ROUTLEDGE HANDBOOK OF CLIMATE CHANGE IMPACTS ON INDIGENOUS PEOPLES AND LOCAL COMMUNITIES

This Handbook examines the diverse ways in which climate change impacts Indigenous Peoples and local communities and considers their response to these changes.

While there is well-established evidence that the climate of the Earth is changing, the scarcity of instrumental data oftentimes challenges scientists' ability to detect such impacts in remote and marginalized areas of the world or in areas with scarce data. Bridging this gap, this Handbook draws on field research among Indigenous Peoples and local communities distributed across different climatic zones and relying on different livelihood activities, to analyse their reports of and responses to climate change impacts. It includes contributions from a range of authors from different nationalities, disciplinary backgrounds, and positionalities, thus reflecting the diversity of approaches in the field. The Handbook is organised in two parts: Part I examines the diverse ways in which climate change – alone or in interaction with other drivers of environmental change – affects Indigenous Peoples and local communities; Part II examines how Indigenous Peoples and local communities are locally adapting their responses to these impacts. Overall, this book highlights Indigenous and local knowledge systems as an untapped resource which will be vital in deepening our understanding of the effects of climate change.

The *Routledge Handbook of Climate Change Impacts on Indigenous Peoples and Local Communities* will be an essential reference text for students and scholars of climate change, anthropology, environmental studies, ethnobiology, and Indigenous studies.

Victoria Reyes-García is an ICREA Research Professor at the Institut de Ciència i Tecnología Ambientals, Universitat Autònoma de Barcelona, Spain.

ROUTLEDGE ENVIRONMENT AND SUSTAINABILITY HANDBOOKS

ROUTLEDGE HANDBOOK OF ENERGY COMMUNITIES AND SMART CITIES
Edited by Maciej M. Sokołowski and Anna Visvizi

ROUTLEDGE HANDBOOK OF GLOBAL LAND AND RESOURCE GRABBING
Edited by Andreas Neef, Chanrith Ngin, Tsegaye Moreda Shegro, and Sharlene Mollett

THE ROUTLEDGE HANDBOOK OF ECOMEDIA STUDIES
Edited by Antonio López, Adrian Ivakhiv, Stephen Rust, Miriam Tola, Alenda Y. Chang and Kiu-wai Chu

ROUTLEDGE HANDBOOK OF URBAN BIODIVERSITY
Edited by Charles H. Nilon & Myla F.J. Aronson

ROUTLEDGE HANDBOOK OF CLIMATE CHANGE IMPACTS ON INDIGENOUS PEOPLES AND LOCAL COMMUNITIES

Edited by Victoria Reyes-García
With Santiago Álvarez-Fernández, Petra Benyei, Laura Calvet-Mir, David García-del-Amo, André B. Junqueira, Xiaoyue Li, Vincent Porcher, Anna Porcuna-Ferrer, Anna Schlingmann and Ramin Soleymani

Routledge
Taylor & Francis Group
LONDON AND NEW YORK
earthscan
from Routledge

Designed cover image: Vincent Porcher and the legend: Betsileo funeral ceremony, Madagascar.

First published 2024
by Routledge
4 Park Square, Milton Park, Abingdon, Oxon OX14 4RN

and by Routledge
605 Third Avenue, New York, NY 10158

Routledge is an imprint of the Taylor & Francis Group, an informa business

Funded by European Research Council (Grant FP7-771056-LICCI).

British Library Cataloguing-in-Publication Data
A catalogue record for this book is available from the British Library

ISBN: 978-1-032-41213-9 (hbk)
ISBN: 978-1-032-41215-3 (pbk)
ISBN: 978-1-003-35683-7 (ebk)

DOI: 10.4324/9781003356837

CONTENTS

FIGURES

TABLES

CONTRIBUTORS

Mariam Abazeri is an Adjunct Professor at the University of Miami, USA.

Ruddy Afriyie is a Postgraduate Researcher at Kwame Nkrumah University of Science and Technology, Ghana.

Santiago Álvarez-Fernández is a Data Analyst at the Institut de Ciència i Tecnología Ambientals, Universitat Autònoma de Barcelona, Spain.

Emmanuel M.N.A.N. Attoh is a Researcher – Climate Adaptation Specialist at International Water Management Institute, Cololmbo, Sri Lanka and a Postdoctoral Researcher at Wageningen University, Netherlands as at the initial stages of this work.

Ayansina Ayanlade is an Associate Professor in the Department of Geography, Obafemi Awolowo University, Ile-Ife, Nigeria and a Researcher and Lecturer in the Department of Geography and Regional Research, University of Vienna, Universitätsstraße 7/5, 1010 Vienna, Austria.

Oluwatoyin Seun Ayanlade is a Senior Research Fellow at the African Institute for Science Policy and Innovation, Obafemi Awolowo University, Ile-Ife, Nigeria.

Dániel Babai is a Senior Research Fellow at the Research Centre for the Humanities in Hungary, and is interested in ethnoecology and extensive grassland management in Central Europe.

Petra Benyei was a Postdoctoral Researcher at the Institut de Ciència i Tecnología Ambientals, Universitat Autònoma de Barcelona, Spain. She is now a Juan de la Cierva Fellow at the Institute of Economy, Geography and Demography, Consejo Superior de Investigaciones Científicas, Spain.

Enoch Bessah is a Lecturer at Kwame Nkrumah University of Science and Technology, Ghana.

Laura Calvet-Mir is a Research Fellow at Institut d'Estudis Regionals i Metropolitans de Barcelona, Spain, and Affiliated Researcher at Urban Transformation and Global Change Laboratory, Universitat Oberta de Catalunya, Spain.

Rosario Carmona is a Postdoctoral Researcher in the Department of Anthropology of the Americas, University of Bonn, Germany, and a Researcher at the Center for Integrated Disaster Risk Management (CIGIDEN), Chile.

Ereck Chakauya is a Network Director at AUDA-NEPAD Agency/Southern Africa Network for Biosciences (NEPAD SANBio), hosted by Council for Scientific and Industrial Research (CSIR), South Africa.

Rumbidzayi Chakauya is a Postdoctoral Researcher in the Department of Environmental Sciences, University of South Africa, South Africa.

Mouna Chambon is a PhD student at the Institut de Ciència i Tecnología Ambientals, Universitat Autònoma de Barcelona, Spain.

Ouerle Chao is a PhD student at the Institut de Ciència i Tecnología Ambientals, Universitat Autònoma de Barcelona, Spain.

Dhrishna Charan is an Assistant Lecturer in Physics in the School of Science and Technology, The University of Fiji, Fiji.

Zhuo Chen is a PhD Researcher at the Faculty of Social Sciences, University of Helsinki, Finland.

Esther Conde is an Independent Researcher working at the Centro Boliviano de Desarrollo Socio-Integral, San Borja, Beni, Bolivia.

Janice Cumberbatch is a Senior Lecturer at the Centre for Resource Management and Environmental Studies (CERMES) at the University of the West Indies, Barbados.

Aida Cuni-Sanchez is an Associate Professor in the Department of International Environmental and Development Studies (NORAGRIC), Norwegian University of Life Sciences, Norway.

Vincenza Ferrara is an Early-Stage Researcher in the Department of Archaeology and Ancient History (Uppsala University) and the Department of Human Geography (Stockholm University).

Anna Fuchs is an Early-Stage Researcher at the Institute of Organic Farming, University of Natural Resources and Life Sciences, Vienna.

Eranga K. Galappaththi is an Assistant Professor in the Department of Geography at Virginia Polytechnic Institute and State University (Virginia Tech), USA.

David García-del-Amo is a Postdoctoral Researcher at the Institut de Ciència i Tecnologia Ambientals, Universitat Autònoma de Barcelona, Spain.

Drew Gerkey is an Associate Professor in the Department of Anthropology and the School of Language, Culture and Society, Oregon State University, USA.

Patrícia Gonçalves is a fisherwoman and Master candidate researcher at NOVA University Lisbon, School of Science and Technology – Campus Caparica, Portugal.

Théo Guillerminet is an Early-Stage Researcher at AGAP Institut, CIRAD, Montpellier, France.

Catrina Hinds is a Project Coordinator at the Centre for Resource Management and Environmental Studies (CERMES) at the University of the West Indies, Barbados.

Eric Hirsch is an Assistant Professor of Environmental Studies at Franklin & Marshall College, USA.

Tomás Huanca is an Independent Researcher working at the Centro Boliviano de Desarrollo Socio-Integral, San Borja, Beni, Bolivia.

Andrea E. Izquierdo is a Researcher at the Multidisciplinary Institute of Plant Biology (IMBIV), of the National University of Córdoba and National Scientific and Technical Research Council (CONICET), Argentina.

Obert Jiri is a Professor and Chief Director of Agricultural and Rural Development Advisory Services at the Ministry of Lands, Agriculture, Fisheries, Water and Rural Development, Zimbabwe.

Leneisja Jungsberg is a Staff Manager at Nordregio, an international research centre for regional development and planning.

André B. Junqueira is a Postdoctoral Researcher at the Institut de Ciència i Tecnología Ambientals, Universitat Autònoma de Barcelona, Spain.

Benjamin Klappoth is an Independent Researcher working at the intersection between organic farming, agroecology, and climate change.

Julia A. Klein is a Professor in the Department of Ecosystem Science and Sustainability, Colorado State University, USA.

Gordana Kranjac-Berisavljevic is a Professor at the University for Development Studies, Ghana.

Xiaoyue Li is a Postdoctoral Researcher at the Institut de Ciència i Tecnología Ambientals, Universitat Autònoma de Barcelona, Spain.

Johan Lindberg is a mechanical engineer from the Royal Institute of Technology (KTH, Sweden). He is an expert in traction motors, drive systems, programming, and statistics.

Simona Lippi is an Independent Conservation Biologist Consultant with a solid experience in participatory natural resources management approaches.

Fulco Ludwig is a Professor at Wageningen University, Netherlands.

Machate Machete is a Professor in the Department of Environmental Sciences, University of South Africa, South Africa.

Simeon Albert Materechera is a Professor at the Faculty of Natural and Agricultural Sciences, North-West, University, South Africa.

Giulia Mattalia is a Postdoctoral Researcher at the Institut de Ciència i Tecnología Ambientals, Universitat Autònoma de Barcelona, Spain and at the New York Botanical Garden, United States.

Patrick McConney is a Senior Lecturer at the Centre for Resource Management and Environmental Studies (CERMES) at the University of the West Indies, Barbados.

Mônica Mesquita is a Senior PhD Researcher at MARE Centre at NOVA University Lisbon, School of Science and Technology – Campus Caparica, Portugal.

Peter Graham Mortyn is a Professor in the Department of Geography, Universitat Autònoma de Barcelona, Spain.

Ben Mwangi is a Researcher at the Institute of Climate and Adaptation, University of Nairobi, Kenya.

Lydia A. Olaka is a Senior Lecturer at The Technical University of Kenya, Kenya.

Daniel Oliveira is a PhD candidate researcher at MARE Centre at NOVA University Lisbon, School of Science and Technology – Campus Caparica, Portugal.

Isaac Ayo Oluwatimilehin is a PhD Researcher in the Department of Geography, Obafemi Awolowo University, Ile-Ife, Nigeria, Obafemi Awolowo University, Ile-Ife, Osun State, Nigeria.

Hazel A. Oxenford is a Professor at the Centre for Resource Management and Environmental Studies (CERMES) at the University of the West Indies, Barbados.

Maria Pena is a Project Officer at the Centre for Resource Management and Environmental Studies (CERMES) at the University of the West Indies, Barbados.

Vincent Porcher is a PhD Researcher at the Institut de Ciència i Tecnología Ambientals, Universitat Autònoma de Barcelona, Spain.

Anna Porcuna-Ferrer is a PhD Researcher at the Institut de Ciència i Tecnología Ambientals, Universitat Autònoma de Barcelona, Spain.

Kerala Railoa is a Project Manager in the School of Science and Technology at the University of Fiji. She is managing a USAID-funded project focussing on Aquaculture for Fijian communities.

Victoria Reyes-García is an ICREA Research Professor at the Institut de Ciència i Tecnología Ambientals, Universitat Autònoma de Barcelona, Spain.

Carla Roncoli is Associated Faculty in the Departments of Anthropology and Environmental Sciences at Emory University, United States.

Massimiliano Sanfilippo is an Independent Natural Climate Solutions and Participatory Forest Management Consultant.

Anna Schlingmann is a PhD researcher at the Institut de Ciència i Tecnología Ambientals, Universitat Autònoma de Barcelona, Spain.

Daniela Schulman is Lab Manager in the Energy and Environment Transitions Lab, Political Science Department, University of California, Santa Barbara.

Christoph Schunko is an Associate Professor at the Institute of Organic Farming, University of Natural Resources and Life Sciences, Vienna.

Victoria N. Sharakhmatova is a Postdoctoral Scholar in the Department of Geography and ARCTICenter, University of Northern Iowa, USA and Associate Professor at the Russian Foreign Trade Academy, Far Eastern Branch, Russia.

Ashneel A. Singh is a Lecturer in Biology in the Discipline of Biological and Chemical Sciences, School of Agriculture, Geography, Environment, Ocean and Natural Sciences (SAGEONS), The University of the South Pacific, Fiji.

Priyatma Singh is a Lecturer in Biology in the School of Science and Technology, The University of Fiji. She is also pursuing her PhD in Environmental Science from Linkoping University, Sweden.

Ramin Soleymani is a Software Developer at the Institut de Ciència i Tecnología Ambientals, Universitat Autònoma de Barcelona, Spain.

Yorgos Stratoudakis is a Researcher at the Portuguese Institute of Sea and Atmosphere, Lisbon, Portugal.

Nataliya Stryamets is a Researcher in the Forestry Faculty, School for Forest Management, Swedish University of Agricultural Sciences and Roztochya Nature Reserve, Ivano-Frankove, Ukraine.

Adrien Tofighi-Niaki is a Project Coordinator at the Institut de Ciència i Tecnología Ambientals, Universitat Autònoma de Barcelona, Spain.

Virginia Toscano Rivalta is a Master's student in the Department of Life Sciences and Systems Biology, University of Turin, Turin, Italy.

Christian R. Vogl is an Associate Professor at the Institute of Organic Farming, University of Natural Resources and Life Sciences, Vienna.

Nicola Wendt-Lucas is a Research Fellow at Nordregio, an international research centre for regional development and planning. The research centre was established by the Nordic Council of Ministers in Stockholm, Sweden.

Noelia Zafra-Calvo is a Researcher at the Basque Centre for Climate Change bc3, Scientific Campus of the University of the Basque Country, Spain.

INTRODUCTION

Understanding climate change impacts on Indigenous Peoples and local communities: A global perspective from local studies

Victoria Reyes-García, Santiago Álvarez-Fernández, Petra Benyei, Laura Calvet-Mir, Mouna Chambon, David García-del-Amo, André B. Junqueira, Xiaoyue Li, Vincent Porcher, Anna Porcuna-Ferrer, Anna Schlingmann, Ramin Soleymani and Adrien Tofighi-Niaki

Why a book on climate change impacts on Indigenous Peoples and local communities?

There is overwhelming evidence that the climate of the Earth is changing (IPCC, 2021), although our understanding of the myriad of ways in which these changes affect local social-ecological systems across the globe is patchier (IPCC, 2022). This is particularly the case in remote and/or marginalized areas of the world, where the scarcity of instrumental data challenges scientists' ability to detect change (Maraun & Widmann, 2017; Rosenzweig & Neofotis, 2013). Moreover, there is also growing evidence that changes in weather patterns have and will continue to have far-reaching and disproportionate impacts upon Indigenous Peoples and local communities (Bardsley & Wiseman, 2016; Galloway-McLean, 2017), including impacts on livelihood activities and health, as well as social and cultural impacts (e.g., Green & Minchin, 2014; Maikhuri et al., 2018; Race et al., 2016). This is largely the case because climate change impacts are socially mediated: those already in social disadvantage, and particularly ethnic and political disadvantages, are likely to find themselves disproportionately affected by climate change impacts.

However, Indigenous Peoples and local communities with a long history of interaction with the environment are not mere victims of climate change impacts; they are also legitimate custodians of knowledge regarding climate change and its impacts and rights-holders to participate in and contribute to climate change decision-making at both local and international levels. Indigenous Peoples and local communities have complex knowledge systems that allow them to detect changes in the local climate and the impacts of such changes in the biophysical systems on which they depend (Reyes-García et al., 2016; Savo et al., 2016), as well as the interactions of climate-driven changes with other drivers of local environmental change (Merten et al., 2020; Smith et al., 2017). Moreover, insights from Indigenous and local knowledge systems can also improve our understanding

DOI: 10.4324/9781003356837-1

of how climate change affects local socioeconomic systems, livelihoods, and cultures, focusing on what matters to local people (Crate & Nuttall, 2016; Sillitoe, 2021). Therefore, Indigenous and local knowledge systems have the untapped potential to contribute to a deeper and more detailed understanding of the myriad effects of climate change on local social-ecological systems around the world, including in very remote areas.

Indigenous Peoples and local communities directly experience climate change impacts, but each group experiences impacts in a different way, not only because such impacts are place-specific but also because climate change impacts on social-ecological systems and responses to those impacts are mediated by local socioeconomic systems and cultural aspects. For example, while sea-level rise is a climate-related phenomenon with potential effects on the millions of people living close to sea level, specific biophysical (e.g., magnitude of tidal influences, geologic subsidence, overall island size, and relief) and socioeconomic conditions (e.g., resources to cope with sea-level rise, livelihood strategy) mediate how different people perceive this change and the extent to which they feel affected by it. In that context, Indigenous peoples and local communities often rely on their knowledge systems to respond to the socially mediated climate change impacts (Schlingmann et al., 2021).

Chapters in this book examine the diverse ways in which climate change -alone or in interaction with other drivers of environmental change- affects Indigenous Peoples and local communities (Part 1) and how Indigenous Peoples and local communities are locally adapting their responses to these impacts (Part 2). Cases featured draw on first-hand information collected from field research in diverse areas of the world and cover different climatic zones and livelihood activities. We argue that the approach is important in three ways. First, examining local perceptions of climate change impacts and local responses to them contributes to return to a human scale a discussion often focused on 'mega-trends' and 'mega-drivers' of change. Second, in generating human-scale data on the impacts of and responses to climate change, the chapters in this book help reframe scholarship on the human impacts of climate change, drawing its social, economic, and political dimensions, thus contributing to a line of scholarship that proposes that the global understanding of climate change impacts means recognizing their historical and political origins. Finally, by focusing on Indigenous Peoples and local communities, we hope to contribute to make climate injustices more visible and to emphasize the need to recognize Indigenous Peoples and local communities' rights in climate research and decision-making.

The next section of this introduction addresses issues referring to authors' common ground and yet diverging positionalities. The following two sections bring together the main findings from chapters in this book and associated research to give a global perspective of the lessons learned from multiple case studies across (1) climate change impacts on Indigenous Peoples and local communities and (2) responses to climate change impacts by Indigenous Peoples and local communities. The next section then presents lessons learned from working with different knowledge systems on the topic of climate change impacts. This general introduction ends with a general policy recommendation derived from our work accompanied by concrete steps for researchers and decision makers to put the general recommendation into action.

The context of the book

Most chapters in this book result from a community of practice organized around LICCI – *Local Indicators of Climate Change Impacts: The contribution of local knowledge to climate change research*, a project funded by the European Research Council designed to bring contextualized information from Indigenous Peoples and local communities to climate change research and policy

(Reyes-García et al., 2019). The editors of the book are the Principal Investigator of the project (ICREA Research Prof. Victoria Reyes-García) and all the core members of the LICCI team (post-doctoral scholars, PhD students, and technicians) who, over the five years of the project, have coordinated the network. Most contributions to this book are from members of the LICCI network, although a few chapters have been written by authors working on related topics who were purposely invited to enlarge the coverage of some regions or topics underrepresented in the LICCI network.

The LICCI network includes about 50 Indigenous and non-Indigenous researchers working with and affiliated with institutions around the world. The common ground of this network is a shared interest in exploring the diverse ways in which climate change affects Indigenous Peoples and local communities and how they respond to it. Members of the network recognize the acute need to increase the transferability, integration, and scalability of Indigenous and local knowledge into climate change research through the creation of a community of practice that uses standardized methods for the collection and coding of locally relevant but cross-culturally comparable data (Reyes-García et al., 2019). All members of the network also gratefully acknowledge and deeply respect the invaluable contributions of the Indigenous Peoples and local communities, whose traditional knowledge and practices have formed the foundation of this research.

Despite this common ground, there are also differences and particularities among network members, which are reflected in the content of the book. Thus, contributions to this book come from authors from different nationalities, disciplinary backgrounds, and positionalities, a diversity that inevitably affects authors' own experiences and engagement with local communities and the way information is analyzed and interpreted. This has resulted in different epistemological stances on the treatment of Indigenous and local knowledge systems across chapters. Importantly, while maintaining a diversity of theoretical approaches, authors have agreed on a common set of concepts that bring some internal coherence to this edited volume (see Glossary).

Chapters in this book also draw from a diversity of methodological tools used to collect and analyze observations of and responses to climate change impacts. Data collection tools range from ethnography and qualitative data collection methods (e.g., semi-structured interviews, focus groups, participant observation) to more systematic forms of data collection (e.g., surveys). Several authors also use secondary sources (e.g., time series of climatic data, historical sources). Data analysis varies from featuring testimonies to statistical analysis, with some chapters being more interpretative than others.

This edited volume does not intend to provide an exhaustive view of climate change impacts on Indigenous Peoples and local communities around the world nor on how people respond to such impacts. Moreover, we acknowledge some geographic biases (e.g., there is only one chapter from the South Pacific, Fiji, none from Australia or New Zealand, and some regions like the Amazon are underrepresented). However, we argue that, despite these biases, bringing together a collection of case studies allows us to identify common trends as well as idiosyncrasies – both of which are important to understand global trends and the myriad of climate change impacts across the world. In the following three sections, we aim to trespass case studies idiosyncrasies, to summarize major trends derived from the work of the LICCI network, with particular emphasis on the chapters published in this book.

Climate change impacts on Indigenous Peoples and local communities

Research by the LICCI network, including chapters in this book, provides three main insights in relation to how climate change impacts Indigenous Peoples and local communities' and how they understand and report such impacts.

Indigenous Peoples and local communities report numerous, ongoing, tangible, and widespread environmental changes

All the contributions to this volume emphasize that Indigenous Peoples and local communities report climate change impacts as currently ongoing, locally contextualized, and affecting diverse elements of local social-ecological systems, including their physical (e.g., atmosphere, rivers, soils) and biological (e.g., trees, crops, fish) components (see also Reyes-García et al., 2024). For example, among many others, impacts recorded include changes in elements of the atmospheric system, such as changes in precipitation patterns (e.g., Babai, 2024) and fog (Mwangi et al., 2024), temperature increase (e.g., Abazeri, 2024; Mattalia et al., 2024), or weather unpredictability (e.g., Hirsch, 2024; Stratoudakis et al., 2024) including changes in seasonality (e.g., Gerkey & Sharakhmatova, 2024; Miara et al., 2022). Reports also feature changes in the physical system, including changes in freshwater availability (e.g., Izquierdo & Schlingmann, 2024) and soil humidity (Estevo et al., 2022), changes in snowfall and snow cover (e.g., García-del-Amo, Mortyn et al., 2023; Jungsberg & Wendt-Lucas, 2024), and changes in sea- and river-ice (e.g., Galappaththi, 2024; Gerkey & Sharakhmatova, 2023). Among the many reports, changes in the water cycle are particularly abundant (e.g., Estevo et al., 2022; Junqueira et al., 2021; Reyes-García et al., 2024).

Indigenous Peoples' and local communities' reports of impacts go beyond changes in the atmospheric and physical systems and include changes in the life system, including wildlife and managed systems. Thus, research conducted by the LICCI network brings to light many reports of changes in abundance and phenology of wild plants and terrestrial and aquatic animals (e.g., Geffner-Fuenmayor, 2022; Schunko et al., 2023). But such reports also include observations of phenological changes in their agricultural calendars (Babai, 2024; Hirsch, 2024), the productivity of traditional and modern crop varieties (e.g., Attoh et al., 2024; Carmona, 2022; Chakauya et al., 2024; Labeyrie et al., 2021), including cash crops such as tea or coffee (Mwangi et al., 2024), the increase of livestock pests (e.g., Attoh et al., 2024; Junqueira et al., 2021), or the newly problematic abundance of previously present, but non-problematic, species (e.g., McConney et al., 2024). Finally, Indigenous Peoples and local communities also report on a variety of impacts on their ways of living, including reduced feasibility to practice traditional hunting and fishing livelihoods (Galappaththi, 2024; Gerkey & Sharakhmatova, 2024), mobility restrictions (e.g., Gerkey & Sharakhmatova, 2024), or damaging of houses and other infrastructure (e.g., Jungsberg & Wendt-Lucas, 2024).

Indigenous Peoples' and local communities' report multiple and synergistic drivers of environmental change

While the LICCI project had an original focus on climate change impacts, a consistent finding in all our study sites is that climate change impacts cannot be fully separated from other changes that are being experienced locally. In fact, this is also a finding currently featured in the scientific literature (e.g., Pörtner et al., 2021). Indeed, the research results of the LICCI network emphasize that Indigenous Peoples' and local communities' reports of environmental changes are interrelated and driven by multiple stressors (see also Li et al., 2021). In particular, our research emphasizes the importance of cascading effects across elements of the social-ecological system and of multiple drivers of change.

On the one hand, our research emphasizes that Indigenous Peoples and local communities identify many relations between elements of their social-ecological systems, pointing at many cascading effects, mainly from changes in the atmospheric and physical systems to changes in the life system (e.g., García-del-Amo, Calvet-Mir et al., 2024; Reyes-García et al., 2024). For example, inhabitants of Sierra Nevada, Spain, report that decreasing rain and snow events diminish rivers' flow, reducing the duration and number of working natural springs, decreasing soil humidity, and increasing soil erosion. Changes in precipitation also affected wild flora, as it reduced the duration and extension of grazing areas, which in turn affected wild fauna's altitudinal distribution, with animals moving closer to the villages looking for food and water. Moreover, the shorter duration and frequency of rainfalls and snowfalls have a direct negative effect on agricultural and livestock activities (including beekeeping), decreasing their productivity and increasing the number of pests and diseases, which are aggravated by the destruction of crops by wild fauna searching for food (García-del-Amo et al., 2024).

On the other hand, our research features how Indigenous Peoples and local communities acknowledge climate change as one, but not necessarily the most important, of the multiple drivers of environmental change (e.g., Izquierdo & Schlingmann, 2024). Across case studies, we noticed that Indigenous Peoples and local communities recognize that environmental change is driven by many inter-related contextual factors, which vary from resource over-extraction (Carmona, 2024b; Estevo et al., 2022; Reyes-García et al., 2024), to development programs (Attoh et al., 2024; Porcuna-Ferrer et al., 2024), creation of new infrastructures (Izquierdo & Schlingmann, 2024; Junqueira et al., 2021), or state policies (e.g., Carmona, 2024b; Chen, 2024; Lippi & Sanfilippo, 2024). Importantly, our research also suggests that, besides these drivers, for many Indigenous Peoples and local communities, an important additional driver of environmental change is the degradation of human relation with the environment, including the loss of caring practices (e.g., Chao et al., 2024; Chen, 2024; Hirsch, 2024; Reyes-García et al., 2024).

Research results of the LICCI network clearly show that Indigenous Peoples and local communities identify numerous dynamic interactions of positive and negative feedback between climate, social, political, ecological, economic, and cultural structures. The recognition of this multifactorial and nested complexity acknowledges that climate change impacts do not come in a vacuum, but often add to situations of colonialism, inequality, and environmental injustice. Within these contexts, climate change, while recognized as having an impact, is often not considered the main topic of concern (e.g., Blanch-Ramirez et al., 2022; Hirsch, 2024; Izquierdo & Schlingmann, 2024; Levy Guillén, 2022; Porcuna-Ferrer et al., 2023). For example, for agriculturalists in the Colca Valley (Peru), climate change impacts are mediated by structures of inequality, violence, and environmental injustice rooted in a highly centralized and urbanized form of capitalism configured by Peru's colonial past (Hirsch, 2024). In the same vein, Koryak, Chukchi, and Even peoples, inhabitants of the Kamchatka Peninsula in Northeast Siberia, Russia, report that climate change impacts are exacerbated by legacies of social transformation derived from Soviet and post-Soviet policies (Gerkey & Sharakhmatova, 2024). Moreover, precisely because of political and economic inequalities, some groups -e.g., Indigenous and local women- may have more at stake in addressing climate change impacts and other sources of environmental change than others (e.g., Abazeri, 2024; Porcuna-Ferrer et al., 2024). Finally, as featured in several chapters of this book, Indigenous Peoples and local communities identify that synergistic interactions between climate change and other drivers of change can amplify (e.g., Carmona, 2023; Hirsch, 2024; Junqueira et al., 2021) or attenuate (e.g., Chen, 2024) the overall impacts of change.

Overall, Indigenous Peoples and local communities' understandings of change provide a holistic, multi-causal, and multi-scalar complex picture of the relations between humans and the environment, entangling ecological observations with socio-economic, cultural and political critiques. In doing so, they provide a political – and not a purely ecological- view of change, highlighting the need to focus on intersecting impacts.

Indigenous Peoples and local communities' reports of environmental change are not uniform

A third finding that derives from research by the LICCI network is that reports of changes vary across geographical areas and social characteristics of groups and individuals. We found variation at different levels. First, as climate-driven environmental change manifests differently across regions, so do reports of these changes (Reyes-García et al., 2024), even at local scales (García-del-amo et al., 2023). Moreover, the same indicator of change exerts different impacts, or impacts of different intensities, across livelihood strategies (e.g., Estevo et al., 2022; García-del-Amo et al., 2022; Reyes-García et al., 2024). For example, weather instability makes agricultural labour increasingly speculative and anxiety-riddled (Hirsch, 2024; Vázquez-Martínez, 2022), and sea-ice-based hunting extremely dangerous (Galappaththi, 2024).

Second, we also found variation associated with whether impacts were perceived as negative. Indeed, reports of climate-driven environmental changes are often perceived as negative, but not always. Some reports of environmental changes are presented as having a positive impact. For example, while most Kolla-Atacameños People and local communities in the Argentine Puna perceive climate change as negative, some people also perceive warmer climates in the local extremely cold weather as positive (e.g., Izquierdo & Schlingmann, 2024). Some people also report as positive changes that permit longer fishing seasons (e.g., Stratoudakis et al., 2024) or that increase the length of agricultural season (e.g., Blanch-Ramirez et al., 2022; Estevo et al., 2022; Fuchs et al., 2024).

Finally, ethnicity or cultural norms (e.g., Mwangi et al., 2024), household context (e.g., Lippi & Sanfilippo, 2024) and individual characteristics, such as gender (Abazeri, 2024), age, or level of relation with the environment (García-del-Amo et al., 2023) are also linked to variation in reports of climate-driven environmental changes. For example, in the mountains of Sierra Nevada, Spain, people whose livelihood directly depends on nature report many more changes and relations among observed changes than people who do not live in close relation with the environment (García-del-Amo et al., 2023), and Kikuyu and Meru smallholder farmers living in mountains in central Kenya report the use of different adaptation strategies, e.g., irrigation is more widespread amongst the Meru than among the Kikuyu, arguably because cultural differences between these groups shape their adaptation strategies (Mwangi et al., 2024).

Indigenous Peoples and local communities' responses to impacts from climate change and associated stressors

Research by the LICCI network also provides main insights on how Indigenous Peoples and local communities respond to the compounded impacts originating from climate change and other stressors.

Indigenous Peoples and local communities respond to climate change impacts in plural ways

The analysis of different case studies clearly shows that Indigenous Peoples and local communities respond to current impacts in diverse ways (see also Schlingmann et al., 2021). Among cases featured by the LICCI network, most documented responses include coping with change (e.g., Carmona, 2022; Porcuna-Ferrer et al., 2024) or the adoption of incremental adaptations, i.e., modifying livelihood strategies (e.g., Ávila et al., 2021; Hirsch, 2024; Labeyrie et al., 2021; Vázquez-Martínez, 2022).

For example, farmers in the Carpathians have changed their mowing, sowing, and harvesting practices to respond to weather unpredictability (Babai, 2024). Transformational adaptations, including migration to cities and changing to non-nature-related jobs, are widespread across geographical areas and livelihood activities, but they are not dominant (Zant et al., in press). Results call for caution when inferring that response options adopted are "adaptive" or "sustainable", as the long-term consequences of responses adopted are often unknown (e.g., Ferrara & Lindberg, 2024). Moreover, some responses, while allowing for immediate coping with climate change impacts, can be detrimental in the long-term, increasing pressure on Indigenous Peoples and local communities to discontinue adaptive practices or to adopt unsustainable relationships with the territory (Zant et al., in press). For example, a review of the literature suggests that switching to off-farm work, while helping to cope with short term climate change impacts, also results in a decline in traditional activities, reduction in food production and self-sufficiency, and higher market dependency, potentially decreasing households' resilience (Galappaththi & Schlingmann, 2023).

Social, political, and cultural barriers hamper Indigenous Peoples' and local communities' adoption of response options

A second finding reflected in research by the LICCI network is that structural social, political, cultural, and other types of barriers hamper Indigenous Peoples' and local communities' adoption of some response options to climate change impacts. For example, Tyrol farmers in Austria are constrained to adopt transformational adaptation options as such responses require community work, which is in shortage because of social barriers (e.g., rural depopulation) (Fuchs et al., 2024). Likewise, farmers in Northern Ghana face financial barriers to adopt technical improvements proposed by government officials to enhance adaptation (Attoh et al., 2024).

Structural barriers are symptomatic of Indigenous Peoples' and local communities' histories of colonialism (Fayazi et al., 2020; Piggott-McKellar et al., 2019), for which they cannot be successfully addressed through technical means, as featured in several chapters of this book. For example, the Mapuche-Pehuenche in Southern Chile are considered vulnerable to climate change, but they express that what has made them vulnerable is the violation of their rights to their territory and way of life through the progressive degradation of their land by logging extraction and their structural marginalization from the decision-making processes (Carmona, 2023). Similarly, Quechua descendant farmers in the Colca Valley, Peru, shifted to market-oriented agriculture under the promise of economic development and the premise of a stable climate. The decrease of agricultural productivity under current market-dependence and unpredictable weather conditions is at the root of their vulnerability (Hirsch, 2024). In that understanding, chapters in this book also highlight how people suffering additional layers of marginalization (e.g., Indigenous women in patriarchal structures) face even further barriers to adaptation options (e.g., Ayanlade et al., 2024; Porcuna-Ferrer et al., 2024).

The adoption of response options might result in trade-offs and unequal distribution of costs and benefits across actors with different roles in society. For example, among Bassari farmers' (Senegal), gender and access to financial, physical, and natural capital define who can implement and benefit from new agricultural technologies. In this case, the introduction of cotton as cash crop has increased men's income but has also resulted in the abandonment of drought-tolerant native crops, increasing women's work burden and the overall household's vulnerability to market uncertainties (Porcuna-Ferrer et al., 2024).

Indigenous and local knowledge systems provide contextualized and suitable response options

Our third important finding in relation to Indigenous Peoples' and local communities' responses to change is that Indigenous and local knowledge systems provide contextualized and suitable options to respond to the simultaneous impacts of climate change and other drivers of global change. Indigenous and local knowledge systems have contributed to climate change response mechanisms that reflect local ontologies and governance systems (Galappaththi, 2024; Reyes-García & Junqueira, 2023). They are contextualized in local livelihoods (Schlingmann et al., 2021) and cultural preferences (Mwangi et al., 2024), rely on local resources and means (e.g., Chakauya et al., 2023), and address the synergistic effects of simultaneous drivers of change (e.g., Abazeri, 2024). For example, Sereer farmers in Senegal shift between crop varieties depending both on climate trends and agricultural extension policies, showing their ability to cope with simultaneous drivers of change (Ruggieri et al., 2021). As Indigenous and local knowledge systems allow the adoption of autonomous responses (e.g., Jungsberg & Wendt-Lucas, 2024), these knowledge systems can be considered as enabling factors in the context of climate change adaptation. Chapters in this book suggest that such bottom-up responses can help manage multiple stressors and minimize vulnerabilities (e.g., Abazeri, 2024; Galappaththi, 2024). For example, among Inuit, the use of co-management to respond to climate change impacts supports community resilience by improving food security, fostering social learning, and co-producing knowledge (Galappaththi, 2024).

Research by the LICCI network also shows that Indigenous and local knowledge can be effective to shape response options in interaction with other bodies of knowledge. For example, using online systems to integrate local and scientific knowledge of communities in coastal areas can serve to assess risk and vulnerability situations and shape adaptation or mitigation strategies to face environmental change and climate risks (Iwama et al., 2023). Also, interacting information from autonomous responses to climate change impacts and national adaptation policies can improve the fit between responses at different scales (Singh et al., 2024) and improve all phases of the policy cycle that generate response options (e.g., Carmona, 2024b; McConney et al., 2024).

Working with different knowledge systems to contribute to climate change research and policy

Beyond research findings, work undertaken by the LICCI network also provides important insights from the research process, in particular, in relation to conducting research with Indigenous Peoples and local communities about climate change impacts. In this section, we outline three main lessons learned regarding our research process.

Indigenous Peoples' and local communities' conceptualizations of climate change are not fully transferable to scientific categories

Indigenous knowledge and local knowledge systems are holistic systems encompassing knowledge (e.g., monitoring changes), practices (e.g., water management), and value systems (e.g., relational values of nature) that do not always have equivalents in scientific categories (see also Orlove et al., 2023). Different members of the LICCI network worked with aspects of Indigenous and local knowledge systems referring to climate change impacts that were encoded in different cultural manifestations, including proverbs (Garteizgogeascoa et al., 2020), agricultural calendars (Miara et al., 2022), rituals and beliefs (Chao et al., 2024), or ways of managing landscapes, ecosystems, or species (Abazeri, 2024; Reyes-García et al., 2023). Some of these elements cannot be fully transferred to scientific categories.

However, in many cases, members of the LICCI network were able to weave elements of different knowledge systems to generate a more enriched view of climate change impacts (Reyes-García et al., 2024) and potential responses (McConney et al., 2024; Singh et al., 2024). For example, weather-related proverbs provide long-term observations of many elements of the atmospheric system and their impact on the biophysical world. In Sierra Nevada, Spain, the reported inaccuracy of weather-related proverbs provides a richer understanding of environmental changes than the limited number of indicators that scientists can locally measure (Garteizgogeascoa et al., 2020). Similarly, regarding response options, the contrast between autonomous and planned responses can guide more comprehensive and less redundant adaptation policies, as shown by the analysis contrasting of local adaptation in iTaukei communities and national adaptation plans in Fiji (Singh et al., 2024).

Understandings of climate change impacts derived from different knowledge systems often, but not always, overlap

A second finding brought from our research process is that, across case studies, reports from different knowledge systems often, but not always, overlap. For example, the Tuareg of Illizi (Algeria) reported changes and irregularities in their ecological calendar, which have impacted their pastoral and semi-pastoral livelihoods. These reports are aligned with scientifically measured climate observations. Paradoxically, although participants recall with detail the climatic disasters that happened in the region over the last century, they do not explicitly report changes in the frequency of extreme events (Miara et al., 2022).

There are many reasons that can explain mismatches between information from different knowledge systems. For example, mismatches might come from tensions between different understandings of the term "climate change." Research among the Bassari of south-eastern Senegal shows how the local interpretation of changing climatic conditions has ontological foundations. The Bassari do not understand 'climate' as something separated, but rather as a phenomenon inextricably linked with socio-economic dynamics and immaterial aspects, such as supernatural forces (e.g., Porcuna-Ferrer et al., 2023). Mismatches between information from different knowledge systems can also come from divergent perceptions of reality due to other changes, such as changes in lifestyle (e.g., houses are now isolated, people wear warmer clothing, or use more efficient technology) (e.g., Ferrara & Lindberg, 2024; Stratoudakis et al., 2024). For example, fishers in Portugal argued that people perceive environmental conditions as milder now than in the past because now they have more robust fishing gears and vessels that allow more secure navigation (Stratoudakis et al., 2024). Moreover, mismatches can also come from a focus on different spatial or temporal scales (e.g., Ferrara & Lindberg, 2024; Mattalia et al., 2024; Reyes-García & Junqueira, 2023). Science often focuses on planetary changes and long (geological) time, whereas Indigenous and local knowledge systems often focus on changes that happen at the local geographical scale and in short (historical) time (e.g., Ferrara & Lindberg, 2024; Reyes-García & Junqueira, 2023). The scale at which information is aggregated (e.g., global vs. local, short-term vs. long-term) might also lead to different conclusions.

Even when reports from different knowledge systems match, Indigenous and local knowledge systems signal new complexities and nuances that might have been previously invisible, often emphasizing aspects that matter to local people. For example, in the Juruá river (Brazil), while climatic models show inconclusive or contradictory precipitation trends, there is a clear local understanding that summers have become wetter (Estevo et al., 2022), thus complementing scientific understandings. Complementarities between knowledge systems can help define research priorities. For example, Meru farmers in Mt. Kenya (Kenya) reported changing rainfall patterns and increased temperatures (matching with weather stations data), which are increasing pests and diseases in the banana fields, their preferred staple food. However, farmers also explained that the impacts are aggravated given the lack of research and access to improved banana varieties, which together reduces banana productivity (Mwangi et al., 2024).

Overall, exploring mismatches and complementarities between knowledge systems, rather than validating information from one or another knowledge system, should lead to an enriched picture of reality and more contextualized research priorities (Tengö et al., 2014).

Current scientific research practices do not uphold Indigenous and local knowledge systems and ignore environmental impacts of research

Our last lesson learned from examining our research process refers to several mismatches noted between the goals of our project and the use of current research practices. In executing this project, several members of the LICCI network signaled that current scientific research practices largely privilege norms and standards that do not encourage the co-construction of new knowledge based on evidence from different knowledge systems (see also Orlove et al. 2024). Moreover, such practices reproduce structural barriers that do not encourage citizen's participation in science (Iwama et al., 2023) and do not acknowledge the power dynamics set by professional researchers, often in privileged positions in the production of knowledge (Reyes-Garcia et al., 2024).

Even more, members of the network also acknowledged that research projects framed under current research "best practices" are often not obliged – and sometimes not even aware- of the need to respect Indigenous Peoples' rights to their knowledge – as enshrined in the United Nations Declaration on the Rights of Indigenous Peoples (UNDRIP) – and consequently to their data, nor are scientific institutions equipped to do so (Reyes-García et al., 2022a).

Finally, members of the network also acknowledged that under current frameworks, researchers are not incentivized to acknowledge and reduce the environmental impact of their research activities, thus -ironically- contributing to aggravate the carbon impact of research activities, which ultimately perpetuates the paradox of aggravating climate change impacts among Indigenous Peoples and local communities. For example, during its initial phase, the LICCI project emitted an estimate of 161 tons CO2-eq, which could have been largely reduced to 92 tons CO2-eq (or 53% of the emissions) by applying a standard set of measures already proposed by scholars aiming to decarbonize research or even to 4 tons CO2-eq (or 2.4% of the estimated emissions) by applying more strict measures aiming to reach carbon neutrality. Most emissions reductions come from reducing travel, which is not likely to happen unless the research sector develops normative standards of scientific research practice that encourage, value or even impose the reduction of carbon emissions (Reyes-García et al., 2022b).

Looking forward: policy recommendations

A growing body of research suggests that Indigenous Peoples and local communities are systematically marginalized in climate research and policy, a status that is linked to their general marginalization in society, current political power structures, and histories of colonization (e.g., Chakraborty & Sherpa, 2021; Conway et al., 2019; Ford et al., 2016; Yap & Watene, 2019). This realization has given rise to several recommendations on how to work with Indigenous Peoples and local communities in climate research and policy (e.g., McElwee et al., 2020; Roue & Nakashima, 2018). Following this trend, in this last section, we provide a main recommendation for partnering with Indigenous and local knowledge holders to engage them and their knowledge systems in climate research and policy. We then present some concrete steps for researchers, research institutions, and decision makers to put the aforementioned recommendation into action. Given the diversity of context in which Indigenous Peoples and local communities are found and the different focus of our research, we also emphasize that the application of any policy recommendation will need to be contextualized.

Policy recommendation: *Recognize Indigenous Peoples and local communities as legitimate custodians of knowledge regarding climate change and its impacts and as key rights-holders to participate in and contribute to climate change decision-making at both local and international levels.*

Concrete steps for ***researchers and research institutions*** to put the aforementioned recommendation into action include

Include representatives from Indigenous Peoples and local communities in scientific committees, national delegations, or advisory/technical bodies dealing with climate and environmental change research.

Establish protocols for validating and endorsing data and information when working with Indigenous knowledge systems in research institutions, to ensure these align with Indigenous Data Sovereignty principles. This should involve including representatives from Indigenous Peoples and local communities in research, ethics, and data regulating bodies and setting up mechanisms and infrastructure that facilitate the respect of these rights.

Recognize and uphold Indigenous Peoples' rights (as enshrined in UNDRIP, 2007) in climate and environmental research, especially in projects that involve Indigenous Peoples' and local communities' knowledge to address climate change impacts. Such rights include, among others, collective rights to resources, land, knowledge and, consequently, data.

Evaluate research that involves Indigenous and local knowledge based on criteria that go beyond academic metrics, including evaluation criteria such as formal recognition of Indigenous Peoples and local communities as knowledge holders, type of engagement with the community, and research benefits to the community.

Create specific research opportunities that allow for the extra time required to prepare proposals for collaborative research.

Provide support, and particularly funding, for Indigenous researchers and Indigenous research institutions.

Foster national and international networks and exchanges of experiences (e.g., via workshops) among Indigenous Peoples and local communities around climate change impacts to ensure a safe space for exchanging and valuing their knowledge and worldviews, without necessarily seeking to transfer it to scientific terms.

Implement a zero-carbon emission policy on climate change research projects (and beyond) to minimize the emissions generated by research activities, which ultimately perpetuate the paradox of aggravating climate change impacts among Indigenous Peoples and local communities.

Concrete steps for ***decision makers*** to put the aforementioned recommendation into action include:

Enhance decision makers' competencies and capacities to receive, comprehend, and value insights from Indigenous peoples and local communities, both at the individual and institutional levels.

Include representatives of Indigenous Peoples and local communities at all levels of climate decision-making. Given their existing management of territories, at the local level, Indigenous Peoples and local communities should be actively included in local climate policy and planning. At the global level, Indigenous Peoples and local communities should be guaranteed permanent, effective, and meaningful participation in intergovernmental processes and bodies.

Incorporate Indigenous and local knowledge into all stages of climate decision-making, from defining climate change impacts, vulnerabilities, and adaptations to designing climate policies and making decisions about climate finance. Specifically, as Indigenous Peoples and local communities must adapt to changes they are not responsible for, and as responding to these changes incurs high costs, governments must cover the increased costs associated with dealing with climate change and its impacts. This includes both economic and non-economic costs associated with loss and damage from climate change impacts.

Uphold Indigenous Peoples' Rights (as enshrined in UNDRIP, 2007) in climate policy, which involves undergoing proper and continuous consultation and Free, Prior Informed Consent (FPIC) before establishing any policies for climate mitigation, adaptation, and reparation.

Ensure *multifoci* and inclusive adaptation policies that consider potential synergies between climatic and socioeconomic factors, which prioritize contextualized and locally implemented solutions, and avoid technological lock-in.

Improve policies based on contextualized intersectional vulnerability assessments that consider the specific vulnerabilities of certain groups (e.g., Indigenous Peoples, ethnic minorities, women, children, landless people) and integrate those factors in climate action plans at all governance levels.

Improve horizontal and vertical coherence in climate adaptation policies to integrate the multiple and simultaneous challenges faced by Indigenous Peoples and local communities across different sectors and geographic scales.

Acknowledgments

Research leading to this paper has received funding from the European Research Council under an ERC Consolidator Grant (FP7-771056-LICCI). We warmly thank all members of the LICCI network and all Indigenous Peoples and local communities who have contributed to frame this research with many inputs over the duration of the project. Special thanks to E. Poncela, for her unconditional help in managing the project and to R. Carmona for comments to a previous version of this work. This work contributes to the "María de Maeztu" Programme for Units of Excellence of the Spanish Ministry of Science and Innovation (CEX2019-000940-M).

References

Abazeri, M. (2024). Strengthening participatory governance to adapt and other sociocultural responses to climate change impacts among farming communities in Kerman, Iran. In V. Reyes-García, S. Álvarez-Fernández, P. Benyei, L. Calvet-Mir, D. García-del-Amo, A. B. Junqueira, X. Li, V. Porcher, A. Porcuna-Ferrer, A. Schlingmann, & R. Soleymani (Eds.), *Routledge Handbook of Climate Change Impacts on Indigenous Peoples and Local Communities*. Oxfordshire & New York: Routledge. Chp 18, pp 303–321.

Attoh, E.M.N.A.N., Afriyie, R., Kranjac-Berisavljevic, G., Bessah, E., & Ludwig, F. (2023). Changing terrain: evidence of climate change impacts and adaptive responses of Dagbani Indigenous communities, northern Ghana. In V. Reyes-García, S. Alvarez-Fernandez, P. Benyei, L. Calvet-Mir, D. García-del-Amo, A. B. Junqueira, X. Li, V. Porcher, A. Porcuna-Ferrer, A. Schlingmann, & R. Soleymani (Eds.), *Routledge Handbook of Climate Change Impacts and Adaptation Strategies of Indigenous Peoples and Local Communities*. Oxfordshire & New York: Routledge. Chp 14, pp 244–259.

Ávila, J. V. da C., Clement, C. R., Junqueira, A. B., Ticktin, T., & Steward, A. M. (2021). Adaptive management strategies of local communities in two amazonian floodplain ecosystems in the face of extreme climate events. *Journal of Ethnobiology*, *41*(3), 409–426. https://doi.org/10.2993/0278-0771-41.3.409

Ayanlade, A., Oluwatilmilehing, I. A., & Ayanlade, O. S. (2024). Climate change impacts on agriculture and barriers to adaptation technologies among rural farmers in Southwestern Nigeria. In V. Reyes-García, S. Alvarez-Fernandez, P. Benyei, L. Calvet-Mir, D. García-del-Amo, A. B. Junqueira, X. Li, V. Porcher, A. Porcuna-Ferrer, A. Schlingmann, & R. Soleymani (Eds.), *Routledge Handbook of Climate Change Impacts and Adaptation Strategies of Indigenous Peoples and Local Communities*. Oxfordshire & New York: Routledge. Chp 15, pp 259–272.

Babai, D. (2024). "The weather is more erratic; it changes faster…" Local perceptions of climate change in the Eastern Carpathians, Romania. In V. Reyes-García, S. Alvarez-Fernandez, P. Benyei, L. Calvet-Mir, D. García-del-Amo, A. B. Junqueira, X. Li, V. Porcher, A. Porcuna-Ferrer, A. Schlingmann, & R. Soleymani (Eds.), *Routledge Handbook of Climate Change Impacts and Adaptation Strategies of Indigenous Peoples and Local Communities*. Oxfordshire & New York: Routledge. Chp 4, pp 74–94.

Bardsley, D. K., & Wiseman, N. D. (2016). Socio-ecological lessons for the Anthropocene: Learning from the remote Indigenous communities of Central Australia. *Anthropocene*, *14*, 58–70. https://doi.org/10.1016/J.ANCENE.2016.04.001

Blanch-Ramirez, J., Calvet-Mir, L., Aceituno-Mata, L., & Benyei, P. (2022). Climate change in the Catalan Pyrenees intersects with socioeconomic factors to shape crop diversity and management. *Agronomy for Sustainable Development*, *42*(5), 1–14. https://doi.org/10.1007/s13593-022-00806-3

Carmona, R. (2022). Resilience requires change: Assessing Pehuenche responses to climate change impacts in Southern Chile. *Environmental Justice*, *15*(3), 185–195. https://doi.org/10.1089/ENV.2021.0044

Carmona, R. (2024). Climate change governance and Indigenous Peoples participation: An analysis from the Chilean case. *Environmental Politics*. https://doi.org/10.1080/09644016.2023.2180911

Carmona, R. (2024). The colonial roots of climate vulnerability. Analysing the case of the Mapuche-Pehuenche People in Southern Chile. In V. Reyes-García, S. Álvarez-Fernández, P. Benyei, L. Calvet-Mir, D. García-del-Amo, A. B. Junqueira, X. Li, V. Porcher, A. Porcuna-Ferrer, A. Schlingmann, & R. Soleymani (Eds.), *Routledge Handbook of Climate Change Impacts on Indigenous Peoples and Local Communities*. Oxfordshire & New York: Routledge. Chp 12, pp 196–207.

Chakauya, R., Materechera, S. A., Jiri, O., Chakauya, E., & Machete, M. (2024). Climate change impacts on agriculture, adaptation and resilience: insights from local farmers in South-East Zimbabwe. In V. Reyes-García, S. Alvarez-Fernandez, P. Benyei, L. Calvet-Mir, D. García-del-Amo, A. B. Junqueira, X. Li, V. Porcher, A. Porcuna-Ferrer, A. Schlingmann, & R. Soleymani (Eds.), *Routledge Handbook of Climate Change Impacts and Adaptation Strategies of Indigenous Peoples and Local Communities*. Oxfordshire & New York: Routledge. Chp 16, pp 273–287.

Chakraborty, R., & Sherpa, P. Y. (2021). From climate adaptation to climate justice: Critical reflections on the IPCC and Himalayan climate knowledges. *Climatic Change*, *167*(3–4). https://doi.org/10.1007/S10584-021-03158-1

Chao, O., Li, X., & Reyes-García, V. (2024). Faith, reciprocity, and balance: Inner Mongolian Ovoo offering ritual and its contribution to climate change adaptation. In V. Reyes-García, S. Alvarez-Fernandez, P. Benyei, L. Calvet-Mir, D. García-del-Amo, A. B. Junqueira, X. Li, V. Porcher, A. Porcuna-Ferrer, A. Schlingman, & R. Soleymani (Eds.), *Routledge Handbook of Climate Change Impacts and Adaptation Strategies of Indigenous Peoples and Local Communities*. Routledge.

Chen, Z. (2024). Roads, riding stables, and highland barley: Livelihood diversification as climate change adaptation among Tibetans in Shangri-La, China. In V. Reyes-García, S. Alvarez-Fernandez, P. Benyei, L. Calvet-Mir, D. García-del-Amo, A. B. Junqueira, X. Li, V. Porcher, A. Porcuna-Ferrer, A. Schlingmann, & R. Soleymani (Eds.), *Routledge Handbook of Climate Change Impacts and Adaptation Strategies of Indigenous Peoples and Local Communities*. Oxfordshire & New York: Routledge. Chp 20, pp 335–345.

Conway, D., Nicholls, R. J., Brown, S., Tebboth, M. G. L., Adger, W. N., Ahmad, B., Biemans, H., Crick, F., Lutz, A. F., De Campos, R. S., Said, M., Singh, C., Zaroug, M. A. H., Ludi, E., New, M., & Wester, P. (2019). The need for bottom-up assessments of climate risks and adaptation in climate-sensitive regions. *Nature Climate Change 2019 9:7*, *9*(7), 503–511. https://doi.org/10.1038/s41558-019-0502-0

Crate, S. A., & Nuttall, M. (2016). *Anthropology and Climate Change: From Actions to Transformations*. Routledge.

Estevo, M. de O., Junqueira, A. B., Reyes-García, V., & Campos-Silva, J. V. (2022). Understanding multidirectional climate change impacts on local livelihoods through the lens of local ecological knowledge: A study in Western Amazonia. *Society and Natural Resources*. 36:3, 232-249, DOI: 10.1080/08941920.2022.2153294

Fayazi, M., Bisson, I.-A., & Nicholas, E. (2020). Barriers to climate change adaptation in indigenous communities: A case study on the Mohawk community of Kanesatake, Canada. *International Journal of Disaster Risk Reduction*, *49*, 101750. https://doi.org/10.1016/j.ijdrr.2020.101750

Ferrara, V., & Lindberg, J. (2024). Climate and environmental change perceptions. A case from rural Sicily, Italy. In V. Reyes-García, S. Álvarez-Fernández, P. Benyei, L. Calvet-Mir, D. García-del-Amo, A. B. Junqueira, X. Li, V. Porcher, A. Porcuna-Ferrer, A. Schlingmann, & R. Soleymani (Eds.), *Routledge Handbook of Climate Change Impacts on Indigenous Peoples and Local Communities*. Oxfordshire & New York: Routledge. Chp 6, pp 109–124.

Ford, J. D., Cameron, L., Rubis, J., Maillet, M., Nakashima, D., Willox, A. C., & Pearce, T. (2016). Including indigenous knowledge and experience in IPCC assessment reports. *Nature Climate Change*, *6*(4), 349–353. https://doi.org/10.1038/nclimate2954

Fuchs, A., Vogl, C. R., & Schunko, C. (2024). Counteracting land abandonment: Local adaptation strategies to climate change impacts of alpine farmers in Eastern Tyrol, Austria. In V. Reyes-García, S. Alvarez-Fernandez, P. Benyei, L. Calvet-Mir, D. García-del-Amo, A. B. Junqueira, X. Li, V. Porcher, A. Porcuna-Ferrer, A. Schlingmann, & R. Soleymani (Eds.), *Routledge Handbook of Climate Change Impacts and Adaptation Strategies of Indigenous Peoples and Local Communities*. Oxfordshire & New York: Routledge. Chp 17, pp 287–302.

Galappaththi, E.K. How can co-management support adaptation to climate change? The case of co-existing fisheries from Pangnirtung, Nunavut, Canada. In V. Reyes-García et al. (Ed.), *Routledge Handbook of Climate Change Impacts on Indigenous Peoples and Local Communities*. Routledge: Oxfordshire & New York. Chp 23, pp 370–388. 2024.

Galappaththi, E., & Schlingmann, A. (2023). The sustainability assessment of Indigenous and local knowledge-based climate adaptation responses in agricultural and aquatic food systems. *Current Opinion in Environmental Sustainability*, *62*, 101276.

Galloway-McLean, K. (2017). *Advance Guard: Climate Change Impacts, Adaptation, Mitigation and Indigenous Peoples: A Compendium of Case Studies*. http://wedocs.unep.org/handle/20.500.11822/17886

García-del-Amo, D., Calvet-Mir, L., Mortyn, G., & Reyes-García, V. (2024). Network analysis of climate change impacts reported by local communities of Sierra Nevada, Spain. In V. Reyes-García, S. Alvarez-Fernandez, P. Benyei, L. Calvet-Mir, D. García-del-Amo, A. B. Junqueira, X. Li, V. Porcher, A. Porcuna-Ferrer, A. Schlingmann, & R. Soleymani (Eds.), *Routledge Handbook of Climate Change Impacts and Adaptation Strategies of Indigenous Peoples and Local Communities*. Oxfordshire & New York: Routledge. Chp 5, pp 95–108.

García-del-Amo, D., Gálvez-García, C., Iniesta-Arandia, I., Moreno-Ortiz, J., & Reyes-García, V. (2022). Local ecological knowledge and the sustainable co-management of Sierra Nevada's Social-Ecological System. *The Landscape of the Sierra Nevada*, 351–367. https://doi.org/10.1007/978-3-030-94219-9_21

García-del-Amo, D., Mortyn, P. G., & Reyes-García, V. (2023). Local reports of climate change impacts in Sierra Nevada, Spain: Sociodemographic and geographical patterns. *Regional Environmental Change*, *23*(1), 1–16. https://doi.org/10.1007/S10113-022-01981-5/FIGURES/2

Garteizgogeascoa, M., García-del-Amo, D., & Reyes-García, V. (2020). Using proverbs to study local perceptions of climate change: A case study in Sierra Nevada (Spain). *Regional Environmental Change*, *20*(2), 1–12. https://doi.org/10.1007/s10113-020-01646-1

Geffner-Fuenmayor, C. (2022). *Artisanal Fishers' Perceptions of Climate and Non-climate Environmental Change* [MSc Thesis]. Universitat Autònoma de Barcelona.

Gerkey, D., & Sharakhmatova, V. (2024). Local observations of climate change and impacts on livelihoods in Kamchatka, Russia. In V. Reyes-García, S. Alvarez-Fernandez, P. Benyei, L. Calvet-Mir, D. García-del-Amo, A. B. Junqueira, X. Li, V. Porcher, A. Porcuna-Ferrer, A. Schlingmann, & R. Soleymani (Eds.), *Routledge Handbook of Climate Change Impacts and Adaptation Strategies of Indigenous Peoples and Local Communities*. Oxfordshire & New York: Routledge. Chp 2, pp 45–60.

Green, D., & Minchin, L. (2014). Living on climate-changed country: Indigenous health, well-being and climate change in remote Australian communities. *EcoHealth*, *11*(2), 263–272. https://doi.org/10.1007/S10393-013-0892-9/METRICS

Hirsch, E. (2024). "Not like it used to be": contending with the altered agricultural calendar in Andean Peru's Colca Valley. In V. Reyes-García, S. Álvarez-Fernández, P. Benyei, L. Calvet-Mir, D. García-del-Amo, A. B. Junqueira, X. Li, V. Porcher, A. Porcuna-Ferrer, A. Schlingmann, & R. Soleymani (Eds.), *Routledge*

Handbook of Climate Change Impacts on Indigenous Peoples and Local Communities. Oxfordshire & New York: Routledge. Chp 11, pp 179–195.

IPCC. (2021). *Climate Change 2021: The Physical Science Basis. Contribution of Working Group I to the Sixth Assessment Report of the Intergovernmental Panel on Climate Change* (V. Masson-Delmotte, P. Zhai, A. Pirani, S. L. Connors, C. Péan, S. Berger, N. Caud, Y. Chen, L. Goldfarb, M. I. Gomis, M. Huang, K. Leitzell, E. Lonnoy, J. B. R. Matthews, T. K. Maycock, T. Waterfield, O. Yelekçi, R. Yu, & B. Zhou, Eds.). Cambridge University Press.

IPCC. (2022). *Climate Change 2022: Impacts, Adaptation, and Vulnerability. Contribution of Working Group II to the Sixth Assessment Report of the Intergovernmental Panel on Climate Change* (H.-O. Pörtner, D. C. Roberts, M. Tignor, E. S. Poloczanska, K. Mintenbeck, A. Alegría, M. Craig, S. Langsdorf, S. Löschke, V. Möller, A. Okem, & B. Rama, Eds.). Cambridge University Press.

Iwama, A. Y., Brañas, F., Nuñez, D., Collao, D., Soleymani-Fard, R., Lanyon, C., Tofighi-Niaki, A., da Silva, L., Benyei, P., Ther, F., & Albaglí, S. (2023). Integrating citizen science observations in climate mapping: Lessons from coastal zone geovisualization in Chilean Patagonia and the Brazilian Southeast. In P. E. Perkins (Ed.), *Climate Justice and Participatory Research: Building Climate-Resilient Commons*. University of Calgary Press.

Izquierdo, A. E., & Schlingmann, A. (2024). Local perceptions of climate change in the context of socioeconomic and political changes in a High Andean community from the Argentine Puna. In V. Reyes-García, S. Alvarez-Fernandez, P. Benyei, L. Calvet-Mir, D. García-del-Amo, A. B. Junqueira, X. Li, V. Porcher, A. Porcuna-Ferrer, A. Schlingmann, & R. Soleymani (Eds.), *Routledge Handbook of Climate Change Impacts and Adaptation Strategies of Indigenous Peoples and Local Communities*. Oxfordshire & New York: Routledge. Chp 8, pp 136–150.

Jungsberg, L., & Wendt-Lucas, N. (2024). Local indicators, adaptation actions, and resilience efforts for a warming climate in North Greenland. In V. Reyes-García, S. Alvarez-Fernandez, P. Benyei, L. Calvet-Mir, D. García-del-Amo, A. B. Junqueira, X. Li, V. Porcher, A. Porcuna-Ferrer, A. Schlingmann, & R. Soleymani (Eds.), *Routledge Handbook of Climate Change Impacts and Adaptation Strategies of Indigenous Peoples and Local Communities*. Oxfordshire & New York: Routledge. Chp 22, pp 356–369.

Junqueira, A. B., Fernández-Llamazares, Á., Torrents-Ticó, M., Haira, P. L., Nasak, J. G., Burgas, D., Fraixedas, S., Cabeza, M., & Reyes-García, V. (2021). Interactions between climate change and infrastructure projects in changing water resources: An ethnobiological perspective from the Daasanach, Kenya. *Journal of Ethnobiology*, *41*(3). https://doi.org/10.2993/0278-0771-41.3.331

Labeyrie, V., Renard, D., Aumeeruddy-Thomas, Y., Benyei, P., Caillon, S., Calvet-Mir, L., M. Carrière, S., Demongeot, M., Descamps, E., Braga Junqueira, A., Li, X., Locqueville, J., Mattalia, G., Miñarro, S., Morel, A., Porcuna-Ferrer, A., Schlingmann, A., Vieira da Cunha Avila, J., & Reyes-García, V. (2021). The role of crop diversity in climate change adaptation: Insights from local observations to inform decision making in agriculture. *Current Opinion in Environmental Sustainability*, *51*, 15–23. https://doi.org/10.1016/J.COSUST.2021.01.006

Levy Guillén, L. (2022). *Percepciones locales de los impactos del cambio climático y adaptaciones entre los agricultores de montaña de Sierra Nevada,* [MSc Thesis]. Universitat Autònoma de Barcelona.

Li, X., Junqueira, A. B., & Reyes-García, V. (2021). At the crossroad of emergency: Ethnobiology, climate change, and indigenous peoples and local communities. *Journal of Ethnobiology*, *41*(3), 307–312. https://doi.org/10.2993/0278-0771-41.3.307

Lippi, S., & Sanfilippo, M. (2024). Smallholders' perceptions of climate change in a post-socialist Albania. In V. Reyes-García, S. Alvarez-Fernandez, P. Benyei, L. Calvet-Mir, D. García-del-Amo, A. B. Junqueira, X. Li, V. Porcher, A. Porcuna-Ferrer, A. Schilngmann, & R. Soleymani (Eds.), *Routledge Handbook of Climate Change Impacts and Adaptation Strategies of Indigenous Peoples and Local Communities*. Oxfordshire & New York: Routledge. Chp 7, pp 124–135.

Maikhuri, R. K., Phondani, P. C., Dhyani, D., Rawat, L. S., Jha, N. K., & Kandari, L. S. (2018). Assessment of climate change impacts and its implications on medicinal plants-based traditional healthcare system in Central Himalaya, India. *Iranian Journal of Science and Technology, Transaction A: Science*, *42*(4), 1827–1835. https://doi.org/10.1007/S40995-017-0354-2/METRICS

Maraun, D., & Widmann, M. (2017). Statistical downscaling and bias correction for climate research. *Statistical Downscaling and Bias Correction for Climate Research*. Cambridge: Cambridge University Press, https://doi.org/10.1017/9781107588783

Mattalia, G., Stryamets, N., Toscano Rivalta, V., & Reyes-García, V. (2024). Correspondence between local and scientific knowledge of climate change: The case of Hutsuls, Northern Romanian Carpathians.

In V. Reyes-García, S. Alvarez-Fernandez, P. Benyei, L. Calvet-Mir, D. García-del-Amo, A. B. Junqueira, X. Li, V. Porcher, A. Porcuna-Ferrer, A. Schlingmann, & R. Soleymani (Eds.), *Routledge Handbook of Climate Change Impacts and Adaptation Strategies of Indigenous Peoples and Local Communities*. Oxfordshire & New York: Routledge. Chp 1, pp 30–44.

McConney, P., Cumberbatch, J., Hinds, C., Oxenford, H. A. & Pena, P. (2024) Sargassum seaweed challenges from local to national level in the Caribbean: a policy cycle perspective. In V. Reyes-García, S. Álvarez-Fernández, P. Benyei, L. Calvet-Mir, D. García-del-Amo, A. B. Junqueira, X. Li, V. Porcher, A. Porcuna-Ferrer, A. Schlingmann, & R. Soleymani (Eds.), *Routledge Handbook of Climate Change Impacts on Indigenous Peoples and Local Communities*. Oxfordshire & New York: Routledge. Chp 3, pp 61–73.

McElwee, P., Fernández-Llamazares, Á., Aumeeruddy-Thomas, Y., Babai, D., Bates, P., Galvin, K., Guèze, M., Liu, J., Molnár, Z., Ngo, H. T., Reyes-García, V., Roy Chowdhury, R., Samakov, A., Shrestha, U. B., Díaz, S., & Brondízio, E. S. (2020). Working with Indigenous and local knowledge (ILK) in large-scale ecological assessments: Reviewing the experience of the IPBES Global Assessment. *Journal of Applied Ecology*, *57*(9), 1666–1676. https://doi.org/10.1111/1365-2664.13705

Merten, J., Stiegler, C., Hennings, N., Purnama, E. S., Röll, A., Agusta, H., Dippold, M. A., Fehrmann, L., Gunawan, D., Hölscher, D., Knohl, A., Kückes, J., Otten, F., Zemp, D. C., & Faust, H. (2020). Flooding and land use change in Jambi Province, Sumatra: Integrating local knowledge and scientific inquiry. *Ecology and Society*, *25*(3), 1–29. https://doi.org/10.5751/ES-11678-250314

Miara, M. D., Negadi, M., Tabak, S., Bendif, H., Dahmani, W., Hammou, M. A., Sahnoun, T., Snorek, J., Porcher, V., Reyes-García, V., & Teixidor-Toneu, I. (2022). Climate change impacts can be differentially perceived across time scales: A study among the Tuareg of the Algerian Sahara. *GeoHealth*, *6*(11), e2022GH000620. https://doi.org/10.1029/2022GH000620

Mwangi, B., Olaka, L. A., Zabra-Calvo, N., Klein, J. A., & Cuni-Sanchez, A. (2024). The role of culture in climate change adaptation: Insights from two mountain regions in Kenya. In V. Reyes-García, S. Alvarez-Fernandez, P. Benyei, L. Calvet-Mir, D. García-del-Amo, A. B. Junqueira, X. Li, V. Porcher, A. Porcuna-Ferrer, A. Schlingmann, & R. Soleymani (Eds.), *Routledge Handbook of Climate Change Impacts and Adaptation Strategies of Indigenous Peoples and Local Communities*. Oxfordshire & New York: Routledge. Chp 19, pp 322–334.

Orlove, B., Dawson, N., Sherpa, P., Adelekan, I., Alangui, W., Carmona, R., Coen, D., Nelson, M., Reyes-García, V., Sanago, G., & Wilson, A. (2023). Placing diverse knowledge systems at the core of transformative climate research. *Ambio*, 52:1431–1447. https://doi.org/10.1007/s13280-023-01857-w

Piggott-McKellar, A. E., McNamara, K. E., Nunn, P. D., & Watson, J. E. M. (2019). What are the barriers to successful community-based climate change adaptation? A review of grey literature. *Local Environment*, *24*(4), 374–390. https://doi.org/10.1080/13549839.2019.1580688

Porcuna-Ferrer, A., Guillermnet, T., Alvarez-Fernandez, S., Calvet-Mir, L., Labeyrie, V., Porcuna-Ferrer, E., & Reyes-García, V. (2023). "So many things have changed": Local understandings of climate change among the Basari. *Environmental Science and Policy*, 148, 103552, SI "The contested nature of climate change: feminist and decolonial approaches for transformative adaptation" I. Iniesta Arandia & F. Ravera. https://doi.org/10.1016/j.envsci.2023.103552

Porcuna-Ferrer, A., Guillerminet, T., Klappoth, B., & Schlingmann, A. (2024). Agricultural adaptation to multiple stressors in a climate change context. A case study in south-eastern Senegal. In V. Reyes-García, S. Alvarez-Fernandez, P. Benyei, L. Calvet-Mir, D. García-del-Amo, A. B. Junqueira, X. Li, V. Porcher, A. Porcuna-Ferrer, A. Schlingmann, & R. Soleymani (Eds.), *Routledge Handbook of Climate Change Impacts and Adaptation Strategies of Indigenous Peoples and Local Communities*. Oxfordshire & New York: Routledge. Chp 13, pp 227–243.

Pörtner, H.-O., Scholes, B., Agard, J., Archer, E., Arneth, A., Bai, X., Barnes, D., Burrows, M., Chan, L., Cheung, W. L. (William), Diamond, S., Donatti, C., Duarte, C., Eisenhauer, N., Foden, W., Gasalla, M. A., Handa, C., Hickler, T., Hoegh-Guldberg, O., … Ngo, H. (2021). *Scientific Outcome of the IPBES-IPCC Co-sponsored Workshop on Biodiversity and Climate Change*. https://doi.org/10.5281/ZENODO.4659159

Race, D., Mathew, S., Campbell, M., & Hampton, K. (2016). Understanding climate adaptation investments for communities living in desert Australia: Experiences of indigenous communities. *Climatic Change*, *139*(3–4), 461–475. https://doi.org/10.1007/S10584-016-1800-4/METRICS

Reyes-García, V., Benyei, P., Junqueira, A.B., Huanca, T., Conde, E. (2024). A complex matrix. Reports of environmental change and its drivers by the Tsimane', Bolivian Amazon. In V. Reyes-García, S. Álvarez-Fernández, P. Benyei, L. Calvet-Mir, D. García-del-Amo, A. B. Junqueira, X. Li, V. Porcher,

A. Porcuna-Ferrer, A. Schlingmann, & R. Soleymani (Eds.), *Routledge Handbook of Climate Change Impacts on Indigenous Peoples and Local Communities*. Oxfordshire & New York: Routledge. Chp 10, pp 164–178.

Reyes-García, V., Cámara-Leret, R., Halpern, B. S., O'Hara, C., Renard, D., Zafra-Calvo, N., & Díaz, S. (2023). Biocultural vulnerability exposes threats of culturally important species. *Proceedings of the National Academy of Sciences of the United States of America, 120*(2), https://doi.org/10.1073/pnas.2217303120

Reyes-García, V., Fernández-Llamazares, Á., Guèze, M., Garcés, A., Mallo, M., Vila-Gómez, M., & Vilaseca, M. (2016). Local indicators of climate change: The potential contribution of local knowledge to climate research. *Wiley Interdisciplinary Reviews: Climate Change, 7*(1), 109–124. https://doi.org/10.1002/wcc.374

Reyes-García, V., García-del-Amo, D., Benyei, P., Fernández-Llamazares, Á., Gravani, K., Junqueira, A. B., Labeyrie, V., Li, X., Matias, D. M., McAlvay, A., Mortyn, P. G., Porcuna-Ferrer, A., Schlingmann, A., & Soleymani-Fard, R. (2019). A collaborative approach to bring insights from local observations of climate change impacts into global climate change research. *Current Opinion in Environmental Sustainability, 39*, 1–8. https://doi.org/10.1016/j.cosust.2019.04.007

Reyes-García, V., Graf, L., Junqueira, A. B., & Madrid, C. (2022b). Decarbonizing the academic sector: Lessons from an international research project. *Journal of Cleaner Production, 368*, 133174. https://doi.org/10.1016/J.JCLEPRO.2022.133174

Reyes-García, V., & Junqueira, A. B. (2023). Climatic and ecological change in the Americas: A perspective from historical ecology. In J. A. Whitaker, C. G. Armstrong, & G. Odomme (Eds.), *Climatic and Ecological Change in the Americas: A Perspective from Historical Ecology. Postface*. Routledge.

Reyes-García, V., Tofighi-Niaki, A., Austin, B. J., Benyei, P., Danielsen, F., Fernández-Llamazares, Á., Sharma, A., Soleymani-Fard, R., & Tengö, M. (2022a). Data sovereignty in community-based environmental monitoring: Toward equitable environmental data governance. *BioScience, 72*(8), 714–717. https://doi.org/10.1093/BIOSCI/BIAC048

Rosenzweig, C., & Neofotis, P. (2013). Detection and attribution of anthropogenic climate change impacts. *Wiley Interdisciplinary Reviews: Climate Change, 4*(2), 121–150. https://doi.org/10.1002/wcc.209

Roue, M., & Nakashima, D. (2018). Indigenous and local knowledge and science: From validation to knowledge coproduction. *The International Encyclopedia of Anthropology*, 1–11. https://doi.org/10.1002/9781118924396.wbiea2215

Ruggieri, F., Porcuna-Ferrer, A., Gaudin, A., Faye, N. F., Reyes-García, V., & Labeyrie, V. (2021). Crop diversity management: Sereer smallholders' response to climatic variability in Senegal. *Journal of Ethnobiology, 41*(3), 389–408. https://doi.org/10.2993/0278-0771-41.3.389

Savo, V., Lepofsky, D., Benner, J. P., Kohfeld, K. E., Bailey, J., & Lertzman, K. (2016). Observations of climate change among subsistence-oriented communities around the world. *Nature Climate Change, 6*(5), 462–473. https://doi.org/10.1038/nclimate2958

Schlingmann et al., 2024 A., Graham, S., Benyei, P., Corbera, E., Martinez Sanesteban, I., Marelle, A., Solemany-Fard, R., & Reyes-García, V. (2021). Global patterns of adaptation to climate change by Indigenous Peoples and local communities. A systematic review. *Current Opinion in Environmental Sustainability, 51*, 55–64. https://doi.org/10.1016/J.COSUST.2021.03.002

Sillitoe, P. (2021). The anthroposcene of weather and climate: Ethnographic contributions to the climate change debate. *The Anthroposcene of Weather and Climate: Ethnographic Contributions to the Climate Change Debate*, 1–345. https://doi.org/10.3167/9781800732315

Singh, P., Schulman, D., Charan, D., Reyes-García, V., Schlingmann, A., Singh, A. A., & Railoa, K. (2024). Climate change adaptation in Fiji: Local adaptation strategies to enhance national policy. In V. Reyes-García, S. Alvarez-Fernandez, P. Benyei, L. Calvet-Mir, D. García-del-Amo, A. B. Junqueira, X. Li, V. Porcher, A. Porcuna-Ferrer, A. Schlingmann, & R. Soleymani (Eds.), *Routledge Handbook of Climate Change Impacts and Adaptation Strategies of Indigenous Peoples and Local Communities*. Oxfordshire & New York: Routledge. Chp 24, pp 389–408.

Smith, B. M., Basu, P. C., Chatterjee, A., Chatterjee, S., Dey, U. K., Dicks, L. V., Giri, B., Laha, S., Majhi, R. K., & Basu, P. (2017). Collating and validating indigenous and local knowledge to apply multiple knowledge systems to an environmental challenge: A case-study of pollinators in India. *Biological Conservation, 211*, 20–28. https://doi.org/10.1016/J.BIOCON.2017.04.032

Stratoudakis, Y., Gonçalvez, P., Oliveira, D., & Mesquita, M. (2024). Settlement, way of life and worldviews: How socio-environmental changes impact and are interpreted by artisanal fishing communities in Portugal. In V. Reyes-García, S. Alvarez-Fernandez, P. Benyei, L. Calvet-Mir, D. García-del-Amo, A.

B. Junqueira, X. Li, V. Porcher, A. Porcuna-Ferrer, A. Schlingmann, & R. Soleymani (Eds.), *Routledge Handbook of Climate Change Impacts and Adaptation Strategies of Indigenous Peoples and Local Communities*. Oxfordshire & New York: Routledge. Chp 9, pp 151–164.

Tengö, M., Brondizio, E. S., Elmqvist, T., Malmer, P., & Spierenburg, M. (2014). Connecting diverse knowledge systems for enhanced ecosystem governance: The multiple evidence base approach. *AMBIO*, *43*(5), 579–591. https://doi.org/10.1007/s13280-014-0501-3

Vázquez-Martínez, L. M. (2022). *Farmers' Perspectives on Changes in Milpa Cropping Systems in Oaxaca, Mexico* [MSc Thesis]. Universitat Autònoma de Barcelona.

Yap, M. L. M., & Watene, K. (2019). The Sustainable Development Goals (SDGs) and Indigenous Peoples: Another missed opportunity? *Journal of Human Development and Capabilities*, *20*(4), 451–467. https://doi.org/10.1080/19452829.2019.1574725

Zant, M., Schlingmann, A., Reyes-García, V., & García-del-Amo, D. (2024). Incremental and transformational adaptation to climate change among Indigenous Peoples and local communities: A global review. *Mitigation and Adaptation Strategies for Global Change*.

PART I

Impacts of climate change and other drivers of global change on local social-ecological systems

INTRODUCTION TO PART I

Impacts of climate and global change on local social-ecological systems

Learning from Indigenous and local knowledge

David García-Del-Amo and André B. Junqueira

Climate change is a global threat affecting all ecosystems and societies in the world (IPCC 2022). Temperature rise, changes in precipitation patterns, and increases in the frequency and intensity of extreme events such as droughts, floods, and storms have been extensively documented by scientists in the last decades, and are a reality across most of the planet (Gulev et al. 2021; Seneviratne et al. 2021). Changes in weather and atmospheric systems are linked to multiple other changes in aquatic and terrestrial physical systems, such as the melting of glaciers and permafrost, sea level rise, ocean acidification, and soil erosion (Fox-Kemper et al. 2021). These changes in atmospheric and physical systems have large and diverse impacts in biodiversity, affecting species distribution, composition, behaviour, reproduction, diseases, as well as the dynamics and functioning of entire ecosystems (Armstrong Mckay et al. 2022; Cooley et al. 2022; Parmesan et al. 2022). Altogether, these biophysical changes have large consequences for socio-ecological systems, as they directly affect resource use and management systems, food, water and economic security, infrastructure and human physical and mental health (Xu et al. 2009; Cissé et al. 2022; García-del-Amo et al. 2022).

Impacts of climate change on socio-ecological systems are very diverse, heterogeneous, and show an uneven geographic distribution. In regions characterised by extreme environmental conditions such as polar regions, deserts, coastal areas, small islands, and mountains, climate change impacts are manifesting with higher intensity (IPCC 2019a, 2019b, 2021). Also, while climate change affects societies all over the world, Indigenous Peoples and local communities are considered particularly vulnerable to its impacts, mostly due to their high dependence on nature and their historical social and economic marginalisation, often due to colonialism (Fayazi et al. 2020; Sultana 2022; Carmona 2024, Chapter 12). Indeed, there are many well-documented examples of Indigenous Peoples and local communities harshly affected by climate change impacts, with effects that compromise different elements of their strongly nature-based livelihoods and culture (Crate 2011). For example, in Chile, Mapuche communities express concern about climate change impacts on araucaria forests (*Araucaria araucana*) as, due to the cultural and economic importance of the species for their society, such impacts might become a threat to Mapuche culture and livelihoods (Carmona 2024, Chapter 12). Similarly, Indigenous Peoples and local communities living in mountainous regions with perpetual snow relate the disappearance of the snowpack to the loss of their local identity, beyond the loss of water resources for their survival (Konchar et al. 2015).

At the same time, Indigenous Peoples and local communities are increasingly recognized as key actors in the challenge of understanding climate change impacts and adapting to them.

 DOI: 10.4324/9781003356837-3

Through their historical and intimate relationship with local ecosystems, Indigenous Peoples and local communities have developed a rich and detailed body of knowledge, based on which they manage nature and identify subtle changes, being the first to react to them (Reyes-García et al. 2016; Savo et al. 2016). In addition, many of them live in remote areas, where historical records and scientific studies are scarce (Rosenzweig and Neofotis 2013). The recognition of the potential of Indigenous knowledge and local knowledge systems in identifying and adapting to climate change impacts is reflected by the recent exponential increase in the number of publications on the topic (Reyes-García et al. 2019; Petzold et al 2020), and by the increasing, although incipient, inclusion of Indigenous Peoples and local communities into climate action (McElwee et al. 2020; Pörter et al. 2021; Schipper et al. 2022; Ford et al. 2016; David-Chavez and Gavin 2018).

As research and policy initiatives on the interface among Indigenous Peoples and local communities, their knowledge systems, and climate change have evolved in the last decades, new research topics, perspectives, and refined understandings also have emerged (see also Orlove et al. 2023). Chapters in this first part of the book bring insights from Indigenous Peoples and local communities experiencing climate change impacts around the world, represented by 12 case studies located in Albania, Italy, Portugal, Romania, Russia, Spain, Argentina, Bolivia, Chile, Peru, and the Caribbean sea region. The case studies featured cover a wide environmental and cultural diversity, including Indigenous Peoples and local communities who practise different livelihood activities (e.g., agriculture, fishing, hunting-gathering) and inhabit different ecosystems such as mountains, polar regions, coastal areas, tropical forests. In this introduction, we discuss insights and advances brought by these chapters on the following three main topics: (1) the place-specific, nuanced, and patterned nature of local knowledge about climate change impacts; (2) the holistic and interconnected understanding of climate change impacts from the perspective of Indigenous knowledge and local knowledge systems, and (3) the local understanding of interactions between climate change and other drivers of socio-environmental change.

Climate change impacts: place-specific, socially and environmentally patterned

While Indigenous knowledge and local knowledge of climate change may have initially been approached as a static and uniform entity, there is increasing evidence unveiling how this knowledge is nuanced and patterned, contingent on socio-economic, cultural and environmental factors (Armah et al. 2015). Echoing patterns found for Indigenous knowledge and local knowledge in general, local understandings about climate change impacts vary across livelihood activities, gender, age, ethnicity, and are influenced by external factors such as access to information or policy initiatives (Crona et al. 2013; Mwangi et al. 2024, Chapter 19). For example, in Chapter 5, García-del-Amo et al. (2024), working with local communities from the mountains of Sierra Nevada, Spain, show how local ecological knowledge about climate change impacts is contingent on the level of engagement with agricultural and livestock activities. In the same line, in Chapter 6, Ferrara and Lindberg (2024), working with local communities in Sicily, show how communities' experiences of 'changing weather' or 'changing climate' are partially culturally constructed and influenced by changes in lifestyles. The way in which Indigenous Peoples and local communities experience climate change impacts is also historically and socially constructed, as demonstrated by Carmona (2024) in Chapter 12 with the case of Mapuche communities in Chile. Similarly, Gerkey and Sharakhmatova (2024), working with local communities of the Kamchatka Peninsula in Russia, show in Chapter 2 how perceived impacts of climate change are aggravated and influenced by the socio-political context where they take place – in their case, by social changes that took place in the post-Soviet era.

This refined understanding of Indigenous knowledge and local knowledge is crucial to leverage the inclusion of Indigenous Peoples and local communities into policy agendas, as well as to foster dialogues between Indigenous knowledge and local knowledge and scientific knowledge (Ford et al. 2016; Xu et al. 2021). Indeed, some of the chapters in this part allude to how a deeper, thorough understanding of Indigenous and local knowledge can favour bridging knowledge systems and hence leverage the development of climate policies and adaptation strategies. For example, in Chapter 1, Mattalia et al. (2024) explored correspondences between climate knowledge from Hutsul communities in Romania and instrumental records, highlighting how the complementarity between knowledge systems can provide a robust basis for co-developing effective climate policies. Comparing evidence derived from local knowledge with instrumental records is a common practice in studies focusing on Indigenous Peoples, local communities and climate change (e.g., Savo et al. 2016; Fernández-Llamazares et al. 2017; Chen and Chen 2020; Paudel et al. 2020). Importantly, and as shown in examples from this volume where local knowledge has been compared with instrumental data (Babai 2024, Chapter 4; Ferrara & Lindberg 2024, Chapter 6; Lippi & Sanfilippo 2024, Chapter 7; Mattalia et al. 2024, Chapter 1), rather than validating one knowledge system with information from another one, this approach should foster identifying complementarities and favouring dialogues between knowledge systems.

While exploring synergies and complementarities between different knowledge systems allows building an 'enriched picture' (Tengö et al. 2014) of climate change impacts, there are still many barriers to the actual engagement of Indigenous and local knowledge into climate policy and action. For example, in the third chapter, McConney et al. (2024) analyse the potential contribution of local knowledge to address an environmental problem derived from climate change impacts in Caribbean countries. During the last decade, massive influxes of pelagic sargassum seaweed have intermittently inundated the Caribbean Sea coasts, disrupting coastal and marine social-ecological systems. The authors show how, in spite of the lack of institutionalised mechanisms for influencing policy, local knowledge has the potential to inform how coastal area ecosystems and livelihoods have been impacted by the sargassum influxes and to guide local coping measures, especially in small-scale fisheries.

The holistic and interconnected understanding of climate change impacts

Through their intimate and historical connections with territories, Indigenous Peoples and local communities have developed holistic worldviews and knowledge systems, in which elements of the natural, human and spiritual world are seen as indissociable, interconnected and interdependent. This understanding is also reflected in how they experience and interpret climate change impacts (Crate 2011; Nakashima and Krupnik 2018). While local knowledge systems may indeed be capable of identifying a change in isolation, for example, a decrease in rainfall, they can also be powerful to map the interconnections between this and other changes, as well as its cascading impacts that extend to different elements of local livelihoods (Junqueira et al. 2021; García-del-Amo et al. 2024, Chapter 5). Cascading effects are a common phenomenon in complex systems, where all components are interconnected, and it is also a fundamental topic in resilience, sustainability and social-ecological frameworks (Xu et al. 2009; Turner et al. 2016; Burton et al. 2020; Haro-Monteagudo et al. 2020). Cascading effects of climate change have been demonstrated (e.g. Bahlai et al. 2021; Rocha et al. 2018), with research showing that they could be able to trigger the extinction of species (Folke et al. 2004; Krönke et al. 2020; Armstrong Mckay et al. 2022). Some of the contributions in this part of the book document reports of cascading effects of climate change derived from Indigenous and local knowledge. In Chapter 4, Babai (2024) examines perceptions of farmer communities in the Eastern Carpathians of Romania, describing how changes in the atmospheric system (in particular changes in the temporal distribution of rainfall, mean temperature, seasonality, and drought intensity) lead to changes in crop productivity, phenology and quality, and trigger local responses such as shifts in the agricultural calendar and practices. In Chapter 7, Lippi and Sanfilippo (2024) document that smallholder farmers in Albania also report atmospheric changes leading to cascading effects in agricultural systems and triggering adaptation responses.

Indigenous and local knowledge can also help unveiling other imbricate and complex connections between climate change impacts. In Chapter 5, García-del-Amo et al. (2024) used a network analysis to investigate climate change impacts reported by local communities from the mountains of Sierra Nevada, Spain, identifying triggering impacts, cascading effects, and mapping the network of interconnections between changes. Likewise, Reyes-García et al (2024), in Chapter 10, unveil how the Tsimane′ understanding of environmental change in Bolivian Amazonia involves a complex network of interconnected social and environmental changes and drivers. Altogether, these examples demonstrate how Indigenous Peoples and local communities′ holistic understanding of nature is also expressed in their understanding of climate change impacts. This holistic understanding unveils the complexity of climate change impacts and their interrelations, allowing the identification of overseen impacts of climate change, of their consequences on the functioning of social-ecological systems, and provides a strong basis for designing local adaptation plans.

The local understanding of interactions between climate change and other drivers of socio-environmental change

While climate change is indeed a major societal challenge and one that is disproportionately impacting Indigenous Peoples and local communities all over the world, its actual impacts on socio-ecological systems depend on interactions with other environmental and socio-political factors. During the last decades, researchers have demonstrated the substantial influence of human activities on ecosystems through direct and indirect drivers of change (Eyring et al. 2021). Land-use change, overexploitation of nature, pollution of waters and soils, or emission of GHG are direct drivers of change affecting ecosystems (Arneth et al. 2020). Indirect drivers of change include rural exodus, economic pressures, policy regulations, and other socio-political factors (Díaz et al. 2019). Together, these drivers are leading to changes in ecosystems, lifestyles, and societies in several aspects and at a pace never seen before in human history (Butchart et al. 2010; Steffen et al. 2015; Díaz et al. 2019; IPBES 2019).

Several of the contributions presented in this volume demonstrate how understanding socio-environmental change from the perspective of Indigenous and local knowledge systems can help in unveiling the complexities of the interactions between multiple drivers of change – including climate change –, and how they are expressed in local socio-ecological systems. Combining information from local knowledge with meteorological records and a long-term reconstruction of climatic and societal past in Sicily, Italy, in Chapter 6, Ferrara and Lindberg (2024) show how environmental changes have been the norm rather than the exception in the region. Such changes are driven by a combination of climatic fluctuations, human agency, and long-term ecological dynamics. In Chapter 8, Izquierdo and Schlingmann (2024), working in the Argentinian Puna, show how local communities see climate change as one among multiple drivers of socio-environmental change, and also how certain drivers (such as the arrival of new technologies and the improvement of transportation conditions) are associated with positive changes, while others (e.g., climate change, migration) are mostly associated with negative ones. Similarly, Stratoudakis et al. (2024) in Chapter 9, demonstrate how fishing communities on the coast of Portugal understand the changes that have affected their fishing practices in the last decades as the product of interactions among multiple socio-economic and environmental drivers, including climate change. In Bolivian Amazonia, Reyes-García et al. (2024) show in Chapter 10 how the Tsimane' complex understanding of environmental change acknowledges the importance of the interactions between different drivers, including social, political, ecological and spiritual elements.

Importantly, many of the socio-environmental changes and drivers acknowledged by Indigenous Peoples and local communities are rooted in the colonialism and/or marginalisation to which they have been historically exposed (Whyte 2017, 2020). For example, Lippi and Sanfilippo (2024), in Chapter 7, demonstrate how smallholder farmers in Albania acknowledge that their livelihoods have been deeply affected by socio-economic changes following the collapse of the communist regime (e.g., land fragmentation, lack of government support, migration). This situation also poses severe barriers to their capacity of implementing adaptation responses to climate change. Likewise, Hirsch (2024) in Chapter 11 discusses how Quechua farmers in the Peruvian Andes have been shifting to market-oriented agriculture stimulated by decades of development projects, which has resulted in reduced food and economic sovereignty, now further aggravated by the challenges of an unpredictable climate. In a similar tone, Carmona (2024) in Chapter 12 debates how the current vulnerability of Mapuche-Pehuenche communities in the Chilean Andes to climate change is 'socially constructed', rooted in a historical process involving changes in land use, pressures for the extraction of forestry resources, and the implementation of public policies. Taken together, these examples bring important insights, rooted on Indigenous and local knowledge, about how social and environmental changes, with the complexity of their multiple drivers, historical and political dimensions, affect socio-ecological systems.

Based on examples from diverse environmental and socio-cultural contexts, the contributions presented in Part I advances the understanding of how climate change and other drivers of global change interact, are expressed in local socio-ecological systems, and are experienced by Indigenous Peoples and local communities. They also demonstrate the potential of Indigenous and local knowledge to disentangle the nuanced and interconnected nature of climate change impacts and drivers. As climate change impacts aggravate the historical vulnerability of Indigenous Peoples and local communities (Birkmann et al. 2022), these advances are timely and crucial to identify pathways for the development of effective climate policies and adaptation strategies.

References

Armah, F. A., Yengoh, G. T., Luginaah, I., Chuenpagdee, R., Hambati, H., & Campbell, G. (2015). Monitored versus experience-based perceptions of environmental change: evidence from coastal Tanzania. *Journal of Integrative Environmental Sciences*, *12*(2), 119–152. https://doi.org/10.1080/1943815X.2015.1017505

Armstrong McKay, D. I., Staal, A., Abrams, J. F., Winkelmann, R., Sakschewski, B., Loriani, S., ... & Lenton, T. M. (2022). Exceeding 1.5 C global warming could trigger multiple climate tipping points. *Science*, *377*(6611), eabn7950.

Arneth, A., Shin, Y. J., Leadley, P., Rondinini, C., Bukvareva, E., Kolb, M., ... & Saito, O. (2020). Post-2020 biodiversity targets need to embrace climate change. *Proceedings of the National Academy of Sciences*, *117*(49), 30882–30891. https://doi.org/10.1073/pnas.2009584117

Babai, D. (2024). "The weather is more erratic, it changes faster…" Local perceptions of climate change in the Eastern Carpathians. In V. Reyes-García, S. Álvarez-Fernández, P. Benyei, L. Calvet-Mir, D. García-del-Amo, A. B. Junqueira, X. Li, V. Porcher, A. Porcuna-Ferrer, A. Schlingmann, & R. Soleymani (Eds.), *Routledge Handbook of Climate Change Impacts on Indigenous Peoples and Local Communities*. Oxfordshire & New York: Routledge. Chp 4, pp 74–94.

Bahlai, C. A., Hart, C., Kavanaugh, M. T., White, J. D., Ruess, R. W., Brinkman, ... & Thompson, J. R. (2021). Cascading effects: insights from the U.S. long term ecological research network. *Ecosphere, 12*(5). https://doi.org/10.1002/ecs2.3430

Birkmann, J., Liwenga, E., Pandey, R., Boyd, E., Djalante, R., Gemenne, F., ... & Wrathall, D. (2022). Poverty, livelihoods and sustainable development. In Pörtner, H.-O., et al. (eds). *Climate Change 2022: Impacts, Adaptation, and Vulnerability. Contribution of Working Group II to the Sixth Assessment Report of the Intergovernmental Panel on Climate Change.* Cambridge University Press, Cambridge and New York, pp. 1171–1274, https://doi.org/10.1017/9781009325844.010

Burton, P. J., Jentsch, A., & Walker, L. R. (2020). The ecology of disturbance interactions. *BioScience, 70*(10), 854–870. https://doi.org/10.1093/biosci/biaa088

Butchart, S. H., Walpole, M., Collen, B., Van Strien, A., Scharlemann, J. P., Almond, R. E., … & Watson, R. (2010). Global biodiversity: indicators of recent declines. *Science, 328*(5982), 1164–1168. https://doi.org/10.1126/science.1187512

Carmichael, B., Wilson, G., Namarnyilk, I., Nadji, S., Cahill, J., & Bird, D. (2017). Testing the scoping phase of a bottom-up planning guide designed to support Australian Indigenous rangers manage the impacts of climate change on cultural heritage sites. *Local Environment, 22*(10), 1197–1216. http://dx.doi.org/10.1080/13549839.2017.1332018

Carmona, R. (2023). The colonial roots of climate vulnerability. Analysing the case of the Mapuche-Pehuenche People in Southern Chile. In V. Reyes-García, S. Álvarez-Fernández, P. Benyei, L. Calvet-Mir, D. García-del-Amo, A. B. Junqueira, X. Li, V. Porcher, A. Porcuna-Ferrer, A. Schlingmann, & R. Soleymani (Eds.), *Routledge Handbook of Climate Change Impacts on Indigenous Peoples and Local Communities*. Oxfordshire & New York: Routledge. Chp 12, pp 196–207.

Chen, T. L., & Cheng, H. W. (2020). Applying traditional knowledge to resilience in coastal rural villages. *International Journal of Disaster Risk Reduction, 47*, 101564. https://doi.org/10.1016/j.ijdrr.2020.101564

Cissé, G., McLeman, R., Adams, H., Aldunce, P., Bowen, K., Campbell-Lendrum, D., … & Tirado, M. C. (2022). Health, wellbeing, and the changing structure of communities. In Pörtner, H.-O., et al. (eds). *Climate Change 2022: Impacts, Adaptation, and Vulnerability. Contribution of Working Group II to the Sixth Assessment Report of the Intergovernmental Panel on Climate Change*. Cambridge University Press, Cambridge and New York, pp. 1041–1170. https://doi.org/10.1017/9781009325844.009

Cooley, S., Schoeman, D., Bopp, L., Boyd, P., Donner, S., Ito, S. I., … & Yool, A. (2022). Ocean and coastal ecosystems and their services. In Pörtner, H.-O., et al. (eds). *Climate Change 2022: Impacts, Adaptation, and Vulnerability. Contribution of Working Group II to the Sixth Assessment Report of the Intergovernmental Panel on Climate Change.* Cambridge University Press, Cambridge and New York, pp. 379–550. https://doi.org/10.1017/9781009325844.005

Crate, S. A. (2011). Climate and culture: anthropology in the era of contemporary climate change. *Annual Review of Anthropology, 40*, 175–194. https://doi.org/10.1146/annurev.anthro.012809.104925

Crona, B., Wutich, A., Brewis, A., & Gartin, M. (2013). Perceptions of climate change: linking local and global perceptions through a cultural knowledge approach. *Climatic Change, 119*, 519–531. https://doi.org/10.1007/s10584-013-0708-5

David-Chavez, D. M., & Gavin, M. C. (2018). A global assessment of Indigenous community engagement in climate research. *Environmental Research Letters, 13*(12), 123005. https://doi.org/10.1088/1748-9326/aaf300

Díaz, S., Settele, J., Brondízio, E. S., Ngo, H. T., Agard, J., Arneth, A., … & Zayas, C. N. (2019). Pervasive human-driven decline of life on Earth points to the need for transformative change. *Science*, *366*, eaax3100. https://doi.org/10.1126/science.aax3100

Eyring, V., Gillett, N. P., Achutarao, K., Barimalala, R., Barreiro Parrillo, M., Bellouin, N., … & Sun, Y. (2021). Human influence on the climate system. In Masson-Delmotte, V., et al. (eds). *The Physical Science Basis. Contribution of Working Group I to the Sixth Assessment Report of the Intergovernmental Panel on Climate Change.* Cambridge University Press, Cambridge and New York, pp. 423–552, https://doi.org/10.1017/9781009157896.005

Fayazi, M., Bisson, I.-A., & Nicholas, E. (2020). Barriers to climate change adaptation in indigenous communities: a case study on the Mohawk community of Kanesatake, Canada. *International Journal of Disaster Risk Reduction*, *49*, 101750. https://doi.org/10.1016/j.ijdrr.2020.101750

Fernández-Llamazares, Á., Garcia, R. A., Díaz-Reviriego, I., Cabeza, M., Pyhälä, A., & Reyes-García, V. (2017). An empirically tested overlap between indigenous and scientific knowledge of a changing climate in Bolivian Amazonia. *Regional Environmental Change, 17*, 1673–1685. https://doi.org/10.1007/s10113-017-1125-5

Ferrara, V., & Lindberg, J., (2023). Climate and environmental change perceptions. A case from rural Sicily (Italy). In V. Reyes-García, S. Álvarez-Fernández, P. Benyei, L. Calvet-Mir, D. García-del-Amo, A. B. Junqueira, X. Li, V. Porcher, A. Porcuna-Ferrer, A. Schlingmann, & R. Soleymani (Eds.), *Routledge Handbook of Climate Change Impacts on Indigenous Peoples and Local Communities*. Oxfordshire & New York: Routledge. Chp 6, pp 109–124.

Folke, C., Carpenter, S., Walker, B., Scheffer, M., Elmqvist, T., Gunderson, L., & Holling, C. S. (2004). Regime shifts, resilience, and biodiversity in ecosystem management. *Annual Review of Ecology, Evolution, and Systematics*, *35*, 557–581. https://doi.org/10.1146/annurev.ecolsys.35.021103.105711

Ford, J. D., Cameron, L., Rubis, J., Maillet, M., Nakashima, D., Willox, A. C., & Pearce, T. (2016). Including indigenous knowledge and experience in IPCC assessment reports. *Nature Climate Change*, *6*(4), 349–353. https://doi.org/10.1038/nclimate2954

Fox-Kemper, B., Hewitt, H. T., Xiao, C., Aðalgeirsdóttir, G., Drijfhout, S. S., Edwards, T. L., ... & Yu, Y. (2021). Ocean, cryosphere and sea level change. In Masson-Delmotte, V., et al. (eds). *The Physical Science Basis. Contribution of Working Group I to the Sixth Assessment Report of the Intergovernmental Panel on Climate Change.* Cambridge University Press, Cambridge and New York, pp. 1211–1362. https://doi.org/10.1017/9781009157896.011

García-del-Amo, D., Calvet-Mir, L., Mortyn, P. G., & Reyes-García, V. (2024). Network analysis of climate change impacts reported by local communities of Sierra Nevada, Spain. In V. Reyes-García, S. Álvarez-Fernández, P. Benyei, L. Calvet-Mir, D. García-del-Amo, A. B. Junqueira, X. Li, V. Porcher, A. Porcuna-Ferrer, A. Schlingmann, & R. Soleymani (Eds.), *Routledge Handbook of Climate Change Impacts on Indigenous Peoples and Local Communities.* Oxfordshire & New York: Routledge. Chp 2, pp 45–60.

García-del-Amo, D., Gálvez-García, C., Iniesta-Arandia, I., Moreno-Ortiz, J., Reyes-García, V. (2022). Local ecological knowledge and the sustainable co-management of Sierra Nevada's social-ecological system. In Zamora, R., & Oliva, M. (eds). *The Landscape of the Sierra Nevada.* Springer, Cham. https://doi.org/10.1007/978-3-030-94219-9_21

García-del-Amo, D., Mortyn, P. G., & Reyes-García, V. (2023). Local reports of climate change impacts in Sierra Nevada, Spain: sociodemographic and geographical patterns. *Regional Environmental Change, 23*(1), 1–16. https://doi.org/10.1007/s10113-022-01981-5

Gerkey, D., & Sharakhmatova, V. (2023). Local observations of climate change and impacts on livelihoods in Kamchatka, Russia. In V. Reyes-García, S. Álvarez-Fernández, P. Benyei, L. Calvet-Mir, D. García-del-Amo, A. B. Junqueira, X. Li, V. Porcher, A. Porcuna-Ferrer, A. Schlingmann, & R. Soleymani (Eds.), *Routledge Handbook of Climate Change Impacts on Indigenous Peoples and Local Communities.* Oxfordshire & New York: Routledge. Chp 2, pp 45–60.

Gulev, S. K., Thorne, P. W., Ahn, J., Dentener, F. J., Domingues, C. M., Gerland, S., ... & Vose, R. S. (2021). Changing state of the climate system. In Masson-Delmotte, V., et al. (eds). *The Physical Science Basis. Contribution of Working Group I to the Sixth Assessment Report of the Intergovernmental Panel on Climate Change.* Cambridge University Press, Cambridge and New York, pp. 287–422. https://doi.org/10.1017/9781009157896.004

Haro-Monteagudo, D., Palazón, L., & Beguería, S. (2020). Long-term sustainability of large water resource systems under climate change: a cascade modeling approach. *Journal of Hydrology, 582*, 124546. https://doi.org/10.1016/j.jhydrol.2020.124546

Hirsch, E. (2024). "Not like it used to be": contending with the altered agricultural calendar in Andean Peru's Colca Valley. In V. Reyes-García, S. Álvarez-Fernández, P. Benyei, L. Calvet-Mir, D. García-del-Amo, A. B. Junqueira, X. Li, V. Porcher, A. Porcuna-Ferrer, A. Schlingmann, & R. Soleymani (Eds.), *Routledge Handbook of Climate Change Impacts on Indigenous Peoples and Local Communities.* Oxfordshire & New York: Routledge. Chp 11, pp 179–195.

IPBES (2019). *Global Assessment Report of the Intergovernmental Science-Policy Platform on Biodiversity and Ecosystem Services.* Brondízio, E. S., Settele, J., Díaz, S., & Ngo, H. T. (eds). IPBES Secretariat, Bonn, 1144 pages. ISBN: 978-3-947851-20-1

IPCC (2019a) *IPCC Special Report on the Ocean and Cryosphere in a Changing Climate.* Pörtner, H-O, Roberts, D. C, Masson-Delmotte, V., Zhai, P., Tignor, M., et al. (eds). Cambridge University Press, Cambridge and New York, p. 755. https://doi.org/10.1017/9781009157964

IPCC (2019b) *Climate Change and Land: An IPCC Special Report on Climate Change, Desertification, Land Degradation, Sustainable Land Management, Food Security, and Greenhouse Gas Fluxes in Terrestrial Ecosystems.* [P.R. Shukla, J. Skea, E. Calvo Buendia, V. Masson-Delmotte, H.-O. Pörtner, D. C. Roberts, P. Zhai, R. Slade, S. Connors, R. van Diemen, M. Ferrat, E. Haughey, S. Luz, S. Neogi, M. Pathak, J. Petzold, J. Portugal Pereira, P. Vyas, E. Huntley, K. Kissick, M. Belkacemi, J. Malley, (eds.)], p. 906.

IPCC (2021) *Climate Change 2021: The Physical Science Basis. Contribution of Working Group I to the Sixth Assessment Report of the Intergovernmental Panel on Climate Change.* Masson- Delmotte, V., Zhai, P., Pirani, A., Connors, S. L, Péan, C., et al. (eds). Cambridge University Press, Cambridge and New York. https://doi.org/10.1017/9781009157896

IPCC (2022) *Climate Change 2022: Impacts, Adaptation, and Vulnerability. Contribution of Working Group II to the Sixth Assessment Report of the Intergovernmental Panel on Climate Change.* Pörtner, H.-O., Roberts, D. C., Tignor, M., Poloczanska, E. S., Mintenbeck, K., et al. (eds). Cambridge University Press, Cambridge and New York, p. 3056. https://doi.org/10.1017/9781009325844

Izquierdo, A., & Schlingmann, A. (2024). Local knowledge of climate change in the context of social, political and technological changes in a High Andean community from the Argentine Puna. In V. Reyes-García, S. Álvarez-Fernández, P. Benyei, L. Calvet-Mir, D. García-del-Amo, A. B. Junqueira, X. Li, V. Porcher,

A. Porcuna-Ferrer, A. Schlingmann, & R. Soleymani (Eds.), *Routledge Handbook of Climate Change Impacts on Indigenous Peoples and Local Communities*. Oxfordshire & New York: Routledge. Chp 8, pp 136–150.

Junqueira, A. B., Fernández-Llamazares, Á., Torrents-Ticó, M., Haira, P. L., Nasak, J. G., Burgas, D., ... & Reyes-García, V. (2021). Interactions between climate change and infrastructure projects in changing water resources: an ethnobiological perspective from the Daasanach, Kenya. *Journal of Ethnobiology*, *41*(3), 331–348. https://doi.org/10.2993/0278-0771-41.3.331

Konchar, K. M., Staver, B., Salick, J., Chapagain, A., Joshi, L., Karki, S., ... & Ghimire, S. K. (2015). Adapting in the shadow of Annapurna: a climate tipping point. *Journal of Ethnobiology*, *35*(3), 449-471. https://doi.org/10.2993/0278-0771-35.3.449

Krönke, J., Wunderling, N., Winkelmann, R., Staal, A., Stumpf, B., Tuinenburg, O. A., & Donges, J. F. (2020). Dynamics of tipping cascades on complex networks. *Physical Review E, 101*(4), 1–9. https://doi.org/10.1103/PhysRevE.101.042311

Lippi, S., & Sanfilippo, M. (2024). Smallholders perceptions of climate change in a post-socialist Albania. In V. Reyes-García, S. Álvarez-Fernández, P. Benyei, L. Calvet-Mir, D. García-del-Amo, A. B. Junqueira, X. Li, V. Porcher, A. Porcuna-Ferrer, A. Schlingmann, & R. Soleymani (Eds.), *Routledge Handbook of Climate Change Impacts on Indigenous Peoples and Local Communities*. Oxfordshire & New York: Routledge. Chp 7, pp 124–135.

Mattalia, G., Stryamets, N., Toscano-Rivalta, V., & Reyes-García, V. (2024). Correspondence between local and scientific knowledge of climate change: the case of Hutsuls of Northern Romanian Carpathians. In V. Reyes-García, S. Álvarez-Fernández, P. Benyei, L. Calvet-Mir, D. García-del-Amo, A. B. Junqueira, X. Li, V. Porcher, A. Porcuna-Ferrer, A. Schlingmann, & R. Soleymani (Eds.), *Routledge Handbook of Climate Change Impacts on Indigenous Peoples and Local Communities*. Oxfordshire & New York: Routledge. Chp 1, pp 30–44.

McConney, P., Cumberbatch, J., Hinds, C., Oxenford, H. A., Pena, M. (2024). Sargassum seaweed challenges from local to national level in the Caribbean: a policy cycle perspective. In V. Reyes-García, S. Álvarez-Fernández, P. Benyei, L. Calvet-Mir, D. García-del-Amo, A. B. Junqueira, X. Li, V. Porcher, A. Porcuna-Ferrer, A. Schlingmann, & R. Soleymani (Eds.), *Routledge Handbook of Climate Change Impacts on Indigenous Peoples and Local Communities*. Oxfordshire & New York: Routledge. Chp 3, pp 61–73.

McElwee, P., Fernández-Llamazares, Á., Aumeeruddy-Thomas, Y., Babai, D., Bates, P., Galvin, K., ... & Brondízio, E. S. (2020). Working with Indigenous and local knowledge (ILK) in large-scale ecological assessments: reviewing the experience of the IPBES Global Assessment. *Journal of Applied Ecology*, *57*(9), 1666–1676. https://doi.org/10.1111/1365-2664.13705

Mwangi, B., Olaka, L. A., Zafra-Calvo, N., Klein, J. A., and Cuni-Sanchez, A., (2024). The role of culture in climate change adaptation: insights from two mountain regions in Kenya. In V. Reyes-García, S. Álvarez-Fernández, P. Benyei, L. Calvet-Mir, D. García-del-Amo, A. B. Junqueira, X. Li, V. Porcher, A. Porcuna-Ferrer, A. Schlingmann, & R. Soleymani (Eds.), *Routledge Handbook of Climate Change Impacts on Indigenous Peoples and Local Communities*. Oxfordshire & New York: Routledge. Chp 19, pp 322–335.

Nakashima, D., & Krupnik, I. (Eds.) (2018). *Indigenous Knowledge for Climate Change Assessment and Adaptation* (Vol. 2). Cambridge: Cambridge University Press.

Orlove, B., Dawson, N., Sherpa, P., Adelekan, I., Alangui, W., Carmona, R., Coen, D., Nelson, M., Reyes-García, V., Sanago, G., Wilson, A. (2023). Placing diverse knowledge systems at the core of transformative climate research. *Ambio*, 52, 1431–1447. https://doi.org/10.1007/s13280-023-01857-w

Parmesan, C., Morecroft, M. D., Trisurat, Y., Adrian, R., Anshari, G. Z., Arneth, A., ... & Talukdarr, G. H. (2022). Terrestrial and freshwater ecosystems and their services. In Pörtner, H.-O., et al. (eds). *Climate Change 2022: Impacts, Adaptation, and Vulnerability. Contribution of Working Group II to the Sixth Assessment Report of the Intergovernmental Panel on Climate Change.* Cambridge University Press, Cambridge and New York, pp. 197–377. https://doi.org/10.1017/9781009325844.004

Paudel, B., Zhang, Y., Yan, J., Rai, R., Li, L., Wu, X., ... & Khanal, N. R. (2020). Farmers' understanding of climate change in Nepal Himalayas: important determinants and implications for developing adaptation strategies. *Climatic Change, 158*, 485–502. https://doi.org/10.1007/s10584-019-02607-2

Petzold, J., Andrews, N., Ford, J. D., Hedemann, C., & Postigo, J. C. (2020). Indigenous knowledge on climate change adaptation: a global evidence map of academic literature. *Environmental Research Letters*, *15*(11), 113007. https://doi.org/10.1088/1748-9326/abb330

Pörtner, H. O., Scholes, R. J., Agard, J., Archer, E., Arneth, A., Bai, X., ... & Ngo, H. T. (2021). *IPBES-IPCC Co-Sponsored Workshop Report on Biodiversity and Climate Change; IPBES and IPCC.* https://doi.org/10.5281/zenodo.4782538

Reyes-García, V., Benyei, P., Junqueira, A.B., Huanca, T., Conde, E. (2024). A complex matrix. Reports of environmental change and its drivers by the Tsimane', Bolivian Amazon. In V. Reyes-García, S. Álvarez-Fernández, P. Benyei, L. Calvet-Mir, D. García-del-Amo, A. B. Junqueira, X. Li, V. Porcher, A. Porcuna-Ferrer, A. Schlingmann, & R. Soleymani (Eds.), *Routledge Handbook of Climate Change Impacts on Indigenous Peoples and Local Communities*. Oxfordshire & New York: Routledge. Chp 10, pp 164–179.

Reyes-García, V., Fernández-Llamazares, Á., Guèze, M., Garcés, A., Mallo, M., Vila-Gómez, M., & Vilaseca, M. (2016). Local indicators of climate change: the potential contribution of local knowledge to climate research. *Wiley Interdisciplinary Reviews: Climate Change*, *7*(1), 109–124. https://doi.org/10.1002/wcc.374

Reyes-García, V., García-del-Amo, D., Benyei, P., Fernández-Llamazares, Á., Gravani, K., Junqueira, A. B., … & Soleymani-Fard, R. (2019). A collaborative approach to bring insights from local observations of climate change impacts into global climate change research. *Current Opinion in Environmental Sustainability*, *39*, 1–8. https://doi.org/10.1016/j.cosust.2019.04.007

Rocha, J. C., Peterson, G., Bodin, Ö., & Levin, S. (2018). Cascading regime shifts within and across scales. *Science*, *362*(6421), 1379–1383. https://doi.org/10.1126/science.aat7850

Rosenzweig, C., & Neofotis, P. (2013). Detection and attribution of anthropogenic climate change impacts. *Wiley Interdisciplinary Reviews: Climate Change, 4*(2), 121–150. https://doi.org/10.1002/wcc.209

Savo, V., Lepofsky, D., Benner, J. P., Kohfeld, K. E., Bailey, J., & Lertzman, K. (2016). Observations of climate change among subsistence-oriented communities around the world. *Nature Climate Change*, *6*(5), 462–473. https://doi.org/10.1038/nclimate2958

Schipper, E. L. F., Revi, A., Preston, B. L., Carr, E. R., Eriksen, S. H., Fernandez-Carril, L. R., … & Singh, P. K. (2022). Climate resilient development pathways. In Pörtner, H.-O., et al. (eds). *Climate Change 2022: Impacts, Adaptation, and Vulnerability. Contribution of Working Group II to the Sixth Assessment Report of the Intergovernmental Panel on Climate Change.* Cambridge University Press, Cambridge and New York, pp. 2655–2807, https://doi.org/10.1017/9781009325844.027

Seneviratne, S. I., Zhang, X., Adnan, M., Badi, W., Dereczynski, C., Di Luca, A., … & Zhou, B. (2021). Weather and climate extreme events in a changing climate. In Masson-Delmotte, V., et al. (eds). *The Physical Science Basis. Contribution of Working Group I to the Sixth Assessment Report of the Intergovernmental Panel on Climate Change.* Cambridge University Press, Cambridge and New York, pp. 1513–1766, https://doi.org/10.1017/9781009157896.013

Steffen, W., Richardson, K., Rockström, J., Cornell, S. E., Fetzer, I., Bennett, E. M., … & Sörlin, S. (2015). Planetary boundaries: guiding human development on a changing planet. *Science*, *347*(6223), 1259855. http://dx.doi.org/10.1126/science.1259855

Stratoudakis, Y., Gonçalves, P., Oliveira, D., & Mesquita, M. (2024). Settlement, way of life and worldviews: how socio-environmental changes impact and are interpreted by artisanal fishing communities in Portugal. In V. Reyes-García, S. Álvarez-Fernández, P. Benyei, L. Calvet-Mir, D. García-del-Amo, A. B. Junqueira, X. Li, V. Porcher, A. Porcuna-Ferrer, A. Schlingmann, & R. Soleymani (Eds.), *Routledge Handbook of Climate Change Impacts on Indigenous Peoples and Local Communities*. Oxfordshire & New York: Routledge. Chp 9, pp 151–164.

Sultana, F. (2022). The unbearable heaviness of climate coloniality. *Political Geography, 99*, 102638. https://doi.org/10.1016/j.polgeo.2022.102638

Tengö, M., Brondizio, E. S., Elmqvist, T., Malmer, P., & Spierenburg, M. (2014). Connecting diverse knowledge systems for enhanced ecosystem governance: the multiple evidence base approach. *Ambio*, *43*, 579–591. https://doi.org/10.1007/s13280-014-0501-3

Turner Ii, B. L., Esler, K. J., Bridgewater, P., Tewksbury, J., Sitas, N., Abrahams, B., … & Mooney, H. (2016). Socio-Environmental Systems (SES) research: what have we learned and how can we use this information in future research programs. *Current Opinion in Environmental Sustainability*, *19*, 160–168. https://doi.org/10.1016/j.cosust.2016.04.001

Whyte, K. (2017). Indigenous climate change studies: indigenizing futures, decolonizing the Anthropocene. *English Language Notes*, *55*(1), 153–162.

Whyte, K. (2020). Too late for indigenous climate justice: ecological and relational tipping points. *Wiley Interdisciplinary Reviews: Climate Change, 11*(1), e603. https://doi.org/10.1002/wcc.603

Xu, H., Cao, Y., Yu, D., Cao, M., He, Y., Gill, M., & Pereira, H. M. (2021). Ensuring effective implementation of the post-2020 global biodiversity targets. *Nature Ecology & Evolution*, *5*(4), 411–418. https://doi.org/10.1038/s41559-020-01375-y

Xu, J., Grumbine, R. E., Shrestha, A., Eriksson, M., Yang, X., Wang, Y. U. N., & Wilkes, A. (2009). The melting Himalayas: cascading effects of climate change on water, biodiversity, and livelihoods. *Conservation Biology, 23*(3), 520–530. https://doi.org/10.1111/j.1523-1739.2009.01237.x

1

CORRESPONDENCE BETWEEN LOCAL AND SCIENTIFIC KNOWLEDGE OF CLIMATE CHANGE

The case of Hutsuls, Northern Romanian Carpathians

Giulia Mattalia, Nataliya Stryamets, Virginia Toscano Rivalta and Victoria Reyes-García

Introduction

Mountain regions provide crucial nature's contributions to highland communities, but also to lower land inhabitants (Schirpke et al. 2019). Despite their essential contributions, mountain areas are particularly vulnerable due to the combination of direct (e.g., extreme weather events, including a steady increase of precipitation) and indirect climate change impacts (e.g., landslides) (Chakraborty 2021; Li et al. 2022). Nevertheless, detecting those impacts is often challenging and efforts based on instrumental data mostly result in coarse predictive models. This is partly so because, in addition to the large heterogeneity in environmental conditions observed in mountain areas, these areas suffer from meteorological data paucity (Gelfan et al. 2017; Hawkins and Sutton 2009; Viviroli et al. 2011).

In the quest to understand climate change impacts in mountain areas, scientists have recently highlighted the potential contributions of local communities, which can provide valuable information (Chaudhary & Bawa 2011; Reyes-García et al. 2016, 2019; Savo et al. 2016). Indeed, the scientific community has started to acknowledge that the perspectives of local communities (defined here as people who have interacted with the surrounding environment for long periods) could provide crucial information on local climate change impacts (Madhanagopal & Pattanaik 2020 and the references within). For instance, climate observations by Ecuadorian mountain communities have helped to select suitable datasets for climatological analysis, providing data otherwise not available (Kieslinger et al. 2019). Local reports of climate change impacts can also enhance our understanding of cascading effects of changes in elements of the atmospheric system (e.g., Garcia-del-Amo 2023; 2024 (2024 refers to chapter 5); Gentle & Thwaites 2016; Ingty 2017; Xenarios et al. 2019). For instance, mountain herders in Nepal note that the scarcity of pastures is generated by extended droughts and lack of snowfall (Aryal et al. 2014). Similarly, Indian mountain farmers,

DOI: 10.4324/9781003356837-4

who mostly rely on rain-fed agriculture, note a reduction in crop productivity due to an overall decrease in precipitation (Shukla et al. 2019).

Despite its richness, to date, local communities' knowledge is rarely taken into account in science and policy fora, for reasons that range from limited accessibility in terms of funding to language barriers (Aryal et al. 2016; Comberti et al. 2016; Matera 2020; Negi et al. 2017). Among the studies on the topic, some have compared climate impacts derived from local knowledge with reports from instrumental recordings, with contrasting results. For example, a study conducted among nomadic Mongolian pastoralists found an alignment between reports of local knowledge and instrumental recordings, local knowledge documenting climate extremes not recorded by instrumental records (Tumenjargal et al. 2020). Conversely, Rakhmanova et al. (2021) found that most of the climate change impacts perceived by the rural dwellers of Western Siberia do not match the measurements by weather stations, arguably because instrumental records are not as specific as local observations. Finally, Shrestha et al. (2019) found a partial overlap between information from the two knowledge systems in Nepal, with convergence on temperature shifts but not on precipitations.

Among the different world regions, understanding of local communities' knowledge on climate change impact is very much neglected in Europe. Indeed, most of the scarce work on local observations of climate change impacts conducted in European mountain ranges has focused on users of ski resorts (e.g., Trawöger 2014 & Wolfsegger et al. 2008 in Austria; Dinca et al. 2014 in Romania, Cholakova & Dogramadjeva 2019 in Bulgaria; Prettenthaler et al. 2022 for a pan-European study), thus ignoring the body of knowledge hold by communities with a long history of interaction with the environment (see García-del-Amo et al. 2023).

To enrich our understanding of climate change impacts in European mountain areas, in this chapter, we explore observations reported by a small ethnolinguistic community and compare them to information obtained from instrumental records in the past 60 years. Specifically, the aims of this work are (1) to document climate change impacts observed by Hutsuls, a mountain community living in Romanian Carpathians; (2) to describe the climatic trends detected through a close-by meteorological station; and (3) to compare data from the local community with those reported by the meteorological station.

Materials & methods

We conducted research in the Carpathians, the second-largest mountain range in Europe, where we worked with the Hutsuls. Recently, the climate of the Carpathians has been examined by several scholars, and particularly through the project "CARPATCLIM, Climate of the Carpathian region", which has analyzed data from hundreds of local meteorological stations until 2010.

The Hutsuls

We conducted research among the Hutsuls, a mountain ethnolinguistic minority living in the Carpathians, largely on Ukrainian territory, although there are also approximately 15,000 Hutsuls in Suceava County, Romania (Figure 1.1). Hutsuls speak a language of Slavic origin, also named Hutsul, as well as the national languages of the states where they live (Ukrainian or Romanian).

Traditionally, Hutsuls relied on the local environment for their livelihood, which heavily depended on small-scale forestry and farming (Huţuleac 2014; Mattalia et al. 2022; Saghin et al. 2017). In spring (April and May) Hutsuls' main activity is preparing homegardens, and especially seeding potato patches. In summer (June–August), the most important activity for Hutsul communities is haymaking, an activity which they complement with wild berries, mushrooms, and

Figure 1.1 Map of the study area: position of (a) Romania within Europe, (b) Suceava County within Romania, and (c) the upper Suceava Valley and the Selyatin meteorological station in Ukraine, close to the Romanian border. Own modification from https://www.mapchart.net/.

medicinal herbs foraging. Autumn (September and October) is the time for preparing food and feeding reservoirs for winter times while harvesting the last mushrooms. During winter (November–March), Hutsuls slaughter pigs, rarely sheep and calves, and prepare their meat. Despite the importance of these traditional activities, nowadays, young Hutsuls increasingly rely on seasonal jobs abroad, as there is a lack of job opportunities in the upper Suceava Valley (especially in industry and services) and the European Union citizenship has facilitated migration to other countries.

Data collection methods

This work uses both secondary and primary information. Secondary information includes data collected at the meteorological station of Selyatin by two different providers. Primary information includes data obtained from semi-structured face-to-face interviews and focus group discussions (FGD) following the research protocol of the Local Indicators of Climate Change Impacts (LICCI) research project (Reyes-García et al. 2023). Before any data were collected, we requested the Free, Prior, and Informed Consent of all participants (CEEAH 4781).

Climatological data were obtained from two sources. For the period 1961–2009, meteorological records were obtained from CARPATCLIM (through its Ukrainian representative). The project ran between 2010 and 2013 and involved institutions from nine countries of the Carpathian area (see Szalai et al. 2013 for details). Among the selected stations, there was Selyatin, the closest meteorological station to the study area (despite its location in the Ukrainian territory; 47°52′ N lat., 25°13′ E long.; 762 m). In our analysis, we integrated the data obtained from CARPATCLIM and data from an online dataset (https://rp5.ua/Weather_archive_in_Selyatin), which contained data from Selyatin meteorological station for the period 2010–2019. We combined information on the two data sources to generate a unique dataset for the period 1961–2019. Both datasets include three variables: daily temperature (avg, in °C), daily precipitation (sum, in mm), and daily snow cover (sum, in cm). Unfortunately, the second dataset (2010–2019) presents incomplete data about snow cover, thus we could not use them for completing the desired time frame (1961–2019).

Twenty in-depth semi-structured interviews were carried out in June and July 2019 in the upper Suceava Valley (Figure 1.1). Quota sampling was used to select informants from different gender and life stages. Our sample includes 13 women and seven men ranging between 37 and 92 years of age. During interviews, we asked participants to report changes observed in the environment. We started by asking to describe any environmental change observed and then we continued by asking them more precisely about specific elements (e.g., forest, flora; for the full protocol, see

Reyes-García et al. 2023). To avoid biases, the expression "climate change" was not used, but rather we referred to "changes in the environment". The term "weather" was mentioned to elicit changes regarding seasons, temperature, or precipitation. Although interviews covered many environmental changes, here we focus on observations of changes in elements of the atmospheric system (i.e., temperature, precipitation, and seasonal events).

Three FGD with people who did not take part in the semi-structured interviews were organized to validate the information mentioned during semi-structured interviews (Reyes-García et al. 2023). In FGD, we presented the list of changes compiled during semi-structured interviews and asked participants to tell us if they had also observed such changes. After the discussion, for each observed change, the researcher moderating the FGD noted her appreciation of whether participants (1) completely agreed; (2) agreed after the debate; (3) disagreed after the debate; or (4) completely disagreed with the statement proposed by the interviewer.

Interviews and FGD were conducted in Romanian by the first author with the help of a native facilitator. We did not record the interviews but took detailed notes of responses.

Data analysis methods

Secondary data were used to calculate (i) daily and (ii) monthly average temperature, (iii) daily temperature range, (iv) monthly precipitation, and (v) annual snow coverage.

Observations of environmental changes noted during interviews were categorized according to the hierarchical classification proposed by Reyes-García et al. (2023). Specifically, we classified observations of environmental changes into Local Indicators of Climate Change Impacts (LICCI), which were then organized into "impacted elements" (e.g., mean temperature) grouped by subsystems affected (e.g., temperature).

We compared observations of changes in elements of the atmospheric system with climatic trends derived from instrumental data from the Selyatin meteorological station following the correspondence in Table 1.1.

Table 1.1 Correspondence between climate trends (a) mentioned by Romanian Hutsuls and (b) derived from instrumental data (Selyatin station), for the period 1961–2019

Climate trend reported by Hutsuls	*Climate trend derived from instrumental data*	
	Original variable	*Trend calculation*
Before there was not such a hot temperature	Daily max temperature (°C)	Days/year with a maximum temperature above 25.4°C (equivalent to the 95th percentile).
Winters are milder	Daily temperature (°C)	Average temperature during winter (avg. daily, to avg. monthly, to avg. winter temperature).
Winter can be colder [than it used to be]	Daily min temperature (°C)	Number of days with minimum temperatures <−16.6°C (equivalent to the fifth percentile).
Temperature changes abruptly	Daily max and min temperature (°C)	Number of days with a temperature variation (max temp-min temp) >21°C (equivalent to the 95th percentile).
Temperature is more extreme	Daily max and min temperature (°C)	We defined extreme cold days, days when minimum temperature <−16.6°C (equivalent to the fifth percentile) and extreme hot when maximum temperature >25.4°C (equivalent to the 95th percentile). We calculated the total number of days per decade with extreme cold or hot days.

(Continued)

Table 1.1 (Continued)

Climate trend reported by Hutsuls	*Climate trend derived from instrumental data*	
	Original variable	*Trend calculation*
Some days are too cold, some days are too hot	Daily avg temperature (°C)	Unusual months (i.e., "too cold" or "too hot") were months with temperature variation over the 95th percentile calculated for every month. We calculated the monthly average temperature, then the average for every month, and finally, the variation for each month.
It snows less	Snow cover sum	Sum of cm of snowfall.
It rains more often	Number of days with recorded daily precipitation	We calculated the days in which rain precipitations occurred per year (excluding the winter months, as at this time, it snows).
Seasons do not exist anymore. Now there are only two seasons [Seasons do no longer present the characteristics they used to have]	Daily temperature; daily precipitation	We defined unusual months (i.e., "too cold" or "too warm"), when temperature variations were over the 95th percentile calculated for every month. To determine "unusual months", we calculated the monthly average temperature, then we calculated the average for every month and the variation for each month of each year 1961–2019. We also calculated seasonal precipitation sum for the years to determine anomalies in it.
Spring comes earlier	Daily snow cover	Date of the last snowfall recorded for each winter.

We are aware that concepts derived from local knowledge may not necessarily overlap with concepts derived from scientific knowledge due to the multifaceted nature of knowledge systems (Barnhardt & Oscar Kawagley 2005; Silvano & Valbo-Jørgensen 2008). Despite these difficulties, and in line with what Tengö et al. (2014) and Fassnacht et al. (2018) have proposed, we argue that the effort to compare evidence derived from different knowledge systems can generate new and important insights to understand the complexity of climate change impacts.

Results

Hutsuls reported several observations of changes in elements of the atmospheric system, including changes in temperature, precipitation, and seasonal events (Table 1.2). In what follows, we describe ten observations of climate change reported by Hutsuls and corresponding trends detected by the meteorological station of Selyatin. We then discuss the level of similarity between data from the two sources.

Changes in temperature

We documented six different observations referring to changes in temperature. The six observations referred either to warmer weather, temperature fluctuations, or more frequent occurrences of extreme temperatures than in the past (see Table 1.2).

Extreme hot temperatures were reported to occur during longer summer periods than in the past. One informant mentioned that *"Before, there were not such hot temperatures [during summer]"* (interview with an Hutsul elder, July 2019). Two out of three FGD agreed on this observation.

Table 1.2 Observations of change in elements of the atmospheric system reported by Hutsuls, level of agreement in Focus Group Discussion (FGD), and matching with records from Selyatin meteorological station

Subsystem	*Local observation*[a]	*FGD agreement*[b]	*Matching with instrumental recordings*
Temperature	Before there was not such a hot temperature.	2/3	Yes
	Winters are milder; there are no cold winters anymore.	2/3	Yes
	Winter can be colder [than it used to be]	3/3	No
	Temperature changes abruptly	3/3	No
	Temperature is more extreme	3/3	Partial
	Some days are too cold, some days are too hot	3/3	Yes
Precipitation	It snows less	3/3	N.D.
	It rains more often	3/3	N.D.
Seasonal events	Seasons do not exist anymore; now there only two seasons	3/3	Yes
	Spring comes earlier	1/3	Yes

[a] Local observation refers to the textual observation reported by Hutsuls.
[b] FGD: Focus group discussion carried out with Hutsuls.

This observation also partially matches with instrumental data, as the meteorological station of Selyatin recorded that the number of days with temperatures above 25.4°C is increasing (Figure 1.2A). For instance, if we compare the first (1961–1980) and the last two decades of analysis (2000–2019), we observe an increase of 0.4°C (from 27°C to 27.4°C) in the average temperature of the hot days. We also observe an increase in the number of days characterized by extreme temperature, from 279 days during the 1961–1980 period to 419 days during the 2000–2019 period (+33%).

Despite the overall trend, the last decade is characterized by alternating hot summers with few days with temperatures above 25.4°C. For instance, in 2012 and 2015, the meteorological station recorded 35 and 33 days, respectively, with a temperature above 25.4°C, while in 2014, only three days with temperatures in that range were recorded. In 2018 no day with a temperature above 25.4°C was recorded.

A commonly mentioned change, validated in two of three FGD, was the general temperature increase during winter. In informants' words: "*Winters are milder*". The Hutsuls call "winter" the period of the year when the ground is covered by snow. Typically, winter was considered to start in November and end in late March or early April (although snowfall can occur from the beginning of October to late April). Informants reported that winters were more rigid in the past, with frosting for several months. The trend of wilder winters reported by Hutsuls corresponds with the trend found when analyzing instrumental data. Thus, according to data from the meteorological station, during the period 1961–2019, there was a temperature increase of 2°C during the months of November–March (being 1961–1970 and 2010–2019 avg winter temperatures respectively equal to −3.9 and −1.9°C). The increase has been more remarkable since the 1990s (Figure 1.2B), with unusually warmer winters in 2019 and 2014.

Hutsuls also mentioned that, although winters are generally milder, there are also some peaks of unusually cold temperatures with some extremely cold winter days, even colder than in the past. Despite these observations, the analysis of instrumental data reveals that extremely low temperatures in winter are not as common as in the past. For instance, in the last decade (2010–2019), there

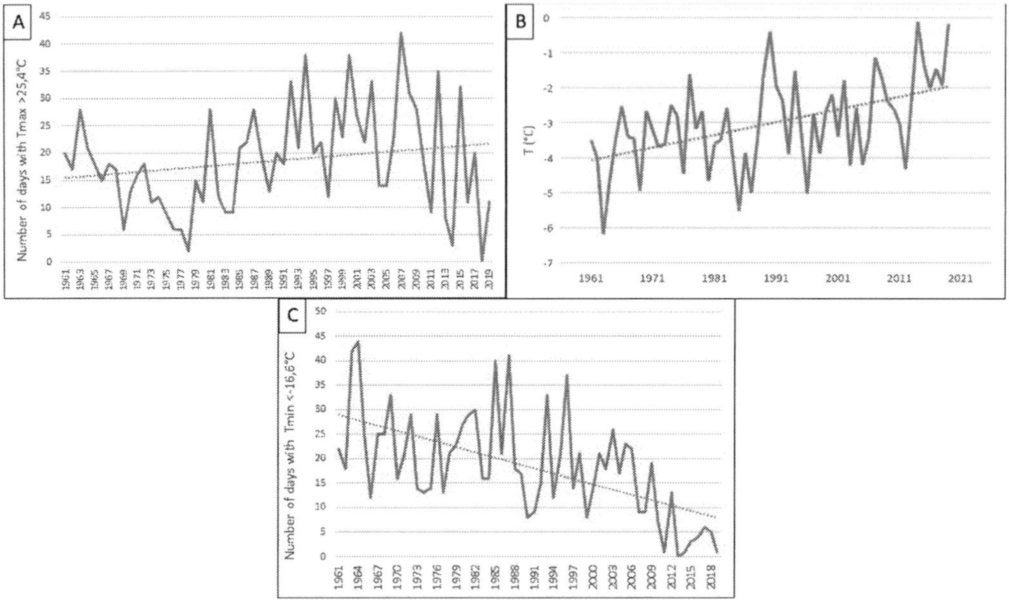

Figure 1.2 Changes in temperature during the period 1961–2019, derived from instrumental data. (A) Number of days/year with extremely high temperatures. (B) Average temperature during winter. (C) Number of days with extremely low temperatures.

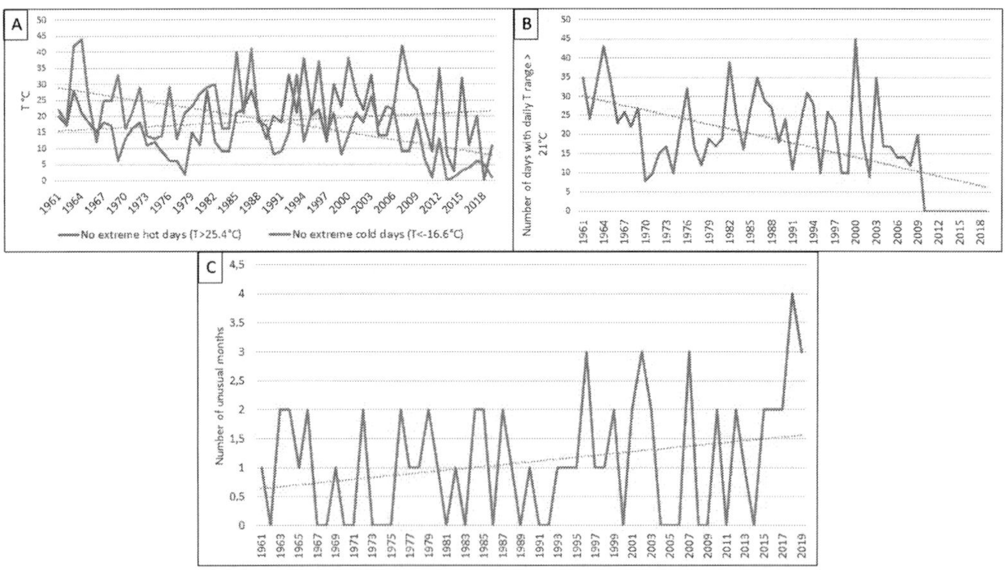

Figure 1.3 Changes in temperature during the period 1961–2019, derived from instrumental data. (A) Number of days with extreme temperature per decade. (B) Number of months with extreme average monthly temperature per decade. (C) Number of months with unusual temperature. See Table 1.1 for definitions.

were 137 fewer days (about 85%) characterized by temperature below −16.6°C than in the period 1961–1970 (Figure 1.2C).

All FGD agreed on the observation that temperature changes more abruptly now than in the past. This observation is relevant because it affects Hutsuls' ability to forecast weather. In their own words: "The weather is becoming increasingly strange". This observation matches with results of the analysis of instrumental data, which show an increase in extremely hot days after the 1961–1979 period and a decrease in extremely cold days during the 2010–2019 period, which has been also unusual in relation to the low number of extremely hot days (Figure 1.3A).

Another change mentioned by Hutsuls during semi-structured interviews, and validated by all FGD, was that nowadays extremely hot and extremely cold days are more frequent than in the past. In their own words: "*Temperature is now more extreme*". The analysis of instrumental data suggests that the last decade (2010–2019) has the lowest number of days with extremely cold temperatures ($n = 41$) recorded during the period. Indeed, instrumental data suggest that the last decade has experienced the lowest number of days with extreme temperatures since the 1960–1969 decade (Figure 1.3B). The 1990–1999 decade had the highest number of extremely hot days ($n = 273$). We consider that there is a partial overlap between Hutsul reports and instrumental records. Indeed, on the longer term (e.g. for older Hutsul participants) this represents an increasing trend of the number of days with extreme temperatures compared to the period before 1990.

In the same line, Hutsuls also reported that "S*ometimes is too cold and sometimes is too hot*", an observation that matches Hutsul's collective memory regarding the weather and the perception of the current climate. All FGD agreed on this observation. By this observation, the analysis of monthly temperature trends for the period 1961–2019 reveals a decadal increase in the number of months characterized by unusual temperatures (Figure 1.3C). Overall, instrumental data shows that the decade 2010–2019 registered the highest number of months with unusual temperatures within our temporal series ($n = 18$).

Precipitation

Hutsuls reported two observations regarding changes in precipitation, one referring to snowfall and the other to rainfall. Hutsuls reported that snowfall is less abundant than it used to be, an observation that was validated in the three FGD. Although the instrumental recordings that would allow a comparison with this observation are incomplete, particularly in the years 2008–2019 (Figure 1.4A), they suggest a decreasing trend, which will be in the same line as the local observations.

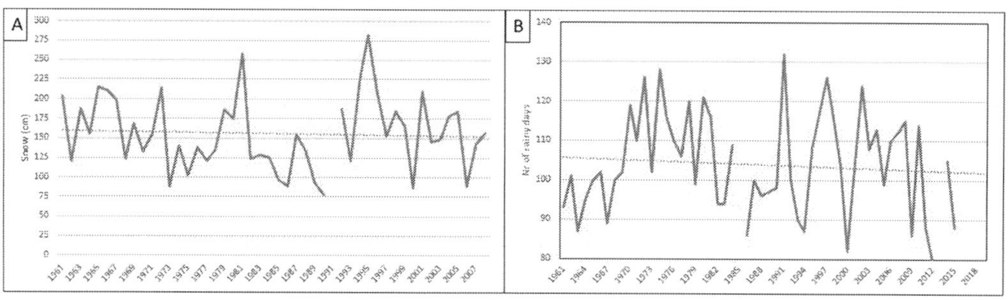

Figure 1.4 Changes in precipitation during the period 1961–2019, derived from instrumental data. (A) Snowfall during winter months. (B) Number of rainy days per year.

Hutsuls also reported that rainfall is more frequent nowadays than in the past, an observation also agreed in the three FGD. Again, data from the meteorological station of Selyatin were insufficient to determine the number of rainy days during the period 2010–2019 as data from the years 2013, 2016, 2017, and 2018 were not complete. However, the general trend seems to be decreasing, although the number of rainy days appears to be increasingly variable since the 1990s (Figure 1.4B).

Seasonal events

During semi-structured interviews, Hutsuls reported two main changes in seasonal events, namely that seasons, as they used to distinguish them, do not exist anymore and that spring comes earlier now than in the past.

During interviews, respondents mentioned that "*Seasons do not exist anymore*". Other people said that "*Now, there are only two seasons*", referring to the difference between summer and winter, and implying that the seasons shift abruptly between these two, with spring and autumn not being well-defined anymore. The three FGD agreed on this observation, which also seems to be in consonance with changes in temperature and precipitation documented through instrumental recordings. As mentioned, instrumental data show that winter temperature (avg T −3°C) is increasing and that snow precipitations are decreasing (Figure 1.5A). In the dataset analyzed, 2011 was characterized by the driest winter since 1961.

Spring average temperature (7.8°C) was relatively stable over the period 1961–2019, although precipitations experienced a slightly increasing trend. Spring precipitation presented an unusual peak in 2012, corresponding to the heavy rains recorded on May 27th, which resulted in 968 mm of rain in 12 hours (Figure 1.5B). Summer trends (avg T 14.6°C) indicate a slight increase in temperatures,

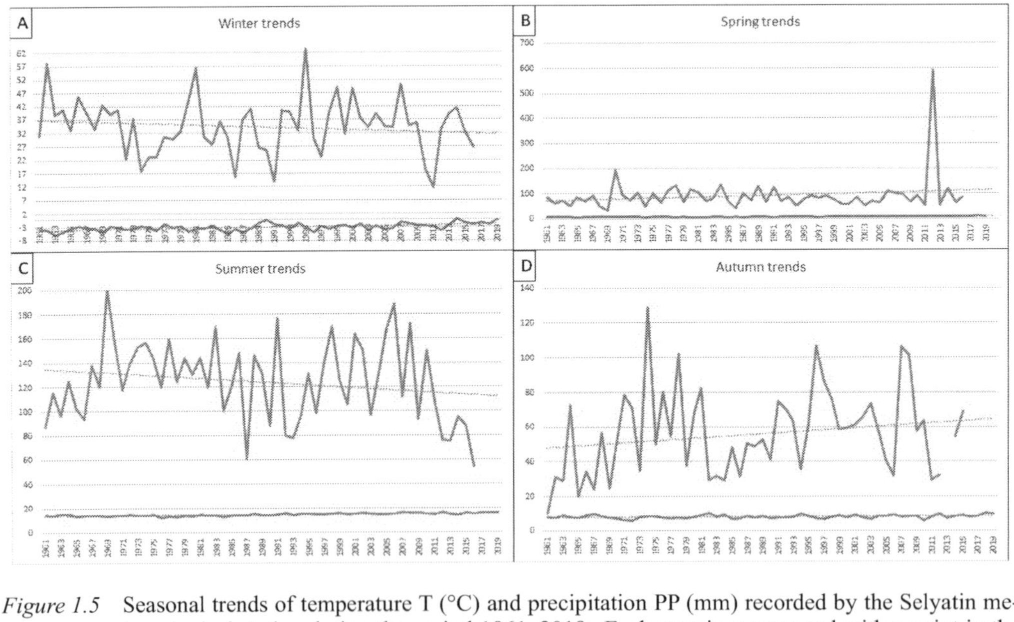

Figure 1.5 Seasonal trends of temperature T (°C) and precipitation PP (mm) recorded by the Selyatin meteorological station during the period 1961–2019. Each year is represented with a point in the graph corresponding to the monthly average seasonal temperature (lower blue lines) and precipitation sum (upper orange lines).

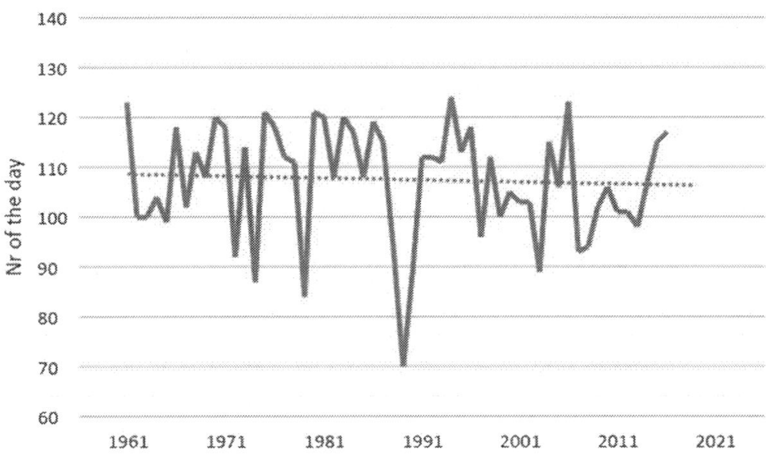

Figure 1.6 Latest day of snow cover according to the Selyatin meteorological station. Day 100 corresponds to April 9th.

with the summer of 2012 characterized by an exceptionally hot temperature and a severe decline in summer precipitations (Figure 1.5C). Autumn trends (avg T 8.1°C) show slightly increasing temperatures and decreasing rainfall, thus resembling more to summer trends (Figure 1.5D). Beyond general trends, in the last decade, the meteorological station of Selyatin recorded three unusual warm seasons (2012, 2018, 2019). Thus, changes in the main characteristics of each season detected by the closest meteorological station are in line with Hutsuls' observations that seasons are now blurred and sometimes hard to be distinguished. For instance, winters are getting milder thus resembling autumns, and autumns are becoming warmer and drier, like late summers.

Informants participating in semi-structured interviews also mentioned that "*Spring comes earlier*", although there was a certain level of disagreement with this trend. Indeed, only participants in one FGD agreed on the trend and several informants mentioned that spring can come earlier or later than in the past, depending on the year. However, the observation overlaps with instrumental data, which shows an advancement of the last day of snow cover, which corresponds to the beginning of spring (Figure 1.6).

Discussion

We present ten changes in elements of the atmospheric system reported by Hutsuls. Out of these, six showed similar trends to data gathered in the closest meteorological station, two were not aligned, and two could not be assessed due to gaps in instrumental data. In this discussion, we comment on this comparison and discuss the potential to combine Hutsul knowledge with instrumental data for enriching our understanding of climate change impacts in European mountain areas.

Before discussing these results, we want to highlight two caveats that might affect them. First, we acknowledge that we used convenient sampling to select study participants, for which our sample might not be representative of the entire Hutsul population of the upper Suceava Valley. Second, the instrumental dataset used is not fully complete (especially concerning precipitation), limiting our ability to make comparisons.

Climate change impacts reported by Hutsuls resemble those reported by other mountain communities

The first important finding of this work is that changes reported by Hutsuls (e.g., warmer temperatures, precipitation decrease, changes in seasonal events) have also been observed in other mountain areas of the world (see, for instance, Babai 2024; Fuchs et al. 2024; Garcia-del-Amo 2024). For example, Cuni-Sanchez et al. (2018) in Kenya and Pandey (2019) in Nepal find a general increase in the mean temperature and a decrease in rainfall, and farmers in Himalayan and South African highlands mountain report an increase in winter temperature (Basannagari and Kala 2013; Rankoana 2016; Sujakhu et al. 2016). Increased frequency of cold and hot temperatures was also found among farmers in the Ecuadorian Andes (Córdova et al. 2019; Postigo 2014), although several other studies only report an increase in extreme heat – but not cold – events (e.g., Meena et al. 2019). As Hutsuls, Ethiopian mountain cereal farmers, Pakistani yak and goat herders, and Mexican horticulturalists have also noticed unusual hot temperatures (Joshi et al. 2013; Kassie et al. 2013; Sánchez-Cortés & Chavero 2011). Changes in precipitation have also been mentioned by subsistence Himalayan and Andean farmers (Gurgiser et al. 2016; Joshi et al. 2019). Changes in seasonality have been reported by potato growers in Bolivia and by mixed farmers in India (Boillat & Berkes, 2013; Meena et al. 2019).

We derive an important insight from such similarities. Climate change might result in similar impacts across mountain ecosystems of the Earth. These impacts seem to be generally characterized by increased temperatures and decreased precipitations. This is in line with the ideas that mountains are particularly affected by climate change impacts (Schirpke et al. 2019) and calls for urgent management strategies (Chakraborty 2021; Li et al. 2022; Lurgi et al. 2012). Considering the high climatic variability and instrumental data paucity in mountain areas, the changes perceived by local populations can inform the development of resource managers on tailoring management strategies to face the current climatic crisis.

Local climate change impacts reported by Hutsuls mostly align with instrumental data

The second main finding of this work refers to the general match between Hutsuls' reports of changes in elements of the atmospheric system and reports from data from the meteorological station. Six out of eight observations of climate change impacts reported by Hutsuls are similar to those recorded by the meteorological station of Selyatin. Moreover, Hutsuls' observations of extreme temperatures are also in line with findings from several studies analyzing instrumental data in the Carpathian Mountains (Bartholy & Pongrácz 2007; Birsan et al. 2019; Spinoni et al. 2015). This match reinforces the idea that communities depending on nature are particularly prone to perceive the impacts of climate change, and thus provide essential knowledge, particularly in context where no other data sources are available (e.g., Gentle & Maraseni 2012; Pearce et al. 2012; Shukla et al. 2019). Instrumental recordings can provide precise and frequent data on temperature and precipitation for long periods. However, they are potentially subject to data gaps, such as those derived from technical or human errors, or breakage of instruments due to weather conditions or wild fauna and flora interactions. For example, in the Selyatin station there were gaps concerning rainfall amount and, in a couple of cases, also concerning temperature. While we do not know the exact reasons for these data gaps, the case exemplifies the importance of finding alternative data sources in mountain regions.

It is interesting to notice that while Hutsuls' reports of increased incidence of unusual hot periods, milder winters, and decreased snowfall amount are aligned with scientific observations and projections (Birsan et al. 2014; Micu 2009), the reports of an increase in extremely cold events differ with scientific records, which show a robust reduction of such events (Birsan et al. 2014). We can think of several explanations for this mismatch. For example, the average annual temperature rise may have affected the overall perception of extreme "cold" temperatures. Another possible reason for the mismatch could be the different scales and sensibility of the two knowledge systems. Considering Hutsuls' main activity (mixed farming), they may be more sensitive to the climate in certain periods of the year. Thus, it is possible that Hutsuls particularly noticed an increase in the frequency of the rains during spring and summer, which are crucial for haymaking, which would contrast with instrumental data that records an overall decrease in rainy days on a monthly or annual basis. These local reports are crucial as climate policies are often planned at national and global scales, neglecting the peculiarity of specific areas and particularly the heterogeneity of mountain regions.

Conclusions

We found several instances of overlap between Hutsuls' reports of climatic changes and meteorological records, particularly concerning changes in temperature and seasonal events. This overlap supports the idea that different knowledge systems bring synergic insights to our understanding of local climate change impacts for which considering insights of both knowledge systems could lead to better contextualized, and potentially more efficient, climate adaptation plans. Following previous work on the topic, we propose a blended approach, combining both data from meteorological stations and mountain communities' reports of climate change as the basis for co-developing climate policies.

Acknowledgments

We thank the participants of the first LICCI partner workshop: Rosario Carmona, Daniel Babai, Leneisja Jungsberg, and Priyatma Singh for supporting and fostering the writing process of this chapter. We are grateful to Yurii Nabyvanets and CARPATCLIM for sharing the dataset of Selyatin and to Angela for her linguistic facilitation in the field. We also thank Camilla Dibari, Marco Moriondo and Giacomo Trombi for their suggestions for data analysis. This project received funding from the European Research Council under a Consolidator Grant (FP7-771056-LICCI) and contributed to the "María de Maeztu Unit of Excellence" (CEX2019-000940-M).

References

Aryal, A., Brunton, D., & Raubenheimer, D. (2014). Impact of climate change on human-wildlife-ecosystem interactions in the Trans-Himalaya region of Nepal. *Theoretical and Applied Climatology, 115*(3), 517–529.

Aryal, S., Cockfield, G., & Maraseni, T. N. (2016). Perceived changes in climatic variables and impacts on the transhumance system in the Himalayas. *Climate and Development, 8*(5), 435–446.

Babai, D, (2024). "The weather is more erratic; it changes faster..." Local perceptions of climate change in the Eastern Carpathians, Romania. Reyes-García, S. Álvarez-Fernández, P. Benyei, L. Calvet-Mir, D. García-del-Amo, A. B. Junqueira, X. Li, V. Porcher, A. Porcuna-Ferrer, A. Schlingmann, & R. Soleymani (Eds.), Routledge Handbook of Climate Change Impacts on Indigenous Peoples and Local Communities. Oxfordshire & New York: Routledge. Chp 4, pp 74–94.

Barnhardt, R., & Oscar Kawagley, A. (2005). Indigenous knowledge systems and Alaska Native ways of knowing. *Anthropology & Education Quarterly, 36*(1), 8–23.

Bartholy, J., & Pongrácz, R. (2007). Regional analysis of extreme temperature and precipitation indices for the Carpathian Basin from 1946 to 2001. *Global and Planetary Change, 57*(1–2), 83–95.

Basannagari, B., & Kala, C. P. (2013). Climate change and apple farming in Indian Himalayas: A study of local perceptions and responses. *PLoS One, 8*(10), e77976.

Birsan, M. V., Dumitrescu, A., Micu, D. M., & Cheval, S. (2014). Changes in annual temperature extremes in the Carpathians since AD 1961. *Natural Hazards, 74*(3), 1899–1910.

Birsan, M. V., Micu, D. M., Nita, I. A., Mateescu, E., Szep, R., & Keresztesi, A. (2019). Spatio-temporal changes in annual temperature extremes over Romania (1961–2013). *Romanian Journal of Physics, 64*(7–8), 816.

Boillat, S., & Berkes, F. (2013). Perception and interpretation of climate change among Quechua farmers of Bolivia: Indigenous knowledge as a resource for adaptive capacity. *Ecology and Society, 18*(4), 21.

Chakraborty, A. (2021). Mountains as vulnerable places: A global synthesis of changing mountain systems in the Anthropocene. *GeoJournal, 86*(2), 585–604.

Chaudhary, P., & Bawa, K. S. (2011). Local perceptions of climate change validated by scientific evidence in the Himalayas. *Biology Letters, 7*(5), 767–770.

Cholakova, S., & Dogramadjeva, E. (2019). Climate change and ski industry in Pamporovo resort, Bulgaria: An exploratory study on the tourists' perceptions. *European Journal of Tourism Research, 22*, 166–192.

Comberti, C., Thornton, T., & Korodimou, M. (2016). Addressing Indigenous peoples' marginalisation at international climate negotiations: Adaptation and resilience at the margins. Working paper, ECI, University of Oxford.

Córdova, R., Hogarth, N. J., & Kanninen, M. (2019). Mountain farming systems' exposure and sensitivity to climate change and variability: Agroforestry and conventional agriculture systems compared in Ecuador's indigenous territory of Kayambi people. *Sustainability, 11*(9), 2623.

Cuni-Sanchez, A., Omeny, P., Pfeifer, M., Olaka, L., Mamo, M. B., Marchant, R., & Burgess, N. D. (2018). Climate change and pastoralists: Perceptions and adaptation in montane Kenya. *Climate and Development, 11*(6), 513–524.

Dinca, A. I., Surugiu, C., Surugiu, M., & Frent, C. (2014). Stakeholder perspectives on climate change effects on tourism activities in the northern Romanian Carpathians: Vatra Dornei resort case study. *Human Geographies, 8*(1), 27.

Fassnacht, S. R., Allegretti, A. M., Venable, N. B., Fernández-Giménez, M. E., Tumenjargal, S., Kappas, M., Laituri M. J., Batbuyan B., & Pfohl, A. K. (2018). Merging indigenous knowledge systems and station observations to estimate the uncertainty of precipitation change in Central Mongolia. *Hydrology, 5*(3), 46.

Fuchs, A., Vogl C. R., & Schunko, C. (2024). Counteracting land abandonment: Local adaptation strategies to climate change impacts of alpine farmers in Eastern Tyrol, Austria. In V. Reyes-García, S. Álvarez-Fernández, P. Benyei, L. Calvet-Mir, D. García-del-Amo, A. B. Junqueira, X. Li, V. Porcher, A. Porcuna-Ferrer, A. Schlingmann, & R. Soleymani (Eds.), *Routledge Handbook of Climate Change Impacts on Indigenous Peoples and Local Communities*. Oxfordshire & New York: Routledge. Chp 17, pp 287–302.

García-del-Amo, D., Calvet-Mir, L., Mortyn, P.G., Reyes-García V. (2024). Network analysis of climate change impacts reported by local communities of Sierra Nevada, Spain. In V. Reyes-García, S. Álvarez-Fernández, P. Benyei, L. Calvet-Mir, D. García-del-Amo, A. B. Junqueira, X. Li, V. Porcher, A. Porcuna-Ferrer, A. Schlingmann, & R. Soleymani (Eds.), Routledge Handbook of Climate Change Impacts on Indigenous Peoples and Local Communities. Oxfordshire & New York: Routledge. Chp 5, pp 95–108.

García-del-Amo, D., Mortyn, P.G., & Reyes-García, V. (2023). Local reports of climate change impacts in Sierra Nevada, Spain: Sociodemographic and geographical patterns. *Regional Environmental Change*, 23, 14.

Gelfan, A., Gustafsson, D., Motovilov, Y., Arheimer, B., Kalugin, A., Krylenko, I., & Lavrenov, A. (2017). Climate change impact on the water regime of two great Arctic rivers: Modelling and uncertainty issues. *Climatic Change, 141*(3), 499–515.

Gentle, P., & Maraseni, T. N. (2012). Climate change, poverty and livelihoods: Adaptation practices by rural mountain communities in Nepal. *Environmental Science & Policy, 21*, 24–34.

Gentle, P., & Thwaites, R. (2016). Transhumant pastoralism in the context of socioeconomic and climate change in the mountains of Nepal. *Mountain Research and Development, 36*(2), 173–182.

Gurgiser, W., Juen, I., Singer, K., Neuburger, M., Schauwecker, S., Hofer, M., & Kaser, G. (2016). Comparing peasants' perceptions of precipitation change with precipitation records in the tropical Callejón de Huaylas, Peru. *Earth System Dynamics, 7*(2), 499–515.

Hawkins, E., & Sutton, R. (2009). The potential to narrow uncertainty in regional climate predictions. *Bulletin of the American Meteorological Society, 90*(8), 1095–1108.

Huţuleac, D. (2014). Despre originea, limba şi cultura etniei huţule în contextul Globalizării. *Memoria ethnologica, 14(50/51)*, 74–85.

Ingty, T. (2017). High mountain communities and climate change: Adaptation, traditional ecological knowledge, and institutions. *Climatic Change, 145*(1–2), 41–55.

Joshi, A., Farquhar, S., Assareh, N., Dahlet, L., & Landahl, E. (2019). Climate change in Lamjung District, Nepal: Meteorological evidence, community perceptions, and responses. *Environmental Research Communications*, *1*(3), 031004.

Joshi, S., Jasra, W. A., Ismail, M., Shrestha, R. M., Yi, S. L., & Wu, N. (2013). Herders' perceptions of and responses to climate change in northern Pakistan. *Environmental Management*, *52*(3), 639–648.

Kassie, B. T., Hengsdijk, H., Rötter, R., Kahiluoto, H., Asseng, S., & Van Ittersum, M. (2013). Adapting to climate variability and change: Experiences from cereal-based farming in the Central Rift and Kobo Valleys, Ethiopia. *Environmental Management*, *52*(5), 1115–1131.

Kieslinger, J., Pohle, P., Buitrón, V., & Peters, T. (2019). Encounters between experiences and measurements: The role of local knowledge in climate change research. *Mountain Research and Development, 39*(2), R55–R68.

Li, B. V., Jenkins, C. N., & Xu, W. (2022). Strategic protection of landslide vulnerable mountains for biodiversity conservation under land-cover and climate change impacts. *Proceedings of the National Academy of Sciences, 119*(2), e2113416118.

Lurgi, M., Lopez, B. C., & Montoya, J. M. (2012). Climate change impacts on body size and food web structure on mountain ecosystems. *Philosophical Transactions of the Royal Society B: Biological Sciences, 367*(1605), 3050–3057.

Madhanagopal, D., & Pattanaik, S. (2020). Exploring fishermen's local knowledge and perceptions in the face of climate change: The case of coastal Tamil Nadu, India. *Environment, Development and Sustainability, 22*(4), 3461–3489.

Matera, J. (2020). Bridging traditional and scientific knowledge of climate change: Understanding change through the lives of small island communities. *Human Ecology*, *48*(5), 529–538.

Mattalia, G., Stryamets, N., Balazsi, A., Molnar, G., Gliga, A., Pieroni, A.,... & Reyes-García, V. (2022). Hutsuls' perceptions of forests and uses of forest resources in Ukrainian and Romanian Bukovina. *International Forestry Review, 24*(3), 393–410.

Meena, R. K., Verma, T. P., Yadav, R. P., Mahapatra, S. K., Surya, J. N., Singh, D., & Singh, S. K. (2019). Local perceptions and adaptation of indigenous communities to climate change: Evidences from High Mountain Pangi valley of Indian Himalayas. *Indian Journal of Traditional Knowledge, 18*(1), 58–67.

Micu, D. (2009). Snow pack in the Romanian Carpathians under changing climatic conditions. *Meteorology and Atmospheric Physics*, *105*(1–2), 1–16.

Negi, V. S., Maikhuri, R. K., Pharswan, D., Thakur, S., & Dhyani, P. P. (2017). Climate change impact in the Western Himalaya: People's perception and adaptive strategies. *Journal of Mountain Science, 14*(2), 403–416.

Pandey, R. (2019). Farmers' perception on agro-ecological implications of climate change in the Middle-Mountains of Nepal: A case of Lumle Village, Kaski. *Environment, Development and Sustainability, 21*(1), 221–247.

Pearce, T., Ford, J. D., Caron, A., & Kudlak, B. P. (2012). Climate change adaptation planning in remote, resource-dependent communities: An Arctic example. *Regional Environmental Change, 12*(4), 825–837.

Postigo, J. C. (2014). Perception and resilience of Andean populations facing climate change. *Journal of Ethnobiology*, *34*(3), 383–400.

Prettenthaler, F., Kortschak, D., & Woess-Gallasch, S. (2022). Modelling tourists' responses to climate change and its effects on alpine ski tourism–A comparative approach for European Regions. *Journal of Outdoor Recreation and Tourism 39*, 100525.

Rakhmanova, L., Kolesnichenko, L., Kuzhevskaya, I., Kolesnichenko, I., Vorobev, R., Tyulyupo, S., Drozdov, V., & Shaduyko, O. (2021). Perspectives of climate change: A comparison of scientific understanding and local interpretations by different Western Siberian communities. *Ambio*, *50*, 2072–2089.

Rankoana, S. A. (2016). Perceptions of climate change and the potential for adaptation in a rural community in Limpopo Province, South Africa. *Sustainability*, *8*(8), 672.

Reyes-García, V., Álvarez-Fernández, S., Benyei, P., García-del-Amo, D., Junqueira, A. B., Labeyrie, V., ... & Soleymani, R. (2023). Local indicators of climate change impacts described by indigenous peoples and local communities: Study protocol. *PLoS One*, *18*(1), e0279847.

Reyes-García, V., García-del-Amo, D., Benyei, P., Fernández-Llamazares, Á., Gravani, K., Junqueira, A. B., & Mortyn, P. G. (2019). A collaborative approach to bring insights from local observations of climate change impacts into global climate change research. *Current Opinion in Environmental Sustainability, 39*, 1–8.

Reyes-García, V., Fernández-Llamazares, Á., Guèze, M., Garcés, A., Mallo, M., Vila-Gómez, M., & Vilaseca, M. (2016). Local indicators of climate change: the potential contribution of local knowledge to climate research. *Wiley Interdisciplinary Reviews: Climate Change*, *7*(1), 109-124.
Saghin, D., Matei, E., Cojocaru, I., Manea, G., & Cocoş, O. (2017). The Hutsuls in South Bukovina: From rural tradition to sustainable development. *Eastern European Countryside, 23*(1), 119–145.
Sánchez-Cortés, M. S., & Chavero, E. L. (2011). Indigenous perception of changes in climate variability and its relationship with agriculture in a Zoque community of Chiapas, Mexico. *Climatic Change*, *107*(3–4), 363–389.
Savo, V., Lepofsky, D., Benner, J. P., Kohfeld, K. E., Bailey, J., & Lertzman, K. (2016). Observations of climate change among subsistence-oriented communities around the world. *Nature Climate Change, 6*(5), 462–473.
Schirpke, U., Tappeiner, U., Tasser, E. (2019). A transnational perspective of global and regional ecosystem service flows from and to mountain regions. *Scientific Reports*, *9*(1), 1–11.
Shrestha, U. B., Shrestha, A. M., Aryal, S., Shrestha, S., Gautam, M. S., & Ojha, H. (2019). Climate change in Nepal: A comprehensive analysis of instrumental data and people's perceptions. *Climatic Change, 154*(3–4), 315–334.
Shukla, R., Agarwal, A., Sachdeva, K., Kurths, J., & Joshi, P. K. (2019). Climate change perception: An analysis of climate change and risk perceptions among farmer types of Indian Western Himalayas. *Climatic Change*, *152*(1), 103–119.
Silvano, R. A., & Valbo-Jørgensen, J. (2008). Beyond fishermen's tales: Contributions of fishers' local ecological knowledge to fish ecology and fisheries management. *Environment, Development and Sustainability,* 10(5), 657–675.
Spinoni, J., Lakatos, M., Szentimrey, T., Bihari, Z., Szalai, S., Vogt, J., & Antofie, T. (2015). Heat and cold waves trends in the Carpathian Region from 1961 to 2010. *International Journal of Climatology, 35*(14), 4197–4209.
Sujakhu, N. M., Ranjitkar, S., Niraula, R. R., Pokharel, B. K., Schmidt-Vogt, D., & Xu, J. (2016). Farmers' perceptions of and adaptations to changing climate in the Melamchi valley of Nepal. *Mountain Research and Development*, *36*(1), 15–30.
Szalai, S., Auer, I., Hiebl, J., Milkovich, J., Radim, T. Stepanek, P., Zahradnicek, P., Bihari, Z., Lakatos, M., Szentimrey, T., Limanowka, D., Kilar, P., Cheval, S., Deak, Gy., Mihic, D., Antolovic, I., Mihajlovic, V., Nejedlik, P., Stastny, P., Mikulova, K., Nabyvanets, I., Skyryk, O., Krakovskaya, S.Vogt, J., Antofie, T., & Spinoni, J. (2013). Climate of the Greater Carpathian Region. Final Technical Report.www.carpatclim-eu.org.
Tengö, M., Brondizio, E. S., Elmqvist, T., Malmer, P., & Spierenburg, M. (2014). Connecting diverse knowledge systems for enhanced ecosystem governance: The multiple evidence based approach. *Ambio, 43*(5), 579–591.
Trawöger, L. (2014). Convinced, ambivalent or annoyed: Tyrolean ski tourism stakeholders and their perceptions of climate change. *Tourism Management*, *40*, 338–351.
Tumenjargal, S., Fassnacht, S. R., Venable, N. B., Kingston, A. P., Fernández-Giménez, M. E., Batbuyan, B., Laituri, M.J., Kappas M. & Adyabadam, G. (2020). Variability and change of climate extremes from indigenous herder knowledge and at meteorological stations across central Mongolia. *Frontiers of Earth Science*, 14, 286–297.
Viviroli, D., Archer, D. R., Buytaert, W., Fowler, H. J., Greenwood, G. B., Hamlet, A. F.,... & Lorentz, S. (2011). Climate change and mountain water resources: Overview and recommendations for research, management and policy. *Hydrology and Earth System Sciences*, *15*(2), 471–504.
Wolfsegger, C., Gössling, S., & Scott D (2008) Climate change risk appraisal in the Austrian ski industry. *Tourism Review International, 12*(1), 13–23.
Xenarios, S., Gafurov, A., Schmidt-Vogt, D., Sehring, J., Manandhar, S., Hergarten, C., Shigaeva J. & Foggin, M. (2019). Climate change and adaptation of mountain societies in Central Asia: Uncertainties, knowledge gaps, and data constraints. *Regional Environmental Change, 19*(5), 1339–1352.

2

LOCAL OBSERVATIONS OF CLIMATE CHANGE AND IMPACTS ON LIVELIHOODS IN KAMCHATKA, RUSSIA

Drew Gerkey and Victoria N. Sharakhmatova

Introduction

Indigenous communities throughout the Arctic are confronting unprecedented challenges arising from anthropogenic climate change. Although climate change is evident throughout the globe, the pace of change in the Arctic has been much faster. Surface temperatures have risen twice as fast in the Arctic, relative to lower latitudes, a phenomenon known as "Arctic amplification" (Previdi et al 2021). Rising temperatures interact with other components of climatic and biological systems at both local and global scales in complex ways, bringing profound changes in sea ice, snow, rain, permafrost, wildfire, extreme weather, erosion, and phenology to the Arctic (Druckenmiller et al 2022). Recent climate model simulations suggest these extreme changes in climatic and biological systems will become routine, with a "new Arctic" climate emerging during the remaining decades of the 21st century (Landrum & Holland 2020). This "new Arctic" will continue to alter interactions between people and their environments, linking the impacts of climate change with ongoing social dynamics of colonialism and Indigenous resurgence to create a "total environment of change" for Arctic communities (Moerlein & Carothers 2012).

Despite increasing investments in scientific research on climate change, our ability to understand the changes that are occurring now, predict the changes that will unfold in the future, and anticipate the impacts these changes will have on Arctic social-ecological systems remains limited. Challenges include temporal and spatial gaps in instrumental and proxy data on climatic and biological systems, as well as theoretical and methodological obstacles that obscure links within social-ecological systems. Expanding efforts to document local observations of climate change and its impacts can help address these challenges (Reyes-García et al 2020). Collaborations between social scientists and communities can generate unique, insightful data on changes in climatic and biological systems, as well as their interactions with human systems at temporal and spatial scales that match the impacts of climate change on communities, supporting efforts for mitigation and adaptation. This approach draws on a strong foundation of Indigenous and local knowledge, developed through generations of accumulated individual and collective experience via livelihoods that are inextricably linked to social-ecological systems.

DOI: 10.4324/9781003356837-5

Efforts to integrate insights on climate change manifestations and impacts from Indigenous and local knowledge with insights from scholars working in the social and natural sciences are ongoing throughout the Arctic (Alessa et al 2015; Fienup-Riordan & Rearden 2013; Gearheard et al 2017; Jungsberg & Wendt-Lucas 2024; Krupnik & Jolly 2002; Marino 2015). Similar collaborations between social scientists and communities in Siberia are relatively less common than in other parts of the Arctic, with some notable exceptions (Alessa et al 2015; Crate 2022; Forbes et al 2009). Research that brings together Indigenous and local knowledge with scientific research in Siberia is particularly important, considering the geographic scope, demographic base, and cultural diversity in this part of the Arctic. Over half of the Arctic coastline and half of terrestrial lands are located in Siberia, along with nearly three-quarters of the Arctic's human population, including 40 distinct Indigenous peoples (Glomsrød & Wei 2021). The geographic, demographic, and socio-cultural prominence of Siberia in the Arctic exists alongside the unique, unprecedented changes in Siberia's climate relative to the rest of the Arctic (Hantemirov et al 2022), highlighting the importance of this region for research on observations and impacts of climate change.

The research featured in this chapter is a modest contribution to this effort, featuring local observations of climate change and its impacts on a predominantly Indigenous community in Northeastern Siberia, Russia. Data were collected as part of the Local Indicators of Climate Change Impacts project (LICCI), using a common protocol (Reyes-Garcia et al 2023). Specifically, we used semi-structured interviews and focus groups to document local reports of social and environment change, identify their impacts on people's lives, and discuss the strategies people use to adapt to these changes.

Case study

We conducted our research in Khailino, a predominantly Indigenous community of about 600 people, located in the northern part of the Kamchatka Peninsula in Northeast Siberia, Russia. The community lies along the Tylovayam River (also known as the Tylgovayam), near its confluence with the Vyvenka River, a major watershed that flows into the Pacific Ocean. The landscape surrounding the community includes a mix of tundra and taiga, with mountains running along both sides of the watershed. The area has a subarctic climate, with long, cold winters and short, warm to cool summers, and is coded "Dfc" in the Köppen-Geiger climate classification system (Beck et al 2018).

Khailino's Indigenous inhabitants include Koryak, Chukchi, and Even peoples, whose traditional livelihoods focus primarily on reindeer herding, but also include fishing, hunting, and foraging. Along with reindeer, salmon represent a second keystone in traditional livelihoods, with abundant summer harvests providing a substantial source of food and income that lasts throughout the year. Since the arrival of non-Indigenous settlers from Russia and other parts of the former Soviet Union, people in Khailino have also come to rely on crops of potatoes and vegetables, which grow during the short, light-intensive summer season, and provide food for the rest of the year. Accordingly, most Indigenous and non-Indigenous residents of Khailino today have livelihoods based on a mixed cash-subsistence economy. Households combine income from wage labor, pensions, and the informal cash economy with food produced through fishing, hunting, herding, foraging, and gardening (Gerkey 2010).

Khailino is part of the Oliutorskii Raion, a district that encompasses the Northeastern part of the peninsula with an administrative center in Tilichiki, a coastal town located about 70 km away. The Oliutorskii Raion is part of the Kamchatka Krai, a subject of the Russian Federation whose regional capital, Petropavlovsk-Kamchatskii, is located about 1,000 km to the South and is home to about 69% of the peninsula's total population. The concentration of settlements in and around the regional capital, combined with the large size of the peninsula, results in very low population densities in Kamchatka, especially in the northern regions, where Khailino is located.

The village of Khailino was founded in 1932, as part of the Soviet government's effort to colonize Siberia and the Far East. Indigenous Koryak, Chukchi, and Even communities whose traditional territories include present-day Khailino were encouraged to settle in the village through a combination of collaboration and coercion, following processes of settler colonialism that had been enacted elsewhere in Siberia (Slezkine 1994). Some of the most dramatic changes for Khailino's Indigenous communities arose during the period of collectivization in the 1930s and 1940s (Antropova 1971; Bilibin 1933), when reindeer herding, salmon fishing, and hunting were integrated with the Soviet economy through the formation of state farms (*sovkhoz*, singular) and collective farms (*kolkhoz*, singular). Prior to collectivization, Indigenous communities were composed of extended family units (*obshchina*, singular) that migrated throughout their territories according to seasonal rounds linked to subsistence activities (Jochelson 1908). After collectivization, most people began to live in the village, allowing children to enroll in school and adults to work in jobs associated with the village economy, public works, and administration. Many adults remained employed in reindeer herding through the sovkhoz, which included 10–12 "brigades" of herders, each responsible for managing 1,000–2,000 reindeer, which were almost entirely government property. During this period, total herd sizes in Khailino increased to as many as 15,600 reindeer, which represented about 43% of all the domestic reindeer in the Oliutorskii Raion from 1975 to 1985 (Korchmit 2001). People in Khailino recall this period fondly as a time when herding was one of the most respected, well-compensated professions in the village and served as a cultural and economic foundation for the community.

After nearly six decades of collectivization, the collapse of the Soviet economy in the 1990s initiated a long period of chaos and uncertainty (King 2003). Although many state farms throughout Siberia were privatized or liquidated entirely, Khailino's sovkhoz continued as a government enterprise, managed primarily through coordination between local and regional governments in Kamchatka. However, access to government subsidies and resources was severely disrupted during this period, playing a key role in dramatic declines in herd sizes. By 2008, there were two "brigades" of herders working for the sovkhoz under the auspices of a regional government organization called "Koryakolenprom," with a total herd size that fluctuated between 2,000 and 3,000 reindeer. By this time, Koryakolenprom also managed reindeer herds in four other communities in the Oliutorskii Raion and neighboring Penzhinskii Raion, with about 13,400 reindeer in Oliutorskii Raion. Although the number of reindeer in the region has increased to 15,500 by 2021 (Kalitin 2021), herd sizes are still less than half what they were in the 1980s, prior to the collapse of the Soviet economy. Following the declines in reindeer herding, salmon fishing has become central in people's livelihoods in the post-Soviet era. In addition to providing food that can be consumed all summer and preserved to last year-round, salmon also provide limited income through the sale of caviar made from salmon roe. However, the reliance on salmon caviar for income has also created social and environmental challenges, including poaching and conflicts over territories and quotas (Gerkey 2011, 2016; Sharakhmatova 2012).

Research design and methods

The research presented here builds on previous research conducted separately by both Gerkey and Sharakhmatova, dating back to 2007. Gerkey has been studying the social and ecological dynamics of Khailino's mixed cash-subsistence economy, focusing on Indigenous knowledge and cultural practices, post-Soviet institutions, and networks of social support through the lens of cooperation and collective action. During multiple research trips between 2007 and 2019, Gerkey gathered data through a combination of participant observation, interviews, surveys, and behavioral economics. This research provides a foundation for understanding contemporary livelihoods in Khailino and how they are associated with the observations and impacts of climate change we report in this chapter.

Sharakhmatova has also previously conducted research in Khailino, including a 2010 project that used semi-structured interviews and surveys to document observations of climate change from 153 people living throughout Kamchatka, including participants from Khailino. Sharakhmatova (2011) found that 82% of survey participants have observed changes in weather and climate conditions in places where they practice traditional subsistence activities and that 50% of respondents have shifted locations for these activities in response. Sharakhmatova's analysis of the direct and indirect impacts of climate change on traditional subsistence activities throughout Kamchatka supported recommendations to create a monitoring network that would involve Indigenous communities in gathering observations about climate change and its impacts on reindeer herding, fishing, hunting, gathering, and other traditional subsistence activities (Sharakhmatova 2014).

Following the protocol for the LICCI project (Reyes-García et al 2023), Gerkey and Sharakhmatova worked together to design a sampling strategy, assemble a timeline of significant historical events, and outline key aspects of local livelihoods to ensure participants in our study reflected variation in the lived experiences and perspectives of people in Khailino. Once Gerkey obtained human subjects research approval from the Oregon State University Institutional Review Board (#2021-1163), our data collection proceeded in two steps. First, Sharakhmatova conducted semi-structured interviews with 15 people identified as particularly experienced and knowledgeable, either from our previous research in the community or based on recommendations from our local research assistant (Table 2.1). To ensure the data we gathered reflected different perspectives in the community, we chose participants based on their economic livelihoods, age, and gender. Second, Sharakhmatova conducted two focus groups with eight participants each, including four people who participated in semi-structured interviews and four who did not (Table 2.1). We used these focus groups to refine our understanding of local observations of climate change from the interviews. Specifically, we developed a list of changes identified by interview participants, presented these to the focus group, one at a time, for discussion, and assessed levels of consensus regarding the types of change that are occurring, the timing of these changes, their impacts on people, and the adaptive strategies people are using in response. These discussions also explored potential drivers underlying changes and suggested possible links to changes across physical, biological, and human systems. Together, these two steps were intended to ensure that we document the range of knowledge and experience that different people in the community have about climate change, while also identifying variation in consensus within the community about how, when, and why these changes are occurring.

In the sections that follow, we present our results, starting with observations of change identified during the interviews and focus groups, followed by people's explanations for how these changes are impacting their livelihoods, and strategies they have begun developing to adapt.

Table 2.1 Demographic and socio-economic background of participants in semi-structured interviews and focus groups in Khailino, Russia

Interview	*Focus group*	*Gender*	*Age*	*Ethnic group*	*Work*	*Salmon fishing*	*Reindeer herding*	*Hunting*	*Gathering*	*Winter fishing*	*Gardening*
X	X	Woman	48	Koryak	Education, Artist	X			X	X	X
X		Man	70	Even	Pensioner	X				X	X
X		Man	54	Nymylan	Unemployed	X	X		X	X	X
X		Woman	44	Chukchi	Education	X				X	X
X	X	Woman	72	Layrovetlan	Reindeer herder (retired)	X	X		X	X	X
X		Woman	61	Koryak	Childcare	X			X	X	X
X		Man	62	Chukchi	Construction	X			X	X	X
X		Woman	64	Chukchi	Pensioner	X	X		X	X	X
X		Man	39	Non-indigenous	Transportation technician		X	X	X	X	
X		Woman	34	Non-indigenous	Veterinarian						X
X	X	Woman	61	Chukchi	Pensioner	X			X	X	X
X	X	Woman	48	Koryak	Education				X		X
X		Man	60	Koryak	Dog-sled musher	X		X	X	X	
X		Man	50	Koryak	Entrepreneur	X			X	X	
X		Man	48	Koryak	Public works	X		X	X	X	X
	X	Woman	58	Koryak	Education	X			X		X
	X	Woman	36	Chukchi	Secretary	X			X		X
	X	Woman	62	Even	Education	X			X		X
	X	Woman	59	Non-indigenous	Education				X		X

Local observations of climate change

Through our semi-structured interviews, we documented observations of change in physical, biological, and human systems, including many reportedly driven by climate change. We summarize these observations by coding observed changes based on their similarity across interviews, noting the number of people who reported each type of change, and the level of consensus from focus group discussions (Table 2.2). We limit our description and analysis to the most frequently observed changes with connections to climate change and the highest levels of consensus.

Warmer temperatures, less snow, softer snow in winter

A clear consensus emerged from our interviews and focus groups that winters in Khailino are warmer and with less snow than in the past. Multiple people linked warmer temperatures to the seasonal transition between fall and winter, specifically the time when the large rivers surrounding the village freeze and become safe for travel by snowmachines, dog sleds, sport-utility vehicles, and heavy machinery, including freight trucks and all-terrain, tracked vehicles that provide the village with coal, groceries, and other supplies. Typically, rivers are frozen by mid to late October and remain so until spring, but recently people noted that ice can remain unstable into November, sometimes freezing sufficiently to allow travel for a few weeks, only to thaw again, making travel unsafe and unreliable. Warmer temperatures also change the quality of the snow, with a consensus emerging from our focus groups that snow has become much softer, which also makes transportation more difficult, particularly for smaller, personal vehicles like snowmachines.

Participants in interviews and focus groups have also noticed fewer strong blizzards and extreme cold events. The Kamchatka Peninsula is surrounded by the Sea of Okhotsk to the West and the North Pacific Ocean to the East, generating cyclones throughout winter, during which large amounts of snow fall continuously for several days and strong winds create large snowdrifts. One Indigenous Koryak woman in her 50s, who was born in Khailino and has lived there most of her

Table 2.2 List of observations of change from semi-structured interviews and focus groups, sorted by level of agreement developed during focus group discussions and the number of people who observed the change (frequency)

Observed change	*Description*	*Frequency*	*Focus group agreement*
Increase in frequency of problems with transportation	Winter road established earlier, but thaws during winter. It is more difficult to get on the winter road in the off-season. Often in winter, the road is bad.	19	Full agreement
Decrease in the amount of snowfall	There is less snow in winter	11	Full agreement
Decrease in ice stability/resistance	The river freezes and then thaws and then freezes	10	Full agreement
Decrease in the speed of ice melting or break-up	There is practically no ice break-up. The ice just melts and disappears	10	Full agreement
Later timing of ice formation	The river freezes later	10	Full agreement

(*Continued*)

Table 2.2 (Continued)

Observed change	*Description*	*Frequency*	*Focus group agreement*
Earlier crop harvesting time	Harvest potatoes earlier	7	Full agreement
Decrease in freshwater transparency/ increase in concentration of dissolved particles	The river is silting up	7	Full agreement
Increase in the frequency of freeze events	Frosts appear earlier in fall/late summer, unpredictable	7	Full agreement
Earlier timing of ice melting or break-up; increased variation in ice melting or breaking patterns in lakes or rivers	Ice break-up time has changed	7	Full agreement
Increase in frequency of problems with transportation	Soft snow interferes with snowmachines	5	Full agreement
Increase in the amount of rainfall in winter	It rains in the winter months	5	Full agreement
Softer physical structure and texture of snow	Snow is softer in warmer weather	5	Full agreement
Decrease in the frequency of frost days	Fewer hard frost in winter	4	Full agreement
Decrease in the frequency of cold days	Less cold days in winter time	11	Agreement after debate
Increased variation in the location of river bank erosion	The river changes its course, erodes the river bank on fishing grounds	10	Agreement after debate
Increase in the amount of rainfall in summer	Rainier August	6	No agreement after debate
Increase in wildfire frequency	There are more fires	6	No agreement after debate
Decrease in the mean temperature in summer	Summer temperature is colder	6	No agreement after debate
Reduced productivity of wild plants and fungi	There are fewer mushrooms and berries in the vicinity of the village. We have to move farther away from the village	4	No agreement after debate
Increased difficulties for domestic animal ability to forage due to environmental/climatic conditions	Difficult for the reindeer to forage	2	Not discussed
Decreased temperature of river water	Water temperature in river is colder in summer	2	Not discussed
Increase in wind strength or speed	Winds are stronger than before	2	Not discussed
Decrease in social and economic viability of traditional livelihoods	The profession of reindeer herder has ceased to be prestigious	10	Not discussed
Decrease in levels of water in lakes and ponds	A lake dried up not far from the village	5	Not discussed
Changes in the physical structure and texture of snow	Tundra ice-crusted in winter	2	Not discussed
Declines in abundance of fish	There are less Chinook salmon (*Oncorhynchus tshawytscha* or "*chavycha*")	1	Not discussed

life, explained "*We used to have very strong blizzards, with snow up to the second floor of our buildings, but now it doesn't happen often*" (personal interview, November 11, 2021). Although the proximity to the ocean keeps temperatures higher in Kamchatka than in other parts of Siberia, inland communities like Khailino typically experience multiple extreme cold events where temperatures drop as low as −40 degrees C. While such events do still occur, participants in our interviews and focus groups suggest that these extreme cold events are less frequent in recent years than they used to be in the past.

Another indicator of warmer temperatures and less snow in winter is an increase in rain during winter months. As one of the people we interviewed explained, "*I don't even remember in the past when we had rain in winter. Elders say it was very rare, and now we have rain every year. Two years ago, we had rain every month, even in winter!*" (personal interview, November 9, 2021). Similarly, the participants in one of our focus groups mentioned that they celebrated the 2021 New Year holiday with a mix of snow and rain, which they considered quite atypical for that time of year. This combination of snow and rain has also apparently resulted in an increase in "rain-on-snow" events, where rain falls on a bed of snow, followed by lower temperatures that freeze the rain, leaving a hard crust of snow that poses particular problems for reindeer trying to feed on lichen below the snow.

Seasonal transitions and spring river ice

In addition to changes in seasonal transitions from fall to winter, about half of the participants in our interviews reported that the conditions of river ice have been changing during the transition from winter to spring. In the past, as temperatures began to rise in late winter and early spring, river ice would start to break up, then, in about a week or two, large chunks of ice would flow downstream, leaving the river mostly clear of ice. Although the timing of this event varied from one year to the next, the rapid break-up and departure of river ice was a prominent marker of the onset of spring (Figure 2.1). Recently, however, people have noticed that the river ice simply melts in the river, leaving fewer large chunks of ice to float downstream. Both of our focus groups discussed this observed change extensively. People in one group suggested that in the 1980s and 1990s, the break-up period lasted as little as three to four days, while in recent years, it is hard to even pinpoint a specific time because the ice gradually disappears into the river. Our second focus group made similar observations, discussing the increased variability in the timing of seasonal transitions from one year to the next, including large differences between 2020 and 2021 alone.

Individual interviews echoed these observations. One adult man in his 40s, a Russian who was born in Khailino and has lived there his entire life, recalled that during his childhood ice break-ups usually occurred in late May or early June, but now the timing is closer to early May. As an active fisher and hunter who travels frequently beyond the village, he explained that ice was typically strong in April and could even be safe in May if one was careful. In other words, as explained, "*April was winter!*" (personal interview, November 9, 2021). His assessment was confirmed in the first focus group, which noted that in recent years the snow has mostly disappeared during the first week of May and the ground has been sufficiently warm and dry that people have begun planting their gardens. People in that focus group also recalled that in the past there was sufficient snow to use dog sleds up to June.

More than half of the participants in our interviews suggested that levels of erosion along river banks have been increasing in recent years, though there was some disagreement about this observation during our focus group discussions. We never arrived at a consensus about the drivers of increased erosion, whether due to changes in summer precipitation, spring river ice break-up, or

Figure 2.1 Khailino youth gather at the river edge during spring break-up, May 12, 2008. Photo by Drew Gerkey.

some other aspect of weather in winter and spring. However, there was a general consensus that levels of silt in the river are increasing as a result of erosion.

Unpredictable summers (temperature, precipitation)

Increased variability in weather during summer was also noted by multiple participants in our interviews and confirmed during our focus groups. People have observed more frequent freeze events in late summer than in years past, particularly in August. While these temperatures are not sufficiently low to speed up the formation of ice on rivers, they do disrupt subsistence food production via fishing and gardening at a crucial time. Their unpredictability is particularly challenging because summer temperatures are typical in some years, while in others freezing events arise just as people are finishing preparing fish and harvesting their gardens.

The increased variability in summer weather may have made it more difficult to arrive at consensus about the changes people are observing during this season. About a third of the people in our interviews reported that summer temperatures have been colder and wetter than in the past, especially in August. Similarly, a third also suggested that wildfires have increased in summer and that the productivity of berries and mushrooms has decreased, conditions typically associated with warm, dry weather. However, participants in our focus groups did not agree on these changes. Our first focus group discussed large swings in daytime temperatures that have become more common in summer over the last four to five years, with very hot weather when it is sunny and very cold weather when it is cloudy. They explained that these changes are especially noticeable when they are working to harvest salmon at their fish camps because of the large amount of time spent working outside. Perhaps, this explains the mix of observations related to warmer, drier weather along with observations related to colder, wetter weather that are both apparently occurring in summer.

Climate change impacts on livelihoods

Problems with transportation

Many of the changes people have observed in temperature, precipitation, and seasonal transitions from fall to winter and from winter to spring create significant problems with transportation between Khailino and Tilichiki, the administrative center. Although helicopter flights are available to and from Khailino year-round, these flights are expensive, frequently disrupted by inclement weather, and limited to transporting people and small amounts of personal belongings, rather than heavy freight. Most of the groceries and supplies the village relies upon arrive via ground-transportation, where there are several modes that depend on the season and weather conditions.

In winter, spring, and fall, the only vehicle that can reliably transport people and freight to and from Khailino is a large, tank-like all-terrain vehicle called a "*vezdekhod*." A vezdekhod is capable of floating across large rivers, crossing soft, boggy expanses of tundra, and pushing through thick stands of brush, though progress is often slow and difficult. In winter, a vezdekhod becomes a more effective mode of transportation, moving quickly on both un-groomed expanses of snow and established winter roads. While the primary cargo is typically groceries and other supplies for village stores, people also frequently travel via vezdekhod, either riding inside the cargo hold or outside on the roof. Although the trip takes longer and is less comfortable than flying by helicopter, the increased frequency and affordability of trips make a vezdekhod a useful option for passengers (Figure 2.2). However, because the priority is always freight and space for passengers is limited, people may struggle to find a vezdekhod for days or weeks at a time. Alternatives expand considerably in winter, once temperatures are cold enough to freeze the large rivers and establish a solid base of ice and snow for the winter road between Khailino and Tilichiki. Once the winter road is formed, large freight trucks begin to transport coal for the village's centralized heating system, while large-wheeled buses run scheduled trips for passengers between the two villages. Although the road remains quite uneven, some people also choose to travel in personal sport-utility vehicles, though these trips can be slow and challenging when conditions are not ideal. Finally, with sufficient snow and good weather, people may travel by snowmachine in winter.

Altogether, the multiple modes of transportation available during winter greatly enhance people's ability to travel beyond Khailino, which is increasingly important for a variety of reasons. Khailino has limited banking, internet, and cell phone services, and no hospital, so people often travel to Tilichiki to access these services. Moreover, as the administrative center, people sometimes need to travel to Tilichiki to complete bureaucratic and legal tasks. Finally, Tilichiki is the gateway to air travel to the regional capital, Petropavlovsk-Kamchatskii, where people frequently travel to visit relatives, pursue higher education, seek employment opportunities, or fly to other regions of Russia and beyond.

Changes in temperature and precipitation during the seasonal transition from fall to winter and from winter to spring disrupt ground transportation in several ways. First, the increases in time to freeze-up in fall and break-up in spring make travel difficult during these periods, even for a vezdekhod. When river ice is not sufficiently frozen to support a vezdekhod's weight, it can be difficult, dangerous, or even impossible to complete the multiple river crossings necessary to travel between Khailino and Tilichiki. As a result, most vezdekhod drivers simply wait until the ice is

Figure 2.2 A vezdekhod travels back to Khailino after visiting the reindeer herds to provide supplies and assist with the spring corral, March 17, 2008. Photo by Drew Gerkey.

fully formed in fall or entirely clear in spring, bringing all trips to a halt in the meantime. While these periods without ground transportation might last a week or so in the past, it is now common for them to extend for several weeks, and even months at a time. Second, warmer temperatures and increased rain throughout winter undermine the integrity of the winter road, making it more difficult for buses and sport-utility vehicles to transport passengers. Similarly, people explained that softer snow also creates problems for snowmachines, clogging tracks and making progress slow, which requires more gasoline and increases the cost of trips. These circumstances make a vezdekhod the only alternative to helicopter flights. Our first focus group explained that in the previous year, the passenger bus was only able to complete three trips on the winter road before shutting down entirely in early March. In fact, these conditions disrupted our data collection for this project in 2021. Sharakhmatova was delayed for nearly a week in traveling from Tilichiki to Khailino by warm temperatures and rain that affected the formation of river ice and the winter road. The trip had to be postponed until mid-November.

Aside from seeking alternative forms of transportation that are less reliable, more expensive, or more dangerous, there is little people in Khailino can do to adapt to these difficulties in transportation. People can delay their travel plans or try to build in more flexibility to account for the uncertainty, but either way, the result is often a substantial increase in the costs of travel outside the community. Some people appear to respond to these impacts by reducing their travel and trying to ground their livelihoods within the community, so that travel is less essential. Others see transportation difficulties as a reason to re-locate to Tilichiki or Petropavolvsk-Kamchatskii, which reduces the need to travel but can increase social and emotional stress that comes from losing connections to the community.

Problems with subsistence

Changes in winter weather—particularly increases in rain that lead to more frequent "rain-on-snow" events—create substantial challenges for reindeer herders. During winter, reindeer use their disk-shaped hooves to dig through snow to reach their primary food, lichen (Figure 2.3). When rain falls on top of snow and then freezes with decreasing temperatures, a hard crust forms on the surface that is difficult for the reindeer to penetrate with their hooves. As a result, herders need to move the herd to new pastures, where conditions are more favorable, or provide an alternative food source. Migrating to new pastures costs valuable time and energy during an already difficult, stressful time of year for both people and animals, and herders may need to travel far to find pastures unaffected by weather patterns that created the rain-on-snow event.

This impact was only mentioned by a few of the participants in our interviews and focus groups, but this is likely because most people involved in reindeer herding are based in the tundra with the reindeer herds, often located far outside the village. The two people who mentioned this in our interviews provide technical and administrative support to the reindeer herders, including one who regularly drives a vezdekhod to bring them supplies. This man, a Russian who has lived in Khailino his entire life, shared insights from the herders about the effects of rain-on-snow events on the reindeer. For example, he explained that when these events happen later in winter or early spring, during the calving season, pregnant female reindeer are particularly vulnerable:

> They lose a lot of mass and become weak. They need fat. When the mother is weak, she gives birth to weak calves, and many of them die. When we have a good winter with snow, the reindeer do better and sometimes give birth to twins.
>
> (personal interview, November 9, 2021)

Figure 2.3 Reindeer from one of Khailino's herds forage for lichen underneath the snow, March 15, 2008. Photo by Drew Gerkey.

A successful calving season, where many young reindeer survive, is critically important for the reindeer herders. It is one of the only ways they can maintain and increase herd sizes, which remain substantially smaller than they were during the height of the Soviet era, largely due to the chaotic collapse of the Soviet economy and the transition to a post-Soviet, market-based economic system.

The same changes in seasonal transitions and the conditions of river ice and snow that disrupt transportation between Khailino and Tilichiki also affect efforts to maintain supply connections between reindeer herders and the village. Two years prior to our interview with him, the vezdekhod driver explained that he and his team were delayed in traveling to the reindeer herds due to excessive rain. Once they finally arrived, they were not able to hold the fall corral and complete the annual harvest of reindeer due to rain-on-snow events that made conditions difficult. As a result, they were stranded with the herds for an entire month, unable to complete their work or travel back to the village.

Subsistence livelihoods are also disrupted by changes in weather during summer, particularly the earlier onset of frosts in late summer and early fall. Multiple interview respondents mentioned several occasions in recent years when an early frost wiped out their entire potato crop. They explained that if potatoes are not removed from the ground and stored safely before the first frost, they spoil. Although the timing of the first frost varies from year-to-year, people have noticed unexpected frost events occurring much earlier than in the past, which makes it hard for them to adjust the timing of their harvest accordingly. The problem is compounded by the short, intense duration of the growing season in Khailino, where gardens cannot be planted until late May or June and must be harvested in August, but can be successful due to the relatively long periods of summer daylight in northern latitudes.

Intersections of climate change and social change in the post-Soviet era

It is important to understand the observations and impacts of climate change described in this chapter in relation to ongoing forms of cultural, economic, and political change in the post-Soviet era. While many of the observations we documented relate to physical and biological changes in the environment, people's explanations of the impacts these changes have on their ways of life establish clear connections with human systems, particularly transportation infrastructure and traditional subsistence livelihoods. Our analysis suggests that the significance of these impacts for people in Khailino cannot be understood fully without an appreciation for the dramatic transformations that have occurred during the pre-Soviet and Soviet era, as well as ongoing tensions between colonialism and Indigenous resurgence in the post-Soviet era.

Disruptions to transportation infrastructure caused by increasing temperatures and rain in winter, along with associated changes in seasonal transitions during fall freeze-up and spring break-up of river ice, negatively impact people's ability to move beyond Khailino. Decreased mobility in turn affects people's livelihoods, personal relationships, and quality of life in a variety of ways. Ever since the Soviet era, when Indigenous communities that previously migrated throughout the area were compelled to reside in permanent settlements like Khailino, transportation infrastructure has been essential.

Reindeer herding provides a clear example. Prior to collectivization, reindeer herds were managed by extended family units (*obshchina*) that migrated with their herds to seasonal pastures. During the Soviet era, these family units were replaced by "brigades" of workers who were usually located hundreds of kilometers outside Khailino, often separated from their families. However, Soviet era investments in collective institutions like Khailino's "Korfskii Sovkhoz" included regular helicopter flights and vezdekhod trips that maintained regular flows of people and supplies between the herds and the village. When the Soviet economy collapsed, the subsidies that maintained this transportation infrastructure diminished, reducing the frequency of trips to and from the herds. In the post-Soviet era, herders are often isolated from the village for months at a time, relying on a few trips a year that typically coincide with key events, like the corrals in fall and spring, which are important for harvesting and managing the reindeer herd. As we described above, the timing of these trips is often disrupted by the increasingly unpredictable environmental conditions that affect snow and ice. In addition to the practical and logistical impacts on reindeer herding, these changes also place heavy personal burdens on reindeer herders, who are forced to spend longer periods of time away from their family and friends back in the village.

Reduced mobility beyond Khailino also impacts people who are not directly involved in reindeer herding. Soviet era investments in transportation infrastructure were also sharply reduced in the post-Soviet era, making it more difficult and expensive to travel to the administrative center, Tilichiki, and beyond to the regional capital, Petropavlovsk-Kamchatskii. Interview participants explained that during the Soviet era, Khailino maintained a runway that could accommodate small airplanes as well as helicopters, increasing the number of flights to and from the village. Moreover, ticket subsidies during the Soviet era made these flights much more affordable. As a result, people became accustomed to traveling frequently outside the village to pursue professional opportunities or maintain personal relationships. Doing so is more difficult now, forcing people to spend more time and money to travel or to relocate temporarily or permanently outside the community. Perhaps this is one factor contributing to recent declines in Khailino's population, which also carries emotional weight for those who remain.

In these ways and many others, the impacts of climate change on people in Khailino are exacerbated by the legacies of social transformation from collectivization and colonialism during the Soviet and post-Soviet eras. Appreciating the complex connections between climate change and social change is essential for those seeking to contribute to efforts to document ongoing forms of climate change, anticipate future changes, and take practical steps to mitigate or adapt.

Acknowledgments

We thank the people of Khailino, particularly those who participated in this project, as well as those who have contributed to our previous research in the community. We are grateful to Anna A. Uvarova, who provided research assistance during data collection. Our research for this project received financial support from the LICCI project (European Research Council # 771056-LICCI). We thank Dr. Victoria Reyes-García (Principal Investigator), Dr. Xiaoyue Li (Postdoctoral Researcher), Encarna P. Fernández (Project Manager), and the rest of the LICCI Core Team for their assistance and insight during all phases of the project.

References

Alessa, L., Grace Beaujean, Leah Bower, Iver Campbell, Olga Chernenko, Patricia Cochran, Margie Coopchiak, Maryann Fidel, Uliana Fleener, Jim Gamble, Arlene Gundersen, Verna Immingan, Lisa Jackson, Alice Kalmakoff, Andrew Kliskey, Sharon Merculief, Delbert Pungowiyi, Olga (Olia) Sutton, Eddie Ungott, Joni Ungott, & Jessica Veldstra (2015) *Bering Sea Sub-Network: Project Summary Report 2015 (Aleut International Association)*. CAFF Monitoring Series. Akureyri: Conservation of Arctic Flora and Fauna.

Antropova, V. V. (1971). *Kultura i Byt Koryakov*. [*Culture and Daily Life of the Koryaks*]. Leningrad, RU: Izdaltel'stvo Nauka.

Beck, H. E., Zimmermann, N. E., McVicar, T. R., Vergopolan, N., Berg, A., & Wood, E. F. (2018). Present and future Köppen-Geiger climate classification maps at 1-km resolution. *Scientific Data*, *5*(1), 1–12.

Bilibin, N. (1933). *Klassovoe Rassloenie Kochevykh Koryakov.* [*Class Stratification among Nomadic Koryaks*]. Vladivostok, RU: Dal'giz.

Crate, S. A. (2022). *Once upon the Permafrost: Knowing Culture and Climate Change in Siberia.* Tucson, AZ: University of Arizona Press.

Druckenmiller, M. L., R. L. Thoman, & T. A. Moon, Eds. (2022). *Arctic Report Card 2022.* https://doi.org/10.25923/yjx6-r184.

Fienup-Riordan, A., & Rearden, A. (2013). *Ellavut/Our Yup'ik World and Weather: Continuity and Change on the Bering Sea Coast.* Seattle, WA: University of Washington Press.

Forbes, B. C., Stammler, F., Kumpula, T., Meschtyb, N., Pajunen, A., & Kaarlejärvi, E. (2009). High resilience in the Yamal-Nenets social–ecological system, west Siberian Arctic, Russia. *Proceedings of the National Academy of Sciences*, *106*(52), 22041–22048.

Gearheard, S., Holm, L. K., Huntington, H., Leavit, J. M., Mahoney, A. R., Opie, M.,... & Sanguya, J. (2017). The meaning of ice. *People and Sea Ice in Three Arctic Communities.* International Polar Institute Press. ISBN: 0996193855

Gerkey, A. P. (2010). *From State Collectives to Local Commons: Cooperation and Collective Action among Salmon Fishers and Reindeer Herders in Kamchatka, Russia* (Doctoral dissertation, Rutgers University-Graduate School-New Brunswick).

Gerkey, D. (2011). Abandoning fish: The vulnerability of Salmon as a cultural resource in a post-Soviet commons. *Anthropology of Work Review*, *32*(2), 77–89.

Gerkey, D. (2016). The emergence of institutions in a post-Soviet commons: Salmon fishing and reindeer herding in Kamchatka, Russia. *Human Organization*, *75*(4), 336–345.

Glomsrød, S., & Wei, T. (2021). Comparative analysis of Arctic economies from a macro level perspective. pp. 39–50 in *The Economy of the North*, Glomsrød, S., G. Duhaime and I. Aslaksen (eds.). ECONOR 2020. ISBN 978-82-587-1275-3-A4

Hantemirov, R. M., Corona, C., Guillet, S., Shiyatov, S. G., Stoffel, M., Osborn, T. J.,... & Fonti, P. (2022). Current Siberian heating is unprecedented during the past seven millennia. *Nature Communications*, *13*(1), 1–8.

Jochelson, V. (1908). *The Koryak.* Memoirs of the American Museum of Natural History 10. Publications of the Jesup North Pacific Expedition 6. Leidon: E. J. Brill.

Jungsberg, L. & Wendt-Lucas, N. (2024). Local indicators, adaptation actions and resilience efforts for a weaming climate in North Greenland. In V. Reyes-García, S. Álvarez-Fernández, P. Benyei, L. Calvet-Mir, D. García-del-Amo, A. B. Junqueira, X. Li, V. Porcher, A. Porcuna-Ferrer, A. Schlingmann, & R. Soleymani (Eds.), *Routledge Handbook of Climate Change Impacts on Indigenous Peoples and Local Communities.* Oxfordshire & New York: Routledge. Chp 22, pp 356–369.

Kalitin, R. R. (2021). *The Current Conditions and Problems of Northern Domesticated Reindeer Herding and How to Solve Them.* [Современное состояние, проблемы северного домашнего оленеводства и пути их решения]. Rossiiskaya Arktika. No. 15, pp 28–39. https://doi.org/10.24412/2658-4255-2021-4-28-39

King, A. D. (2003). Without deer there is no culture, nothing. *Anthropology and Humanism, 27*(2), 133–164.

Korchmit, V. A. (2001). *Olenevodstvo Koryakii: Etapi, razvitia, perspektivy vozrozhdenia.* [*Reindeer Herding in Koryakia: Periods, Development, Perspectives of Revival.*] Petropavlovsk-Kamchatskii, RU: Kamchatskii Gosudarstvennyi Tekhnicheskii Universitet.

Krupnik, I., & Jolly, D. (2002). *The Earth Is Faster Now: Indigenous Observations of Arctic Environmental Change. Frontiers in Polar Social Science.* Fairbanks, AK: Arctic Research Consortium of the United States.

Landrum, L., & Holland, M. M. (2020). Extremes become routine in an emerging new Arctic. *Nature Climate Change*, *10*(12), 1108–1115.

Marino, E. (2015). *Fierce Climate, Sacred Ground: An Ethnography of Climate Change in Shishmaref, Alaska.* Fairbanks, AK: University of Alaska Press.

Moerlein, K. J., & Carothers, C. (2012). Total environment of change: Impacts of climate change and social transitions on subsistence fisheries in northwest Alaska. *Ecology and Society*, *17*(1).

Previdi, M., Smith, K. L., & Polvani, L. M. (2021). Arctic amplification of climate change: A review of underlying mechanisms. *Environmental Research Letters*, *16*(9), 093003.

Reyes-García,V., Álvarez-Fernández, S., Benyei, P., García-del-Amo, D., Junqueira, A. B., Labeyrie, V., et al. (2023) Local indicators of climate change impacts described by indigenous peoples and local communities: Study protocol. *PLoS One 18*(1), e0279847. https://doi.org/10.1371/journal.pone.0279847

Reyes-García, V., Fernández-Llamazares, Á., García-del-Amo, D., Cabeza, M. (2020). Operationalizing Local Ecological Knowledge in Climate Change Research: Challenges and Opportunities of Citizen Science. In: Welch-Devine, M., Sourdril, A., Burke, B. (Eds.). *Changing Climate, Changing Worlds. Ethnobiology*. Cham: Springer. https://doi.org/10.1007/978-3-030-37312-2_9

Sharakhmatova, V. N. (2011). *Observation of Climate Change by Kamchatka Indigenous Peoples*. Petropavlovsk-Kamchatsky: Kamchatpress, p. 78.

Sharakhmatova, V. N. (2012). Indigenous Peoples' Traditional Fishing in Kamchatka and Local Community Development Concept Based on Sustainable Use of Fish Resources: Problems and Solutions. pp. 89–108 in Colombi, B. J., & Brooks, J. (Eds.). *Keystone Nations: Indigenous Peoples and Salmon across the North Pacific*. Santa Fe, NM: School for Advanced Research Press.

Sharakhmatova, V. N. (2014). Adaptation to climate change and the importance of traditional knowledge of the indigenous peoples of the north in the international projects and negotiation processes. *At the Crossroad of Continent: Materials of the XXXI Krasheninnikov Reading*; Publishing house of the Information-publishing center of the Kamchatka Regional Scientific Library named after S.P. Krasheninnikov; Petropavlovsk-Kamchatsky, Russia, 2014, pp. 363–370. Available online: www.kamlib.ru/resourses/sharakhmatova.htm (accessed on December 12, 2022).

Slezkine, Y. (1994). *Arctic Mirrors: Russia and the Small Peoples of the North*. Ithaca, NY: Cornell University Press.

3

SARGASSUM SEAWEED CHALLENGES FROM LOCAL TO NATIONAL LEVEL IN THE CARIBBEAN

A policy cycle perspective

Patrick McConney, Janice Cumberbatch, Catrina Hinds, Hazel A. Oxenford and Maria Pena

Introduction

Climate change and climate variability are current areas of concern in small-scale fisheries globally. Understanding climate dynamics is benefitting from more and better scientific and local knowledge despite persistent uncertainties (Barange et al. 2018). Improvements in marine climate change awareness and action are required in the Caribbean (Monnereau and Oxenford 2017) within regional climate contexts and policy (CSGM 2020). An area in which the need is most apparent is the complex issue of recent large sargassum seaweed influxes.

Since 2011, massive influxes of holopelagic sargassum seaweed have inundated and been stranded upon the beaches of the coasts of countries in and bordering the Caribbean Sea. The "Great Atlantic sargassum belt" can extend over 8,000 km and convey 10–20 million tons of seaweed biomass (Wang et al. 2019). The floating mats of brown algae comprise mainly morphotypes of *Sargassum fluitans* and *S. natans*. These are not invasive alien species, but typically occurring Atlantic and Caribbean species now seen in atypically large quantities. In typical quantities, floating sargassum mats are ecological assets providing mobile habitats harbouring high biodiversity and sought by commercial small-scale fisheries (Oxenford et al. 2021). Before 2011, the typical quantities reaching Caribbean coasts were small enough to be removed manually, if at all, not harming tourism or coastal transportation, and being traditionally used in household and small-scale agriculture as fertiliser and mulch (Desrochers et al. 2020). Now, the amount of seaweed in the largest sargassum influxes creates a biological hazard, prompting some island states to declare national emergencies when their beaches and response systems are overwhelmed.

The sargassum influxes originate in the North Equatorial Recirculation Region of the tropical Atlantic Ocean before being transported by ocean currents north and westward into the Caribbean Sea with growth aided by riverine nutrients (Johnson et al. 2020) as shown in Figure 3.1. There is significant interannual variability in the sizes of the seaweed blooms and variability in the patterns of where and when sargassum makes landfall, mainly on windward coasts. Sargassum influxes

 DOI: 10.4324/9781003356837-6

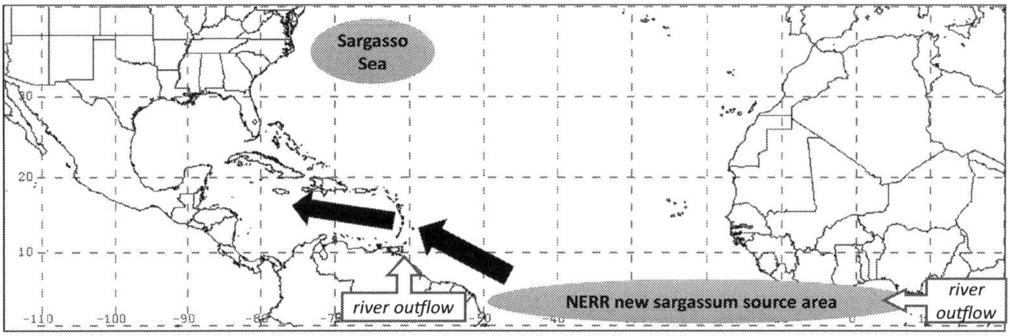

Figure 3.1 New sargassum source area and ocean current transport into the Caribbean Sea (by authors).

have been linked to climate change and climate variability, although the physical climate-related drivers and ecosystem mechanisms are not fully understood (Skliris et al. 2022). Even less is documented about the human dimensions of sargassum-impacted social-ecological systems, despite the significant intersectoral socio-economic disruptions that they cause (Fraga and Robledo 2022).

The sargassum social-ecological system links the seaweed to impacts on small-scale fisheries, tourism, recreation, energy, transportation, and other economic sectors. In the climate-vulnerable small island developing states (SIDS) of the Caribbean, sargassum is emerging as a wicked problem, meaning that the problem is difficult to specify and understand, while having no clear solution (Jentoft and Chuenpagdee 2009; Rittel and Webber 1973). Noting the challenges of the many uncertainties of climate science, the limited capacities of SIDS, struggling small blue economies, and the quest for sustainable development, solutions to sargassum problems will require a combination of strategies drawn from disaster risk management, climate change adaptation, and commercial innovation and entrepreneurship (Oxenford et al. 2021). Practical action towards adaptation is being taken from local to national levels of governance in Caribbean SIDS. State and non-state actors are beginning to collaborate in *ad hoc* sargassum problem-solving and in generating new multi-level opportunities for sargassum use (Desrochers et al. 2020). Communication among these stakeholders at different levels of governance has generally not yet been through institutions such as learning networks that optimise information exchange and knowledge mobilisation in marine resource governance (Van der Plank et al. 2020). Although scientific and local knowledge are both being used to cope with sargassum, there has been little effort to institutionalise, integrate, and formalise their use. To examine relationships between local knowledge and policy, in this chapter we consider two questions applied to each stage of a generic policy cycle:

- What types of local knowledge are available and used in sargassum management?
- How can the use of local knowledge in policy cycles for sargassum be enhanced?

Context

Within this century, the Caribbean is expected to experience even warmer air and sea temperatures, sea level rise, more extreme weather events, ocean acidification, intense rainfall as well as more drought, and changing physical and biological oceanographic conditions (Cox et al. 2021; CSGM 2020). Fisherfolk in the region have already observed changes in the marine environment that some attribute to climate change (James 2009; Kinch and Oxenford 2019). Climate change adaptation is now becoming incorporated into fisheries and marine ecosystem-based management

(Fanning et al. 2011). Improving global climate models, down-scaling climate projections to island size, and implementing mitigation measures will be necessary but not sufficient climate action (Nurse 2011). Adaptation to climate change will be essential in the Caribbean (Turner et al. 2020).

Sargassum influxes add to the complexity of marine resource governance under climate change and variability. Interactions among winds, currents, upwellings, nutrient transport from continental rivers, sargassum growth and mortality rates, and other factors that vary seasonally, but are less predictable due to climate change and climate variability, challenge our understanding of the ocean environment (Skliris et al. 2022). Natural science has dominated sargassum research with less attention to social science and socio-economic aspects (Ramlogan et al. 2017). Caribbean countries have focused on solutions such as sargassum removal from beaches and its use as raw material in products such as fertilisers (Rosellón-Druker et al. 2022). In small-scale fisheries and other arenas of marine resource governance in the Caribbean Large Marine Ecosystem (CLME), there have been multi-level governance initiatives to link local organisations and communities to national and transboundary institutional arrangements (Fanning et al. 2009). Such conduits for local knowledge are less evident in the case of sargassum (Van der Plank et al. 2022).

For all marine resources, it has been a challenge to network stakeholders into effective governance arrangements to tackle issues at appropriately matched social and ecological scales (Fanning et al. 2021; Mahon and Fanning 2019). Governance concepts have only recently been applied to sargassum (Van der Plank et al. 2022) but few initiatives have yet focused on the links among the actors at national to local levels (McConney and Compton 2020). A common thread running through the governance narratives of climate, fisheries and sargassum is the need to have functional policy cycles that link stakeholders to form networks. Caribbean marine resource governance literature has used a generic five stage policy cycle as a means of assessing how state and non-state actors interact and exchange information to decide upon and achieve shared objectives or solve problems (Fanning et al. 2009, 2013). For example, multi-level, nested, policy cycles have featured prominently in the large marine ecosystem governance framework used for the Caribbean Sea and North Brazil Shelf Large Marine Ecosystems (Fanning et al. 2013). Figure 3.2 is a generic example of a policy cycle.

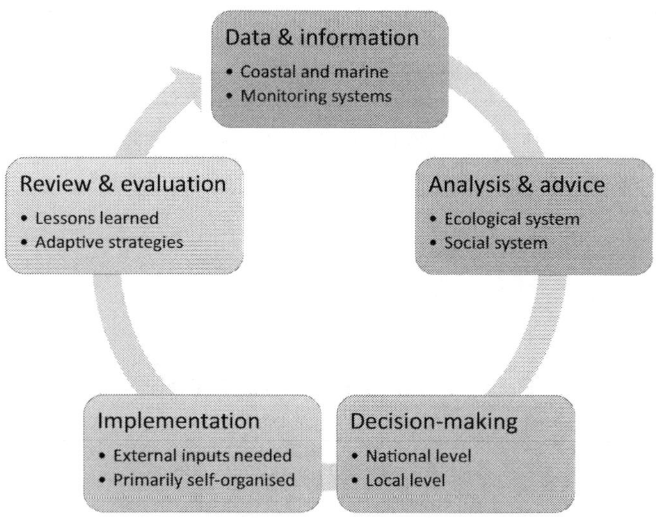

Figure 3.2 Generic policy cycle showing examples of factors relevant to sargassum (by authors).

Activities, actors, policy instruments, and their interactions are dynamic in policy cycle stages, but the data and information stage can be an entry point for including local knowledge. Analysis and advice stage benefits from both social and ecological local knowledge of the social-ecological system. Decision-making ideally incorporates local knowledge in multi-level governance for community to national level interventions. At implementation, local knowledge should feature prominently in enabling self-organisation for climate change adaptation. Review and evaluation processes feature local knowledge in lessons learned and consequent adaptation strategies.

Little formal integration of local knowledge typically occurs at Caribbean marine science policy interfaces (McConney et al. 2016). Deficiencies in obtaining, integrating and using the best available science and local knowledge, such as advocated for climate change adaptation (Kettle et al. 2014), are evident. Scientific and local knowledge are constrained by massive sargassum influxes having no precedent, and so no records or memory to call upon. Science and local knowledge are very slowly informing Caribbean sargassum policy (McConney and Oxenford 2020). Also see Singh et al. (2024) for a comparable Pacific Island situation.

Methods

Using a hypothetical generic sargassum national policy cycle as an analytical framework, we examine the actual and potential use of local knowledge to inform policy. Sargassum formal and informal practices, protocols, adaptive strategies, and management plans in Caribbean countries are reviewed. The qualitative analysis posed questions such as those presented in Table 3.1.

Identifying data limitations helped to filter out unsuitable methods. From prior research, it was known that the use of online literature databases and search engines was not a feasible approach. There are few peer-reviewed publications on Caribbean sargassum policy and governance. Except for natural science journal articles, regional documentation is mainly in the grey literature of project and meeting reports, and draft or restricted manuscripts that are not publicly accessible. Mass media lacks reports on sargassum institutional arrangements. For this exploratory investigation, a convenience sample of peer-reviewed publications and grey literature was drawn from the document collections of the authors and the specialised Sargassum Reference Repository of the Centre for Resource Management and Environmental Studies (CERMES), with which all authors are affiliated. Documents were supplemented by author's participant observation in regional and national sargassum initiatives (e.g., research, outreach, consultation).

Table 3.1 Key considerations for integrating local knowledge into a sargassum policy cycle

Stage of the policy cycle	*Considerations for integration of local knowledge (LK)*
Data and information	What science gaps and uncertain areas can LK fill? How can science data and LK best be integrated?
Analysis and advice	Do LK sources understand the science being used? Which LK citizens or sources do policy-makers trust?
Decision-making	Which holders of LK need to be engaged in local decisions? What decision institutions and processes fit local needs?
Implementation	Can LK be used directly in fieldwork through collaboration? Can applied science add value to the practical use of LK?
Review and evaluation	How does LK evolve through collective learning-by-doing? Are learning networks useful for scaling up LK in reviews?

Five recent journal articles were most relevant to local knowledge in sargassum policy cycles (e.g., Van der Plank et al. 2022), but other articles informed analysis and interpretation. Around 20 items of regional grey literature comprising mainly technical reports, conference proceedings, and sargassum management plans (e.g., CERMES and MMABE 2021) provided the bulk of documentation on local knowledge in policy cycles. No document addressed local knowledge formally used in a full sargassum policy cycle, as one does not exist, but aspects of sargassum management (e.g., participatory beach stranding site profiling) were relevant. Content within documents was linked to each stage by qualitative judgement. Methods reflect a preliminary study of how local knowledge of sargassum is currently used, and how it can be formally incorporated into full sargassum policy cycles whenever they are developed in the future.

Results and discussion

This section presents and discusses findings by stage in the policy cycle on local knowledge availability and current use, with perspectives on how local knowledge may be improved in future policy cycles. Caribbean sargassum natural and social science research relevant to local knowledge has been linked more to harvest sector coastal operations in small-scale fisheries (e.g., Speede et al. 2019) than to climate change or marine resource policy, an exception being Cox et al. (2021) which tackled wider local knowledge integration and use.

Data and information

This policy cycle stage is about gathering scientific data and the best available information. The latter can include local knowledge. Even if information exchange is not formalised, it is customary for Caribbean state officials, academic researchers, and others to engage resource users to obtain data and information. This is especially so where institutional capacity for data collection is low due to inadequacies in staffing, budget, equipment or skills and where the few available marine resource data sets are inadequate (Cox and Oxenford 2018).

Fishers' local knowledge from observations at sea over the decade of sargassum influxes (e.g., on associated biodiversity and impacts on fishing) assist in validating and interpreting official fisheries data for trends and patterns linked to sargassum (Oxenford et al. 2019). For example, Cox et al. (2021) combine Caribbean fishers' observations, perceptions, and fishing practices with science to understand changing climate impacts and environmental risks in fisheries. Their analyses include information on climate and sargassum coping and adaptive strategies in fisheries. They claim that "mobilising this local knowledge for evidence-based policy-making is essential for more effective and sustainable implementation of climate change adaptation strategies" (Cox et al. 2021: 26), as seen in later stages of the policy cycle.

Due to the influxes being as much a surprise to science as to citizens, communication of all types in support of better exchange of information on sargassum has been more challenging than in climate science generally (McConney and Oxenford 2020). Few fisheries resource users or managers participate in sargassum email lists or online webinars. Fisherfolk and fisheries authority information exchanges tend to be via social media or in person and linked to projects (Oxenford et al. 2019). Caribbean SIDS have not yet pursued citizen science to monitor sargassum.

In the future, sharing fishers' observations could become a routine procedure while protecting data sovereignty (Reyes-García et al. 2022). Improving sargassum oceanographic and landfall model forecasts and refining sargassum early warning systems relies on large observational data sets at fine scale. For example, fishers with mobile phones at sea could submit locations, photographs, and other information on sargassum to authorities or researchers via a smart phone application. Text or voice notes would add local knowledge to data for interpretation. Larger countries already incorporate citizen science into sargassum initiatives (Arellano-Verdejo et al. 2020; Collado-Vides et al. 2018; Iporac et al. 2020). Reyes-García et al. (2020) emphasise the ways in which operationalising local knowledge and utilising citizen science can be both challenging and rewarding for informing climate adaptation. New technologies simplify data and knowledge sharing by fisherfolk and others (FAO and WorldFish 2020).

Currently, local knowledge is used to document sensitive coastal ecological areas (e.g., sea turtle nests), timelines of sargassum impacts and responses, assessments of site vulnerability, local sargassum uses, and other factors in sargassum management (CERMES and MMABE 2021). Fisherfolk, watersport operators, tourism workers, coastal residents, environmental groups, and others share field observations to improve socio-economic data and information (Ramlogan et al. 2017). Sargassum action learning groups of stakeholders are promoted by the Caribbean Natural Resources Institute (CANARI) in several countries, but information exchange remains *ad hoc* and without sufficient feedback to the providers of local knowledge regarding sargassum and climate science (McConney and Oxenford 2020).

A future approach to practical integration of local and scientific knowledge is learning-by-doing, ranging from *ad hoc* collaborative fieldwork to full-time citizen science programmes (Collado-Vides et al. 2018). Sargassum monitoring can be a means of achieving knowledge integration. Initially, sargassum protocols emphasised state authority (e.g., CRFM 2016). Recent protocols by Alleyne et al. (2022), Baldwin et al. (2022), and Small et al. (2022) set out field techniques that could engage citizen groups or schools, and improve data augmented by local knowledge while protecting data sovereignty (Reyes-García et al. 2022).

For the future, incorporating more local knowledge on the human dimensions of climate change impacts, and sargassum influxes specifically, could be done through the global initiative known as Socio-economic Monitoring for Coastal Management (SocMon), designed for local level engagement (Edwards et al. 2019). According to these authors, SocMon has been used for Caribbean small-scale fisheries and marine protected areas to contribute socio-economic data for adaptive management decision-making. SocMon incorporates climate change indicators which can be adapted to add sargassum local knowledge.

Analysis and advice

Local knowledge is often sought in the data and information stage, especially in data-poor situations, but ideally there will be a little separation between that stage and the analysis and advice stage. This chapter's authors have encountered useful time series of sargassum data stored in government offices with no system for analysis and advice. Now, local knowledge rather than accumulated data tends to be sought and analysed to generate urgent advice. The means of selecting local knowledge sources and verifying information is therefore of utmost importance. Local knowledge from social networks or formal channels must be understood to find reliable advice (Ramlogan et al. 2017). Statutory institutions such as Fisheries Advisory Committees are a means by which individual fisherfolk and their organisations make a direct input into policy. However, few Caribbean SIDS currently have fully functional Fisheries Advisory Committees. Addressing this deficiency is critical in drafting sargassum adaptive management strategies that incorporate all stages of the policy cycle (CERMES and MMABE 2021).

Currently, there are few examples of co-development of knowledge products with fisherfolk that demonstrate the useful outputs of local knowledge integration with science to enhance analysis and advice. For example, Speede et al. (2019) validated the best practices of fishers and boat owners from both practical and scientific angles. They show how innovations in fishing vessels and fishing gear relate to ecological features of sargassum. Although the good practices are simply advisory, they may have implications for formal use of local knowledge if used in new standards for fishing equipment and practices.

Currently, fisherfolk have limited knowledge of climate science with which to contextualise or interpret their observations and upon which to base advice in formal processes (Cox et al. 2021). Caribbean authorities and mass media often fall short of fully informing the public on environmental matters, a situation that applies to marine science in general (McConney et al. 2016). Offering advice on ecosystem-based management, an ecosystem approach to fisheries, integrated coastal management, and the like requires close attention not only to sargassum, but also to the environmental, economic, and social aspects of sargassum interactions including in tourism, renewable energy, and coastal marine transportation. The scope of analysis and advice based upon local knowledge is important and needs to be broadened.

In the future, much of the human dimension sargassum advice may be derived from local knowledge combined with science and technology (e.g., Hinds et al. 2016). Ramlogan et al. (2017) provided an integrated socio-economic analysis framed by resilience thinking in which local knowledge was used to advise on how fishers were likely to cope with or adapt to sargassum influxes. It can document practices and guide ways to deal with sargassum such as by changing fishing practices, adding homemade technological adaptations to engines and hulls, and other innovations not recorded in official documents (Speede at al. 2019). Advice on what is feasible and practical at the level of the individual, fishing enterprise, household, and organisation often relies – at least partially – upon local knowledge. Such advice must be aggregated in the future for application in national policy cycles, but will not necessarily be homogenous across local communities due to their diversity in exposure and sensitivity to climate change impacts.

A quarterly Sargassum Sub-regional Outlook Bulletin is currently disseminated to around 1,000 diverse subscribers, representing a special case of analysis and advice (CERMES 2022). This bulletin informs regional fisheries and tourism stakeholders of a three-month Caribbean forecast for sargassum quantities, distribution, and possible impacts through ocean science advice that can be easily understood and combined with local knowledge by readers to fit site-specific situations. Compiled by sargassum researchers, the bulletin could serve as a time-limited policy brief for multi-stakeholder, evidence-based, decision-making, and elicit contributions of local knowledge. However, this needs to be further strengthened.

The recent sargassum monitoring protocols and best practices, for example, Alleyne et al. (2022), Baldwin et al. (2022), and Small et al. (2022), open new opportunities for developing citizen science approaches that are simple but scientifically robust. These protocols can encourage marine resource stewardship by actively engaging local resource users and their households in analysis and advice. They may also foster trust and respect amongst the stakeholders, which would also be beneficial for the remainder of the policy cycle.

Decision-making

At this stage of the policy cycle, local knowledge is useful for deciding on matters such as feasible community level subsidiarity to manage sargassum, public resources to deploy, and budget allocations if sargassum policy is to enable self-organisation and empowerment. In the absence of fully functional sargassum policy cycles, or formal institutional arrangements that would support them, no clear cases of local knowledge use in formal decision-making were found in the literature. However, one could envision a village committee guided by using sargassum local knowledge in its decision-making procedures. Applying principles of good governance in decision-making (e.g., inclusiveness, participation) facilitates the incorporation of local knowledge through equal opportunity for diverse sources to be heard (Compton et al. 2020). For example, one would need to know the interests and key roles of various stakeholders (Fanning et al. 2021), how to ensure accountability and transparency through effective communication (McConney and Oxenford 2020), and other structural and functional features of governance arrangements tailored to local circumstances. Without good governance, power asymmetries and other factors may intentionally or inadvertently stifle local knowledge at this stage.

Other features that are relevant to all stages of a policy cycle become more prominent in the decision-making stage. Gender mainstreaming is an example (Compton et al. 2020). Gender is important throughout the policy cycle as local knowledge of women may differ from men's knowledge. Moreover, women's knowledge is often critical in household adaptation measures and socio-economic aspects of livelihoods (Edwards et al. 2019). These aspects can include knowledge of public or women's health, and other social impacts of sargassum that men may downplay due to differences in gendered perceptions of risk (Fraga and Robledo 2022).

In the future, local knowledge can also be used to decide on no-regrets adaptation interventions at the community level more than at the national level of the policy cycle. Governance arrangements such as Fisheries Advisory Committees and Ocean Governance Committees provide national intersectoral coordination mechanisms (NICs) for integrating different knowledge types to inform adaptation to climate change and variability (Compton et al. 2020). Collective action on sargassum coordinated across sectors at local, national, and regional levels will be critical for successful decision-making on climate issues (CSGM 2020).

Implementation

Implementation of site level sargassum adaptations and interventions is currently informed by local knowledge, but significant improvements are warranted. For example, some sites use manual labour with hand tools rather than heavy machinery to responsibly remove sargassum in ecologically sensitive areas or to reach off-road areas (Sealys and Felix 2017). Residents are guides and resource persons to implement such measures. Nearshore sargassum removal requiring boats and fisherfolk knowledge of sea conditions, bathymetry, ecology, and weather is another case in point. In the future, local knowledge that was useful in analysis and advice can also be made more useful in site level activities when shared by the knowledge holders on the coast in real time during interventions such as sargassum beach clean-ups.

Local knowledge may also be important in the future for adding sargassum to local economic opportunities for innovation, investment, and entrepreneurship as private sector engagement will be essential to turn sargassum from being a liability into an asset. Potential sargassum uses include fertiliser, for bioenergy, in construction, and for cosmetics (Desrochers et al. 2020). However,

Oxenford et al. (2021) urge caution in proceeding with some products due to high heavy metal and arsenic levels in sargassum. It will be important to tap into local knowledge on uses to apply public health and safety standards to innovation.

Fisheries learning exchanges for knowledge mobilisation between neighbouring islands with different experiences of sargassum, have been an effective means of sharing local knowledge among fisherfolk through peer-to-peer interaction (Cox and McConney 2021). Higher-level exchanges should be used in implementation to facilitate regional cooperation among policy actors as to date this has been attempted with only moderate success (CRFM 2016).

Although not explicitly mentioning policy cycles, researchers have recommended that "engagement of fisherfolk should be improved to co-produce adaptation strategies that ensure appropriate adaptation measures which complement locally established coping strategies" (Cox et al. 2021: 25). This acknowledges that local knowledge should be an essential contributor to the implementation of any sargassum policy cycle in the future.

Review and evaluation

Finally, in the policy stage of review and evaluation local knowledge can be useful for social learning and adaptation . This role can be within the institutions for participatory monitoring and evaluation commonly used in externally funded projects and programmes. As noted in the methods section, seldom is open data or publicly accessible literature available on sargassum local knowledge. A culture of critical analysis and constructive criticism of sargassum initiatives must utilise exchanges between local and scientific knowledge, leading to improved knowledge integration that does not subordinate local knowledge.

On reviewing draft national sargassum management plans (e.g., Ince 2017; Sealys and Felix 2017; Williams 2017), Cox et al. (2019) found that the plans and the protocol on which they were based (CRFM 2016) ignored climate change adaptation. Cox et al. (2019) urged more attention to collaborative planning for developing adaptive capacity. Subsequent national sargassum adaptive management strategies (CERMES 2021a–d) incorporated sargassum local management plans. This hierarchy was intended to infuse more use of local knowledge at the community or site level, while facilitating aggregation of knowledge at the national level. In the future, multi-level adaptation should better address influxes that vary annually in landfall location, severity, ratios of sargassum species and morphotypes. However, strategies and plans are only useful if tested and treated as living documents, with continuous review and evaluation, as knowledge is created locally from experience and from scientific investigation.

In the future, each policy cycle stage should use more sargassum local knowledge. During the COVID-19 pandemic, regional online forums on sargassum catalysed local knowledge exchange across the Caribbean. Appreciation for review and evaluation was apparent as sargassum stakeholders determined critical gaps in data and action. For example, the use of drones for remote and sophisticated sargassum monitoring became a focus in the multi-level diffusion of technology, as team fieldwork was constrained (Baldwin et al. 2022). With COVID-19 constraints easing, responses to sargassum are being reviewed and evaluated to incorporate new technologies. Local knowledge plays a role in these, particularly as participatory methods are now being used again (Baldwin et al. 2022).

The key points set out above from the findings in all stages of the sargassum policy cycle on present and potential future use of local knowledge are summarised in Figure 3.3.

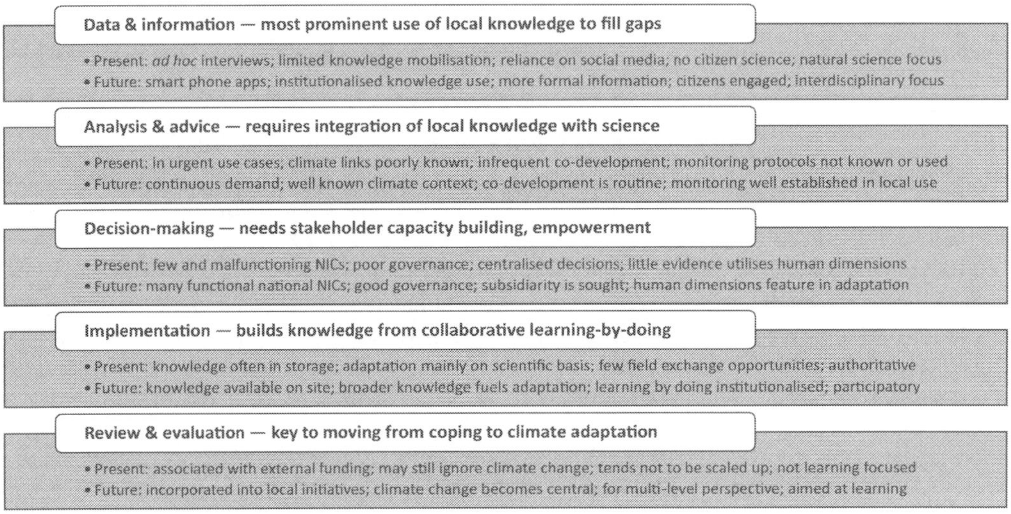

Figure 3.3 Summary of present and potential future use of local knowledge in the policy cycle.

Conclusion

The dilemma of sargassum seaweed being a social-ecological system asset in small to moderate quantities, but a climate-linked threat to Caribbean countries in the massive influxes experienced since 2011, is a unique case of climate change and climate variability. The influxes were a surprise to science and no local knowledge of them existed. In the last decade, both science and local knowledge have resulted in improved understanding of the influxes and ability to cope, but successful adaptation to sargassum influxes is still elusive.

An important issue that has received little attention in climate adaptation is the role of local knowledge in creating and operating complete, multi-level sargassum policy cycles in the affected countries. Comprehensive policy cycles build resilience by ensuring that well-forged links from policy to practise guide climate action. In the case of the sargassum influxes, the exploratory investigation in this chapter found that there are no complete sargassum policy cycles, and that the availability and use of local knowledge are issues to be addressed.

In the scarce literature and the author's own applied research experience, we found mainly examples of *ad hoc* policy-related use of local knowledge in sargassum management. Local community input informed how coastal area ecosystems and livelihoods have been impacted by the influxes. Such knowledge has guided local coping measures, especially in small-scale fisheries, but there are no institutionalised mechanisms for bringing this knowledge to policy.

Deficiencies in the availability and use of local knowledge were evident in all stages of a generic sargassum policy cycle assembled from research findings. Components of policy cycle stages exist, but these were not yet linked to create a coherent whole. The findings illustrate several future opportunities for active learning networks and other beneficial institutional arrangements to incorporate local knowledge at each stage of a policy cycle. There is a need for citizen engagement and local knowledge, especially to convert sargassum from a threat to an opportunity. A sargassum policy cycle will need to be incorporated into climate adaptation measures with support for reducing vulnerability and building resilience.

This transformation in use of local knowledge and climate change adaptation requires major reform of marine resource and climate science-policy interfaces (McConney et al. 2016) with special attention to guidance available on good governance (Compton et al. 2020). Lack of national intersectoral coordination mechanisms for marine governance currently stifle use of local knowledge as these national governance arrangements are crucial for policy success (McConney and Compton 2020). Recent interest in coastal and marine spatial planning could offer avenues for citizen science and engagement in policy cycle stages with a concomitant increase in the availability and use of local knowledge to positively influence climate policy.

Acknowledgements

The authors wish to acknowledge support from the Food and Agriculture Organization and the Caribbean Biodiversity Fund Ecosystem based Adaptation Facility with financing from the International Climate Initiative (IKI) of the German Federal Ministry for Environment, Nature Conservation, and Nuclear Safety through Credit Institute for Reconstruction (KfW).

References

Alleyne, K., Irvine, J. and Oxenford, H.A. (2022). *Sargassum Monitoring Protocol: Monitoring Sargassum Species Morphotypes.* SargAdapt Good Practice Guide Series: 3. Barbados: UWI Centre for Resource Management and Environmental Studies.

Arellano-Verdejo, J. and Lazcano-Hernandez, H.E. (2020). Crowdsourcing for Sargassum monitoring along the beaches in Quintana Roo. In: Mata-Rivera, M.F., Zagal- Flores, R., Arellano Verdejo, J., Lazcano Hernandez, H.E. (Eds.), *GIS LATAM, Communications in Computer and Information Science*. Cham: Springer International Publishing, pp. 49–62.

Baldwin, K., Oxenford, H.A., Weekes, J., Small, M., Irvine, J. and Desai, A. (2022). *Sargassum Monitoring Protocol: Monitoring Sargassum Abundance Using Drones*. SargAdapt Good Practice Guide Series 1. Barbados: UWI Centre for Resource Management and Environmental Studies.

Barange, M., Bahri, T., Beveridge, M.C.M., Cochrane, K.L., Funge-Smith, S. and Poulain, F. (eds.) (2018). *Impacts of Climate Change on Fisheries and Aquaculture: Synthesis of Current Knowledge, Adaptation and Mitigation Options*. FAO Fisheries and Aquaculture Technical Paper No. 627. Rome: FAO.

CERMES. (2021a). *Draft Grenada Sargassum Adaptive Management Strategy.* Volumes 1 and 2. Barbados: UWI Centre for Resource Management and Environmental Studies.

CERMES. (2021b). *Draft Saint Kitts and Nevis Sargassum Adaptive Management Strategy.* Volumes 1 and 2. Barbados: UWI Centre for Resource Management and Environmental Studies.

CERMES. (2021c). *Draft Saint Lucia Sargassum Adaptive Management Strategy.* Volumes 1 and 2. Barbados: UWI Centre for Resource Management and Environmental Studies.

CERMES. (2021d). *Draft St Vincent and the Grenadines Sargassum Adaptive Management Strategy.* Volumes 1 and 2. Barbados: UWI Centre for Resource Management and Environmental Studies.

CERMES. (2022). *Sargassum Sub-Regional Outlook Bulletin.* UWI Centre for Resource Management and Environmental Studies. Available online at: www.cavehill.uwi.edu/cermes/projects/sargassum/outlook-bulletin.aspx

CERMES and MMABE. (2021). *Draft Barbados Sargassum Adaptive Management Strategy.* Volumes 1 and 2. Barbados: UWI Centre for Resource Management and Environmental Studies and the Ministry of Maritime Affairs and Blue Economy.

Collado-Vides, L., van Tussenbroek, B.I., Garcia, M., Rodríguez-Martínez, R., Cassano, V., Oliveira, M.C., Toyota Fujii, M., Oxenford, H. and Krueger-Hadfield, S. (2018). A network to develop a taxonomic, monitoring, and citizen's participation program for Sargassum landings in Florida, Mexico, Barbados, and Brazil. *Proceedings of the Gulf and Caribbean Fisheries Institute* 71:319–324.

Compton, S., McConney, P., Monnereau, I., Simmons, B. and Mahon, R. (2020). *Good Practice Guidelines for Successful National Intersectoral Coordination Mechanisms*. 2nd Edition. CERMES Technical Report. No 88. Barbados: UWI Centre for Resource Management and Environmental Studies.

Cox, S. and McConney, P. (2021). *Perfecting the Art of Fisheries Learning Exchanges in the Eastern Caribbean.* FAO Fisheries and Aquaculture Circular No. 1222. Bridgetown: FAO.

Cox, S. and Oxenford, H.A. (2018). *Summary Report on the Available Catch and Fishing Effort Data for Flyingfish and Dolphinfish in the Eastern Caribbean.* Barbados: UWI Centre for Resource Management and Environmental Studies.

Cox, S., Oxenford, H.A. and McConney, P. (2019). *Summary Report on the Review of Draft National Sargassum Plans for Four Countries in the Eastern Caribbean*. Barbados: UWI Centre for Resource Management and Environmental Studies.

Cox, S., Desai A., Granderson A., Albert D., and Ramkissoon C. (2021). *A Preliminary Look at Integrating Local and Scientific Knowledge for Climate Change Adaptation in the Easter Caribbean Fisheries Sector*. Barbados: UWI Centre for Resource Management and Environmental Studies.

CRFM. (2016). *Model Protocol for the Management of Extreme Accumulations of Sargassum on the Coasts of CRFM Member States*. CRFM technical and advisory document series number 2016/5. Belize: Caribbean Regional Fisheries Mechanism (CRFM) Secretariat.

CSGM (Eds.). (2020). *The State of the Caribbean Climate*. Produced by the Climate Studies Group Mona for the Caribbean Development Bank. Barbados: CDB.

Desrochers, A. Cox S., Oxenford H.A. and van Tussenbroek B. (2020). *Sargassum Uses Guide: A Resource for Caribbean Researchers, Entrepreneurs and Policy Makers*. CERMES Technical Report No. 97. Barbados: UWI Centre for Resource Management and Environmental Studies.

Edwards P., Pena M., Medeiros R.P. and McConney P. (2019). Socioeconomic Monitoring for Sustainable Small-Scale Fisheries: Lessons from Brazil, Jamaica, and St. Vincent and the Grenadines. In: Salas S., Barragán-Paladines M., Chuenpagdee R. (Eds.), *Viability and Sustainability of Small-Scale Fisheries in Latin America and the Caribbean*. MARE Publication Series, Vol 19. Cham: Springer.

Fanning, L., Mahon R. and McConney P. (2009). Focusing on Living Marine Resource Governance: The Caribbean Large Marine Ecosystem and Adjacent Areas Project. *Coastal Management* 37: 219–234.

Fanning, L., Mahon R. and McConney P. [Eds]. (2011). *Towards Marine Ecosystem-Based Management in the Wider Caribbean*. Amsterdam: Amsterdam University Press.

Fanning, L., Mahon R. and McConney P. (2013). Applying the Large Marine Ecosystem (LME) Governance Framework in the Wider Caribbean Region. *Marine Policy* 42: 99–110.

Fanning, L., Mahon R., Compton S., Corbin C., Debels P., Haughton M., Heileman S., Leotaud N., McConney P., M. Moreno P., Phillips, T. and Toro, C. (2021). Challenges to Implementing Regional Ocean Governance in the Wider Caribbean Region. *Frontiers in Marine Science* 8: 286.

FAO and WorldFish. (2020). *Information and Communication Technologies for Small-Scale Fisheries (ICT4SSF) – A Handbook for Fisheries Stakeholders*. In support of the implementation of the Voluntary Guidelines for Securing Sustainable Small-Scale Fisheries in the Context of Food Security and Poverty Eradication. Bangkok.

Fraga, J. and Robledo, D. (2022). Covid-19 and Sargassum Blooms: Impacts and Social Issues in a Mass Tourism Destination (Mexican Caribbean). *Maritime Studies* 21:159–171.

Hinds, C., Oxenford H.A., Cumberbatch J., Fardin F., Doyle E. and Cashman A. (2016). *Golden Tides: Management Best Practices for Influxes of Sargassum in the Caribbean with a Focus on Clean-Up*. Barbados: UWI Centre for Resource Management and Environmental Studies.

Ince, D. (2017). *Protocol for the Management of the Extreme Accumulations of Sargassum on the Coast of Grenada*. Grenada: Fisheries Division.

Iporac, L.A.R., Olszak, S., Burkholder, D. and Collado-Vides, L. (2020). Lessons and Challenges in Piloting "Sargassum Watch," A Citizen Science Program to Monitor Pelagic Sargassum Landings in South Florida. *Proceedings of the Gulf and Caribbean Fisheries Institute* 72: 246–252.

James, P. (2009). An Assessment of Potential Impacts of Climate Change and Climate Variability on Small-scale Fisheries in the Eastern Caribbean. *Proceedings of the Gulf and Caribbean Fisheries Institute* 61: 139–147.

Jentoft, S. and Chuenpagdee, R. (2009). Fisheries and Coastal Governance as a Wicked Problem. *Marine Policy* 33: 553–560.

Johnson, D.R., Franks, J.S., Oxenford, H.A. and Cox, S. (2020). Pelagic Sargassum Prediction and Marine Connectivity in the Tropical Atlantic. *Gulf and Caribbean Research* 31: 20–30.

Kettle, N.P., Dow, K., Tuler, S., Webler, T., Whitehead, J. and Miller, K.M. (2014). Integrating Scientific and Local Knowledge to Inform Risk-Based Management Approaches for Climate Adaptation. *Climate Risk Management* 4: 17–31.

Kinch, A. and Oxenford H.A. (2019). Fishers' Observations of Climate Change Impacts on the Flyingfish Fishery in Barbados. *Proceedings of the Gulf and Caribbean Fisheries Institute* 71: 146–147.

Mahon, R. and Fanning, L. (2019). Regional Ocean Governance: Integrating and Coordinating Mechanisms for Polycentric Systems. *Marine Policy* 107: 103589.

McConney, P. and Compton, S. (2020). National Ocean Governance as a Foundation for Blue Economic Development. The Caribbean Blue Economy. In: P. Clegg, R. Mahon, P. McConney and H.A. Oxenford. (Eds.), *The Caribbean Blue Economy*. Oxford: Routledge, pp. 92–103.

McConney, P. and Oxenford, H.A. (2020). Caribbean Sargassum Phenomenon: Complexities of Communicating. *Journal of Caribbean Environmental Sciences and Renewable Energy* 3: 10–14.

McConney, P., Fanning, L., Mahon, R. and Simmons, B. (2016). A First Look at the Science-Policy Interface for Ocean Governance in the Wider Caribbean Region. *Frontiers in Marine Science* 2: 119.

Monnereau, I. and Oxenford, H.A. (2017). Impacts of Climate Change on Fisheries in the Coastal and Marine Environments of Caribbean Small Island Developing States (SIDS). Commonwealth Marine Economies Programme. *Caribbean Marine Climate Change Report Card: Science Review* 2017: 124–154.

Nurse, L. (2011). The Implications of Global Climate Change for Fisheries Management in the Caribbean. *Climate and Development* 3: 228–241.

Oxenford, H.A., Johnson, D., Cox, S. and Franks, J. (2019). *Report on the Relationships between Sargassum Events, Oceanic Variables and Dolphinfish and Flyingfish Fisheries*. Barbados: UWI Centre for Resource Management and Environmental Studies.

Oxenford, H.A. and Cox, S., van Tussenbroek, B.I. and Desrochers, A. (2021). Challenges of Turning the Sargassum Crisis into Gold: Current Constraints and Implications for the Caribbean. *Phycology* 1: 27–48.

Ramlogan, N.R., McConney, P. and Oxenford, H.A. (2017). *Socio-Economic Impacts of Sargassum Influx Events on the Fishery Sector of Barbados*. CERMES Technical Report No. 81. Barbados: UWI Centre for Resource Management and Environmental Studies.

Reyes-García, V., Fernández-Llamazares, Á., García-del-Amo, D. and Cabeza, M. (2020). Operationalizing Local Ecological Knowledge in Climate Change Research: Challenges and Opportunities of Citizen Science. In: Welch-Devine, M., Sourdril, A., Burke, B. (Eds.), *Changing Climate, Changing Worlds. Ethnobiology*. Cham: Springer.

Reyes-García, V., Tofighi-Niaki, A., Austin, B. J., Benyei, P., Danielsen, F., Fernández-Llamazares, Sharma, A., Soleymani-Fard, R. and Tengö, M. (2022). Data Sovereignty in Community-Based Environmental Monitoring: Toward Equitable Environmental Data Governance. *BioScience* 72: 714–717.

Rittel, H.W. and Webber, M.M. (1973). Dilemmas in a General Theory of Planning. *Policy Sciences* 4: 155–169.

Rosellón-Druker, J., Calixto-Pérez, E., Escobar-Briones, E., González-Cano, J., Masiá-Nebot, L. and Córdova-Tapia, F. (2022). A Review of a Decade of Local Projects, Studies and Initiatives of Atypical Influxes of Pelagic Sargassum on Mexican Caribbean Coasts. *Phycology* 2: 254–279.

Sealys, C. and Felix M. (2017). *Saint Lucia National Strategy for the Management of Sargassum Influxes on Beaches, Bays and Small Harbours*. St. Lucia: Fisheries Department.

Singh, P., Schulman D., Charan D., Reyes-García V., Schlingmann A., Ajay Singh A. and Railoa K. (2024). *Climate Change Adaptation in Fiji: Local Adaptation Strategies to Enhance National Policy*. In V. Reyes-García, S. Álvarez-Fernández, P. Benyei, L. Calvet-Mir, D. García-del-Amo, A. B. Junqueira, X. Li, V. Porcher, A. Porcuna-Ferrer, A. Schlingmann, & R. Soleymani (Eds.), *Routledge Handbook of Climate Change Impacts on Indigenous Peoples and Local Communities*. Oxfordshire & New York: Routledge. Chp 24, pp 389–408.

Skliris, N., Marsh, R., Appeaning Addo, K. and Oxenford, H. (2022). Physical Drivers of Pelagic Sargassum Bloom Interannual Variability in the Central West Atlantic over 2010–2020. *Ocean Dynamics* 72: 383–404.

Small, M., Irvine, J., Baldwin, K., Weekes, J. and Oxenford, H.A. (2022). *Sargassum Monitoring Protocol: Monitoring Sargassum Abundance Using Beach Surveys*. SargAdapt Good Practice Guide Series: 2. Barbados: UWI Centre for Resource Management and Environmental Studies.

Speede, R.F. Cox, S. and Oxenford, H.A. (2019). *Best Practices for Fisherfolk Coping with Sargassum*. Barbados: UWI Centre for Resource Management and Environmental Studies.

Turner, R., McConney P. and Monnereau I. (2020). Climate Change Adaptation and Extreme Weather in the Small-Scale Fisheries of Dominica. *Coastal Management* 48: 436–455.

Van der Plank, S., Corbett, J., Cumberbatch, J., Thomas, B. and Tompkins, E. (2020). *Management of Sargassum Influxes in the Caribbean: National and Regional Governance of a Transboundary Marine Species*. SARTRAC Working Paper 1.

Van der Plank, S., Cox, S., Cumberbatch, J., Mahon, R., Thomas, B., Tompkins, E. L. and Corbett, J. (2022). Polycentric Governance, Coordination and Capacity: The Case of Sargassum Influxes in the Caribbean. *Coastal Management* 50: 285–305.

Wang, M., Hu, C., Barnes, B.B., Mitchum, G., Lapointe, B. and Montoya, J.P. (2019). The Great Atlantic Sargassum Belt. *Science* 365(6448): 83–87.

Williams, O. (2017). *Plan for the Management of the Accumulations of Sargassum on the Coastal and Marine Ecosystem of St. Kitts and Nevis*. St. Kitts and Nevis: Fisheries Department.

4

"THE WEATHER IS MORE ERRATIC; IT CHANGES FASTER…" LOCAL PERCEPTIONS OF CLIMATE CHANGE IN THE EASTERN CARPATHIANS, ROMANIA

Dániel Babai

Introduction

High mountain regions are home to numerous subsistence-oriented communities maintaining highly valued cultural landscapes. These social-ecological systems are directly and indirectly affected by climatic, ecological, sociocultural, economic, and political changes (Savo et al., 2016). Climate change affects mountainous landscapes above the global averages (e.g., Chaudhary & Bawa, 2011; Gobiet et al., 2014). Climate change influences not only the local weather patterns, but also the different elements of the ecosystem, including the vertical and horizontal distribution and phenology of plant and animal species (Feehan et al., 2009; Hansen et al., 2012) and the composition of mountain plant associations (Feehan et al., 2009). Thereby, subsistence-oriented, rural mountain communities operating extensive systems are also seriously affected by climate-related changes (Savo et al., 2016).

Because of the complex topography of mountain areas, it is difficult to exactly determine and model climate change impacts (Elsen et al., 2020). To overcome such difficulties, first-hand and detailed local observations and reports are becoming increasingly valuable in climate research (Reyes-García et al., 2019). Observations made by local communities can reveal more about ongoing trends than models interpolated on the basis of instrumental measurements by geographically distant meteorological stations (Savo et al., 2016; Reyes-García et al., 2019). Although local communities' reports can be distorted by the variability of the weather, short-term trends, extreme weather events, and the public media (Zaval et al., 2014; Reyes-García et al., 2016), research shows that local observations are often concordant with the results of instrumental measurements (Lehner & Stocker, 2015; Savo et al., 2016). Local observations are partly based on community members' own life experience (individual observations), but there is also a multi-generational component (collective observations), which allows for the monitoring of long-term, trend-like changes (Hansen et al., 2012).

DOI: 10.4324/9781003356837-7

Local communities in Central and Eastern Europe are underrepresented in the documentation of local observations of climate change impacts, despite the fact that the region is facing considerable changes, especially regarding average temperatures and precipitation patterns (Spinoni et al., 2015). Within this context, the aim of this study was to document the changes experienced by local people in a mountain community practising extensive farming in the Eastern Carpathians of Romania.

Specifically, the objectives of the study are:

1 To document the local observation of changes in elements of the atmospheric system (temperature, precipitation, air movement, seasonality).
2 To document the direct impacts of changes in elements of the atmospheric system on extensive farming.
3 To document local responses to climate change impacts in farming.

Study area

Research was conducted in Romania, in the Eastern Carpathians, in the village of Lunca de Jos (Gyimesközéplok), within the area of Valea Rece (Hidegségpataka). The village is located in the valley of the Hidegség stream (800–900 MSL) (Figure 4.1a). The height of the surrounding mountains ranges from 900 to 1,500 m.a.s.l. The municipality of Lunca de Jos has an area of 138.3 km^2, of which 30.5% is covered by hay meadows, 36.4% by pasture, 30.3% by forest, 1.1% by arable

Figure 4.1 Study area. (a) The village is nestled in the valley bottom. Between the houses, there are arable fields and hay meadows. (b) Inner hay meadows stretch across the valley floor and foothills. (c) Beyond the inner hay meadows are the mountain pastures and the outer hay meadows. (d) Extensively cultivated hay meadows are particularly species-rich. Photos: Dániel Babai.

land, and 1.7% is built-up area, including infrastructure (Sólyom et al., 2011). Hay meadows and pastures are partly mesophilic, partly mountain meadows formed on nutrient-poor, acid soil, and to a lesser extent species-poor *Nardus* grasslands (Figure 4.1b–d) (Babai et al., 2014). The forest area is primarily spruce forest.

The mountain area has a moderately humid alpine-boreal climate with strong continental influences. The average annual temperature is 4–6°C. The hottest month is July and the coldest is January. Average annual precipitation is 700–800 mm (Ilyés, 2007). Across the Carpathians, most of the precipitation tends to fall in June (12–16% of the annual precipitation). The second peak in precipitation tends to occur in late autumn and early winter (8%) (Micu et al., 2015). Average winter precipitation is 106 mm. The first snowfall usually occurs in November and continuous snow cover lasts over 100 days, approximately until March. The prevailing wind direction is from the west.

The study area has a population of 2,328 people, mostly Hungarians (only 0.4% are Romanians). Around 85–90% of the population is engaged in extensive farming as full-time farmers or as a supplementary source of income (personal observation). The average farm size is 3.8 hectares (Sólyom et al., 2011). Family farms tend to rely on dairy cattle farming (Babai et al., 2014). The management of hay meadows focussed on winter fodder production and on the elimination of the impacts of natural and anthropogenic disturbances and promoting the regeneration of mountain grasslands in order to stabilise long-term hay yields. Landscape and ecological features, together with the political, ethnic, and economic contexts, have contributed to the preservation of an extensive, traditional way of farming, especially in terms of grassland management (Babai et al., 2014, 2021). Semi-subsistence family farms also grow arable crops on small plots. The main crop grown is potatoes (*Solanum tuberosum*). Small-scale fruit and vegetable production (cabbage, onions, apples, plums) are also typical. Edible wild plants (spring vegetables and forest fruits), mushrooms, and medicinal herbs play an important role in the diet and everyday life of many families (Babai et al., 2014).

Methods

Data collection

This study uses the term "climate change" to refer to changes in the state of the climate for which the mean and/or variability of each parameter can be well identified and persist over a long period of time (Reyes-García et al., 2016; IPCC, 2021). The exploration of local observations of farmers in Gyimes related to climate change was based on 23 qualitative, semi-structured interviews conducted in Valea Rece. Key informants were interviewed, identified on the basis of former studies on traditional ecological knowledge of local farmers in Gyimes (see, e.g., Babai et al., 2014). During the sampling, the recommendations of the LICCI protocol were followed (Reyes-García et al., 2023). The interview questions covered changes in the landscape, the natural environment, and the climate, as well as the drivers of change during the period from 1990 to 2020. Among the changes mentioned during the interviews, I took into account observations and reports related to long-term climate trends, primarily those which were considered by the local farmers affecting the flora, vegetation, and land-use practices in Gyimes.

Two focus group discussions were organised with six and five experienced farmers. To validate the data collected during interviews, focus group participants were different from those interviewed. The participants discussed the climate trends that came up in the interviews and either confirmed or refuted them.

Fieldwork was conducted in Hungarian and lasted 21 days distributed between 2019 and 2020. Fieldwork received ethical clearance from the Universitat Autònoma de Barcelona (UAB) (CEEAH 4781). Interviewees were informed in advance, through a prior informed consent form, about the aims of the research, including publication.

Data analysis

All the statements concerning climate change, namely changes related to the atmospheric, physical, biological, and human systems were analysed (for more details, see Reyes-García et al., 2023). I coded the statements related to changes made in the interviews using the categorisation system developed by the Local Indicators of Climate Change Impacts (LICCI) project (Reyes-García et al., 2023). For changes mentioned in the interviews, I documented the direction of the trend (increasing, decreasing, variable) and the degree of consensus, based on the number of interviewees who mentioned the change and the number of interviewees who had similar perceptions of the direction of the trend. Climate-related trends were examined in the context of temperature, precipitation, air movement, and seasonality, as local farmers reported changes only in these categories.

Whenever possible, local reports of observations were supplemented by instrumental data of mesoclimatic studies of the Carpathians (e.g., Micu et al., 2015; Spinoni et al., 2015). Information from the closest meteorological station in the Ciuc basin was not used, because the climate of the basin, and the climate of the study area differ significantly (Petres et al., 2017). Similarly, the nearest meteorological station located at a similar elevation (Ceahlau-Toaca) is approximately 188 km away. Given the geographical distance and different climates in the two locations, data from both stations are not suitable for a comparison with changes reported in Valea Rece.

Results

Local inhabitants from Valea Rece reported several climate change impacts that have been affecting them for some decades: "*Do you know that the weather has changed? In the past, spring used to start somewhat later. Now, for the past two or three years, it has started earlier*" (Jánó Mária – 1944; 28 10 2019). Local farmers in Gyimes mentioned changes categorised in 36 indicators related to elements of the atmospheric system, which were confirmed during the focus group discussions. The most frequently observed changes were related to the temperature (ten indicators, e.g., increasing mean temperature), precipitation (16 indicators, e.g., shifting temporal patterns of precipitation), air movement (7 indicators, e.g., decreasing number of windy days), and seasonality (3 indicators, e.g., changes in the onset of seasons) (Tables 4.1–4.4). The level of consensus among the participants on the direction of the trends of the indicators of climate change impacts reached 100% for 28 indicators (e.g., for trends related to seasonality). The lowest degree of consensus was 40% (i.e.., changes in the frequency of lightning and thunder). The average level of consensus related to the reported trends' directions in the case of the 36 coded indicators was 92%.

Table 4.1 Local observations of climate change impacts perceived by Csángó farmers in Gyimes. Observations of changes in temperature

Indicators	*Number of interviewees*	*Direction*	*Quotation*	*Instrumental data*	*C*	*Literature*
Changes in the temperature during the night	2/2	▼	"At night, the air is much cooler than before. On summer nights it actually used to be possible to be outside in short sleeves, but now it cools down…" (Tankó, László – 1986; 29 10 2019)	Minimum night temperatures increased significantly across the Carpathians, and the frequency of tropical nights (temperature is more than 30°C) increased, with a peak in July.	Ø	Micu et al. (2015)
Changes in temperature fluctuation	6/6 100%	▲	"The heat waves are more intense. There are periods with sudden changes, more erratic weather. The weather changes in a matter of seconds." (Prezsmer, Károly – 1980; 29 10 2019)	No data	–	–
Changes in the frequency of days with extreme temperatures	9/9 100%	▲	"Now when it's so hot, really, this heat is just unheard of…" (Máris, Éva – 1979; 17 11 2019)	Occurrence of extremely warm days and maximum temperature in summer increased across the Romanian Carpathians.	Ok	Spinoni et al. (2015) and Micu et al. (2015)
Changes in the length/ duration of heat waves	3/3	▲	"For a day or two, but not like this, with this heat lasting a whole week and drying everything up. This is a little strange." (László, Csaba – 1964; 18 11 2019)	Higher incidence of warm spells, particularly in areas below 1,600 m is characteristic.	Ok	Birsan et al. (2014), Dumitrescu et al. (2015) and Micu et al. (2015)

Changes in the length/duration of cold waves	5/4 80%	▼	"The cold lasted longer. Now it lasts two or three days, but back then it lasted a week or two." (Máris, Éva – 1979; 17 11 2019)	The duration of cold spells is significantly decreasing, but only over limited areas in the Carpathians.	Ok	Lakatos et al. (2013), Birsan et al. (2014) and Dumitrescu et al. (2015)
Changes in the intensity of heat waves	3/3	▲	"I mean, there were heat waves, but it was always windy up in the mountains, so it was hot, but it wasn't this stifling heat that makes it hard to breathe." (Bodor, László – 1976; 16 11 2019)	Intensity of warm spells has increased (maximum temperature is above 25°C) especially since 2000.	Ok	Micu et al. (2015)
Changes in the intensity of cold waves	4/3	▼	"I'm not saying that it doesn't get cold or snow anymore, just that it used to last longer. Now it'll get cold for a week, but back then it would be 20 to 25 degrees below zero for the whole month of December…" (László, Csaba – 1964; 18 11 2019)	Winter minimum temperatures indicated a stronger warming in the Eastern Carpathians (>5°C).	Ok	Busuioc et al. (2006)
Changes in the frequency of unusual temperatures in a given season	2/2	▲	Extremely warm days in winter: "There have even been floods in January. It would snow one day and warm up so much the next day that it would flood, with ice blocks floating in the water." (Kondor, Béla – 1954; 12 11 2019) Extremely cold days in summer: "It's so uncommon to have frost and snow in the summer and for the crops to freeze. That didn't use to happen." (Tankó, Anna – 1934; 14 11 2019)	No data	–	–

(Continued)

Table 4.1 (Continued)

Indicators	*Number of interviewees*	*Direction*	*Quotation*	*Instrumental data*	*C*	*Literature*
Changes in the mean temperature in a given season	15/15 100%	▲	Spring: "The weather gets warmer in April already too." (Tankó, Erzsébet – 1968; 17 11 2019) Summer: "The summers, they'll keep getting hotter. It's a lot hotter than it used to be." (Jánó, Erika – 1992; 28 10 2019) Autumn: "Autumn is longer and a lot warmer than it used to be… Now autumn competes with summer." (Jánó, Erika – 1992; 28 10 2019) Winter: "Winter was forever the same. Now there aren't such awful cold spells; it's bearable." (Jánó, Mária – 1944; 28 10 2019) (Figure 4.2)	Mean temperature trends have increased in winter, spring and summer, while they are completely absent in autumn (it is a stable season from a thermal point of view). The most extended increase occurs in the late spring and summer months (May–August).	Ok[a]	Birsan et al. (2014), Cheval et al. (2014), Spinoni et al. (2015), Dumitrescu et al. (2015) and Micu et al. (2015)
Changes in sunshine intensity	4/4	▲	"It's much hotter in the sun; when it's shining, it's so hot you can feel your head melting." (Tankó, Erzsébet – 1968; 17 11 2019)	A brightening in the 1990s and 2000s was observed, while the relative sunshine duration shows an increase in spring and summer.	Ok	Spinoni et al. (2015)

Number of interviewees – Number of respondents who perceived changes/number of participants who agreed with the direction of the trend. Legend – direction of trend: ▲: the trend shows an increase, ▼: the trend shows a decrease, ◀▶: the direction of the trend varies (the change cannot be expressed as an increase or decrease). C – correspondence between the reports of local observations and instrumental data (OK: locally reported and mesoclimatic instrumental data are similar, Ø: there is no consensus among local observations and instrumental data).

[a] There is a correspondence regarding the winter, spring, and summer mean temperature increase, but not regarding the mean temperature in autumn (local reports highlight the warming trends of the temperature in autumn, while instrumental data shows the opposite).

Figure 4.2 Due to the increase in mean winter temperatures, snow often melts as early as mid-February on slopes with southern exposure. Photo: Dániel Babai.

Temperature

Temperature-related observations reported by local farmers in Gyimes referred to changes in the annual, seasonal, or even daily variation in mean temperature (Table 4.1). According to farmers' observations, the mean temperature during the winter months has increased significantly (Figure 4.2), while the length and intensity of winter cold spells have decreased. The mean temperature of the summer months has also increased, as have the daily maximum temperatures. The intensity and length of summer heat waves have also increased.

Precipitation

Local observations of changes in precipitation mainly highlight changes in the temporal distribution and intensity of rainfall (Table 4.2). Csángó farmers in Gyimes have observed a significant change in the temporal distribution of summer rainfall, with rainfall being less frequent but more intense than in the past. There are more frequent and longer dry spells and even droughts in the summer, which has a serious impact on the vegetation and freshwater supplies in the studied landscape (Figure 4.3a–4.3b). Meanwhile, there has also been an increasing intensity of rainfall events that cause flash floods and landslides (Figure 4.3c–4.3d).

Air masses

Locally perceived changes in air movement are less significant and trends are not particularly obvious, so they were mentioned less frequently during interviews. However, the decreasing number of windy days and the changing frequency of hailstorms were often mentioned (Table 4.3).

Table 4.2 Local observations of climate change impacts perceived by Csángó farmers in Gyimes.

Indicators	*Number of interviewees*	*Direction*	*Quotation*	*Instrumental data*	*C*	*Literature*
Changes in mean rainfall (not further specified)	4/4	▼	"Well, it rained less this year; there used to be more rain." (Jánó, Erika – 1992; 28 10 2019) (Figure 4.3a,b)	A transition towards a drier mountain climate became visible, and the decline of precipitation amount is present in winter, summer and spring.	Ok	Micu et al. (2015)
Changes in the number of days with rainfall/ rainy days (not further specified)	5/5 100%	▼	"It used to rain more often, and not as much. If we had a week without rain, then it would rain the next week. Now a month or two will go by without it raining at all." (Főcze, Péter – 1942; 15 11 2019)	The frequency of precipitation has decreased since the mid-1980s and early 1990s. The probability of wet day occurrences is on the decrease at most Carpathian sites.	Ok	Micu et al. (2015)
Changes in the intensity of rainfall (not further specified)	4/4	▲	"In the past when it rained, you could even go out and mow. Now when the rain comes, it pours so much that it sweeps away the whole village." (FGD#1, Jánó Béla et al. 19 01 2020)	The average intensity of precipitation were slightly increasing (more rainfall showers in some areas).	Ok	Micu et al. (2015)
Changes in the intensity/ strength of heavy rainfall events	9/8 89%	▲	"It used to be that we would have a week of calm rain, but now we'll get a huge rainstorm, with sudden rainfall coming in and causing destruction." (Prezsmer, Károly – 1980; 29 10 2019)	Indicators related to the intensity of heavy rainfall events did not support the observation that intensity of precipitation events increased over the 1961–2010 period.	Ø	Micu et al. (2015)
Changes in the frequency of heavy rainfall events	7/5 71%	▲	"There have never been so many downpours." (Tankó, Ilona – 1954; 12 11 2019)	The number of days with precipitation above 10 and 20 mm increased, while the annual frequency of heavy rain showers increased slightly.	Ok	Pongrácz and Bartholy (2006), Bartholy and Pongrácz (2010), Dumitrescu et al. (2015) and Micu et al. (2015)

Changes in the frequency of flash floods	4/4	▲	"We used to have more frequent but lighter rainfall, but now when it rains, it pours. When there's heavy rain, there's flooding." (Jánó, Erika – 1992; 28 10 2019) (Figure 4.3c)	Frequency of flash floods increased.	Ok	Palamarchuk et al. (2005), Birsan et al. (2013) and Micu et al. (2015)
Changes in the intensity of flash floods	2/2	▲	"We had a flood in '70 that swept everything away. But it seems like the one we had now was even bigger." (Pasi, Olga – 1945; 14 11 2019)	No data	–	–
Changes in temporal distribution of rainfall	9/9 100%	◀▶	"It's completely different. It used to rain two or three times a week, but - after it rained - the weather was good, everyone got on with their work." (Jánó, Erika – 1992; 28 10 2019)	Intensity of extreme rainfall events increased, coupled with a lower frequency.	Ok	Bartholy and Pongrácz (2007)
Changes in the length/duration of rainfall events	2/2	▲	"It didn't use to be like this. When it rained in the afternoon, it wouldn't rain at night. The weather was good during the day. If there was thunder and lightning during the day, there was none at night." (Pasi, Olga – 1945; 14 11 2019)	No data	–	–
Changes in the predictability of rainfall	10/10 100%	▼	"Now we can't plan the way we were taught as children. We used to be able to tell in the morning if it was going to rain in the afternoon. But not anymore." (Ferenc, Bálint; 15 11 2019)	No data	–	–
Changes in the amount of rainfall in a given season	6/6 100%	▼	"In summer, it rained at least once a week. Now it's really starting to shift; there's less rainfall in the summer than there used to be." (Tankó, László – 1986; 29 10 2019)	Generally, winters, springs, summers and even the extended cold season (November-April), grew drier, while autumns showed predominant increasing trends.	Ok	Dumitrescu et al. (2015) and Micu et al. (2015)

(Continued)

Table 4.2 (Continued)

Indicators	*Number of interviewees*	*Direction*	*Quotation*	*Instrumental data*	*C*	*Literature*
Changes in the intensity/ strength of rainfall in a given season	3/3	▲	"It used to rain more frequently but not as much, but now, when it rains, it pours. Then it sweeps everything away." (Jánó, Erika – 1992; 28 10 2019) (Figure 4.3d)	No data	–	–
Changes in the length/duration of drought	5/5 100%	▲	"It rained more often and not so much. If it didn't rain for a week, then it would rain the next week. Now we'll have a month or two where it doesn't rain at all." (Tankó Tímár, Attila – 1971; 27 10 2019) (Figure 4.3b)	It seems that in the last 15–20 years the droughts have also been longer and more intense than in the past and this is probably due to the temperature rise in the Carpathians.	Ok	Spinoni et al. (2013, 2015) and Micu et al. (2015)
Changes in the frequency of fog or misty days	4/4	▼	"There's not much fog either. There used to be more in the mornings. Spring and autumn were a bit foggier." (Jánó, Erika – 1992; 28 10 2019)	No data	–	–
Changes in the frequency of cloudy days	2/2	▼	"It was so rare to have a completely cloudless sky. In the summer maybe once or twice, but now when it happens, it's like that for months." (Tankó, Erzsébet – 1968; 17 11 2019)	No data	–	–
Changes in the frequency of frost days	4/4	▼	"Well, it used to be that there was frost; we would be sliding down the hillside during the second harvest. But the weather's not like that anymore. It's warmer; there's no frost." (Jánó, Mária – 1944; 28 10 2019)	The decreasing number of frost days is obvious.	Ok	Micu et al. (2015)

Figure 4.3 (a) As a result of drought periods, water yields from springs and wells have fallen significantly. Locals try to reduce the draining effect of streams by building weirs. (b) As a result of drought periods, springs that had never dried before have dried up several times in recent years. (c) Rainwater rushing down the steep slopes during periods of more intense rainfall causes flash floods. (d) High-intensity rainfall causes landslides on hay meadows located on steep slopes. Photos: Dániel Babai.

Seasonality

Local observations made by Csángó farmers in Gyimes indicate drastic changes in the seasons. Most of the observations relate to changes in the seasons and in seasonality (Table 4.4). Observations show variability in the length of the seasons (e.g., winter getting shorter), but also a marked tendency for the transitional seasons of spring and autumn to become shorter or even disappear. At the same time, farmers observe significant shifts in the onset and end of seasons, which may start either earlier or later.

Climate change impacts in farming and farmers' responses

Rising average temperatures, temporal changes in precipitation patterns, and the increasing unpredictability of seasonality and weather all have an impact on agricultural activities, making it difficult to plan grassland management and crop production. According to informants, the flowering of vegetation in hay meadows and pastures is affected by the earlier onset of spring. This is particularly true for grass species such as Poaceae spp., for which one informant mentioned:

Table 4.3 Local observations of climate change impacts perceived by Csángó farmers in Gyimes. Observations of changes in air masses. See notes in Table 4.1

Indicators	*Number of interviewees*	*Direction*	*Quotation*	*Instrumental data*	*C*	*Literature*
Changes in wind strength or speed	4/2	◀▶	"I don't remember such strong winds. When it howls like this, it's as if the sky were falling, tearing down willow branches and everything." (Máris, Éva- 1979; 17 11 2019)/"It's not as windy now as it used to be, when the wind would blow houses down…" (Daradics, Gizella – 1934; 12 11 2019)	The annual wind speed decreased significantly in most areas of the Romanian Carpathians. Eastern Carpathians experienced the most significant weakening of wind speed.	Ok[a]	Cheval et al. (2014), Spinoni et al. (2015), Dumitrescu et al. (2015) and Micu et al. (2015)
Changes in the number of windy days	6/6 100%	▼	"There are fewer now, fewer [windy days]. It used to be windier." (Jánó, Erika – 1992; 28 10 2019)	There is an increased frequency of atmospheric calm because of the strengthening of anticyclonic activity over the last decades.	Ok	Micu et al. (2015)
Change in the frequency of storms (not further specified)	2/1	▲	"There tends to be stronger winds and storms than before. We had them in the past, but now they're more frequent." (Főcze, Péter – 1942; 15 11 2019)	No data	–	–
Changes in the intensity of windstorms	2/2	▲	"The wind, it's brutal. We never really had winds strong enough to blow the tiles off the houses." (Főcze, Péter – 1942; 15 11 2019)	More intense wind events can be expected over the Central Europe in the next decades.	Ok	Leckebusch et al. (2008)
Changes in the frequency of lightning and thunder	5/2 40%	◀▶	"I might be wrong to think that there used to be more thunder and lightning than there is now." (Tankó, Ilona – 1954; 12 11 2019)/"These huge lightning strikes – there are more than there used to be." (Máris, Éva – 1979; 17 11 2019)	No data	–	–
Changes in the frequency of hailstorms	5/3 60%	◀▶	"There's more hailstorms. As I recall, there didn't use to be so much hail that it would stay on the ground and you could gather it up, and it would be white in the summer." (Máris, Éva – 1979; 17 11 2019)	No data	–	–
Changes in the intensity of hailstorms	3/3	▲	"Last year we had ice, ice the size of your palm, the likes of which no one here has ever seen." (Bukhel, Antal – 1941; 11 11 2019)	No data	–	–

[a] Local farmers disagreed about the change in the wind speed, older informants reported about a decreasing trend, while younger participants observed an increasing speed of wind.

Table 4.4 Local observations of climate change impacts perceived by Csángó farmers in Gyimes. Observations of changes in seasonality. See notes in Table 4.1

Indicators	*Number of interviewees*	*Direction*	*Quotation*	*Instrumental data*	*C*	*Literature*
Changes in the length/duration/ disappearance of seasons	9/9 100%	◀▶	Spring: "Spring used to last 2–3 months; now it barely lasts one." (Jánó, Erika; 28 10 2019) Autumn: "When I was a child, we didn't have such long autumns. At this time of the year there would already be snow up to our waists." (László, Csaba – 1964; 18 11 2019)	No data	–	–
Changes in the timing (onset or end) of seasons	12/12 100%	◀▶	Winter: "Now the real winter doesn't start until January. In the past it would snow at the beginning of December, and it wouldn't melt." (Bodor, László – 1976; 16 11 2019) Spring: "Spring comes earlier. The way I remember it, there used to be so much snow on March 15th that people would go up into the woods on sleds. Now it's getting more and more uncommon for there to be snow on the ground for that long." (Tankó, Erzsébet - 1968, 17 11 2019)	In spring, there is a generalized trend towards earlier snowmelts all over the Carpathians. The growing season starts earlier.	Ok	Lakatos et al. (2013) and Micu et al. (2015)
Changes in the timing (onset or end) of seasons	5/5 100%	◀▶	"I've noticed that the autumn lasts longer. Winter tends to extend into springtime. It didn't use to be like that." (Hajnal, János – 1973; 28 10 2019)	No data	–	–

"*Spring comes earlier. The grass matures sooner too.*" [FGD#2- Bartos, János et al.; 26 01 2020]). It is also the case for Fabaceae species that are important for fodder quality. For example, when talking about *Onobrychis viciifolia, one informant mentioned "It bloomed much earlier than it should have. It usually blooms in June, but this time it was in bloom at the end of May. This is influenced by the weather.*" [Tankó, Júlia – 1977; 11 17 2019]. Rising average temperatures and daily maximum temperatures reduce primary biomass production. In the words of one informant: "*There's also less hay because of the heat*" (FGD#1, Jánó, Béla et al.; 19 01 2020). Changes in the temporal distribution of precipitation in summer also result in agricultural drought, as illustrated by one informant: "*When it's hot for a month, the water evaporates quickly from the soil. Here, the perfect thing for the poor soil would be to have a good rain at least once a week*" (Tankó Tímár, Attila – 1971; 27 10 2019). As the intensity of summer rainfall increases, a smaller proportion of rainwater infiltrates into the soil for use by vegetation, thus exacerbating the effects of agricultural drought: "*It's not that soft kind of rain that the ground can soak up; instead, it all pours down at once, and then it* [the rainwater] *runs off. Everything dries out of the earth*" (Kondor, Béla – 1954; 12 11 2019).

To respond to these changes, farmers have changed the timing of grassland management practices. Grazing animals are put out to pasture earlier in the year. One informant said: "*We put them* [the animals] *out to pasture at the end of May. Now we put them out on May 10th, if there's grass*" (Kondor, Béla – 1954; 12 11 2019). The haymaking and aftermath harvesting on hay meadows also happen earlier in the year, as illustrated by one informant: "*Spring starts earlier. Once it has grown, it starts to flower, and after that it loses its fodder value. All that's left is the chaff*" (Kondor, Béla – 1954; 12 11 2019). As vegetation development accelerates, hay quality rapidly deteriorates: "*It* [the biomass] *turns brown quickly. Because it's hot. Brown grass is like weak straw*" (FGD#2- Bartos, János et al.; 26 01 2020). Timely mowing of Fabaceae species, particularly *Onobrychis viciifolia*, is important for fodder quality: "*Because if it blooms earlier, the flowers drop sooner, and if you don't get out there in time to mow it, it'll get old and not be as good quality.*" (Tankó, Júlia – 1977; 11 17 2019).

In arable crop production, agrophenological changes are primarily caused by rising mean annual temperatures, rising mean winter temperatures, and the early onset of spring. Planting of potatoes starts earlier: "*We used to plant the potatoes between May 10th and 15th. Now we plant them in mid-April or even sooner. (…) Spring starts sooner*" (Prezsmer, Béla – 1961; 15 11 2019). Soil moisture also affects the time of planting:

> *Now we have to do it sooner, weather permitting, because then the droughts come. In the past, we wouldn't plant the potatoes until after May 10th. But now, we already start thinking about it after mid-April. Albeit the ground is still cold, but it's damp as well. Later it gets warmer, but then the plants don't get enough moisture.*
>
> (Kondor, Béla – 1954; 12 11 2019)

Potato production is also hampered by the emergence of new pathogens and pests. One informant complained: "*Some kind of disease hits* [the potatoes]. *They say it's the blight* [*Alternaria solani*]. *There's a sort of spray against it. (…) The potatoes don't grow*" (Főcze Péter – 1942; 15 11 2019). Milder winters facilitate the survival of another potato pest which is well-known in the Carpathian basin, namely the Colorado potato beetle (*Leptinotarsa decemlineata*). One informant narrated the evolution of this pest "*These were unheard of in our childhood. They've appeared in the past few years and multiplied. If there's not a harsh winter, they don't freeze*" (Tankó, Erzsébet – 1968; 17 11 2019). Potato growth is negatively affected by changes in the weather and pests: "*The potatoes don't get rain when they need it. By that time, the blight and the potato bugs will have finished them off…*" (FGD#1, Jánó, Béla et al.; 19 01 2020). The crop yield has also fallen:

> *[The potato yield] is much poorer. On our land we used to get twenty sacks of potatoes, sometimes more, and now we get half of that. Ten or twelve. Now if you don't use chemicals, you hardly get anything. [In the past] Nobody used to use chemicals, yet they would still get potatoes.*
>
> (Jánó, Erika – 1992; 28 10 2019)

Work also starts earlier in kitchen gardens, as one informant mentioned: "*We used to do the planting in April, but now we do it in March, if it* [the weather] *is like that*" (László, Csaba – 1964; 18 11 2019). New species and varieties have also appeared: "*Cucumbers, my grandparents always used to buy them, but now we can grow a lot of them. The weather can be that much warmer. Because you know, cucumbers freeze easily*" (Hajnal, János – 1973; 28 10 2019). Some people are even experimenting with frost-sensitive fruits: "*The priest planted peaches, and they ripened. That had never happened before*" (Máris, Éva – 1979; 17 11 2019) (Figure 4.4).

Figure 4.4 The parish priest is experimenting with the planting of new fruit species in his well-kept orchard on the south-facing hillside with its warm microclimate. Photo: Dániel Babai.

Discussion

Subsistence-oriented communities with a direct connection to the environment perceive local climate change impacts (Reyes-García et al., 2016, 2023). Based on the observations of Csángó farmers, it can be established that the weather in Gyimes has changed significantly in the last three decades. The increase of the average temperature, especially in summer and in the winter, the decreasing predictability and changing temporal distribution of the precipitation, and the increase of the intensity of heavy rainfall events coupled with shifting seasonality (e.g., shorter winter, earlier onset of spring) were the most often mentioned indicators of change. All these changes, especially phenological and agrophenological changes, have an essential impact on agricultural activities. Many local observations are consistent with the trends of instrumental measurements published about the Carpathians (cf. Micu et al., 2015; Spinoni et al., 2015). Moreover, the three main climate change impacts mentioned by Csángó farmers in Gyimes (i.e., increase in mean temperatures, changes in the spatio-temporal distribution, intensity, and predictability of rainfall events, and seasonality) are also reported by other communities in the Carpathians (Mattalia et al, 2024, Chapter 1, this volume) and by numerous subsistence-oriented communities living in different biomes (Savo et al., 2016).

The increase in mean temperatures, in particular in summer, and the extremes of daily maxima are trends consistently observed by subsistence-oriented communities worldwide (Nakashima & Krupnik, 2018; Reyes-García et al., 2019). The trends, like increasing daily highs, increasing lengths, and intensity of warm spells reported by Csángó farmers have been also detected by instrumental measurements in several regions of the Carpathians (Busuioc et al., 2010; Melo et al., 2013; Micu et al., 2015). Csángó farmers report that the minimum temperatures of winter cold waves have become drastically milder, with shorter cold wave duration and reduced frequency, while daily peak temperatures well above freezing point (even up to 10°C) are becoming increasingly frequent during the winter resulting in sudden melting of snow. Snow cover develops later than usual, usually in the second half of December instead of in November, and the snow melts earlier. Similar changes are also happening in other regions of the Carpathians (Micu et al., 2015; Mattalia et al. 2024, Chapter 1, this volume), in the mountain regions of Bulgaria (Brown & Petkova, 2007), and in the Alps (Marty, 2013). Permanent snow cover has also disappeared, replaced in several instances by snow cover of shorter duration. This phenomenon has a significant impact on the timing of the onset of spring and the start of the growing season (Menzel et al., 2006; Marty, 2013).

Changes in the temporal distribution of precipitation and increases in precipitation intensity are also global phenomena locally observed in many parts of the world (Sillmann & Roeckner, 2008; Savo et al., 2016). These trends have been documented in mountain regions of Europe (Melo et al., 2013), and in many regions of Central Europe (Didovets et al., 2019), and they are also predominant in Gyimes. Local observations indicate that the temporal distribution and the intensity of precipitation have changed significantly, while the predictability of precipitation has decreased. According to Csángó farmers' observations, the number of days with precipitation has decreased, thus shortening rainy periods and giving rise to periods of drought. Summer rain, which has for the most part remained unchanged in terms of quantity, falls less frequently but with greater intensity, thus increasing the frequency and destructiveness of flash floods after major storms and the occurrence of agricultural (vegetation) droughts with longer rain-free periods.

Some of the farmers interviewed in Gyimes suspect a cyclical nature behind the pattern of precipitation distribution and quantity. "*There were a few years of normal rainfall and heat, then it turned dry, then rainy again. So they alternate,*" said one of our informants. Instrumental measurements suggest a similar cyclical change of wetter and drier four-to-five-year periods in the rest of Carpathians as well (Micu et al., 2015).

The changing and shifting of the seasons is a globally perceived phenomenon (Savo et al., 2016; IPCC, 2021). Changes in temperature and precipitation have also considerably altered the normal course of the seasons in Gyimes. Among the most significant changes were the prolongation of mild autumn, the later onset of winter, and the early onset of spring. These changes have an impact on the development of vegetation and thus on the (earlier) start of the growing season in spring (Menzel et al., 2006; Cleland et al., 2007).

Changes in the weather are a daily topic of conversation among farmers in Gyimes, which might explain the general consensus in the local reports of the direction of trends. The perceived impacts affect the development of vegetation and thus the timing and organisation of agricultural work, which is also a topic of everyday conversations. With weather becoming less predictable, the adaptation of agricultural activities has become crucial (Schlingmann et al., 2021). Locally observed climate-related trends generate significant phenological changes and thus affect the cropping calendar (Menzel et al., 2006; Cleland et al., 2007; Savo et al., 2016), as well as weed and pest species pool and abundance (Chaudhary & Bawa, 2011). Rising mean temperatures in the Carpathians do not yet have a clear documented effect on the phenological state of vegetation (Micu et al., 2015), however, the observations made by Csángó farmers in Gyimes show the impact of climate change on vegetation phenology in this area: keystone species such as grasses (Poaceae) and *Onobrychis viciifolia* bloom earlier, thus bringing forward the time of mowing (Babai et al., 2021).

Climate change impacts on seasonality also affects the agrophenology of crop production in both field and horticultural cultivation. The timing of important agricultural activities, such as tillage or harvesting, changes (Menzel et al., 2006; Lavalle et al., 2009). In Gyimes, it is mainly the sowing and the harvesting dates that have changed for potatoes and cabbage (*Brassica oleracea* convar. *capitata*).

Climate change, especially increasing droughts, both globally and locally, negatively affect biomass quality and production (Sherwood & Fu, 2014). Farmers in Gyimes reported similar experiences, specifically with regard to the yield of the hay meadows. Globally, there is a trend of declining crop production, partly due to the effects of climate change, especially in rain-fed agriculture (Savo et al., 2016). Increases in mean temperatures may also translate into longer growing seasons and the cultivation of new varieties (e.g., Labeyrie et al., 2021). However, this can also have a positive effect on pests (Bale & Hayward, 2010): populations of previously present pests increase or new pest species emerge (Lama & Devkota, 2009). In Gyimes, the increase in mean winter temperatures helps pests to overwinter more successfully (e.g., the Colorado potato beetle) (Reedy et al., 2014). Furthermore, as in other regions of Romania (Fleşeriu et al., 2013), an increase in mean summer temperature or a change in the distribution of precipitation has also resulted in the emergence of *Alternaria solani* and a drastic reduction in potato yield. As most Csángó farmers reject chemical treatments, there is currently no solution to prevent and treat the disease, and local farmers have to buy potatoes on the market. The fluctuating trend of the yield is in line with the increasing variability in crop yields throughout Europe, as a consequence of extreme climatic events (Lavalle et al., 2009).

Conclusion

Local observations of climate change impacts on elements of the atmospheric system in the subsistence-oriented community of Valea Rece mainly refer to changes in mean temperature, temporal distribution and intensity of precipitation, and seasonality. These observations of key indicators are globally observed by local communities and appear to be common, which may indicate either the culture-independent nature of communities' cognitive perceptions of climate change or the biased nature of climate change research. However, trends in observations are often consistent with reports from instrumental measurements, suggesting that observations from communities in direct contact with the natural environment are reliable, which provide a strong basis for adaptive responses.

Observations made by local communities in mountain regions and their knowledge of local ecology help in understanding climate change and its consequences, especially with regard to the management of plant and animal species, habitats, and resources of importance to the community. This empirical knowledge will support subsistence-oriented communities' coping strategies and adaptive responses to the challenges created by climate change.

Acknowledgements

The author would like to thank all the participants who, through semi-structured interviews or focus group discussions, supported the research by contributing their knowledge.

The author's work was supported by ERC Grant No. 771056 LICCI (Local Indicators of Climate Change Impacts – The Contribution of Local Knowledge to Climate Change Research), and by the Hungarian Academy of Sciences (MTA) Grant LENDULET_2020–56.

References

Babai, D., Molnár, Á., & Molnár, Zs. (2014). *Traditional Ecological Knowledge and Land Use in Gyimes (Eastern Carparhians).* MTA BTK Institute of Ethnology – MTA RE Institute of Ecology and Botany: Budapest-Vácrátót, Hungary.

Babai, D., Jánó, B., & Molnár, Zs. (2021). In the trap of interacting indirect and direct drivers: The disintegration of extensive, traditional grassland management in Central and Eastern Europe. *Ecology and Society, 26*(4). https://doi.org/10.5751/ES-12679-260406

Bale, J. S., & Hayward, S. A. L. (2010). Insect overwintering in a changing climate. *Journal of Experimental Biology, 213*(6), 980–994. https://doi.org/10.1242/jeb.037911

Bartholy, J., & Pongrácz, R. (2007). Regional analysis of extreme temperature and precipitation indices for the Carpathian Basin from 1946 to 2001. *Global and Planetary Change, 57*(1), 83–95. https://doi.org/10.1016/j.gloplacha.2006.11.002

Bartholy, J., & Pongrácz, R. (2010). Analysis of precipitation conditions for the Carpathian Basin based on extreme indices in the 20th century and climate simulations for 2050 and 2100. *Physics and Chemistry of the Earth A/B/C, 35*(1), 43–51. https://doi.org/10.1016/j.pce.2010.03.011

Birsan, M. V., Zaharia, L., Chendes, V., & Branescu, E. (2013). Seasonal trends in Romanian streamflow. *Hydrological Processes, 28*, 4496–4505. https://doi.org/10.1002/hyp.9961

Birsan, M. V., Dumitrescu, A., Micu, D. M., & Cheval, S. (2014). Changes in annual temperature extremes in the Carpathians since AD 1961. *Natural Hazards, 74*(3), 1899–1910. https://doi.org/10.1007/s11069-014-1290-5

Brown, R. D., & Petkova, N. (2007). Snow cover variability in Bulgarian mountainous regions, 1931–2000. *International Journal of Climatology, 27*, 1215–1229. https://doi.org/10.1002/joc.1468

Busuioc, A., Giorgi, F., Bi, X., & Ionita, M. (2006). Comparison of regional climate model and statistical downscaling simulations of different winter precipitation change scenarios over Romania. *Theoretical and Applied Climatology, 86*(1), 101–123. https://doi.org/10.1007/s00704-005-0210-8

Busuioc, A., Caian, M., Cheval, S., Bojariu, R., Boroneanţ, C., Baciu, M., Dumitrescu, A. (2010). *Variabilitatea şi schimbarea climei în România (Variability and Climate Change in Romania).* Bucureşti: Pro Universitaria.

Chaudhary, P., & Bawa, K. S. (2011). Local perceptions of climate change validated by scientific evidence in the Himalayas. *Biological Letters, 7*, 767–770. https://doi.org/10.1098/rsbl.2011.0269

Cheval, S., Birsan, M. V., & Dumitrescu, A. (2014). Climate variability in the Carpathian Mountains Region over 1961–2010. *Global and Planetary Change, 118*, 85–96. https://doi.org/10.1016/j.gloplacha.2014.04.005

Cleland, E. E., Chuine, I., Menzel, A., Mooney, H. A., & Schwartz, M. D. (2007). Shifting plant phenology in response to global change. *Trends in Ecology and Evolution, 22*, 357–365. https://doi.org/10.1016/j.tree.2007.04.003

Didovets, I., Krysanova, V., Bürger, G., Snizhko, S., Balabukh, V., & Bronstert, A. (2019). Climate change impact on regional floods in the Carpathian region. *Journal of Hydrology: Regional Studies, 22*, article number: 100590. https://doi.org/10.1016/j.ejrh.2019.01.002

Dumitrescu, A., Bojariu, R., Birsan, M. V., Marin, L., & Manea, A. (2015). Recent climatic changes in Romania from observational data (1961–2013). *Theoretical and Applied Climatology, 122*(1), 111–119. https://doi.org/10.1007/s00704-014-1290-0

Elsen, P. R., Monahan, W. B., & Merenlender, A. M. (2020). Topography and human pressure in mountain ranges alter expected species responses to climate change. *Nature Communications, 11*, article number: 1974. https://doi.org/10.1038/s41467-020-15881-x

Feehan, J., Harley, M., & van Minnen, J. (2009). Climate change in Europe. 1. Impact on terrestrial ecosystems and biodiversity. A review. *Agronomy for Sustainable Development, 29*(3), 409–421. https://doi.org/10.1051/agro:2008066

Fleşeriu, A., Oroian, I. G., Mălinaş, C., Braşovean, I., & Bordeanu, B. (2013). Study upon the Alternaria solani Sorauer attack degree on potato cultures function of climatic conditions from Transylvania. *ProEnvironment, 6*, 615–618.

Gobiet, A., Kotlarski, S., Beniston, M., Heinrich, G., Rajczak, J., & Stoffel, M. (2014). 21st century climate change in the European Alps – a review. *Science of the Total Environment, 493*, 1138–1151. https://doi.org/10.1016/j.scitotenv.2013.07.050

Hansen, J., Sato, M., & Ruedy, R. (2012). Perception of climate change. *Proceedings of the National Academy of Sciences (PNAS), 109*(37), E2415–E2423. https://doi.org/10.1073/pnas.1205276109

Ilyés, Z. (2007). *A tájhasználat változásai és a történeti kultúrtáj 18–20. századi fejlődése Gyimesben* [*Landscape Changes and the 18–20. Century Development of the Historical Cultural Landscape in Gyimes*]. Eger: Eszterházy Károly Főiskola.

IPCC (2021). *Climate Change 2021: The Physical Science Basis*. Contribution of working group I to the sixth assessment report of the intergovernmental panel on climate change. V. Masson-Delmotte, P. Zhai, A. Pirani, S. L. Connors, C. Péan, S. Berger, N. Caud, Y. Chen, L. Goldfarb, M. I. Gomis, M. Huang, K. Leitzell, E. Lonnoy, J. B. R. Matthews, T. K. Maycock, T. Waterfield, O. Yelekçi, R. Yu, & B. Zhou (eds.), Cambridge University Press. https://doi.org/10.1017/9781009157896

Labeyrie, V., Renard, D., Aumeeruddy-Thomas, Y., Benyei, P., Caillon, S., Calvet-Mir, L., Carrière, S., Demongeot, M., Descamp, E., Junqueira, A. B., Li, X., Loqueville, J., Mattalia, G., Miñarro, S., Morel, A., Porcuna-Ferrer, A., Schlingmann, A., Da Cunha, J.V., & Reyes-García, V. (2021). The role of crop diversity in climate change adaptation: Insights from local observations to inform decision-making in agriculture. *Current Opinion in Environmental Sustainability, 51*, 15–23. https://doi.org/10.1016/j.cosust.2021.01.006

Lakatos, M., Szentimrey, T., Bihari, Z., & Szalai, S. (2013). Creation of a homogenized climate database for the Carpathian region by applying the MASH procedure and the preliminary analysis of the data, *Időjárás, 117*(1), 143–158.

Lama, S., & Devkota, B. (2009). Vulnerability of mountain communities to climate change and adaptation strategies. *Journal of Agriculture and Environment, 10*, 76–83. https://doi.org/10.3126/aej.v10i0.2133

Lavalle, C., Micale, F., Houston, T. D., Camia, A., Hiederer, R., Lazar, C., Conte, C., Amatulli, G., & Giampiero, G. (2009). Climate change in Europe. 3. Impact on agriculture and forestry. A review. *Agronomy for Sustainable Development, 29*, 433–446. https://doi.org/10.1051/agro/2008068

Leckebusch, G. C., Weimer, A., Pinto, J. G., Reyers, M., & Speth, P. (2008). Extreme wind storms over Europe in present and future climate: A cluster analysis approach. *Meteorologische Zeitschrift, 17*(1), 67–82. https://doi.org/10.1127/0941-2948/2008/0266

Lehner, F., & Stocker, T. F. (2015). From local perception to global perspective. *Nature Climate Change, 5*(8), 731–734. https://doi.org/10.1038/nclimate2660

Marty, C. (2013). Climate change and snow cover in the European Alps. In C. Rixen, & A. Rolando (eds.), *Impacts of Skiing and Related Winter Recreational Activities on Mountain Environments*. 30–44. Oak Park, IL: Bentham, Science Publishers.

Melo, M., Lapin, M., Kapolková, H., Pecho, J., & Kružicová, A. (2013). Climate trends in the Slovak part of the Carpathians. In J. Kozak, K. Ostapowicz, A. Bytnerowicz, & B. Wyżga (eds.), *The Carpathians: Integrating Nature and Society Towards Sustainability*. 131–150. Springer, Berlin, Heidelberg, Germany.

Menzel, A., Sparks, T. H., Estrella, N., Koch, E., Aasa, A., Ahas, R., Alm-Kübler, K., Bissolli, P., Braslavská, O., Briede, A., Chmielewski, F. M., Crepinsek, Z., Curnel, Y., Dahl, Å., Defila, C., Donnelly, A., Filella, Y., Jatczak, K., Måge, F., Mestre, A., Nordli, Ø., Peñuelas, J., Pirinen, P., Remišová, V., Scheifinger, H., Striz, M., Susnik, A., van Vliet, A. J. H., Wielgolaski, F., Zach, S., & Zust, A. (2006). European phenological response to climate change matches the warming pattern. *Global Change Biology, 12*(10), 1969–1976. https://doi.org/10.1111/j.1365-2486.2006.01193.x

Micu, D. M., Dumitrescu, A., Cheval, S., & Birsan, M. (2015). *Climate of the Romanian Carpathians. Variability and Trends.* Springer International Publishing. https://doi.org/10.1007/978-3-319-02886-6

Nakashima, D., Krupnik, I., & Rubis J. T. (eds.) (2018). *Indigenous Knowledge for Climate Change Assessment and Adaptation* (Vol. 2). Cambridge University Press and UNESCO: Cambridge and Paris.

Palamarchuk, L., Shpyg, V., & Krakovskaia, S. (2005). Floods in the Carpathians: Synoptic analysis and numerical modeling. *Geophysical Research Abstract, 7*, 00967.

Petres, S., Korodi, A., Keresztes, R., & Szép, R. (2017). Tendencies and particularities in thermic inversion episodes in the Ciuc Basin–Eastern Carpathians, Romania. *International Multidisciplinary Scientific GeoConference, 17*(1.4), 445–452. https://doi.org/10.5593/sgem2017/14

Pongrácz, R., & Bartholy, J. (2006). Extreme climate analysis using extreme index time series for the Central/Eastern European region. *Geophysical Research Abstract, 8*, 08552.

Reedy, D., Savo, V., & McClatchey, W. (2014). Traditional climatic knowledge: Orchardists' perceptions of and adaptation to climate change in the Campania region (Southern Italy). *Plant Biosystems, 148*, 699–712. https://doi.org/10.1080/11263504.2013.793753

Reyes-García, V., Fernández-Llamazares, Á., Guèze, M., Garcés, A., Mallo, M., Vila-Gómez, M., & Vilaseca, M. (2016). Local indicators of climate change: The potential contribution of local knowledge to climate research. *Wiley Interdisciplinary Reviews: Climate Change, 7*(1), 109–124. https://doi.org/10.1002/wcc.374

Reyes-García, V., García-del-Amo, D., Benyei, P., Fernández-Llamazares, Á., Gravani, K., Junqueira, A. B., Labeyrie, V., Li, X., Matias, D. M. S., McAlvay, A., Mortyn, P. G., Porcuna-Ferrer, A., Schlingmann, A., & Soleymani-Fard, R. (2019). A collaborative approach to bring insights from local observations of climate change impacts into global climate change research. *Current Opinion in Environmental Sustainability, 39*, 1–8. https://doi.org/10.1016/j.cosust.2019.04.007

Reyes-García, V., Álvarez-Fernández, S., Benyei, P., García-del-Amo, D., Junqueira, A. B., Labeyrie, V., Li, X., Porcher, V, Porcuna-Ferrer, A., Schlingmann, A., & Soleymani, R. (2023). Local indicators of climate change impacts described by indigenous peoples and local communities: Study protocol. *PloS One, 18*(1), e0279847. https://doi.org/10.1371/journal.pone.0279847

Savo, V., Lepofsky, D., Benner, J. P., Kohfeld, K. E., Bailey, J., & Lertzman, K. (2016). Observations of climate change among subsistence-oriented communities around the world. *Nature Climate Change, 6*(5), 462–473. https://doi.org/10.1038/nclimate2958

Schlingmann, A., Graham, S., Benyei, P., Corbera, E., Martínez-Sanesteban, I., Marelle, A., Solemany-Fard, R., & Reyes-García, V. (2021). Global patterns of ground-based adaptation to climate change by Indigenous Peoples and local communities. A systematic review. Special edition on Climate-Decision. *Current Opinion in Environmental Sustainability*, *51*, 55–64. https://doi.org/10.1016/j.cosust.2021.03.002

Sherwood, S., & Fu, Q. (2014). A drier future? *Science*, *343*(6172), 737–739.

Sillmann, J., & Roeckner, E. (2008). Indices for extreme events in projections of anthropogenic climate change. *Climatic Change, 86*, 83–104. https://doi.org/10.1007/s10584-007-9308-6

Sólyom, A., Knowles, B., Bogdán, J., Rodics, G., Biró, R., & Nyirő, G. (2011). *Small Scale Farming in the Pogány-Havas Region of Transylvania. Farming Statistics, Agricultural Subsidies, the Future of Farming.* Final Report. Miercuea Ciuc: Pogányhavas Regional Association.

Spinoni, J., Antofie, T., Barbosa, P., Bihari, Z., Lakatos, M., Szalai, S., Szentimrey, T., & Vogt, J. (2013). An overview of drought events in the Carpathian Region in 1961–2010. *Advances in Science and Research, 10*(1), 21–32. https://doi.org/10.5194/asr-10-21-2013

Spinoni, J., & CARPATCLIM project team (39 authors) (2015). Climate of the Carpathian region in 1961–2010: Climatologies and trends of ten variables. *International Journal of Climatology, 35*(7), 1322–1341. https://doi.org/10.1002/joc.4059

Zaval, L., Keenan, E. A., Johnson, E. J., & Weber, E. U. (2014). How warm days increase belief in global warming. *Nature Climate Change, 4*(2), 143–147. https://doi.org/10.1038/nclimate2093

5

NETWORK ANALYSIS OF CLIMATE CHANGE IMPACTS REPORTED BY LOCAL COMMUNITIES OF SIERRA NEVADA, SPAIN

David García-del-Amo, Laura Calvet-Mir, Peter Graham Mortyn and Victoria Reyes-García

Introduction

Mountainous areas occupy more than 30% of the Earth's surface, where 10% of the world's population lived in 2010 (Jones & O'Neill 2016). Mountains are home to a great diversity of ecosystems across different altitude ranges. Mountains also provide significant ecosystem services, including the provision of fresh water and raw materials, with high mountain ranges being essential reservoirs of hydric resources, both locally and for neighbouring regions (Payne et al. 2020; IPCC 2021). In mountain areas, changes in the atmospheric system disproportionally affect the physical, biological, and human systems. Indeed, mountain ranges are considered hot spots of climate change due to the intensity and visibility of its effects, which are further amplified through cascading effects (IPCC 2021; Pörtner et al. 2021).

Cascading effects are a well-known phenomena in ecology and ecosystem services studies (Rocha et al. 2018; Bahlai et al. 2021), as well as in resilience, sustainability and social-ecological frameworks (Burton et al. 2020; Haro-Monteagudo et al. 2020). Cascading effects are dangerous, as they can contribute to reaching tipping points that trigger regime shifts in ecosystems and cause extinctions (Krönke et al. 2020; Armstrong Mckay et al. 2022). Examples of cascading effects are decreases in soil humidity generated by precipitation decrease (Seneviratne et al. 2010), which in turn affects the growing rate of wild plants (Inouye & Wielgolaski 2013), which affects wild fauna, through the reduction of the abundance and quality of pastures (He & Richards 2015).

In the quest to better understand how climate change impacts propagate through the system, researchers have advocated for the need to include Indigenous and local knowledge (ILK) as an alternative, but equally valid, source of knowledge (Ford et al. 2016; Reyes-García et al. 2019; García-del-Amo et al. 2020). Indigenous Peoples and local communities (IPLC) have developed a deep understanding of the ecological relations in the ecosystems that sustain them, which allows them to detect changes in elements of their social-ecological systems (Smith & Sharp 2012; Hu et al. 2020) including changes in extreme temperatures or the frequency and length of hot or cold waves (Plummer et al. 2014), changes in temporal and spatial rainfall distribution (Gurgiser et al. 2016; Fassnacht et al. 2018), or changes in the predictability of rainfalls (Sowman & Raemaekers 2018).

 DOI: 10.4324/9781003356837-8

Moreover, the holistic nature of their knowledge systems allows IPLC to perceive how changes in the atmospheric system propagate to the physical, biological, and human components of their social-ecological system (Weatherhead et al. 2010). Few studies, however, have systematically analysed how IPLC perceive cascading effects originally triggered by climate change. Analysing the relations among locally perceived climate change impacts might allow to identify the most vulnerable elements, but also the triggering impacts of these cascading effects and the intermediate, and sometimes apparently unconnected, cascading impacts that can generate regimen shifts in ecosystems (Rocha et al. 2018; Collins et al. 2019).

To test this idea, in this study we assess: (1) what are the triggering climate change impacts with highest negative repercussions on agropastoral activities reported by local communities of Sierra Nevada, a mountain region of Spain, and (2) what are the cascading effects triggered by these impacts. Given that local knowledge systems are patterned by sociodemographic characteristics of informants (Benyei et al. 2020; García-del-Amo et al. 2023), we compare information reported by people devoted to agropastoral activities with information reported by local inhabitants not fully devoted to the primary sector. We explore our data using network analysis (NA), which has been broadly used in social-ecological systems research (Calvet-Mir & Salpeteur 2016).

Methodology

Sierra Nevada as a social-ecological system

We conducted research in Sierra Nevada, the most southern mountain region of Europe, located in a semi-arid region of Spain. Sierra Nevada displays a cold climate with cold and dry summers, presenting large thermal variations. Rainfall is scarce, with most precipitations concentrated in winter and spring. Historically, Sierra Nevada was characterised by the presence of permanent snow, with several snowfields and permafrost areas on the summits (Oliva et al. 2018). Because of its biophysical characteristics, Sierra Nevada has become one of the most important biodiversity hot-spots of Europe and the Mediterranean basin (Blanca et al. 2001), leading to the recognition of the territory as a Natural Park (1989), a National Park (1999), and/or a Biosphere Reserve (1986), the latter recognizing the importance of local management in biodiversity maintenance.

Indeed, since the Islamic period in the 8th century, local inhabitants have transformed the territory through the construction of a network of water ditches dug on the ground (*acequias*), which collect melting water from the summits and canalise it along the slopes to the villages and agricultural fields (Pulido-Bosch & Sbih 1995; Martos-Rosillo et al. 2019). This water management system increases water infiltration, recharging the aquifers and increasing the availability of water during summer. The system also reduces soil erosion, favours the existence of flora species with higher water requirements, and makes possible agriculture in a slope area (Martos-Rosillo et al. 2019). Communities sharing irrigation sources have created management rules to ensure water equitable distribution and use. Such knowledge is an essential component of the local knowledge system, allowing the formation of a unique social-ecological system (Blondel 2006; Iniesta-Arandia et al. 2014).

As other mountain regions, Sierra Nevada has recently experienced profound economic, social, and ecological changes. Modernization made the traditional mountain agricultural system non-profitable economically. This fact triggered the abandonment of lands and the rural exodus during the second half of the 20th century (García-del-Amo et al. 2022). The declaration of the area as Natural and later as National Park resulted in further land use changes, as these conservation measures restricted local uses (Zamora et al. 2016). Climate change has also affected the region in many ways. Rainfall reduction and temperature increase drastically reduce hydric resources, extending drought periods (Zamora et al. 2017). Snowfields no longer last all summer, permafrost areas have disappeared (Oliva et al. 2018), and several springs and wells dried some years ago. All these changes impact high mountain meadows, the abundance and distribution of species, and phenological cycles, but they also threat the continuity of livelihoods based on agriculture, pastoralism, and beekeeping (García-del-Amo et al. 2022).

Data collection and coding

From June to December 2018, we conducted 238 surveys in 33 villages of Sierra Nevada applying a purposive sampling. Selected villages had similar social and environmental characteristics (villages with more than 300 inhabitants located over 500 metres above sea level). We aimed to compare the network of climate-related cascading effects perceived by people directly involved in agropastoral activities (i.e., farmers, shepherds, ranchers, and beekeepers, hereafter *experts;* $n = 175$) with the network perceived by inhabitants having professions not related with agropastoral activities (hereafter *non-experts;* $n = 63$). We used a combination of snowball and quota sampling techniques (Shively 2011), and we only interviewed people who had lived in the area for at least 25 years.

All participants responded to a survey asking to identify the five changes with largest negative repercussions on their agropastoral activities. Participants selected impacts from a previously prepared list of 95 climate-related environmental changes potentially occurring in Sierra Nevada (see García-del-Amo et al. 2023 for details). In the reference list, climate-related changes were classified considering whether they affected elements of the atmospheric, the physical, the biological, or the human system (Reyes-García et al. 2019). The classification further divides these four systems in subsystems. Thus, the atmospheric system includes four subsystems (i.e., temperature, precipitation, air masses, and seasons). In turn, the temperature subsystem includes three impacted elements (i.e., mean, extreme, and seasonal temperature), and each impacted element is composed of a different number of indicator categories. Similar observations were grouped together and translated into standardised indicator categories, which were called *"Local Indicators of Climate Change Impacts"* (LICCI). For example, "now rain is harder and more dangerous" and "rain is very destructive nowadays" were classified as the LICCI "*Changes in the intensity / strength of heavy rainfall events*". Following the hierarchical categorization, this LICCI was classified into the impacted element "extreme precipitation", which belongs to the subsystem "precipitation", and the system "atmospheric".

For each of the five LICCI with largest negative repercussions on agropastoral activities mentioned, we asked the informant (a) to provide a description of further impacts (i.e., cascading effects) and (b) to name the two main drivers of the impact. In the analysis that follows, we only used information on impacts attributed to climate change.

Descriptions of further impacts generated by the five selected impacts were noted *verbatim* and later coded as LICCI using the same protocol (i.e., Reyes-García et al. 2019). We also coded reported relations between different components of the system and used that information to differentiate between triggering and cascading impacts. We refer to *triggering impact* as the LICCI generating a cascading effect, to *cascading impact* as the LICCI receiving the impact, and to *cascading effect* as the relation between triggering and cascading impacts. As descriptions normally included more than two LICCI, some elements appear both as triggering and cascading impacts. After analysing all responses, we found six indicators of climate change impacts that were not included in the list of 95 LICCI used in the survey, for which our final list for analysis has 101 LICCI.

Data analysis

We used network analysis to explore relations between reported impacts. To ease visual representation, we analysed data at the impacted element level. The 101 LICCI mentioned correspond to 46 impacted elements. We then created a one-mode adjacency matrix in which rows and columns are the list of impacted elements (46 · 46), which represent the nodes of the networks, which we called "*climate change networks*" (CCN). Impacted elements placed in the first cell of the rows represent triggering impacts and impacted elements placed in the first cell of the columns represent cascading impacts. Cells in the matrices contain the number of respondents mentioning a cascading effect between the elements in the corresponding row and column. Those values represent the tie strength between the nodes of the network and capture the direction of the relation. Thus, a tie indicates that a node triggers the impact mentioned in the receiving node. A node is considered "active" if at least one of its cells is different from zero. The thickness of the ties was rescaled to a score from one to three for visualization.

For each network, we calculated four measures. We calculated network density, representing the number of ties (cascading effects) in the network expressed as a proportion of the maximum possible number of ties, where "1" signifies a fully connected network (Borgatti et al. 2018). We also calculated three network centralization measures, that is, network indegree, network outdegree, and network betweenness. Network centralization degrees (i.e., indegree and outdegree) of valued matrices consider the sums of the values of the ties of each node to give the overall network centralization measure. These measures basically quantify how dispersed are the centralities of the network, with higher scores meaning that a few central nodes generate most of the connections by emitting (network outdegree) or receiving (network indegree) ties from the rest of the nodes. Network betweenness quantifies the grade of modularity of the network, that is, how much each small group (or module) contributes to minimize the distance between modules in the network (Borgatti et al. 2018).

To analyse nodes' interactions, we calculated three centrality scores for each node. The *outdegree* score of a node represents the number of outcoming ties from a node; here indicating the number of respondents reporting cascading effects triggered by node. The *indegree* score of a node represents the number of incoming ties received, signifying the number of people reporting cascading effects affecting it. Finally, the *betweenness* score of a node represents its degree of intermediation, that is, the proportion of all the shortest paths between two nodes of a network that pass through the node, indicating the frequency with which an impacted element is included in the description of cascading effects as an intermediate step (Borgatti et al. 2018).

We conducted separate analyses for the group of experts ($n = 175$) and non-experts ($n = 63$). As the two samples differ in sizes, we normalized data to be able to compare results from statistical analysis. To compare the visual representation of the two groups, in the CCN of experts, we only represented relations perceived by three or more people. We compared networks of both groups of participants analysing densities, network centralization measures, and centrality degree scores of their nodes. All statistical analyses were conducted with SPSS version 22 and UCINET version 6.721 (Borgatti et al. 2002). Network diagrams were carried out with NetDraw version 2.168 (Borgatti 2002).

Results

Across survey responses (238 respondents * five impacts), 67.65% of the environmental changes reported were climate related. The group of experts reported 69.7% ($n = 601$) of responses as climate-related, whereas the group of non-experts reported 64.8% ($n = 204$). Considering only impacts perceived as climate-related, the compiled list of triggering and cascading impacts includes 75 (74.3%) of the 101 LICCI considered. Experts included 69 LICCI and non-experts included 52 LICCI, representing 36 and 32 impacted elements, respectively.

The experts' CCN included 36 active nodes and 1,017 cascading effects (ties), corresponding to a normalised score of 50.2 ties, with an average of 5.8 ties/respondent. The non-experts' CCN included 32 active nodes and 276 ties, corresponding to a normalized score of 38.3 ties, with an average of 4.4 ties/respondent. The normalized value of the network density score of experts' CCN (0.491) was higher than the same value for the non-experts′ CCN (0.133), implying that experts reported more connections among climate change impacts than non-experts. Network centralization outdegree scores of experts' (5.19%) and non-experts' CNN (3.97%) were lower than network centralization indegree scores (9.28% and 8.38% respectively), indicating that both groups do not report a clearly defined core group of nodes triggering most cascading effects. Instead, they report a group of nodes receiving a higher concentration of cascading effects. The non-experts' CCN showed a lower network centralization betweenness score (1.64%) than the experts' CCN (8.49%), suggesting that non-experts reported less intermediate cascading impacts than experts.

Regarding node measures, nine out of the ten triggering impacts with the highest outdegree scores, that is, reported by more respondents as triggering cascading effects, were the same in the experts' and non-experts' CCN (Table 5.1). Overall, elements of the atmospheric and physical systems were most often reported as triggering further impacts. These included *Fresh water availability, Mean precipitation, Seasonal precipitation, Extreme temperature, Snowfall and snow cover,* and *Duration and timing of seasons*. The impact of changes in freshwater availability on crop production was the cascading effect most cited by both groups (Figure 5.1). However, both groups also reported that *Cultivated spp-diseases/pest/mortality*, one of the ten nodes most cited, triggered changes on *Cultivated spp productivity and quality*. Some of the triggering impacts with the highest outdegree score are also impacted by other nodes. However, experts reported more triggering impacts simultaneously generating cascading effects than non-experts. For instance, the group of experts reported cascading effects coming from seven different triggering impacts affecting *Wild flora phenology*, while non-experts reported only two (Figure 5.1).

Five of the ten nodes with the highest indegree scores (i.e., listed as cascading impacts) appear in both the expert and the non-expert CCN. Moreover, both experts and non-experts considered *Cultivated species productivity and quality* and *Freshwater availability* as the nodes receiving the most impacts from cascading effects. However, experts included *Wild flora phenology* and seven elements from the human system among the ten most prominent cascading effects. In contrast, non-experts considered four elements of the human system as most prominent and reported *Human health* as the fourth most impacted cascading impact (Table 5.1). Other impacted elements with high indegree scores belong to the physical system and particularly refer to the water cycle (i.e., *Drought, Mean river flow, Snowfall and snow cover*). In the CCN of experts, there is a higher variety of cascading effects on elements of the biological and human systems than in the CNN of non-experts. Moreover, in CCN of non-experts, *Wild flora disease, pest and mortality* and *Terrestrial fauna abundance* are only reported as triggering impacts affecting *Human health*, whereas *Livestock phenology* and *Livestock productivity and quality* are reported only to be affected by *Extreme temperatures* (Figure 5.2).

The experts and non-experts' CCN shared six out of the ten nodes with the highest betweenness scores (i.e., listed as cascading impacts). These included impacted elements related to the water cycle, which also had a high indegree score (i.e., *Drought, Freshwater availability*) (Table 5.1).

Table 5.1 Ranked list of ten impacted elements with the highest node outdegree (i.e., number of impacts triggered by a node), indegree (i.e., number of impacts received by a node), and betweenness scores (i.e., number of times a node rests on a short path connecting two other nodes which are disconnected) in experts' and non-experts' climate change networks (CCN)

Rank	*Node outdegree*		*Node indegree*		*Node betweenness*	
	Experts	Non-experts	Experts	Non-experts	Experts	Non-experts
1	Fresh water availability	Freshwater availability	Cultivated spp.-productivity and quality	Cultivated spp.-productivity and quality	Pasture-availability and productivity	Freshwater availability
2	Mean precipitation	Mean temperature	Freshwater availability	Freshwater availability	Terrestrial fauna-distribution	Snowfall and snow cover
3	Seasonal precipitation	Seasonal precipitation	Pasture-availability and productivity	Cultivated spp.-diseases/pest/mortality	Freshwater availability	Pasture-availability and productivity
4	Extreme temperature	Extreme temperature	Cultivated spp.-diseases/pest/mortality	Human health	Livestock-productivity and quality	Duration and timing of seasons
5	Snowfall and snow cover	Snowfall and snow cover	Livestock-disease/pest/mortality	Duration and timing of seasons	Cultivated spp-productivity and quality	Soil moisture
6	Seasonal temperature	Mean precipitation	Livestock-productivity and quality	Cultivated spp.-phenology	Cultivated spp.-diseases/pest/mortality	Drought
7	Mean temperature	Drought	Drought	Drought	Drought	Extreme temperature
8	Duration and timing of seasons	Mean river flow	Wild flora-phenology	Mean river flow	Seasonal precipitation	Terrestrial fauna-distribution
9	Drought	Cultivated spp.-diseases/pest/mortality	Pasture-composition and quality	Snowfall and snow cover	Wild flora-phenology	Cultivated spp.-diseases/pest/mortality
10	Cultivated spp.-diseases/pest/mortality	Duration and timing of seasons	Cultivated spp.-phenology	Pasture-availability and productivity	Mean river flow	Wild flora-phenology

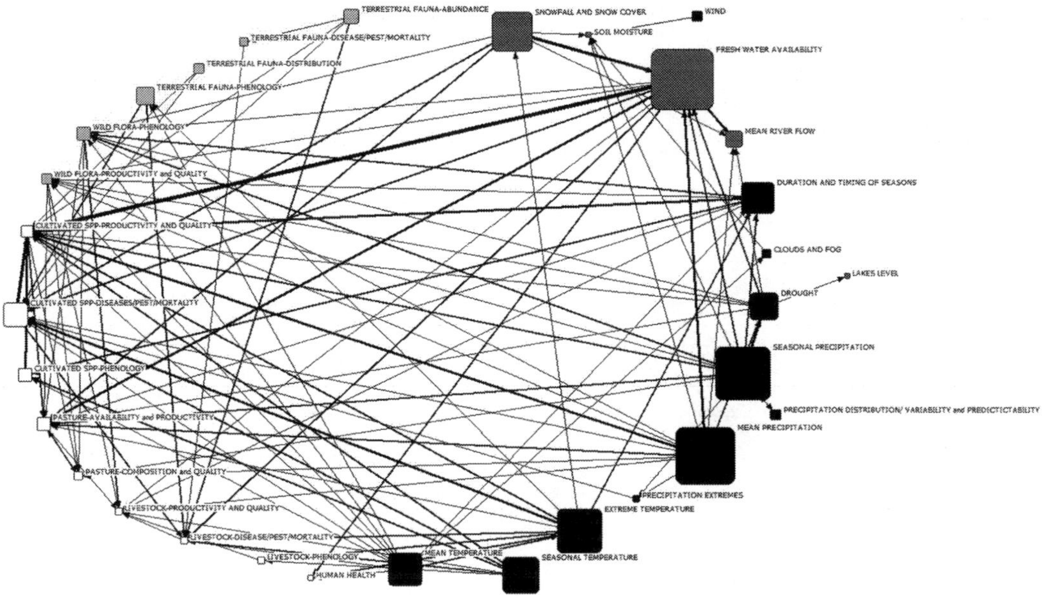

Figure 5.1 Representation of the climate change network (CCN) centralization outdegree of (a) *experts* ($n = 175$) and (b) *non-experts* ($n = 63$).

Note: To balance sample sizes, experts′ CCN only represents cascading effects perceived by at least three respondents. Size of nodes represents the centrality outdegree score of each node. Line thickness represents the strength of the tie based on the number of respondents that mention the cascading effect on a scale from 1 to 3.

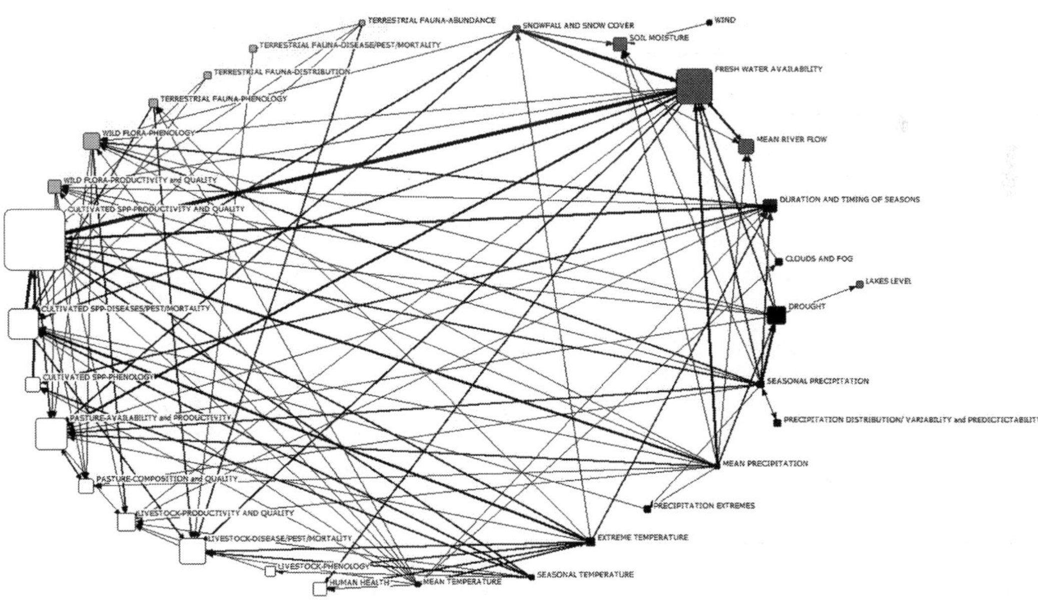

Figure 5.2 Representation of the climate change network (CCN) centralization indegree of (a) *experts* ($n = 175$) and (b) *non-experts* ($n = 63$). See note in Figure 5.1.

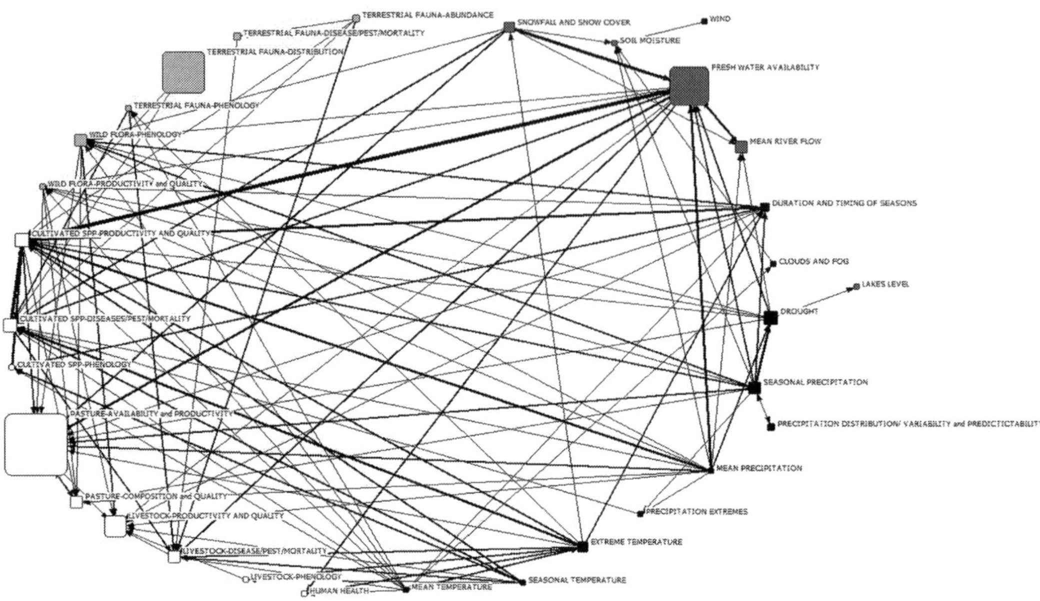

Figure 5.3 Representation of the climate change network (CCN) centralization betweenness of (a) *experts* (n = 175) and (b) *non-experts* (n = 63). See note in Figure 5.1.

Both groups perceived *Pasture availability and productivity* and *Cultivated spp. disease, pest, and mortality* as impacted elements with high betweenness scores, representing a fundamental place in the network due to the several cascading effects that are receiving and their own cascading impacts. Similarly, for the two groups, *Terrestrial fauna distribution* and *Wild flora phenology* were among the ten nodes with the most important intermediate positions in their networks (Figure 5.3). Finally, the group of experts and non-experts respectively reported that seven and six of the nodes with highest betweenness score were also among those with the highest indegree scores. However, experts' CCN shows a higher interconnection among all nodes, with most of them receiving or generating at least two cascading effects. In the experts' CCN, only three nodes are connected to the network through one tie, whereas in non-experts' CCN eight nodes are connected only by one tie, showing a weaker interconnection among network nodes and explaining their lower network centralization betweenness score. Although both groups shared six out of the ten cascading impacts with the highest betweenness score, non-experts emphasized elements of the physical and atmospheric system as fundamental intermediate cascading impacts, being *Freshwater availability* and *Snowfall and snow cover* those with the highest betweenness scores (Figure 5.3). In contrast, the group of experts perceived *Pasture availability and productivity* and *Wild fauna distribution* as those intermediate cascading impacts with the highest betweenness score, thus giving a higher importance to elements of the biological and human systems in these intermediate positions.

Discussion

We derive four main findings from the work presented here. Our first finding is that Sierra Nevada inhabitants report a large number and variety of climate change impacts affecting their social-ecological system and livelihoods. In fact, respondents reported climate change impacts that were not proposed in the survey, showing a deep level of understanding of the social-ecological system. These findings are neither surprising nor new, as it is increasingly acknowledged that people with a long history of interaction with the environment observe detailed changes in their social-ecological system (e.g., Hansen et al. 2016; Díaz et al. 2019; Reyes-García et al. 2019). Nevertheless, they contribute to this research line by emphasizing the value of local knowledge systems.

Second, our methodology allowed us to identify the main climate-related changes observed in Sierra Nevada as triggering additional changes in the social-ecological system. Mainly, elements of the atmospheric and physical system are seen as triggers of changes in elements of the biological and human systems. Indeed, changes in the amount of permanent snow (Uprety et al. 2017; Panikkar et al. 2018; Lippi & Sanfilippo 2024), in the frequency and intensity of snowfalls (Wilson et al. 2015; Babai 2024; Lippi & Sanfilippo 2024), and in freshwater availability (Ingty 2017; Izquierdo & Schlingmann 2024; Mwangi et al. 2024) are climate change impacts recurrently reported in mountain areas because of their cascading effects. As reported by IPLC in other areas of the world (Babai 2024; Carmona 2024; Izquierdo & Schlingmann 2024; Mattalia et al. 2024; Mwangi et al. 2024) and by the scientific community in Sierra Nevada (Hódar & Zamora 2004; Ros-Candeira et al. 2019), our respondents also see temperature increase as a trigger of other changes. Thus, according to our respondents, changes in mean and extreme temperatures affect the quantity, quality, and the survival of crops and orchards, the phenology of crops and wild flora and fauna, and livestock diseases and mortality.

Among the changes that seem to be triggered by a large diversity of impacts, we recorded elements of the human system related to crops, livestock, and pastures, but also elements of the physical system such as changes in freshwater availability and rivers' flow. Changes in productivity, appearance of diseases and pests in crops, pastures and livestock, and phenological changes are climate change impacts also reported by many other communities in mountain areas (Uprety et al. 2017; Babai 2024; Mattalia et al. 2024). Moreover, the group people directly involved in agropastoral activities (i.e., experts) also mentioned changes in wild flora phenology among the most impacted elements, showing high awareness of changes in elements of the biological system. Local knowledge about the relations between atmospheric factors and wild and domesticated species is increasingly evident in the literature, with many examples of IPLC observing phenological changes in their agricultural calendars (Miara et al. 2022), or unusual behaviour of animals and plants to make weather forecasts and predict the intensity or duration of the rainy season or dry periods (Castillo & Ladio 2018; Balehegn et al. 2019).

The third important finding of this work refers to the fact that some of the elements mentioned as being most impacted also play a key intermediate role in maintaining the network structure. Indeed, key intermediate nodes have connections with many other nodes in the network, potentially amplifying cascading effects. This is the case of *Crops, Livestock, and Pasture productivity, Cultivated spp disease, pest, and mortality,* and *Wild flora phenology* but also of *Freshwater availability, Mean river flow and Drought*. For example, shepherds, ranchers, and beekeepers mentioned that changes in temperature, precipitation and snow, and ice duration but also changes in elements of the biological system impact pastures. The rapid disappearance of pastures, in turn, impacts livestock, all together putting at risk the future of pastoralism in Sierra Nevada. In practical terms, this finding calls for attention to the maintenance of traditional irrigation systems, which – for centuries – have allowed the supply of water to agricultural and inhabited areas on the slopes of the mountain, while maintaining the state of the high mountain meadows during the summer, favouring a large biodiversity of wild flora, and offering water for wildlife and livestock of the local population (Pulido-Bosch & Sbih 1995; Martos-Rosillo et al. 2019).

Other elements with an important intermediate key role include changes in *Terrestrial fauna distribution*, which could result in potentially severe consequences for the social-ecological system, if not considered also as a priority. For example, participants mentioned that water scarcity in the high areas of Sierra Nevada is leading wild boars and mountain goats to lower areas, where they invade crop fields to feed and drink. A similar situation has been reported for reindeers in Norway, elephants in India, or wild carnivores in other areas of the world (Pecl et al. 2017). Importantly, this change in wildlife behaviour favours the transmission of diseases to livestock (Martínez-Guijosa et al. 2021), resulting in the slaughtering of livestock avoiding transmission to the human population.

The last important finding of this work is that people directly involved in agropastoral activities (i.e., experts) reported more climate change impacts, cascading effects, and synergistic interactions than people not directly involved (i.e., non-experts). Despite similarities between both groups regarding triggering impacts from the atmospheric and physical systems, experts identified more cascading effects on elements of the biological systems and described more relations between network nodes than non-experts, thus denoting a deeper understanding of ecological functioning of the local ecosystems. In contrast, non-experts focused their attention on elements of the human system, describing fewer cascading effects, directly relating triggering impacts of the atmospheric and physical system to elements of the human system. In addition, non-experts gave much importance to impacts on human health, suggesting a more anthropocentric vision of the Sierra Nevada social-ecological system than experts. This result correlates with a recent study showing that people in Sierra Nevada with life-long bonds with the environment and higher connection and dependence upon ecosystem services report more climate change impacts (García-del-Amo et al. 2023).

Conclusion

Local communities in Sierra Nevada report a great variety of climate change impacts that negatively affect their livelihoods. Beyond adding to the growing number of works pointing at the importance of local knowledge systems, our findings might inform local policies. For example, identifying which climate change impacts generate more cascading effects can help in the design of management plans that minimize the effect of such changes to maintain the resilience of the social-ecological system. We present three specific ways in which our results can contribute to local policies both in Sierra Nevada and beyond.

First, local communities have identified the changes that most strongly impact them. This information can help design priorities in local adaptation plans. For example, changes in freshwater availability consistently appear as the triggering impact with the strongest effect in Sierra Nevada, for which further efforts should be done to protect water networks. Importantly, because Sierra Nevada should be understood as a social-ecological system, protection should include not only the biophysical but also the associated human component, such as irrigation communities.

Second, the network structure found suggests the interrelated nature of impacts across elements and the critical situation described by respondents, as many key intermediate elements are also those receiving more impacts. The holistic nature of local knowledge systems allows a deeper understanding of local climate change impacts. Giving priority to addressing these impacts in the design of adaptation plans could contribute to preserving the sustainability of Sierra Nevada.

Finally, our results also show intracultural differences in local knowledge. Local inhabitants with a dilated experience in agricultural and livestock activities like shepherds, ranchers, beekeepers, and members of irrigation communities are valuable holders of knowledge. Recognizing and integrating them into future co-management climate change research and adaptation plans for the region should be a priority.

Acknowledgements

We thank the collaboration of people from Sierra Nevada who answered our survey and shared with us their valuable knowledge. We acknowledge financial support from the Spanish Ministry of Economy and Competitiveness, through the María de Maeztu Programme for Units of Excellence in R&D (MdM-2015-0552), the project grant CSO2014-59704-P, and funding from the European Research Council (ERC) under grant agreement No 771056-LICCI-ERC-2017-COG. This research received the approval of the Ethics committee of the Autonomous University of Barcelona (CEEAH 4781). This work contributes to the "María de Maeztu" Programme for Units of Excellence (CEX2019-000940-M).

References

Armstrong Mckay, D. I. A., Staal, A., Abrams, J. F., Winkelmann, R., Sakschewski, B., Loriani, S., ... & Lenton, T. M. (2022). Exceeding 1.5°C global warming could trigger multiple climate tipping points. *Science*, 1171. https://doi.org/10.1126/science.abn7950

Babai, D. (2023). "The weather is more erratic, it changes faster..." Local perceptions of climate change in the Eastern Carpathians. In V. Reyes-García, S. Álvarez-Fernández, P. Benyei, L. Calvet-Mir, D. García-del-Amo, A. B. Junqueira, X. Li, V. Porcher, A. Porcuna-Ferrer, A. Schlingmann, & R. Soleymani (Eds.), *Routledge Handbook of Climate Change Impacts on Indigenous Peoples and Local Communities*. Oxfordshire & New York: Routledge. Chp 4, pp 74–94.

Bahlai, C. A., Hart, C., Kavanaugh, M. T., White, J. D., Ruess, R. W., Brinkman, T. J., ... & Thompson, J. R. (2021). Cascading effects: insights from the U.S. Long Term Ecological Research Network. *Ecosphere*, 12(5). https://doi.org/10.1002/ecs2.3430

Balehegn, M., Balehey, S., Fu, C., & Liang, W. (2019). Indigenous weather and climate forecasting knowledge among Afar pastoralists of north eastern Ethiopia: role in adaptation to weather and climate variability. *Pastoralism*, 9(1). https://doi.org/10.1186/s13570-019-0143-y

Benyei, P., Aceituno-Mata, L., Calvet-Mir, L., Tardío, J., Pardo-de-Santayana, M., García-del-Amo, D., ... & Reyes-García, V. (2020). Seeds of change: reversing traditional agroecological knowledge's erosion through a citizen science school program in Catalonia. *Ecology and Society*, 25(2), 19. https://doi.org/ https://doi.org/10.5751/ES-11471-250219

Blanca, G., López-Onieva, M. R., Lorite, J., Martínez-Lirola, M. J., Molero-Mesa, J., Quintas, S., Ruiz-Girela, M., Varo, M. A., & Vidal, S. (2001). *Flora Amenazada y endémica de Sierra Nevada.* Consejería de Medio Ambiente, Junta de Andalucía. https://doi.org/10.1192/bjp.112.483.211-a

Blondel, J. (2006). The "design" of Mediterranean landscapes: a millennial story of humans and ecological systems during the historic period. *Human Ecology*, 34(5), 713–729. https://doi.org/10.1007/s10745-006-9030-4

Borgatti, S. P. (2002). *Netdraw.* Network visualization. Harvard, MA: Analytics Technologies.

Borgatti, S. P., Everett, M. G., & Johnson, J. C. (2018). *Analyzing Social Networks* (J. Seaman (ed.)). SAGE.

Borgatti, S. P., Everett, M. G., & Freeman, L. C. (2002). *Ucinet 6 for Windows: Software for Social Network Analysis.* Harvard, MA: Analytic Technologies.

Burton, P. J., Jentsch, A., & Walker, L. R. (2020). The ecology of disturbance interactions. *BioScience*, 70(10), 854–870. https://doi.org/10.1093/biosci/biaa088

Calvet-Mir, L., & Salpeteur, M. (2016). Humans, plants, and networks: a critical review. *Environment and Society: Advances in Research*, 7(1), 107–128. https://doi.org/10.3167/ares.2016.070107

Carmona, R. (2024). The colonial roots of climate vulnerability. Analysing the case of the Mapuche-Pehuenche people in Southern Chile. In V. Reyes-García, S. Álvarez-Fernández, P. Benyei, L. Calvet-Mir, D. García-del-Amo, A. B. Junqueira, X. Li, V. Porcher, A. Porcuna-Ferrer, A. Schlingmann, & R. Soleymani (Eds.), *Routledge Handbook of Climate Change Impacts on Indigenous Peoples and Local Communities*. Oxfordshire & New York: Routledge. Chp 12, pp 196–207.

Castillo, L., & Ladio, A. (2018). Mammals and birds as ethno-indicators of change: their importance to livestock farmers in Arid Patagonia (Argentina). *Environment, Development and Sustainability*, 20(5), 2161–2179. https://doi.org/10.1007/s10668-017-9983-z

Collins, M., Sutherland, M., Bouwer, L., Cheong, S.-M., Frölicher, T. L., Jacot Des Combes, H., ... & Timmermans, M.-L. (2019). Extremes, abrupt changes and managing risks. *IPCC Special Report on the*

Ocean and Cryosphere in a Changing Climate, 589–655. https://report.ipcc.ch/srocc/pdf/SROCC_FinalDraft_Chapter6.pdf
Díaz, S., Settele, J., Brondízio, E. S., Ngo, H. T., Agard, J., Arneth, A., … & Zayas, C. N. (2019). Pervasive human-driven decline of life on Earth points to the need for transformative change. *Science*, 366(6471). https://doi.org/10.1126/science.aax3100
Fassnacht, S. Allegretti, R., Venable, A. M., Fernández-Giménez, N. B. H., Tumenjargal, M. E., Kappas, S., … & Pfohl, A. K. D. (2018). Merging indigenous knowledge systems and station observations to estimate the uncertainty of precipitation change in central Mongolia. *Hydrology,* 5(3), 1–16. https://doi.org/10.3390/hydrology5030046
Ford, J. D., Cameron, L., Rubis, J., Maillet, M., Nakashima, D., Willox, A. C., & Pearce, T. (2016). Including indigenous knowledge and experience in IPCC assessment reports. *Nature*, 6(4), 349–353. https://doi.org/10.1038/nclimate2954
García-del-Amo, D., Mortyn, P. G., & Reyes-García, V. (2020). Including indigenous and local knowledge in climate research: an assessment of the opinion of Spanish climate change researchers. *Climatic Change*, 160, 67–88. https://doi.org/10.1007/s10584-019-02628-x
García-del-Amo, D., Gálvez-García, C., Iniesta-Arandia, I., Moreno-Ortiz, J., & Reyes-García, V. (2022). Local ecological knowledge and the sustainable co-management of Sierra Nevada social-ecological system. In M. Oliva & R. Zamora (eds.). *The Landscape of the Sierra Nevada. A Unique Laboratory of Global Processes in Spain* (p. 417). Springer Nature Switzerland. https://doi.org/https://doi.org/10.1007/978-3-030-94219-9_21
García-del-Amo, D., Mortyn, P.G. & Reyes-García, V. (2023). Local reports of climate change impacts in Sierra Nevada, Spain: sociodemographic and geographical patterns. *Regional Environmental Change*, 23, 14. https://doi.org/10.1007/s10113-022-01981-5
Gurgiser, W., Juen, I., Singer, K., Neuburger, M., Schauwecker, S., Hofer, M., & Kaser, G. (2016). Comparing peasants' perceptions of precipitation change with precipitation records in the tropical Callejón de Huaylas, Peru. *Earth System Dynamics*, 7(2), 499–515. https://doi.org/10.5194/esd-7-499-2016
Hansen, G., Stone, D., Auffhammer, M., Huggel, C., & Cramer, W. (2016). Linking local impacts to changes in climate: a guide to attribution. *Regional Environmental Change*, 16(2), 527–541. https://doi.org/10.1007/s10113-015-0760-y
Haro-Monteagudo, D., Palazón, L., & Beguería, S. (2020). Long-term sustainability of large water resource systems under climate change: a cascade modeling approach. *Journal of Hydrology*, 582, 124546. https://doi.org/10.1016/j.jhydrol.2020.124546
He, S., & Richards, K. (2015). Impact of meadow degradation on soil water status and pasture management-a case study in Tibet. *Land Degradation and Development*, 26(5), 468–479. https://doi.org/10.1002/ldr.2358
Hódar, J. A., & Zamora, R. (2004). Herbivory and climatic warming: a Mediterranean outbreaking caterpillar attacks a relict, boreal pine species. *Biodiversity and Conservation,* 13(3), 493–500. https://doi.org/10.1023/B:BIOC.0000009495.95589.a7
Hu, Z., Zhang, Y., Gu, F., Li, Y., Shao, H., & Liu, S. (2020). Local residents' perceptions of climate and ecological changes in the eastern Tibetan Plateau. *Regional Environmental Change*, 20(2). https://doi.org/10.1007/s10113-020-01629-2
Ingty, T. (2017). High mountain communities and climate change: adaptation, traditional ecological knowledge, and institutions. *Climatic Change*, 145(1–2), 41–55. https://doi.org/10.1007/s10584-017-2080-3
Iniesta-Arandia, I., Garcia-del-Amo, D., García-Nieto, A. P., Piñeiro, C., Montes, C., & Martín-López, B. (2014). Factors influencing local ecological knowledge maintenance in Mediterranean watersheds: insights for environmental policies. *Ambio*, 44(4), 285–296. https://doi.org/10.1007/s13280-014-0556-1
Inouye, D. W., & Wielgolaski, F. E. (2013). Phenology at high altitudes. In *Phenology: An Integrative Environmental Science* (pp. 249–272). Springer Netherlands. https://doi.org/10.1007/978-94-007-6925-0_14
IPCC. (2021). Climate change 2021: the physical science basis. In V. Masson-Delmotte, et al. (eds.). *Contribution of Working Group I to the Sixth Assessment Report of the Intergovernmental Panel on Climate Change.* Cambridge and New York: Cambridge University Press. https://doi.org/10.1017/9781009157896
Izquierdo, A., & Schlingmann, A. (2024). Local knowledge of climate change in the context of social, political and technological changes in a High Andean community from the Argentine In V. Reyes-García, S. Álvarez-Fernández, P. Benyei, L. Calvet-Mir, D. García-del-Amo, A. B. Junqueira, X. Li, V. Porcher, A. Porcuna-Ferrer, A. Schlingmann, & R. Soleymani (Eds.), *Routledge Handbook of Climate Change Impacts on Indigenous Peoples and Local Communities*. Oxfordshire & New York: Routledge. Chp 8, pp 136–150.
Jones, B., & O'Neill, B. C. (2016). Spatially explicit global population scenarios consistent with the shared socioeconomic pathways. *Environmental Research Letters*, 11(8). https://doi.org/10.1088/1748-9326/11/8/084003

Krönke, J., Wunderling, N., Winkelmann, R., Staal, A., Stumpf, B., Tuinenburg, O. A., & Donges, J. F. (2020). Dynamics of tipping cascades on complex networks. *Physical Review E*, 101(4), 1–9. https://doi.org/10.1103/PhysRevE.101.042311

Lippi, S., & Sanfilippo, M. (2024). Smallholders perceptions of climate change in a post-socialist Albania. In V. Reyes-García, S. Álvarez-Fernández, P. Benyei, L. Calvet-Mir, D. García-del-Amo, A. B. Junqueira, X. Li, V. Porcher, A. Porcuna-Ferrer, A. Schlingmann, & R. Soleymani (Eds.), *Routledge Handbook of Climate Change Impacts on Indigenous Peoples and Local Communities*. Oxfordshire & New York: Routledge. Chp 7, pp 124–135.

Martínez-Guijosa, J., Lima-Barbero, J. F., Acevedo, P., Cano-Terriza, D., Jiménez-Ruiz, S., Barasona, J. Á.,... & Vicente, J. (2021). Description and implementation of an on-farm wildlife risk mitigation protocol at the wildlife-livestock interface: tuberculosis in Mediterranean environments. *Preventive Veterinary Medicine*, 191. https://doi.org/10.1016/j.prevetmed.2021.105346

Martos-Rosillo, S., Ruiz-Constán, A., González-Ramón, A., Mediavilla, R., Martín-Civantos, J. M., Martínez-Moreno, F. J., ... & Durán, J. J. (2019). The oldest managed aquifer recharge system in Europe: new insights from the Espino recharge channel (Sierra Nevada, southern Spain). *Journal of Hydrology*, 578. https://doi.org/10.1016/j.jhydrol.2019.124047

Mattalia, G., Stryamets, N., Toscano-Rivalta, V., & Reyes-García, V. (2024). Correspondence between local and scientific knowledge of climate change: the case of Hutsuls of Northern Romanian In V. Reyes-García, S. Álvarez-Fernández, P. Benyei, L. Calvet-Mir, D. García-del-Amo, A. B. Junqueira, X. Li, V. Porcher, A. Porcuna-Ferrer, A. Schlingmann, & R. Soleymani (Eds.), *Routledge Handbook of Climate Change Impacts on Indigenous Peoples and Local Communities*. Oxfordshire & New York: Routledge. Chp 1, pp 30–44.

Miara, M. D., Negadi, M., Tabak, S., Bendif, H., Dahmani, W., Ait Hammou, M.,... & Teixidor-Toneu, I. (2022). Climate change impacts can be differentially perceived across time scales: a study among the Tuareg of the Algerian Sahara. *GeoHealth*, 6(11). https://doi.org/10.1029/2022GH000620

Mwangi, B., Olaka, L. A., Zafra-Calvo, N., Klein, J. A., & Cuni-Sanchez, A. (2024). The role of culture in climate change adaptation: insights from two mountain regions in Kenya. In V. Reyes-García, S. Álvarez-Fernández, P. Benyei, L. Calvet-Mir, D. García-del-Amo, A. B. Junqueira, X. Li, V. Porcher, A. Porcuna-Ferrer, A. Schlingmann, & R. Soleymani (Eds.), *Routledge Handbook of Climate Change Impacts on Indigenous Peoples and Local Communities*. Oxfordshire & New York: Routledge. Chp 19, pp 322–334.

Oliva, M., Ruiz-Fernández, J., Barriendos, M., Benito, G., Cuadrat, J. M., Domínguez-Castro, F., ... & Vicente-Serrano, S. M. (2018). The little ice age in Iberian mountains. *Earth-Science Reviews*, 177, 175–208. https://doi.org/10.1016/j.earscirev.2017.11.010

Panikkar, B., Lemmond, B., Else, B., & Murray, M. (2018). Ice over troubled waters: navigating the Northwest Passage using Inuit knowledge and scientific information. *Climate Research*, 75(1), 81–94. https://doi.org/10.3354/cr01501

Payne, D., Snethlage, M., Geschke, J., Spehn, E. M., & Fischer, M. (2020). Nature and people in the Andes, East African Mountains, European Alps, and Hindu Kush Himalaya: current research and future directions. *Mountain Research and Development*, 40(2), R1–R14. https://doi.org/10.1659/MRD-JOURNAL-D-19-00075.1

Pecl, G. T., Araújo, M. B., Bell, J. D., Blanchard, J., Bonebrake, T. C., Chen, I. C., ... & Williams, S. E. (2017). Biodiversity redistribution under climate change: impacts on ecosystems and human well-being. *Science*, 355(6332). https://doi.org/10.1126/science.aai9214

Plummer, R., Baird, J., & Pickering, K. (2014). Priming the governance system for climate change adaptation: the application of a Social Ecological Inventory (SEI) to engage actors in Niagara, Canada. *Ecology and Society*, 19(1), 3. https://doi.org/http://dx.doi.org/10.5751/ES-06152-190103

Pörtner, H. O., Scholes, R. J., Agard, J., Archer, E., Arneth, A., Bai, X., ... & Foden, W. (2021). IPBES-IPCC CO-sponsored workshop biodiversity and climate change. *IPBES and IPCC*, 115(4), 608–609. https://doi.org/10.5281/zenodo.4782538.IPBES

Pulido-Bosch, A., & Sbih, Y. Ben. (1995). Centuries of artificial recharge on the southern edge of the Sierra Nevada (Granada, Spain). *Environmental Geology*, 26(1), 57–63. https://doi.org/10.1007/BF00776033

Reyes-García, V., García-del-Amo, D., Benyei, P., Fernández-Llamazares, Á., Gravani, K., Junqueira, A., ... & Soleymani-Fard, R. (2019). A collaborative approach to bring insights from local indicators of climate change impacts into global climate change research. *Current Opinion in Environmental Sustainability*, 39, 1–8. https://doi.org/10.1016/j.cosust.2019.04.007

Rocha, J. C., Peterson, G., Bodin, Ö., & Levin, S. (2018). Cascading regime shifts within and across scales. *Science*, 362(6421), 1379–1383. https://doi.org/10.1126/science.aat7850

Ros-Candeira, A., Pérez-Luque, A. J., Suárez-Muñoz, M., Bonet-García, F. J., Hódar, J. A., Giménez De Azcárate, F., & Ortega-Díaz, E. (2019). Dataset of occurrence and incidence of pine processionary moth in Andalusia, South Spain. *ZooKeys*, 852, 125–136. https://doi.org/10.3897/zookeys.852.28567

Seneviratne, S. I., Corti, T., Davin, E. L., Hirschi, M., Jaeger, E. B., Lehner, I., Orlowsky, B., & Teuling, A. J. (2010). Investigating soil moisture–climate interactions in a changing climate: a review. *Earth-Science Reviews,* 99(3–4), 125–161. https://doi.org/10.1016/J.EARSCIREV.2010.02.004

Shively, G. (2011). Sampling: who, how and how many? In A. Angelsen, H. O. Larsen, & J. F. Lund (eds.). *Measuring Livelihoods and Environmental Dependence: Methods for Research and Fieldwork* (pp. 51–70). Routledge London & Washington: Earthscan.

Smith, H. A., & Sharp, K. (2012). Indigenous climate knowledges. Wiley interdisciplinary reviews. *Climate Change*, 3(5), 467–476. https://doi.org/10.1002/wcc.185

Sowman, M., & Raemaekers, S. (2018). Socio-ecological vulnerability assessment in coastal communities in the BCLME region. *Journal of Marine Systems*, 188, 160–171. https://doi.org/10.1016/j.jmarsys.2018.01.008

Uprety, Y., Shrestha, U. B., Rokaya, M. B., Shrestha, S., Chaudhary, R. P., Thakali, A., Cockfield, G., & Asselin, H. (2017). Perceptions of climate change by highland communities in the Nepal Himalaya. *Climate and Development*, 9(7), 649–661. https://doi.org/10.1080/17565529.2017.1304886

Weatherhead, E., Gearheard, S., & Barry, R. G. (2010). Changes in weather persistence: insight from Inuit knowledge. *Global Environmental Change*, 20(3), 523–528. https://doi.org/10.1016/J.GLOENVCHA.2010.02.002

Wilson, N. J., Todd Walter, M., & Waterhouse, J. (2015). Indigenous knowledge of hydrologic change in the Yukon River basin: a case study of Ruby, Alaska. *Arctic*, 68(1), 93–106. https://doi.org/10.14430/arctic4459

Zamora, R., Pérez-Luque, A. J., Bonet, F. J., Barea-Azcón, J. M., & Aspizua, R. (eds.) (2016). *Global Change Impacts in Sierra Nevada: Challenges for Conservation*. Consejería de Medio Ambiente y Ordenación del Territorio. Granada: Junta de Andalucía.

Zamora, R., Antonio, J. P., & Bonet, F. J. (2017). Monitoring global change in high mountains. *High Mountain Conservation in a Changing World*, 62, 385–413. https://doi.org/10.1007/978-3-319-55982-7

6

CLIMATE AND ENVIRONMENTAL CHANGE PERCEPTIONS

A case from rural Sicily, Italy

Vincenza Ferrara and Johan Lindberg

Introduction

Climate in the Mediterranean region has always been characterised by large spatial and temporal variability (Peyron et al., 2013). Located in such a complex area in terms of climatic regimes, the island of Sicily has not been spared by climate change. Calculated trends in historical rainfall provide evidence of a general decrease in precipitation (Liuzzo et al., 2016) and an increase in warming, with a steepness during the last 25 years (Viola et al., 2014). Such changes are expected to have great impacts on the island. Increasing temperature will bring aridity and an increase in the frequency and intensity of short precipitation extremes (Forestieri et al., 2018). Severe changes are also foreseen in plant phenology and physiology, species distribution, communities' interactions, and structural and functional heterogeneity in ecosystems and agroecosystems. Such ecological responses to climatic changes will be strictly linked to a loss of biocultural diversity, the biophysical traits of an area defined by past and present human practices (Baiamonte et al., 2015; Ferrara et al., 2020). Uncomfortably uncertain about their future, local communities in rural Sicily are thus today in the midst of the progressive degradation of their biocultural heritage, while suffering from serious marginalisation due to contrasting and simultaneous processes of rural depopulation and land abandonment, agricultural intensification (Rühl et al., 2011; Veneziano, 2011; Brandolini et al., 2018; Ferrara et al., 2022), and increasing urbanisation and pollution trends (Rapporto Ambientale – VAS PSR Sicilia, 2014/2020).

What can the analysis of the local impacts of climate and environmental change tell us about these ongoing dynamics? When attempting to understand how climate affects a society, scale matters. We may ask, for instance, what is the influence of current dominant global change drivers (i.e., climate warming and land-use change) on plant communities at a local scale over decennia? Or we could extend the time lags and consider the response of plants to environmental change over a longer time period. Relationships between timescales can provide unique insights into the impacts of climate variability over land. Historical ecology reminds us that downscaling climate to the specific location (place) during the period in which the community under study lived (time) and in the contexts of their experiences (stored in traditional knowledge and genetic evolution) may be a suitable inquiry approach (Rivera-Collazo, 2022; Whitaker et al., 2023). In historical ecology, the timeframe for ecological studies covers different time scales (decennia, centuries, millennia, Vellend et al., 2013; Beller et al., 2020; Crumley, 2021) and past variability in most

 DOI: 10.4324/9781003356837-9

ecosystems (Swetnam et al., 1999) is seen as non-stationary (Safford, 2008). Historical ecology shifts baselines, putting ecosystems and landscapes in a nonequilibrium state, where different temporal and spatial scales co-exist at once (Crumley, 1995). In such a framework, *change* in general, among which climate change, cannot be approached as something happening with the same intensity at different space-time scales, but it needs to be context- and scale-specific. For this reason, a historical ecology approach impels us to cultivate a sense of social and ecological *context.*

This chapter looks at climate and environmental changes as reported by locals in a fluvial system of central Sicily, the Morello Valley, and compares these local reports with temperature and rainfall data collected by meteorological stations *in loco.* Then, inspired by historical ecology, the chapter further frames and discusses the results of a longer-term reconstruction of the climatic, environmental and societal past of this part of Sicily, covering the entire Holocene. Conclusions highlight the importance of multi-causal explanations when approaching climate and environmental change. Such multi-causal explanations should incorporate long-term socially accumulated ecological knowledge with scientific evidence on different timescales, always taking into account how the influence of climate on society strongly depends on a combination of ecological dynamics, relational and societal conditions, place geography, and historical circumstances.

Data and methods

This study is part of a cross-disciplinary research project on the biocultural heritage of Sicilian olive trees, funded by The Swedish Research Council for the period 2020–2024 and aiming to investigate how remnants of ancient agrosystems still present on the island may offer insights to tackle adaptation challenges of agrobiodiversity in the near future. The data presented and analysed in this chapter have been collected within a collaboration with the 'Local Indicators of Climate Change Impacts: the contribution of local knowledge to climate change research' project.

Study area

Fieldwork was conducted by the lead author in the villages of Calascibetta, Villarosa and Villapriolo, located in the Morello valley (Figure 6.1), a fluvial system inhabited since prehistory. Numerous archaeological sites testify sophisticated agricultural and production activities back to the fourth millennium B.C. (Giannitrapani et al., 2014). Water availability thanks to the presence of the Morello river favoured pastoralism and agriculture. Additionally, the central location of the valley in the island facilitated trade and a soil rich in sulphur gave rise to mineral extraction activities (Giannitrapani, 2015).

Early Neolithic groups, mainly dedicated to nomadic pastoralism, developed into complex socio-cultural and economic communities during the Bronze Age (third mill. BC), and their presence continued to be attested throughout Antiquity. Settlements were progressive abandoned during late Antiquity and early Middle Ages due to historical events such as the fall of the Roman Empire, the Muslims invasion and the Norman conquest. During the 9th century AD, one of the three villages in the case study area, Calascibetta, was founded as a Muslim military camp. Successive abandonment of rural sites due to Middle Ages disruptions (e.g., plagues, wars, demographic crises) resulted in the consolidation of extensive feudal estates that, from around the 15th century, focused mainly on cereal production (Pluciennik et al., 2004). A significant colonisation movement in the inner areas accompanied this early modern reorganisation of land use, particularly during the 16th–17th centuries (Davies, 1983; Pluciennik et al., 2004). Many new villages were established, dedicated to cereal production and strictly related to the rise of new land-owning families, who

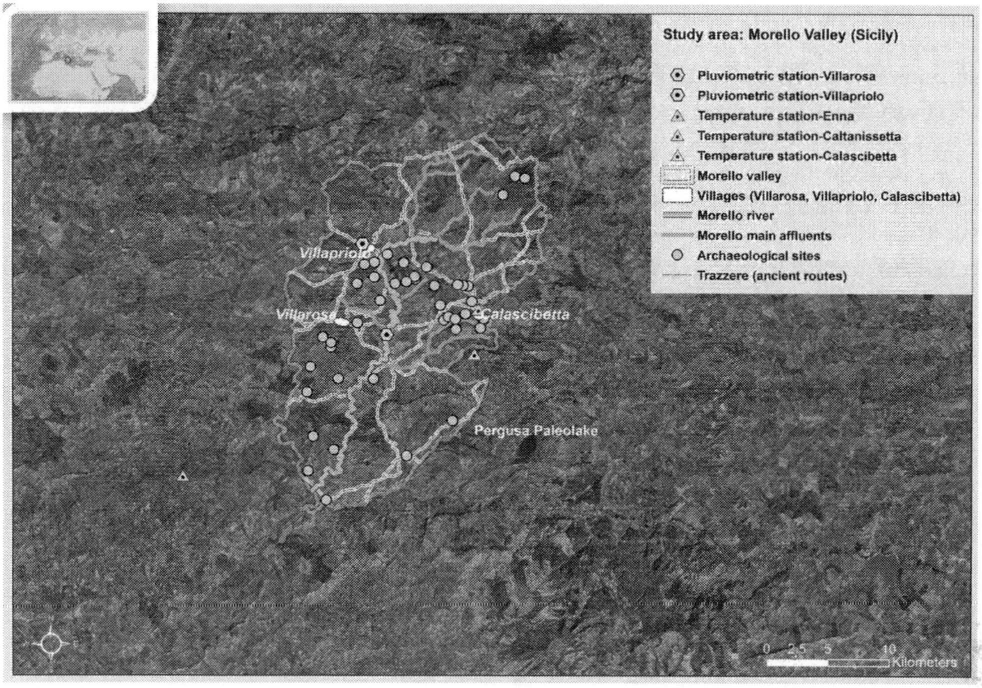

Figure 6.1 Morello Valley study area (Geodata and ortophoto: Sistema Informativo Territoriale – Regione Sicilia; Archaeological dataset: Enrico Giannitrapani; basemap Esri).

controlled the peasant population and their labour, as well as earnings from agriculture and other business (Davies, 1983). The latifundia were banned at the end of the Second World War. Nonetheless, the subsequent redistribution of land to peasants did not help the agricultural development of the area. Devastated by the war and affected by the contemporary collapse of the sulphur mines, the area was hit by a severe migration flow (Aymard & Giarrizzo, 1987) that persists today (e.g., 151.4% is the latest incidence rate for the village of Villarosa, Fondazione Migrantes 2022). Despite the severe abandonment of the rural areas, some small-scale agricultural activities persist, mainly non-irrigated mechanised cereal fields, olive cultivation and livestock (Ferrara et al., 2022).

Data collection and analysis

Data collection and analysis took place in 2020 and 2021, and it involved

1 The collection of local observations of climate and broader environmental changes;
2 The comparison of the local perceptions with trends in time series of weather data (temperature and rainfall) for the period January 1924–July 2021.

Collection of local observations of climate and broader environmental changes

To collect local observations of climate and broader environmental changes in the case study area, we conducted semi-structured interviews and focus groups, following a standardized framework to document local observation of climate and environmental change (Reyes-García et al., 2023).

For the semi-structured interviews, "quota sampling" was used, interviewing different informants in terms of gender, age and livelihood. The semi-structured interviews ($n = 25$) covered seasonal calendars and local ecological knowledge on the main agricultural crops and practices of the area, observed environmental and climatic changes, and their perceived causes and drivers.

Semi-structured interviews were followed by three focus groups discussion (FGD), one in each village, for a total of 18 participants, six on each FGD. The selection of informants was done through convenience sampling, with the aim to interview the most knowledgeable people (local experts), and aiming to capture diversity in terms of livelihood activities and expertise (with special preference to elders). The FGDs aimed to assess the level of group consensus regarding perceived changes and corresponding drivers emerging from the semi-structured interviews, clarify some incomplete information, and identify other changes not listed during interviews. The information coming from the FGDs represents thus the collective perception of change.

The observations of climate and environmental changes, as their corresponding drivers (both generic ones, such as "climate change" or more specific if detailed by the interviewees), have been then grouped after an impact model, according to the system (atmospheric, physical and life systems)[1] in which they have first been observed, and how progressively they have impacted other elements of the system.

Comparison of local perceptions with trends in time series of weather data

Monthly temperature and rainfall data were gathered from the Regional Hydrological Observatory Database[2] and aggregated per year.

The closest meteorological and pluviometric stations to the three villages of the case study area have been chosen for data collection (Figure 6.1). Temperature yearly min and max values covering the period January 1924–July 2021 were collected by the meteorological stations of Enna and Caltanissetta. Due to non-operative periods of these meteorological stations, 0.69% of the temperature monthly data are missing in the time series analysed.

Monthly cumulative rainfall data were collected in the pluviometric stations located in Villarosa (time series: January 1924–December 2002, data gap 17.93 %), Villapriolo (January 1924–December 2015, data gap 13.04%), and Calascibetta[3] (January 1924–July 2021, data gap 2.13%) and aggregated per year. The mean precipitation values in the entire valley were calculated by aggregating the data coming from the three different pluviometric stations.

For both temperature and rainfall data, seasonal estimates were also calculated. Following insights from semi-structured interviews, seasons were defined as follows: autumn included September, October and November; winter included December, January and February; spring included March, April and May; and summer included June, July and August.

Linear regression with t-test was used to assess the linear trend (*sensu* Mudelsee, 2019) and estimate whether there has been a long-term systematic change in the temperature and rainfall mean values over the period considered (Rodrigo & Trigo, 2007; Barnes & Barnes, 2015). In our regression model, the dependent variables were temperature and rainfall values, and the independent variables were the time series. The slopes of the trends were calculated by least-squares linear fitting.

1 https://www.licci.eu/resources/.

2 Osservatorio delle Acque – Agenzia Regionale per i Rifiuti e le Acque (OA-ARRA), Regional Hydrological Observatory Database, https://www.regione.sicilia.it/istituzioni/regione/strutture-regionali/presidenza-regione/autorita-bacino-distretto-idrografico-sicilia/annali-idrologici (last accessed 30 January 2023).

3 Data from the pluviometric station of Calascibetta are available for the period January 2016 – July 2021. To cover the period January 1924 – December 2015, data from the closest pluviometric station (Enna) have been collected.

Results

Local observations of climate and broader environmental changes

The local climate and environmental changes observed in the Morello Valley are presented in Table 6.1. Each observed change is listed under the specific system in which it has been observed (atmospheric, physical and life system). Each change is also associated with the perceived drivers that, in some cases, respondents report in generic terms (such as simply "climate change"), while in other cases are very detailed and specifically correlated to particular changes in both the atmospheric system and broader socio-economic changes.

Observed changes in the atmospheric system

The interviewees observed changes in temperature, seasons, precipitation and storms. The first relevant changes mentioned are the increase in the frequency of hot and warm days, as well as changes in mean temperatures throughout the year. Respondents also reported the increase in abrupt warm events (e.g., heatwaves). These changes, thought to be climate-driven, have led to changes in the length of seasons, perceived as longer and warmer for some seasons (i.e., summer) and shorter and warmer for others (i.e., winter). Spring and fall are reported to be shorter than in the past and, in some cases, nearly disappearing. An indicator of this last climatic variation is spring frost that, according to the respondents, is nearly totally absent today but in the past happened quite often in April. This phenomenon is particularly well remembered as it represented a huge danger for the spike of the cereals, which were already in the growing phase. In the past, this happened quite often and it could represent a loss of even 50% of the harvest for a farmer. Today, cereal farmers are not afraid of frosts or cold, but instead, they are afraid of lack of rain and drought. Overall, nearly all the interviewees mentioned that the system is transitioning to only two main seasons: a long summer followed by a short indefinite season, and a mix of winter, fall and spring. Moreover, they observed a lack of gradual change between seasons. The main perceived drivers of all these climate changes are changes in the mean temperature in a given season.

Precipitation patterns are also perceived as changing. The interviewees have observed an increase in the intensity of summer rains, which they have classified as tropical rains due to their strong intensity and extreme short duration (only 20 or 30 minutes), but recurrent during the entire summer (usually in the afternoon, around 14:00–16:00). These short but intense rainfalls are locally called "water bombs" because they can cause dangerous floods and serious damages to crops, agricultural premises and farm equipment. Together with these sporadic extreme summer rainfalls, informants reported a reduction or even disappearance of precipitations during fall, winter and spring, accompanied by a lack of normal summer precipitations. Spring is also reported as being less rainy today than in the past. This is considered to have a very negative impact on the ecosystem, since nowadays spring does not provide water to sustain crops and wild plants. Moreover, in the past, people could predict rainfalls just by looking at the clouds in the sky, while today rain is totally unpredictable. Nonetheless, people interviewed do not automatically correlate changes in rainfall patterns with temperature increase, but they generally associate causes of such phenomena to broader climatic changes. Last but not least, even changes in snowfall are reported, with an overall perceived reduction in the frequency of snowy days in the winter and early spring.

Observed changes in the physical system

Locals observed changes in the physical system. As a component of the terrestrial physical system, soil is reported to be more impoverished and less fertile than in the past. Furthermore, all

Table 6.1 Local observations of climate and broader environmental changes in the Morello Valley

ATMOSPHERIC SYSTEM		Drivers
Changes in TEMPERATURE	*Increase of warm and hot days in spring* *Increase of abrupt climatic events: heatwaves* *Increase of the frequency of hot and warm days* *Changes in mean temperatures*	Climate change
Changes in PRECIPITATION	*Nearly total lack of summer rain* *Increase of abrupt heavy rain and/or strong hail* *Increase in the intensity of summer rain* *Reduction or disappearance of rainfall in fall, winter and spring* *Reduction of the predictability of rainfall* *Nearly disappearance of spring frost* *Reduction in the frequency of snowy days*	Climate change
Changes in SEASONS	*Changes in the length of the seasons: longer and warmer summers, shorter and warmer winters.* *Shorter (nearly disappearing) springs and autumns shorter.* *No gradual change between the seasons*	Climate change

PHYSICAL SYSTEMS		Drivers
Changes in the SOIL	*Soils are more impoverished and less fertile*	Modernisation in agriculture (intensive mechanisation, use of pesticides and chemical fertilisers, intensification of agricultural cropping patterns)
	Increase of soil erosion	-
	Reduction of soil moisture/humidity	Reduction or disappearance of rainfall in fall, winter and spring
Changes in WILD FIRES	*Increase of seasonal (summer) fires*	Land abandonment, Reduction/disappearance of grazing

LIFE SYSTEM		Drivers
Changes in WILD FAUNA	*Reduction or even disappearance of some wild species (e.g., birds, insects).*	Prolonged use of chemical pesticides and other substances in agriculture, so they also polluted the surrounding environment.
	Increase of the abundance of other wild species (e.g. foxes, wild boars)	-
Changes in WILD FLORA and FUNGI	*Reduction in the abundance of wild flora*	Disappearance of domesticated animals playing seed dispersal role + changes in agricultural practices (mechanisation, intensification, heavy tilling, widespread use of chemical herbicides).
	Reduction in the abundance of wild fungi *Changes in the maturation time of wild fungi* *Changes in the spatial distribution of wild fungi*	Reduction or disappearance of rainfall in fall, winter and spring Increase of abrupt strong hail
Changes in CULTIVATED PLANTS	*Changes in crop flowering time: early flowering* *Changes in crop fruiting time: early fruiting* *Changes in crop maturation time: early maturation time* *Reduction in the amount of fruits produced*	Increase of hot days in spring Increase of abrupt heatwaves, heavy rains and strong hail Reduction or disappearance of rainfall in fall, winter and spring
Changes in CROP WEEDS	*Increase in the frequency or occurrence of weed species considered invasive*	Intensive use of chemical herbicides and intensification of mechanical tilling
Changes in LIVESTOCK	*Increase in the upgrowth and spread of diseases in domesticated animals*	Environmental pollution due to prolonged use of chemical pesticides in agriculture

the informants reported a significant increase in soil erosion and a reduction in soil moisture and humidity. The main driver of these changes is considered to be agricultural modernisation, i.e., mechanisation, use of pesticides and chemical fertilisers, and the intensification of agricultural cropping patterns. Nonetheless, climate change is also seen as a driver of transformations in soil conditions, particularly due to the reduction of rainfall during fall, winter and spring, a change that hampers the maintenance of adequate humidity in the soil.

Wild summer fires are perceived as more frequent and with stronger devastating effects, aggravated by the accumulation of unmowed dry biomass, which is a direct result of land abandonment. Moreover, compared to the past, when domestic animals would eat grass in marginal areas keeping the landscape open, today there are fewer domestic animals around the roads or pasturing freely. The main drivers of increased seasonal fires are thus land abandonment and reduction of goat and sheep herds.

Observed changes in the life system

Within the life system, informants reported changes in both the terrestrial wild fauna, flora and fungi, as well as in cultivated plants and livestock.

When it comes to changes in the abundance of local wild terrestrial animals, informants observed the reduction, or even disappearance, of some species (e.g., wild rabbits, some local species of birds, insects) and the increase of other species (e.g., wild boars, foxes). While people interviewed do not provide any explanation for the increase of some of these wild species, the main driver for the reduction and disappearance of other wild fauna is thought to be the prolonged spread of chemical substances in agriculture during the last decades, which have slowly destroyed their habitats. Wild flora and fungi are also perceived as changing. Wild flora is less abundant now, mainly due to the disappearance of domestic animals along the roads. Domestic animals were used in the past as main means of transport and moved around for pasture or transhumance, thus playing a very important role in seed dispersal. The reduction of wild plants is also associated with changes in agricultural practices (e.g., mechanisation, intensification, heavy tilling, use of herbicides). In the case of wild edible fungi, the interviewees noticed changes in their abundance, spatial distribution and maturation time, correlated to climatic drivers, particularly observed changes in rainfall seasonal patterns and soil humidity.

Cultivated plants are the component of the life system perceived as most affected by changes, both climate and non-climate driven. Respondents reported changes in crop growing patterns, namely early flowering, fruiting, and maturation time. For some years, up to three weeks of anticipation in flowering time have been observed and, generally, early flowering has been observed more often than late flowering. The anticipation of flowering leads to changes in crop fruiting and maturation, with an overall observed tendency towards a reduction in the number of fruits produced. This is the case of olives, for instance, whose successful growth depends on a smooth flowering pattern in late spring. Changes in crop growing patterns have influenced the agricultural calendar, with consequent changes in sowing, planting, and harvesting time. Informants perceive these changes as clearly driven by climatic changes. The increase of warm and hot days in spring cause early flowering and even fruiting, while the increase of abrupt climatic events such as heat waves, strong hail, and heavy rain, perceived as unseasonal, destroys a great share of the flowers or early fruits, reducing the final production of orchards trees. Moreover, interviewees mentioned changes in crop growing patterns as also driven by changes in the number of rainy days, and particularly by the nearly total lack of summer rain.

Within cultivated plants and cropping systems, an increase in the frequency or occurrence of certain weed species considered as invasive has also been observed. This is the case of *Orobanche alsatica* Kirschl. (a weed invasive of *Vicia Faba* L.) and *Centaurea napifolia* L. (a weed of cereal crops). Some of the farmers believe *C. napifolia* comes from California, it is extremely difficult to eradicate (nearly impossible) and they have observed it growing in association with *Hedysarum coronarium* L. According to the informants, the increase of crop weeds is associated with the use of herbicides and mechanisations, which paradoxically have caused some invasive weed species to genetically modify and adapt to the new conditions.

Interviewees have also observed changes in livestock, namely an increase in the upgrowth and spread of diseases for certain domesticated animals. For example, they report the case of young rabbits that cannot grow anymore, becoming sick when separated from the milk of the mother, and a disease in the sheep that makes their milk become more watery. The informants attributed changes to the prolonged use of chemicals and pesticides in agriculture.

The weather as captured by meteorological stations

Results from the analysis of the data collected from local weather stations indicate that there is no statistically significant trend showing an increase or a decrease in the average yearly temperatures. However, for the yearly precipitation, it is possible to see a statistically significant decreasing trend, mainly driven by the decrease in winter precipitation.

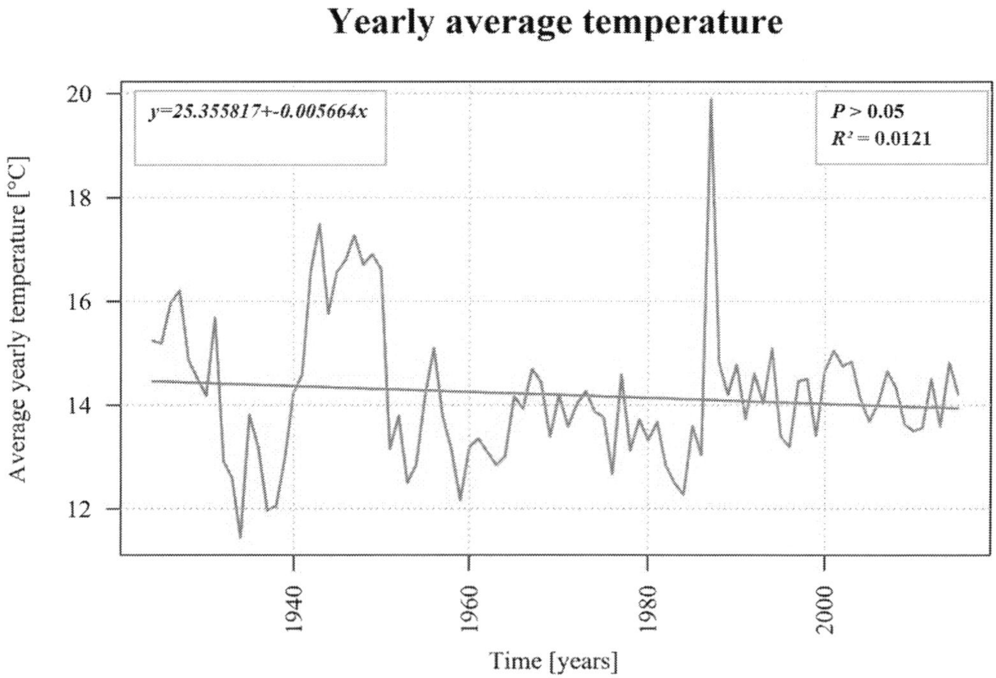

Figure 6.2 Linear trend of mean yearly temperatures in the study area for the period 1924–2021 (Data: Regional Hydrological Observatory Database, Osservatorio delle Acque – Agenzia Regionale per i Rifiuti e le Acque (OA-ARRA), Regione Sicilia).

Temperatures

The analysis of the historical temperature time series gives no statistically significant increasing or decreasing trend ($p > 0.05$). Nonetheless, as it is possible to see from the time series plot in Figure 6.2, temperatures show large fluctuations over the years, with decreasing and increasing peaks.

Results from the analysis of the yearly seasonal temperatures indicate that there is no statistically significant trend showing an increase or a decrease in the average temperatures per season ($p > 0.05$). The time series plot shows significant fluctuations over the years, quite similar across the seasons, with the exception of a steady decrease for the autumn and spring plots in some periods (Figure 6.3). In the time series analysed, after summer, autumn shows to be the warmest season, followed by spring and winter (coldest).

Precipitations

In contrast to temperature, the regression analysis for precipitation shows a statistically significant decreasing trend ($p = 0.02$), with the graphical analysis suggesting large fluctuations over time (Figure 6.4).

The decreasing trend in yearly precipitation is mainly given by the decrease in winter precipitation ($p = 0.006$). From the analysis of the yearly seasonal precipitations, winter is the only season showing a statistically significant trend, while for the other seasons, data do not allow us to assess a significant increase or decrease in the average rainfall pattern. The time series plot shows larger fluctuations over the years and across the seasons, with the highest peaks in autumn-winter and the lowest fluctuations during summer (Figure 6.5).

Discussion

When comparing results from local observations with the evidence from weather stations, what first emerges is the potential of local knowledge to enhance our understanding of complex social-ecological issues, such as the impacts of climate and environmental change on the biophysical context of local communities. Local people recognise many indicators of change that are deeply interwoven and manifested in different elements of their environment, and they emphasise how the observed changes are driven by both climatic and non-climatic factors. Examples include variations in plant phenological behaviour, which locally is attributed to changes in temperatures, or the disappearance of some wild fauna species, attributed to the intensive use of chemicals in farmlands. In other words, for respondents climate and environmental changes are not only caused by external factors (i.e., weather) happening independently from human behaviour, but they have strong anthropogenic causes. Moreover, interviewees stressed how the perception of "changing weather or climate change" can be seen as culturally-constructed and influenced by lifestyles. This is exemplified by the comment of four women, who mentioned that we believe the climate has changed because we have changed our lifestyle. They mentioned, for instance, that today it is normal to go bathing in the sea in May, while this was totally unthinkable in the past. However, they argue that such change did not happen because May is today warmer, but simply because going to the beach is a status symbol, and a way to show off a wealthy socio-economic condition. Another interesting example given by another woman refers to the perception of cold during winters. According to her, today winter is perceived as warmer because houses are now equipped with central heating systems and we have better quality clothes than in the past. In a way, these personal observations, although at odds with the general perception, dovetail with meteorological data showing that there

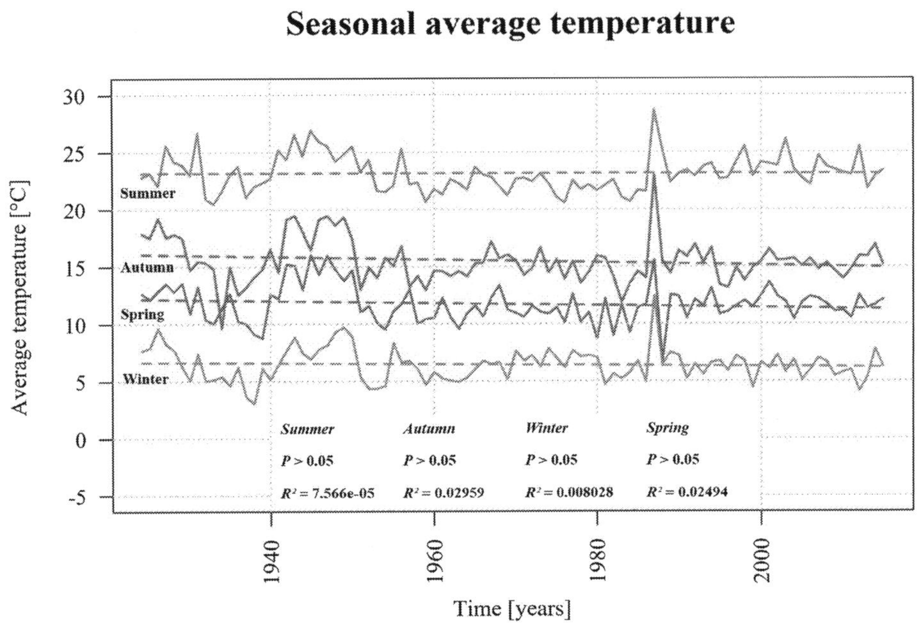

Figure 6.3 Linear trend of seasonal mean yearly temperatures in the study area for the period 1924–2021 (Data: Regional Hydrological Observatory Database, Osservatorio delle Acque – Agenzia Regionale per i Rifiuti e le Acque (OA-ARRA), Regione Sicilia).

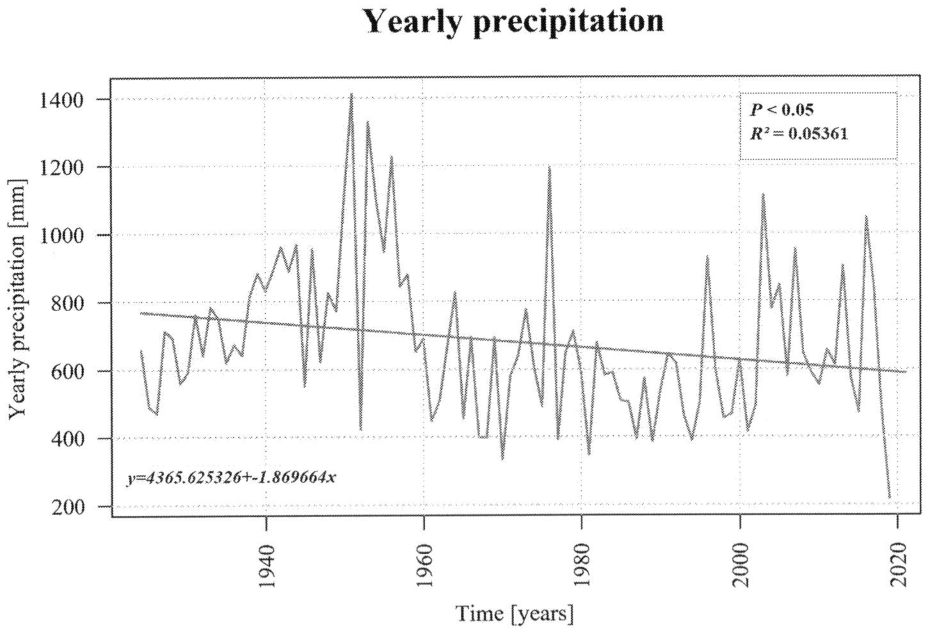

Figure 6.4 Linear trend of mean yearly precipitations in the study area for the period 1924–2021 (Data: Regional Hydrological Observatory Database, Osservatorio delle Acque – Agenzia Regionale per i Rifiuti e le Acque (OA-ARRA), Regione Sicilia).

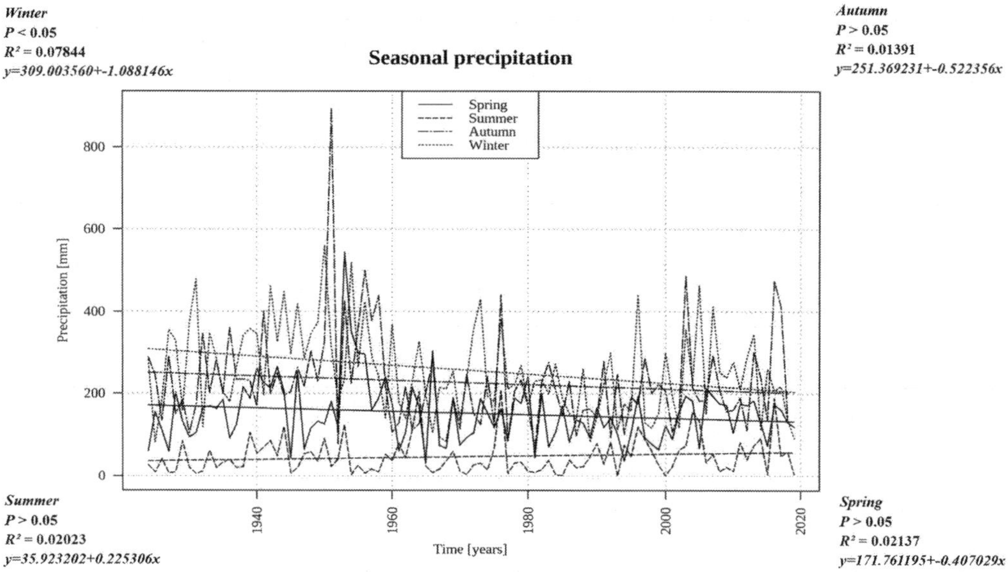

Figure 6.5 Linear trend of mean yearly seasonal precipitations in the study area for the period 1924–2021 (Data: Regional Hydrological Observatory Database, Osservatorio delle Acque – Agenzia Regionale per i Rifiuti e le Acque (OA-ARRA), Regione Sicilia).

is not such a relevant significant increase or decrease in the local climate, but rather, more abrupt but punctual events that might shape perceptions. These accounts provide an invaluable source of information on ongoing ecological change dynamics that cannot be caught by instrumental measurements. The weather station can report an atmospheric event as, for instance, the increase in the intensity of precipitation, but its perception as a "water bomb" with devastating effects on crops can only be captured by local knowledge. People provide more nuanced explanations of changes which reinforce the idea that climate change perceptions are really dependent on the context, including geographical, historical, cultural and epistemological/methodological aspects.

But, what happens if we shift historical baselines and frame these results within a longer temporal perspective? Thanks to palaeoecological and archaeobotanical evidence, we can reconstruct the long-term climatic and environmental history in this part of the island. The Early Holocene period (ca. 9700 BC.) was a relatively humid period, with marked presence of deciduous forests (Sadori et al., 2008, 2013). Winter drought is registered between ca. 5550 and 4550 BC and reduced rainfall around ca. 4050 BC followed by a progressive phase of increasing dryness, which allowed *Olea* and *Pistacia* to expand (Zanchetta et al., 2007; Carroll et al., 2012; Peyron et al., 2013). In response to these climatic changes, both vegetation patterns and human activities underwent abrupt modifications, with intensification of soil exploitation. Anthropogenic impact is also evident from the extinction of numerous large mammals due also to the introduction of domestic species and alteration of vegetation cover (Incarbona et al., 2010). A cooling phase is attested between 2600 and 2000 years BC in connection to a precipitation maximum (cfr. Sadori et al., 2015) reflected in the growth in human settlements with agricultural and specialised pastoralism intensification (Giannitrapani et al., 2014). The increase in sclerophyllous Mediterranean taxa around 1250 BC and the spread of anthropogenic taxa around 650 BC could be an indicator of cropping and burning at a more intensive scale (Sadori et al., 2013). A successive humid period (ca. 450–750 AD) attests to the development of agriculture on a new and larger scale (Sadori et al., 2016). A sudden phase of aridity (ca. 750 AD) is associated with a decrease in synanthropic taxa, a recovery of arboreal vegetation and socio-economic decline (Sadori et al., 2016). As these changes occurred prior to the Arab invasion of Sicily in 827 AD, they could be correlated to the collapse of the Byzantine society and the Muslim conquest of the island. A second period of slightly less warmer climate was during what is known as the *Medieval Climate Anomaly* (ca.1100–1350 AD), characterised by marked temperature and hydroclimatic variability, driven by Atlantic Ocean cycles (Lüning et al., 2019). Finally, the last climatic anomaly before the current era, the *Little Ice Age* (1550–1850 AD), was characterised by cooling and drought (Incarbona et al., 2010).

As we can see, overall, the long-term climatic and environmental history of this area of Sicily shows the occurrence of repeated climatic anomalies throughout the Holocene, correlated with complex patterns of environmental change. The look at the deep past highlights that environmental change has been the norm rather than the exception and that drivers of such changes include climatic fluctuations, human agency and broader ecological dynamics. Furthermore, temporal long-term analysis reveals that the most significant changes in the local history happened during periods of climate-induced environmental change, when both natural and human ecosystems collapsed or developed strategies for their adaptation and survival. In the case of key endemic species (i.e., the olive), human stewardship in the form of traditional ecological memory played a fundamental role in preserving and maintaining this biocultural heritage (Ferrara et al., 2022), dating back to the anthropogenic transformation of natural ecosystems into agro-ecosystems. The combination of different temporal and spatial scales of analysis, as the inclusion of different types of evidence giving voice to local reports and ecological memory can help better contextualise environmental change processes, as understand both macro and micro-climate dynamics.

Conclusion

Many socio-ecological processes unfold within centennial to millennial scales, but what we observe is “the here and now”. By shifting baselines and combining local observations of climate and broader environmental changes with weather data at the micro-scale and a millennial reconstruction of climate and environmental history levels, we show how the perceptions of climate change may be interpreted differently according to scale and context. Longitudinal trends and larger geographical scales give a larger perspective of change. To better understand the correlation between climate dynamics, effects on ecosystems and human behaviour, reductionist oversimplifications about short-term individual causes of change should be avoided. Multi-causal explanations that incorporate long-term socially accumulated ecological knowledge with scientific evidence on different timescales should be favoured instead, taking into account that the influence of climate on society strongly depends on a combination of ecological dynamics, relational and societal conditions, place geography and historical circumstances (Blok, 2010; Marignani et al., 2017; Ljungqvist et al., 2021).

The way we decide to construct, typify and represent (climate) change influences our responses to current and future challenges (Hulme et al., 2009; Alexandra, 2021) and may be our chance to write the next chapter in history.

Acknowledgements

The authors are very grateful to the local communities of the towns of Villarosa, Villapriolo and Calascibetta for being willing to participate in this research and sharing their ecological memory and traditional knowledge. A special thanks to the reviewers for their insightful comments and suggestions to improve the manuscript, and to the LICCI team for collaborations in the framework of the "LICCI. Local Indicators of Climate Change Impacts" project (FP7-771056-LICCI).

Funding information

Research presented in this chapter has received funding by Vetenskapsrådet, under the research project "Oliven och Siciliens bio-kulturella arv", grant number 2020-02625, and by the Swedish Institute of Rome (Lerici grant 2021/2022).

Conflict of interest

The authors declare that they have no conflict of interest.

Compliance with ethical standards

Data collection was realised within the framework of the ERC project "LICCI. Local Indicator of Climate Change Impacts" (FP7-771056-LICCI) and conducted according to Uppsala University and its Department of Archaeology and Ancient History compliance procedures to the *CODEX Rules and Guidelines for Research* (https://www.codex.uu.se/), the Swedish Research Council's (VR) Guidelines for Research Ethics (https://www.vr.se/uppdrag/etik/etik-i-forskningen.html), the Swedish Research Council's Guidelines related to the law "Knowing the ethics of research involving humans" (https://www.riksdagen.se/sv/dokument-och-lagar/dokument/svensk-forfattningssamling/forordning-20071068-med-ins_sfs-2007-1068/). Informed consent was obtained from all the participants in this study.

Data availability statement

The audio recordings of the semi-structured interviews and Focus Groups used in this study are stored in DiVA (Digitala Vetenskapliga Arkivet), and accessible via the following link: (http://uu.diva-portal.org/smash/record.jsf?pid=diva2:1690733). These data are not publicly available to protect the interviewees′ interests, trust and safety.

References

Alexandra, J. (2021). How Do the Cultural Dimensions of Climate Shape Our Understanding of Climate Change? *Climate*, 9, 63. https://doi.org/10.3390/cli9040063

Aymard, M., & Giarrizzo, G. (1987). *La Sicilia*. Giulio Einaudi Editore, Torino.

Baiamonte, G., Domina, G., Raimondo, F.M. et al. (2015). Agricultural Landscapes and Biodiversity Conservation: A Case Study in Sicily (Italy). *Biodiversity and Conservation*, 24, 3201–3216.

Barnes, E.A., & Barnes, R.J. (2015). Estimating Linear Trends: Simple Linear Regression versus Epoch Differences. *Journal of Climate*, 28(24), 9969–9976. https://doi.org/10.1175/JCLI-D-15-0032.1

Beller, E.E., McClenachan, L., Zavaleta, E.S., & Larsen, L.G. (2020). Past Forward: Recommendations from Historical Ecology for Ecosystem Management. *Global Ecology and Conservation*, 21, e00836. https://doi-org/10.1016/j.gecco.2019.e00836

Blok, A. (2010). Topologies of Climate Change: Actor-Network Theory, Relational-Scalar Analytics, and Carbon-Market Overflows. *Environment and Planning D: Society and Space*, 28(5), 896–912. http://dx.doi.org/10.1068%2Fd0309

Brandolini, P., Pepe, G., Capolongo, D., Cappadonia, C., Cevasco, A., Conoscenti, C., Marsico, A., Vergari, F., & Del Monte, M. (2018). Hillslope Degradation in Representative Italian Areas: Just Soil Erosion

Risk or Opportunity for Development? *Land Degradation and Development*, 29, 3050–3068. https://doi.org/10.1002/ldr.2999

Carroll, F.A., Hunt, C.O., Schembri, P.J., & Bonanno, A. (2012). Holocene Climate Change, Vegetation History and Human Impact in the Central Mediterranean: Evidence from the Maltese Islands. *Quaternary Science Reviews*, 52, 24–40. http://dx.doi.org/10.1016/j.quascirev.2012.07.010

Crumley, C.L. (2021). Historical Ecology: A Robust Bridge between Archaeology and Ecology. *Sustainability*, 13, 8210. https://doi.org/10.3390/su13158210

Crumley, C.L. (1995). Heterarchy and the Analysis of Complex Societies. *Archeological Papers of the American Anthropological Association*, 6(1), 1–5. https://doi-org. /10.1525/ap3a.1995.6.1.1

Davies, T. (1983). Changes in the Structure of the Wheat Trade in Seventeenth-Century Sicily and the Building of New Villages. *Journal of European Economic History*, 12, 371–405.

Ferrara, V., Wästfelt, A., & Ekblom, A. (2022). From Landscape as Heritage to Biocultural Heritage in a Landscape. The Ecological and Cultural Legacy of Millennial Land Use Practices for Future Natures. In G. Pettenati (Ed.), *Landscape as Heritage: Critical Perspectives* (pp. 80–90). Routledge, London.

Ferrara, V., Ekblom, A., & Wästfelt, A. (2020). Biocultural Heritage in Sicilian Olive Groves: The Importance of Heterogeneous Landscapes over the Long Term. In M. I. Goldstein & D. A. DellaSala (Eds.), *Encyclopedia of the World's Biomes* (Vol. 5, pp. 135–145). Elsevier, Amsterdam.

Fondazione Migrantes (2022). *Rapporto Italiani nel Mondo 2022. Parte Quarta. Allegati Socio-Statistici*. TAU Editrice, Todi. https://www.migrantes.it/rapporto-italiani-nel-mondo-migrantes-mobilita-italiana-convivere-e-resistere-nellepoca-delle-emergenze-globali/

Forestieri, A., Arnone, E., Blenkinsop, S., Candela, A., Fowler, H., & Noto, L.V. (2018). The Impact of Climate Change on Extreme Precipitation in Sicily, Italy. *Hydrological Processes*, 32, 332–348. https://doi.org/10.1002/hyp.11421

Giannitrapani, E. (2015, May 14–15). *Prehistoric Landscape Dynamics and Social Complexity in the Erei Uplands (Central Sicily) between the 5th and the 3rd Millennium cal. BC* [Paper presentation]. Mediterranean Geoarchaeology Workshop, Cagliari, Italy.

Giannitrapani, E., Ianni, F., Chilardi, S., & Anguilano, L. (2014). Case Bastione: A Prehistoric Settlement in the Erei Uplands (Central Sicily). *ORIGINI*, XXXVI, 181–211.

Hulme, M., Dessai, S., Lorenzoni, I., & Nelson, D.R. (2009). Unstable Climates: Exploring the Statistical and Social Constructions of 'Normal' Climate. *Geoforum*, 40(2), 197–206. https://doi.org/10.1016/j.geoforum.2008.09.010

Incarbona, A., Zarcone, G., Agate, M., Bonomo, S., Di Stefano, E., Masini, F., Russo, F., & Sineo, L. (2010). A Multidisciplinary Approach to Reveal the Sicily Climate and Environment over the Last 20 000 years. *Central European Journal of Geosciences*, 2, 71–82. https://doi-org/10.2478/v10085-010-0005-8

Liuzzo, L., Bono, E., Sammartano, V., & Freni, G. (2016). Analysis of Spatial and Temporal Rainfall Trends in Sicily during the 1921–2012 Period. *Theoretical and Applied Climatology*, 113–129. https://doi-org/10.1007/s00704-015-1561-4

Ljungqvist, F.C., Seim, A., & Huhtamaa, H. (2021). Climate and Society in European History. *WIREs Climate Change*, 12, e691. http://dx.doi.org/10.1002%2Fwcc.691

Lüning, S., Schulte, L., Garcés-Pastor, S., Danladi, I.B., & Gałka, M. (2019). The Medieval Climate Anomaly in the Mediterranean Region. *Paleoceanography and Paleoclimatology*, 34. http://dx.doi.org/10.1029/2019PA003734

Marignani, M., Chiarucci, A., Sadori, L., & Mercuri, A.M. (2017). Natural and Human Impact in Mediterranean Landscapes: An Intriguing Puzzle or Only a Question of Time? *Plant Biosystems*, 151(5), 900–905. https://doi.org/10.1080/11263504.2016.1244121

Mudelsee, M. (2019). Trend Analysis of Climate Time Series: A Review of Methods. *Earth-Science Reviews*, 190, 310–322. https://doi.org/10.1016/j.earscirev.2018.12.005

Peyron, O., Magny, M., Goring, S., Joannin, S., de Beaulieu, J.L., Brugiapaglia, E., Sadori, L., Garfi, G., Kouli, K., Ioakim, C., & Combourieu-Nebout, N. (2013). Contrasting Patterns of Climatic Changes during the Holocene across the Italian Peninsula Reconstructed from Pollen Data. *Climate of the Past*, 9, 1233–1252. https://doi.org/10.5194/cp-9-1233-2013

Pluciennik, M., Mientjes, A., & Giannitrapani, E. (2004) Archaeologies of Aspiration: Historical Archaeology in Rural Central Sicily. *International Journal of Historical Archaeology*, 8(1), 27–65.

Rapporto Ambientale – VAS PSR Sicilia 2014/2020, https://www.psrsicilia.it/wp-content/uploads/2021/11/Allegati_Progarmma_ver10_1.zip

Reyes-García, V., Álvarez-Fernandez, S., Benyei, P., García-del-Amo, D., Junqueira, A.B., Labeyrie, V., Li, X., Porcher, V., Porcuna-Ferrer, A., Schlingmann, A., & Soleymani. R. (2023). Local Indicators of Climate Change Impacts Described by Indigenous Peoples and Local Communities: Study Protocol. *PLoS One*, 18(1), e0279847. https://doi.org/10.1371/journal.pone.0279847

Rivera-Collazo, I. (2022). Environment, Climate and People: Exploring Human Responses to Climate Change. *Journal of Anthropological Archaeology*, 68. https://doi.org/10.1016/j.jaa.2022.101460

Rodrigo, F.S., & Trigo, R.M. (2007). Trends in Daily Rainfall in the Iberian Peninsula from 1951 to 2002. *International Journal of Climatology*, 27, 513–529. http://dx.doi.org/10.1002%2Fjoc.1409

Rühl, J., Caruso, T., Giucastro, M., & La Mantia, T. (2011). Olive Agroforestry Systems in Sicily: Cultivated Typologies and Secondary Succession Processes after Abandonment. *Plant Biosystems*, 145(1), 120–130. https://doi.org/10.1080/11263504.2010.54038

Sadori, L., Giraudi, C., Masi, A., Magny, M., Ortu, E., Zanchetta, G., & Izdebski, A. (2016). Climate, Environment and Society in Southern Italy during the Last 2000 years. A Review of the Environmental, Historical and Archaeological Evidence. *Quaternary Science Reviews*, 136, 173–188. http://dx.doi.org/10.1016/j.quascirev.2015.09.020

Sadori, L., Masi, A., & Ricotta, C. (2015). Climate-Driven Past Fires in Central Sicily. *Plant Biosystems*, 149(1), 166–173. https://doi.org/10.1080/11263504.2014.992996

Sadori, L., Ortu, E., Peyron, O., Zanchetta, G., Vanniere, B., Desmet, M., & Magny, M. (2013). The Last 7 Millennia of Vegetation and Climate Changes at Lago di Pergusa (Central Sicily, Italy). *Climate of the Past*, 9, 1969–1984. https://doi.org/10.5194/cp-9-1969-2013

Sadori, L., Zanchetta, G., & Giardini, M. (2008). Last Glacial to Holocene Palaeoenvironmental Evolution at Lago di Pergusa (Sicily, Southern Italy) as Inferred by Pollen, Microcharcoal, and Stable Isotopes. *Quaternary International*, 181, 4–14. https://doi-org/10.1016/j.quaint.2007.02.024

Safford, H.D. (2008). Land Management in the Anthropocene: Is History Still Relevant? *Eos*, 89, 37.

Swetnam, T., Craig, D.A., & Betancourt, J.L. (1999). Applied Historical Ecology: Using the Past to Manage for the Future. *Ecological Applications*, 9(4), 1189–1206.

Vellend, M., Brown, C., Kharouba, H.M., McCune, J.L., & Myers-Smith, I.H. (2013). Historical Ecology: Using Unconventional Data Sources to Test for Effects of Global Environmental Change. *American Journal of Botany*, 100(7), 1294–1305. https://doi-org/10.3732/ajb.1200503

Veneziano, V. (2011) *Studio ecologico-vegetazionale e cambiamenti ambientali del bacino del fiume Morello (Sicilia centrale) con restituzione cartografica dei tipi vegetazionali* [Doctoral dissertation, Catania University]. Catania University Theses and Dissertations Archive. http://dspace.unict.it/handle/10761/1070

Viola, F., Liuzzo, L., Noto, L.V., Lo Conti, F., & La Loggia, G. (2014). Spatial Distribution of Temperature Trends in Sicily. *International Journal of Climatology*, 34, 1–17. http://dx.doi.org/10.1002%2Fjoc.3657

Whitaker, J.A., Armstrong, C.G., & Odomme, G. (Eds.) (2023). *Climatic and Ecological Change in the Americas: A Perspective from Historical Ecology*. Routledge.

Zanchetta, G., Borghini, A., Fallick, A.E., Bonadonna, F.P., & Leone, G. (2007). Late Quaternary Paleohydrology of Lake Pergusa (Sicily, Southern Italy) as Inferred by Stable Isotopes of Lacustrine Carbonates. *Journal of Paleolimnology*, 38, 227–239. https://doi-org/10.1007/s10933-006-9070-1

7

SMALLHOLDERS PERCEPTIONS OF CLIMATE CHANGE IN A POST-SOCIALIST ALBANIA

Simona Lippi and Massimiliano Sanfilippo

Introduction

In the last decades, Albania has experienced an increase in the frequency of extreme weather events such as floods, heat waves, and drought (Demiraj-Bruçi et al., 2016). Mean annual temperatures have increased since 1960 and the projections indicate that summer temperatures will continue to rise considerably in the following decades. According to the projections provided by the Coupled Model Intercomparison Project Phase 5 (CMIP5) ensemble, mean annual temperatures in 2080–2099 are expected to increase by 2.7°C–6.9°C under the high emission scenario (RCP8.5). Low (RCP2.6) and medium (RCP4.5) emission scenarios also predict an increase in mean annual temperatures for the region. Summer temperatures are expected to rise by 2.4°C–3.1°C during June–August, the highest increase predicted for South Eastern Europe (World Bank, 2021a). Precipitation patterns are expected to change, with a tendency to decreasing precipitation and a contraction of Albania's snowpacks due to higher temperatures (Demiraj-Bruçi et al., 2016). These changes will negatively affect crop yields and livestock production, jeopardising food security and increasing the vulnerability of local communities (Sutton et al., 2013; World Bank, 2021a).

In Albania, 36% of the population is employed in agriculture (World Bank, 2021b) and rural communities depend primarily on agriculture for food security and income. The recent variations in weather patterns and trends represent a significant stressor for their livelihood strategies.

Farmers and livestock keepers are often aware of the impacts of climate change on their livelihood strategies. However, this awareness does not always result in the adoption of effective adaptation strategies (Fierros-Gonzalez and Lopez-Feldman, 2021). A better understanding of farmers' perceptions of climate change impacts and their adaptation strategies is a fundamental step to providing inputs to policymakers and developing tailored policy responses to support farm-level adaptation to climate change (Eitzinger et al., 2018; Zhang et al., 2020).

Small farm size, high rate of deforestation, lack of infrastructure, and poor access to the market exacerbate community vulnerability to climate change impacts (Demiraj-Bruçi et al., 2016). According to the Census of Agricultural Holdings carried out in 2012 (INSTAT, 2012), in Albania more than 98% of agriculture holdings are family farms and up to 86% are small farms with less than 2 ha of land (FAO, 2020).

In 1945, the Albanian Communist Party started a land collectivisation process, which was completed in the 1960s. This process established agricultural production cooperatives and state farms (Cungu and Swinnen, 1999). The nationalisation of lands and collectivisation of agriculture

DOI: 10.4324/9781003356837-10

brought, in a few years, to the creation of Agricultural Production Co-operatives (APCs) and State Farms (SFs) to achieve national self-sufficiency in agriculture (Papajorgji and Alikay, 2016). Until the 1960s, the cultivation of small private plots was allowed, but selling the agricultural products grown on those plots was illegal. The final stage of collectivisation began in 1965; cooperatives and state farms assumed complete control, and family plots were abolished (King and Vullnetari, 2016; Swinnen, 2018).

In 1991, after the collapse of the communist regime, an important land reform was implemented in Albania, with the adoption of Law 7501 "Concerning the Land" dated 19.07.1991. According to this law, collectivised land was redistributed to former cooperative members. To ensure an equitable land redistribution, the farm size allocated to each household was proportional to the number of household members. According to article 6 of the same law, the land was also given to families who were not members of the cooperatives and were not working on state-owned farms and that, as such, had no previous experience in farming.

Land privatisation resulted in a high fragmentation of agricultural holdings, with families owning several non-contiguous plots (Sallaku et al., 2011). The fragmentation resulting from splitting formerly collective land into small private farms is considered one of the main challenges of the Albanian agricultural sector (FAO, 2020).

Perceptions of climate change impacts are influenced by a complex interaction of different factors such as socio-cultural background, knowledge, and individual life experiences (Van der Linden, 2015). Those factors can differ considerably among people living in the same area (Nguyen et al., 2016; Whitmarsh and Capstick, 2018) or even within the same community.

To adapt to climate change, rural dwellers need to gain an accurate understanding of the threats that a changing climate brings to their livelihoods. Furthermore, decision makers need to ensure that local perceptions and cultural factors are taken into due consideration in climate change adaptation policies (Adger et al., 2013; Makuvaro et al., 2018a,b).

Our fieldwork, carried out in the context of the LICCI project (Reyes-García et al., 2023), aimed at investigating (1) how Albanian rural communities perceive the effects of climate change, (2) how they are affected, and (3) how they are adapting and coping with them. Given that climate change is only one of the drivers of change in the study area, our analysis is framed by the understanding of the contextual socio-economic factors that influence Albanian rural communities' livelihood strategies.

Materials and methods

Study area

According to Albanian agroecological zonation (Sutton et al., 2013), our study area is located in the transition between the Southern Highlands and the Northern and Central Mountains zone. The main climatic variables of the municipality are summarised in Table 7.1.

Fieldwork was carried out in three rural areas of Përmet municipality, in the Prefecture of Gjirokastër, in South Albania. The municipality is located in the upper section of the Vjosa river and is characterised by a hilly and mountainous landscape and by the presence of the Bredhi i Hotovës-Dangëlli National Park (WDPA, 2022). The resident population in the municipality was 5,945 inhabitants in 2011 (INSTAT, 2013), when the latest census was conducted.

Cropland covers about 12% (10,844.33 ha) of the total municipal area (Phalke et al., 2017). The production of vegetables in 2020 was equal to 2,538 t, with no production in greenhouses. With only 52 ha destined for fruit production in the municipality and low yields (9.6 kg/tree in 2020)

Table 7.1 Main climatic variables in Përmet municipality (Trendline values)

Year	*1979*	*2019*
Mean annual temperature (°C)	11.3	12.47
Annual average of maximum temperature (°C)	21.64	23.03
Annual average of minimum temperature (°C)	0.09	1.23
Mean annual precipitation (mm)	1,009.97	1,238.58

Source of data: ERA5 – Latest climate reanalysis produced by European Centre for Medium-Range Weather Forecasts/Copernicus Climate Change Service.

Figure 7.1 Kosine Landscape. Photograph by the author.

compared to a national average (23 kg/tree), fruticulture is not an important activity. A similar situation characterises the oliviculture sector: only 35 ha of olive trees exist in the municipality, and with a productivity of 8 kg/trees in 2020, Përmet is among the municipalities with the lowest yields in the country. Relatively more important is the production of grapes, with 261 hectares destined for vineyards.

Goats are the most common type of livestock, with a total of 20,000 heads, 15,000 of which are milk goats. Around 14,000 sheep (Figure 7.1) and 1,000 cows are also kept in the municipality (INSTAT, 2021). Data on municipal beekeeping are unavailable. Nevertheless, in the Prefecture of Gjirokastër, there are around 6,000 beehives and the honey production in 2020 was equal to 214 tons.

We selected three areas with different environmental characteristics in the northern part of the Bredhi i Hotovës-Dangëlli National Park. Two villages, Kosinë (265 m a.s.l.) and Bodar (390 m a.s.l.), are situated at the border of the National Park and count a total of 300 households. The third area is made of a cluster of small settlements (Raban, 720 m a.s.l., Odriçan, 765 m a.s.l. and Pagri, 495 m a.s.l.) totalling about 30 families located entirely within the protected area (Figure 7.2).

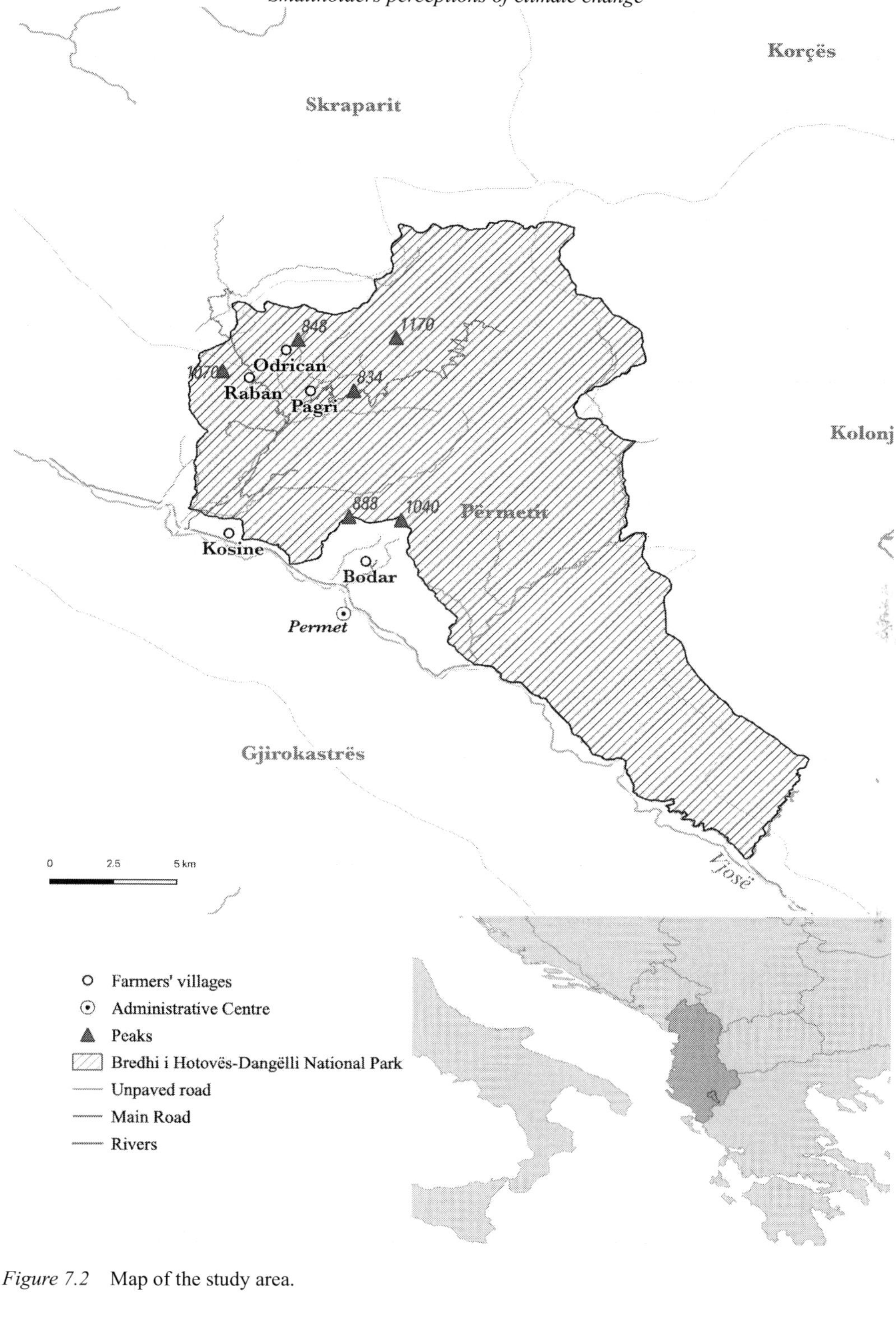

Figure 7.2 Map of the study area.

Table 7.2 Interviewees' livelihood strategies

Livelihood activities	*Description*	*Main activities*
Horticulture	Rainfed agriculture in small plots up to 2 ha. Each family own poultry and one or two heads.	Main crops: Grapes, tomatoes, beans (several varieties), potatoes, peppers, aubergines, onions, tobacco, maize
Livestock farming	Extensive dairy sheep farming	Few livestock keepers own cows. Households also grow horticultural crops in small garden plots and produce fodder.
Beekeeping	Extensive beekeeping inside the national park	Transhumance to higher altitudes is used to ensure three harvests during the year

Those settlements are connected with Kosinë through an unpaved road that crosses the National Park for 13 km. The entire study area is characterised by high biodiversity, making it an important site for the conservation of wild fauna and flora (Keci and Krog, 2014). Family farming and livestock keeping are the area's main economic activities (Table 7.2). A communal water source exists in Kosinë, and community members benefit from a regulated access to the water source for irrigation. On the contrary, households living in Raban, Odriçan, and Pagri rely entirely on rain for their farming activities.

The most common crops in the area are wheat, maize, fodder crops, grapes, potatoes, tomatoes, beans, and tree crops such as olives, walnuts, and figs. The National Park regulation framework shapes the governance of forests and pastures within the protected area. Honey is produced at a considerable scale in the National Park and marketed mainly locally (Keci and Krog, 2014).

Data collection

We conducted face-to-face interviews with 15 farmers and livestock keepers to explore their perceptions of climate change impacts, the repercussions on farming practices, and their responses to these impacts. The interviews were performed in July 2021.

In the first part of the interviews, we explored local livelihoods, seasonal activities, and the natural resources of the surrounding environments. The second part of the interviews consisted of a semi-structured questionnaire to explore farmers' perceptions of changes in weather patterns and the environment. The questionnaire followed the LICCI protocol (Reyes-García et al., 2023). In particular, we focused on the Local indicators of climate change impacts (LICCIs) classification tree to collect farmers' observations on atmospheric, physical, and life system changes .

We used a timeline representation to help interviewees to anchor weather events and key elements to significant historical events and facts in the last 30 years. Timeline is a tool that allows communities to place in chronological order the most important events they experienced in a given period of time (Miara et al., 2022).

Open-ended questions were used to collect information on the impacts of climate change on farming activities and on farmers' strategies to cope with such change. Snowball sampling was used to select the interviewees.

Overall, we interviewed four sheep farmers, nine horticulture farmers that also kept a small number of chickens (one or two), and two beekeepers. The interviewees were eight men and seven women with an average age of 60. Three of the seven interviewed women were widows whowere taking care of the garden and livestock alone. All interviewees were workers of the Agricultural Production Co-operatives (APCs) or State Farms (SFs) during the communist period.

Interviews were conducted with the help of a translator who is currently working as a ranger in the Bredhi i Hotovës-Dangëlli National Park.

Results

The results of the investigation suggest that all informants perceive a change in some key atmospheric variables and that they consider it as a risk to their farming activities.

Perception on climate change

About 65% of interviewees perceived an increase in mean temperatures and heatwave frequency during the summer and an intensification of cold days in winter during the last 15 years. Interviewees indicated an increase in the length of dry spells and heat waves and the occurrence of severely cold winters without snow in 2010, 2013, 2019, 2020, and 2021, with serious negative consequences for agricultural yields (Figure 7.3). One farmer from Pagri produces and markets a distillate grape called Raki. He reported that in 2020, his vineyard did not have any fruits because of a lack of rain, high temperatures, and heatwaves during the summer.

About 65% of interviewees reported decreased snow cover depth and precipitation in the last decade. They perceived significant changes in the length of snow permanence in the previous 15 years. Farmers in Pagri, Raban, and Odriçan said that, in recent years, snowfall has been limited to a few days and only on the mountains' tops. They also reported that the soil remains frozen for longer without snow cover. They consider this event a climate change marker and linked it to the dryer and colder winters.

Moreover, farmers said that the change in snow patterns has a considerable impact on water availability in rivers and springs and on crop pests and soil health. According to their interpretation, the less frequent and less intense snowfall events and the shorter permanence of snow was an important cause of the reduced water availability they are currently experiencing. Moreover, they identified a nexus between the shorter permanence of snow and a higher incidence of crop pests.

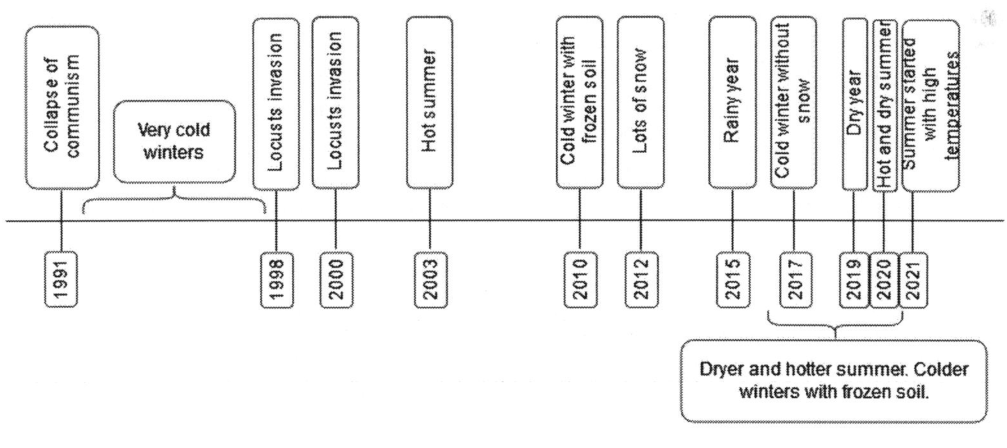

Figure 7.3 Local timeline events reported by respondents during the interviews.

They attributed this change to the fact that reduced winter snow cover can result in more favourable conditions for some species of arthropods.

All informants perceived changes in rainfall patterns, especially during the spring, and 80% of them reported a delay in the start of this season. The spring rainfalls have become less predictable, and this irregularity causes delays or changes in the farming calendar. Farmers and livestock keepers expressed their concern for these irregularities and unpredictable changes. They agreed that the mean temperature and the duration of the summer have increased. Livestock keepers explained that, in the past, they used to grow alfalfa from April to October, which was possible thanks to the rainfall in July and August. Under the current precipitation regime, they cannot harvest alfalfa in summer as the rains in this season have decreased. Nowadays, they are forced to buy extra fodder for their livestock. More than 50% of respondents reported an increase in the frequency of extreme rain days and additional rainfall anomalies throughout the year, making productivity unstable.

In Table 7.3, we summarised farmers' observations of weather patterns. Ninety per cent of the interviewed farmers were conscious of the impacts of climate change on their livelihoods and farming activities. Dwellers of Pagri, Raban, and Odriçan are more dependent on rainwater and natural resources than the interviewees living in Kosinë and Bodar. As a result, they perceive more seriously the risks posed by decreasing rainfall and snowfall, increasing mean temperatures, and higher frequency of dry spells during the summer. They directly attribute to those changes the decreases in productivity they are experiencing. In Kosinë and Bodar, some farmers reported that root pests and diseases have compromised their production of tomatoes and potatoes in the past years. They relate the damages to higher temperatures and erratic rainfall.

Adaptation measures adopted

The results suggest that, although all respondents perceive a higher unpredictability of weather patterns and are aware of the risks that this poses to their livelihood strategies, they are not able to properly cope with these changes. More than 80% of the respondents reported an increase in the use of pesticides as a strategy to cope with more frequent pest outbreaks that they attribute to the reduced snow cover during the winter. In Kosinë e Bodar, farmers manage collective springs, ensuring water availability for crops and livestock throughout the year. The situation is different in Raban, Pagri, and Odriçan where, due to the lack of infrastructure and governmental support, farmers are totally dependent on precipitation. Fifty per cent of livestock keepers reduced the size of their herds in the last few years because of the increasing cost of fodder and the lower summer productivity of pastures. Other livestock keepers decided to specialise in specific phases of the production cycle. For example, one sheep farmer specialised in breeding and marketing lambs and another sheep farming household started to store more hay to cope with the anticipated drying of pastures resulting from the altered rainfall patterns. Similarly, some beekeepers reported that they have recently adopted transhumant beekeeping to cope with decreasing bee fodder availability.

Non-climatic drivers of change

Besides climate change impacts, farmers also mentioned the severe consequences on farmland, productivity, and market economy caused by the socio-economic changes after the collapse of communism. Land fragmentation, the establishment of an open and competitive market, massive migration from rural to urban areas and to foreign countries, insufficient government support, and lack of adequate infrastructures and technologies strongly impact many aspects of farmers'

Table 7.3 Farmers' perceptions in weather patterns according to the LICCI protocol

Variable category – LICCI	*Variable observation*	*Variable description by interviewees*	*n. households*
Changes in the amount of snowfall	Less snowfall in the mountains	The amount of snow has decreased in the last few years. They also experienced more frequent frost without snow	Ten
Changes in the spatial distribution of snowfall	Less snowfall in the mountains	In the past, the snow covered the whole territory including the villages at lower altitudes. Nowadays, only the mountains tops are covered with snow for a few weeks during the winter	Ten
Changes in the variation of rainfall in a given season	Households observed that rainy season changed	Very limited rainfall during the summer. Springs have less water and are drier than in the past	13
Changes in variability of rainfall (not further specified)	Households observed that rainfall patterns changed	Households perceive a decrease in rainfall. Furthermore, they said that during the winter the rainfall is erratic	15
Changes in the frequency of heat waves	Increase of hot days	Most households said that the number of hot days has increased in the last ten years	Ten
Changes in the length/ duration of dry spells	Fodder for the livestock is decreasing due to the spring and summer	Households used to plant alfalfa in April and harvest it every 20 days till October. Nowadays, July is too dry and they are forced to buy new fodder	Ten
Changes in the frequency of cold days	Increasingly dryer and colder winters	Households reported dryer and colder winter in the last six years	Ten
Changes in the length/ duration of heat waves	Higher incidence of pest-related crop damages	Many households think that after 1995, the incidence of damage to crops due to pests and diseases has increased. Tomatoes and potatoes are prone to wilting other pathologies.	Seven
Changes in mean temperature (not further specified)	Lower crops' quality and quantity		Ten

livelihoods. These factors pose important barriers and limitations to farmers' resilience and increase their vulnerability to climate change.

Moreover, all interviewees attributed issues such as a higher incidence of livestock diseases and a general decline in crop health to socio-economic changes rather than climate change. Since 1991, farmers have experienced several challenges in their production system, including difficulties in finding good quality seeds in the market, the misuse of agrochemical inputs, and increased livestock pests and diseases. About 60% of respondents reported that diseases such as bluetongue virus on ruminants, Newcastle disease for poultry, and varroosis for bees entered Albania from neighbouring countries when the communist system collapsed and trade with other countries started.

Discussion

All people interviewed acknowledge that climate change negatively impacts their livelihoods. Most of them agreed that there is a higher variability in annual weather patterns compared to the early 1990s. Changes in rain and snowfall patterns, dry spells' duration, and a shift in the seasons are the most frequently perceived changes in the three study areas. Rural dwellers living in Pagri, Raban, and Odriçan are more dependent on rainfall because they do not have communal water sources for irrigation and, as a result, they are more concerned by the decrease in snow cover and in the number of summer rainy days when compared to farmers living in Kosinë and Bodar.

There is a high degree of agreement among respondents on the climate trends observed in the last 30 years. Recurrent droughts, more erratic rainfall, and a decrease in snowfall were the most commonly perceived changes across the study area.

A comparison of farmers' perceptions with instrumental measures and models reveals an incongruity in the rainfall trends (Table 7.4). Farmers perceive a decreased rainfall in the last ten years, especially in early spring. Nevertheless, according to the European Centre for Medium-Range Weather Forecasts (ECMWF), precipitations slightly increased in the previous decade. This inconsistency between measured precipitation trends and local perceptions has previously been evidenced by other scholars (Simelton et al., 2013; Makuvaro et al., 2018a,b; Madhuri and Sharma, 2020). It is worth noting that farmers accurately reported an increase in the frequency and length of summer dry spells that jeopardised farming productivity. This trend could have influenced their perception of total annual precipitation. On the other hand, farmers' perceptions of temperature trends are more consistent with meteorological data.

Farmers are currently adopting a limited number of coping measures, such as shifting seasonal activities, increasing the use of agrochemical inputs, and buying extra fodder for livestock.

None of the interviewed farmers adopted crop diversification as an adaptation measure. In fact, all of them have been growing the same horticultural crops since 1991. All interviewees buy hybrid seeds from local shops, and only two of them preserve and use local varieties of beans and tomatoes. More than 80% of respondents perceive that the quality of agricultural products

Table 7.4 Comparison between farmers' perceptions and meteorological records for key variables

Farmers perception on climate trends	*Percentage*	*Meteorological records*	*Agreement between perception and measured trends*
Increase in heat waves frequency	65%	Trend line values for the study area show an increase in the frequency of days with maximum temperature >32°C from 4.22 in 2000 to 11.12 in 2017 Raban, Pagri, and Odriçan cluster and from 3.72 in 2000 to 10.66 in Kosinë and Bodar	Y
Intensification and persistence of cold days in winter	65%	The frequency of days with temperature <0°C decreased considerably from 2000 to 2020	N
Decrease in mean annual rainfall	100%	Trendline values for the study area show an increase in mean annual rainfall from 1,009.97 in 1979 to 1,238.58	N

Source of data: ERA5 – Latest climate reanalysis produced by European Centre for Medium-Range Weather Forecasts (ECMWF)/Copernicus Climate Change Service.

has worsened in the last decades, but they attributed this change to the lower quality of imported products and reported higher quality during communism.

Research shows that the adoption of adaptation measures is influenced by many factors (Dang et al., 2019), and results of this study highlight several factors that significantly affect the vulnerability of farming households to climate change. In particular, the interviewees spotlighted how the transformation of the socio-economic system that occurred in the early 1990s still impacts their livelihoods. The decollectivisation of cooperative lands was one of the consequences of the collapse of the communist system (FAO, 2020). The Albanian agricultural land privatisation program that began in 1991 resulted in the distribution of small farms to households. This generated conflicts between those who owned the lands before communism and those who used to work in cooperatives (Zhllima and Imami, 2012). Small-size plots, land fragmentation, limited adoption of crop diversification, and the transition from a communist system to an open market economy resulted in the limited economic sustainability of newly established farms (FAO, 2020). Moreover, farmers' low adaptive capacity is influenced by factors such as lack of technology, poor infrastructure, and limited access to the market (Sutton et al., 2013; Merkoci et al., 2014). These limiting factors have been identified as significant barriers to adopting appropriate adaptive responses (Dang et al., 2019). Furthermore, the lack of infrastructures, such as roads, medical centres, and schools in Raban, Pagri, and Odriçan, led to a massive migration towards urban areas and foreign countries in the last 30 years, further increasing community vulnerability (Zhllima and Imami, 2012; Sutton et al., 2013).

Other factors, such as cultural background, also influence farmers' behaviour (Nguyen et al., 2016). The land reform introduced in 1991 came after 46 years of a centralised agricultural system and resulted in a sudden change in rural dwellers' livelihood strategies. It is worth considering that interviewed farmers and livestock keepers were formerly employed in APCs and SFs, where they were implementing top-down indications and were not allowed to make decisions on agricultural practices and farm management. As a result, they did not acquire a strong legacy in terms of traditional ecological knowledge, which can potentially further reduce their adaptive capacity and resilience.

Conclusions

Our research findings suggest that Albanese farmers who are highly dependent on natural resources have a high awareness of the impact of climate change on their livelihoods. Nevertheless, their adoption of coping and adaptation measures appears to be limited. Rural dwellers seem not particularly inclined to explore suitable adaptive strategies and practices to tackle the impacts of climate change. Socio-economic, political, and cultural factors play an essential role in this respect.

The perception and understanding of changes in weather patterns were widely consistent among interviewed farmers. Moreover, there is a strong convergence in farmers' attitudes and responses to climate change. But in addition to weather patterns, all respondents agree that the new socio-economic system generated disruptive changes in their livelihoods. Despite the fact that interviewees have not clearly identified effective adaptation measures, they believe that without appropriate government support, it is impossible for them to adopt effective measures.

A better understanding of farmers' perceptions and their adaptive attitudes towards climate change is a fundamental step towards developing tailored and effective policies to provide innovative economic and social opportunities to local communities and enhance rural communities' resilience to climate change.

Acknowledgements

The authors are grateful to the local communities of Kosinë, Bodar, Raban, Odriçan, and Pagri. A special thanks to Sokrat Kondili for his availability and translation, to CELIM NGO for their support, and to the Bredhi i Hotovës-Dangëlli National Park.

Conflict of interest. The authors declare that they have no conflict of interest.

Compliance with ethical standards. Data collection was realised within the framework of the ERC project "LICCI. Local Indicator of Climate Change Impacts" (FP7–771056-LICCI).

References

Adger, W. N., Barnett, J., Brown, K., Marshall, N., and O'Brien, K. (2013). Cultural dimensions of climate change impacts and adaptation. *Nature Climate Change*. https://doi.org/10.1038/nclimate1666.

Cungu, A., and Swinnen, J. F. M. (1999). Albania's radical agrarian reform. *Economic Development and Cultural Change*, *47*(3), 605–619. https://doi.org/10.1086/452421.

Dang, H. L., Li, E., Nuberg, I., and Bruwer, J. (2019). Factors influencing the adaptation of farmers in response to climate change: a review. *Climate and Development*. Taylor and Francis Ltd. https://doi.org/10.1080/17565529.2018.1562866.

Demiraj-Bruçi, E., Islami, B., Kamberi, M., Bego, F., Deda, M., Diku, A., Laci, S., Le Tissier, M., Mucaj, L., Mustagi, V., Bogdani-Ndini, M., Pojani, E., Vaso, A., Ylli, A., Zotaj, A., Selfo, G., Gjyli, G., Kamberi, Z., Bekteshi, P., and Bako, A. (2016). *Third National Communication of the Republic of Albania under the United Nations Framework Convention on Climate Change*. Ministry of Environment. https://unfccc.int/sites/default/files/resource/Albania%20NC3_13%20October%202016.pdf.

Eitzinger, A., Binder, C. R., and Meyer, M. A. (2018). Risk perception and decision-making: do farmers consider risks from climate change? *Climatic Change*, *151*(3), 507–524.

FAO (2020). *Smallholders and Family Farms in Albania. Country Study Report 2019*. Budapest. https://doi.org/10.4060/ca7450en.

Fierros-Gonzalez, I., and Lopez-Feldman, A. (2021). Farmers' perception of climate change: a review of the literature for Latin America. *Frontiers in Environmental Science*, *9*, 205.

INSTAT (2012). *Albania – Census of Agricultural Holdings, 2012*. Albanian Institute of Statistics, Tirana.

INSTAT (2013) Population and Housing Census, 2011. Gjirokastër. Institute of Statistics, Tirana. http://www.instat.gov.al/media/3064/6__gjirokaster.pdf.

INSTAT (2021). Regional Statistical Yearbook, 2020. Institute of Statistics, Tirana. http://instat.gov.al/media/8866/regional-statistical-yearbook.pdf.

Keci, E., and Krog, E. (2014). The European Union's IPA 2010 programme for Albania Technical Assistance for Strengthening the Capacity of the Ministry of Environment in Albania for Law Drafting and Enforcement of National Environmental Legislation Activity A.3: Preparation of Strategy and Plans-Protected Areas Management Plans Bredhi i Hotoves – Dangelli National Park Management Plan. https://doi.org/10.13140/2.1.2593.6967.

King, R., and Vullnetari, J. (2016). From shortage economy to second economy: an historical ethnography of rural life in communist Albania. *Journal of Rural Studies*, *44*, 198–207. https://doi.org/10.1016/j.jrurstud.2016.02.010.

Madhuri, and Sharma, U. (2020). How do farmers perceive climate change? A systematic review. *Climatic Change*. Springer Science and Business Media B.V. https://doi.org/10.1007/s10584-020-02814-2.

Makuvaro, V., Murewi, C. T. F., Dimes, J., and Chagonda, I. (2018a). Are smallholder farmers' perceptions of climate variability and change supported by climate records? A case study of Lower Gweru in Semiarid Central Zimbabwe. *Weather, Climate, and Society*, *10*(1), 35–49. https://doi.org/10.1175/WCAS-D-16-0029.1.

Makuvaro, V., Walker, S., Masere, T. P., and Dimes, J. (2018b). Smallholder farmer perceived effects of climate change on agricultural productivity and adaptation strategies. *Journal of Arid Environments*, *152*, 75–82. https://doi.org/10.1016/j.jaridenv.2018.01.016.

Merkoci, A. L., Idrizi, B., and Dvorani, M. (2014). Study on "Vulnerability of agriculture sector in Albania from climate change" powered by the Institute of Energy, Water and Environment. *Revista Shkencore Ndërkombëtare*, *3*, 35–48.

Miara, M. D., Negadi, M., Tabak, S., Bendif, H., Dahmani, W., Hammou, M. A., Sahnoun, T., Snorek, J., Porcher, V., Reyes-García, V., and Teixidor-Toneu, I. (2022). Climate change impacts can be differentially perceived across time scales: a study among the Tuareg of the Algerian Sahara. GeoHealth 6:11 Special collection: rhythms of the earth: ecological calendars and anticipating the anthropogenic climate crisis. https://doi.org/10.1029/2022GH000620.

Nguyen, T. P. L., Seddaiu, G., Virdis, S. G. P., Tidore, C., Pasqui, M., and Roggero, P. P. (2016). Perceiving to learn or learning to perceive? Understanding farmers' perceptions and adaptation to climate uncertainties. *Agricultural Systems*, *143*, 205–216.

Papajorgji E., and Alikay, G. (2016). The function and content of the enterprise during communism in Albania. *Academic Journal of Business, Administration, Law and Social Sciences*. IIPCCL Publishing, Tirana-Albania, *2*(1). https://iipccl.org/wp-content/uploads/2016/03/55-59.pdf.

Phalke, A., Ozdogan, M., Thenkabail, P. S., Congalton, R. G., Yadav, K., Massey, R., Teluguntla, P., Poehnelt, J., Smith, C. (2017). NASA Making Earth System Data Records for Use in Research Environments (MEaSUREs) Global Food Security-support Analysis Data (GFSAD) @ 30-m for Europe, Middle-east, Russia and Central Asia: Cropland Extent Product (GFSAD30EUCEARUMECE). NASA EOSDIS Land Processes DAAC. Retrieved from https://doi.org/10.5067/MEaSUREs/GFSAD/GFSAD30EUCEARUMECE.001.

Republic of Albania (1991) Law 7501 "Concerning the Land" dated 19.07.1991.

Reyes-García, V., Álvarez-Fernandez, S., Benyei, P., García-del-Amo, D., Junqueira, A. B., Labeyrie, V., Li, X., Porcher, V., Porcuna-Ferrer, A., Schlingmann, A., and Soleymani. R. (2023). Local indicators of climate change impacts described by Indigenous Peoples and local communities: study protocol. *PLoS One*, *18*(1), e0279847. https://doi.org/10.1371/journal.pone.0279847.

Sallaku, F., Tota, O., Huqi, B., Jojic, E., Emiri, E., and Fortuzi, S. (2011). Impact of macroeconomic changes and property rights on forest degradation, land use and environmental situation in Albania. In Georgi Zhelezov (Ed.), *Sustainable Development in Mountain Regions* (pp. 203–214). Springer, Dordrecht.

Simelton, E., Quinn, C. H., Batisani, N., Dougill, A. J., Dyer, J. C., Fraser, E. D. G., … Stringer, L. C. (2013). Is rainfall really changing? Farmers' perceptions, meteorological data, and policy implications. *Climate and Development*, *5*(2), 123–138. https://doi.org/10.1080/17565529.2012.751893.

Sutton, W. R., Srivastava J. P., Neumann J. E., Strzepek K. M., and Droogers, P. (2013). *Reducing the Vulnerability of Albania's Agricultural Systems to Climate Change: Impact Assessment and Adaptation Options*. World Bank Study. Washington, DC. https://doi:10.1596/978-1-4648-0047-4.

Swinnen, J. F. M. (2018). *Political Economy of Agrarian Reform in Central and Eastern Europe* (pp. 1–397). Taylor and Francis. https://doi.org/10.4324/9780429446382.

UNEP-WCMC, IUCN (2022). *Protected Planet: The World Database on Protected Areas (WDPA)*. https://www.protectedplanet.net/en.

Van der Linden, S. (2015). The social-psychological determinants of climate change risk perceptions: Towards a comprehensive model. *Journal of Environmental Psychology*, *41*, 112–124.

Whitmarsh, L., and Capstick, S. (2018). Perceptions of climate change. *Psychology and Climate Change*, 13–33. https://doi.org/10.1016/b978-0-12-813130-5.00002-3.

World Bank (2021a). *Climate Risk Profile: Albania*. https://climateknowledgeportal.worldbank.org/sites/default/files/2021-06/15812-Albania%20Country%20Profile-WEB.pdf.

World Bank (2021b). *World Development Indicators*. https://databank.worldbank.org/source/world-development-indicators.

Zhang, C., Jin, J., Kuang, F., Ning, J., Wan, X., and Guan, T. (2020). Farmers' perceptions of climate change and adaptation behavior in Wushen Banner, China. *Environmental Science and Pollution Research*, *27*(21), 26484–26494. https://doi.org/10.1007/s11356-020-09048-w.

Zhllima, E., and Imami, D. (2012). The Albanian land rights security perception and factors influencing it. *Albanian Journal of Agricultural Sciences*, *11*(1), 43–52.

8

LOCAL PERCEPTIONS OF CLIMATE CHANGE IN THE CONTEXT OF SOCIOECONOMIC AND POLITICAL CHANGES IN A HIGH ANDEAN COMMUNITY FROM THE ARGENTINE PUNA

Andrea E. Izquierdo and Anna Schlingmann

Introduction

Indigenous Peoples and local communities in isolated mountainous areas are among the most vulnerable groups to global climate change impacts (Beniston et al., 1997; Beniston, 2003, 2005; IPCC, 2022). This is the case of pastoralist communities in the Argentine Dry Puna, part of the High Andean Plateau at altitudes between 3,200 and 6,700 m. a.s.l. Most climate scenarios for this alpine ecosystem predict a 2°– 4°C temperature increase by the end of the 21st century (Urrutia & Vuille, 2009). While models uncertainties are greater with regard to precipitation trends, the most plausible scenarios predict a decrease in precipitation and cloudiness for the subtropical Andes (Vuille et al., 2008), as well as decreasing water availability and longer dry seasons (Buytaert et al., 2010). This coincides with dendroecological reconstructions of water balance and ecosystem productivity over the past 30 years (Carilla et al., 2013) and centuries (Morales et al., 2015).

At the same time, the Argentine Puna is undergoing rapid socioeconomic changes (Izquierdo & Grau, 2009; Izquierdo et al., 2018a). Historically a very remote area, governmental investments since 1983, when the current democratic period started, have fostered the expansion of transport and communications infrastructure, mainly of road networks and most recently of internet coverage, thereby increasing mobility and access to information, markets, governmental institutions, and labor (Izquierdo et al., 2018a). Extreme environmental conditions in the Puna have determined past patterns of human settlement and land use. Since the establishment of the first inhabitants, estimated about 11,000 years ago, the main subsistence activity in the area has been pastoralism (Martınez, 2018; Olivera, 2018), primarily of sheep and llamas (Quiroga Mendiola & Cladera, 2018). At present, the main growing economic activity is mining, particularly lithium extraction (Izquierdo et al., 2015, 2018a). Lithium reserves in the Argentinian Puna are part of the so-called "lithium triangle"—the largest global reserve of lithium, which extends across northern Chile,

DOI: 10.4324/9781003356837-11

southwestern Bolivia, and northwestern Argentina. Interest in lithium has boomed in recent years due to the rising demand for cell phones and electric vehicle batteries (Dresselhaus & Thomas, 2001; Scrosati & Garche, 2010; Gielen, 2021). While prospects for lithium mining are promising, its environmental consequences are poorly understood (Wanger, 2011; Izquierdo et al., 2015). The future economic, social, and environmental impacts, and the (un)fair distribution of benefits and negative consequences from the lithium industry will highly depend on whether or not local communities will have a say in the decision-making process surrounding lithium extraction activities (Prior et al., 2013; Jerez et al., 2021).

Climatic, socioeconomic, and political factors (e.g., demographic changes, politics, globalization) have widely influenced and transformed local living conditions and livelihoods. In the study area, increased connectivity and access to off-farm work have led to rural–urban migration, a decrease in pastoralist activities, and related land-use changes (Izquierdo & Grau, 2009; Izquierdo et al., 2018a). Climate change adds an additional challenge to the socioeconomic complexity, which offers both opportunities and risks. Climate change and socioeconomic and political changes interact in non-linear ways and can amplify or attenuate the overall impacts on the local environment, and hence, affect local livelihoods and living conditions of local communities and Indigenous Peoples in diverse and unpredicted ways (Pörtner et al., 2021). Vulnerability and resilience to climate change are therefore not only defined by exposure to climate hazards, but also by the changing socioeconomic conditions of the system of concern, which determine its sensitivity and adaptive capacity. Consequently, research on the vulnerability and resilience of social-ecological systems has increasingly recognized the need for multiple stressor approaches by considering all relevant drivers instead of focusing on a single stressor (e.g., Reid & Vogel, 2006; Tschakert, 2007; Räsänen et al., 2016).

Due to their strong dependence on nature through their natural-resource-dependent livelihoods, culture, and place connection, Indigenous Peoples and local communities can be keen observers of environmental changes in their surroundings (Reyes-García et al., 2016). They also know best the context-specific positive and negative impacts of socioeconomic, political, and environmental changes that drive and shape changes in local livelihoods, living conditions, including culture and worldviews (Salick & Ross, 2009). Nonetheless, several studies underline the persistent lack of recognition of Indigenous and local knowledge into research and policy agendas (Ford et al., 2016; Yap & Watene, 2019; Chakraborty & Sherpa, 2021; Reyes-García et al., 2022). The assessment of local reports and evaluations of the relative importance of such drivers of local changes based on their negative and positive impacts is crucial to understanding people's concerns, viewpoints, and well-being aspirations; and – based on that – to improve future adaptation planning by bridging local, Indigenous, and scientific knowledge (Tengö et al., 2017; Reyes-García et al., 2019).

In direct response to this need, this chapter assessed the perceptions of Kolla-Atacameños People and local communities in an area from the Argentine Puna ecoregion. The chapter focuses on understanding the relative importance of climate change in relation to other drivers that shape changes in and imply impact on the local environment, livelihoods, and living conditions. For this, we (1) describe observed changes in the local environment, livelihoods, and living conditions, (2) identify the main drivers of such changes, (3) assess the relative importance of each of the drivers regarding their impacts from a local perspective, and (4) determine the associated positive and negative local impacts of such drivers. The chapter ends with a discussion of the potential differences in prioritization between Indigenous Peoples and the local community in a region of the Argentine Puna and the research and policy agendas.

Methodology

Study area

This study was carried out in an the Argentine Puna in the department of Antofagasta de la Sierra (between 25°10′ and 26°55′S; and 66°30′ and 68°35′ W) in the province of Catamarca (Figure 8.1). The department of Antofagasta de la Sierra is an area of roughly 29,000 km^2 with a population of about 2,000 inhabitants according to the latest published national census (INDEC, 2022). The capital and market center of the department is the homonymous town Antofagasta de la Sierra. In addition to the capital, there are only three more villages in the department: El Peñon, Antofalla, and Los Nacimientos, and some dispersed settlements, permanently or semi-permanently inhabited, locally known as 'puestos'. Puestos are generally associated with 'vegas', a type of key wetland which provides main ecosystem services such as freshwater provision, carbon storage, and pastures for livestock (Izquierdo et al., 2018b).

The Argentine Puna is part of the High Andean Plateau, with an average altitude of 3,500 m.a.s.l. The area is an arid climate region according to the Köppen-Geiger climate classification, with an annual mean temperature of 8°C and annual precipitation between 100 mm in the northeast and 400 mm in the southwest. Most rainfall concentrates in the summer months from December to March (Burkart et al., 1995).

The region of Antofagasta de la Sierra was already inhabited 11,000 years ago by the Diaguitas and Kolla-Atacameños Indigenous peoples (Martinez, 2018). The first inhabitants were hunters and gatherers organized in small groups which concentrated mainly in ravines and meadows. Currently, the area is inahited by Indigenous Peoples who recognize themselves as Kolla-Atacameños and local community who recognizee themselves as "criollos". Their main livelihood is the traditional

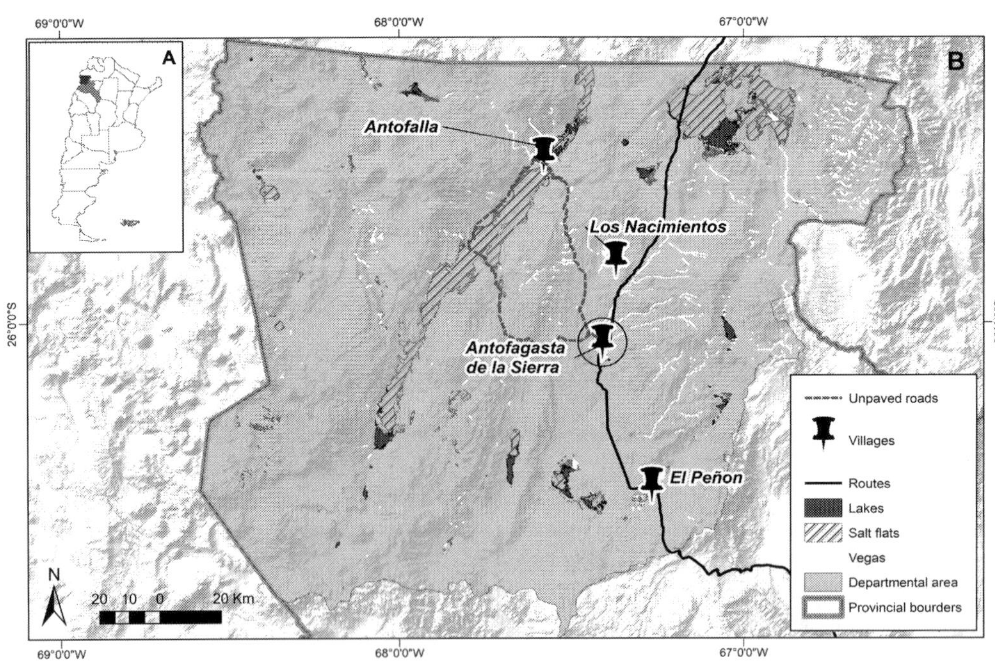

Figure 8.1 Study site location in Argentine and Catamarca borders (A) and in the Antofagasta de la Sierra department (B).

pastoralism of llamas and sheep. Until recently, people in the area practiced transhumant pastoralism, keeping the livestock in pastures at lower altitudes, near the main houses, during the winter and moving it to higher and remote fields during the summer months, a practice well adapted to the extreme climate and environmental conditions of the Puna. However, traditional pastoralism in the Argentine Puna is decreasing. First, current climatic trends toward a decrease in humidity and an increase in drought events imply adverse and threatening climate conditions for traditional livelihoods (Morales et al., 2015, 2023). Second, due to different factors such as increasing employment in mines, easier access to technologies and amenities in urban areas, and resulting rural–urban migration, pastoralism is decreasing, and transhumance is being abandoned (Izquierdo et al., 2018a). Although the region has been important for mining activities since the Jesuits' time (1608–1767), in recent years, the mining industry has experienced an unprecedented boom, especially due to a growing interest in the lithium reserves in the area (Izquierdo et al., 2015, 2018a).

Methods

Data collection took place in the town of Antofagasta de la Sierra, and the three villages El Peñon, Antofalla, and Los Nacimientos. We followed the LICCI protocol (Reyes-García et al., 2023) and divided the research into two phases. In the first phase, from October 2019 to February 2020, we used qualitative data collection methods to identify perceived changes in the local environment, livelihoods and living conditions, and the associated drivers of these changes. We conducted a total of 23 semi-structured interviews (SSI) and three focus group discussions (FGD) with three to seven participants. We used quota sampling for the categories of gender, age group, and livelihood (expertise) to select participants. SSI and FGD lasted between 1 and 1.5 hours and were conducted in Spanish, the national language, spoken by all participants. We first asked generally about any observed changes, and for each reported change, we asked about the potential drivers of such change. We then asked explicitly about changes and their drivers in the local natural environment (atmospheric, physical, and life system, including flora, fauna, and human activities), with a special focus on climate-driven changes. Using information from the interviews, we identified five main drivers of local change: climate change, technology, transport, mining, and migration.

In the second phase, we conducted 78 surveys to collect individual perceptions of positive and negative impacts that arise from the five main drivers of change. All interviewees were household heads. The number of selected household was proportional to the population size of each village. Households were selected based on simple random sampling, while household heads were selected based on convenience quota sampling to achieve an approximately even distribution across gender and age groups. First, we assessed how the interviewees perceived the relative importance of the five main drivers of changes in the local environment, livelihoods, and living conditions. We did so by asking participants to rank the different drivers of change on a scale from 1 (most important) to 5 (least important), according to the impacts they have on their livelihoods and living conditions. Then, we asked if those drivers imply any positive or negative impacts, and which ones.

Results

Observed changes in the local environment, livelihoods, and living conditions, and their main drivers

Through the semi-structured interviews, we identified 31 changes in the local environment, livelihood, and living conditions. Reported changes include changes in precipitation and snowfall (e.g., fewer but more extreme rainfalls), changes in water availability (e.g., a decline in the water volume in streams and wetlands), changes in the pastures (e.g., a decrease in the quality and quantity), changes in the behavior of wild fauna (e.g., increase in the cougar predation on livestock), changes related to invasive species and pests (e.g., the appearance of weeds or insects), and changes related to housings (e.g., increasing roof damages). Twenty-four of those reported changes were directly linked to climate change. For example, a decline in the size of wetlands was directly associated with a decrease in snowfall and rainfall, while damaged roofs were linked to more frequent heavy rainfalls. Other observed changes include higher temperatures and sunshine intensity. Some climate-related changes relate to soil (e.g., greater and faster drying) and ice (e.g., less ice formation or thickness). Informants also reported changes in livestock and crop cultivation, mainly a decrease in pasture and crop productivity as a direct consequence of higher aridity. Seven observed changes were not associated with climate change but with socioeconomic and political drivers such as technology, infrastructure, mining activities, and rural–urban migration. For example, people related new pests and insects with the arrival of trucks that bring products from other areas.

In total, we identified five main drivers of change on the local environment, livelihoods, and living conditions in the study area: (1) the arrival of technology (i.e., electricity, water network) and communications infrastructure (i.e., phone coverage and internet connection) (hereafter technology); (2) improvements in the transport system and road networks (i.e., road paving, traffic) (hereafter transport); (3) extension of mining activities (hereafter mining); (4) rural–urban migration (hereafter migration); and (5) climate change (Table 8.1).

Ranked importance of main drivers of local changes

When asked about the relative importance of each of the five main drivers regarding their impacts on the local environment, livelihoods, and living conditions on a scale from 1 (most important) to 5 (least important), interviewees evaluated changes generated by technology and transport as the most important drivers of change (weighted mean 1.9, respectively), followed by mining (weighted mean 3.4), climate change (weighted mean 3.5), and rural–urban migration (weighted mean 4.4).

Table 8.1 Reported changes in the local environment, livelihoods, and living conditions related to climate change and unrelated to climate change and the number of times each was mentioned in the semi-structured interviews. The direction of change is shown with arrows indicating (increase), (decrease), (later), (temporality changes)

Reported changes in the local environment, livelihoods, and living conditions.	*System*	*Direction*	*Times mentioned*
Climate change related			
Changes in the intensity/strength of rainfall (not further specified).	Atmospheric	↑	12
Changes in the temperature during the day.		↑	9
Changes in sunshine intensity.		↑	6
Changes in the amount of snowfall.		↑	6
Changes in the timing (onset or end) of seasons.		→	5

(*Continued*)

Table 8.1 (Continued)

Reported changes in the local environment, livelihoods, and living conditions.	*System*	*Direction*	*Times mentioned*
Changes in the length/duration of rainfall events.		↓	3
Changes in the amount of rainfall in a given season.		↓	8
Changes in the length/duration of drought.		↑	4
Changes in mean rainfall (not further specified).		↓	7
Changes in temporal distribution of rainfall.		↔	5
Changes in the predictability of rainfall.		↑	8
Changes in mean temperature (not further specified).		↑	7
Changes in freshwater availability.	Physical	↓	9
Changes in freshwater transparency.		↓	1
Changes in river/stream water flow, volume, level, and/or depth.		↓	5
Changes in the thickness of ice in lakes or rivers.		↓	1
Changes in wetland surface.		↓	4
Changes in river/stream water flow, volume, level and/or depth.		↓	4
Changes in soil evaporation.		↑	7
Changes in the abundance of terrestrial fauna (mammals, birds, reptiles, insects, etc.).	Life	↓	2
Changes in pasture productivity.		↓	11
Changes in crop productivity/yield.		↓	5
Changes in the frequency of crop 'pests' (insects, birds, larvae, etc.).		↑	2
Changes in the frequency or occurrence of weed species stated as invasive.		↑	2
Unrelated to climate change			
Changes in infrastructure (routes, transport).	Physical	↑	4
Changes in public services access (electricity, internet, freshwater network).		↑	6
Changes in herbivory interspecific competition (livestock vs. vicuña).	Life	↑	3
Changes in the abundance of terrestrial fauna (mammals, birds, reptiles, insects, etc.).		↑	22
Changes in predation rates (livestock predation by cougar).		↑	13
Demographic changes (rural–urban migration).		↑	7
Changes in availability of paid jobs (mining).		↑	5

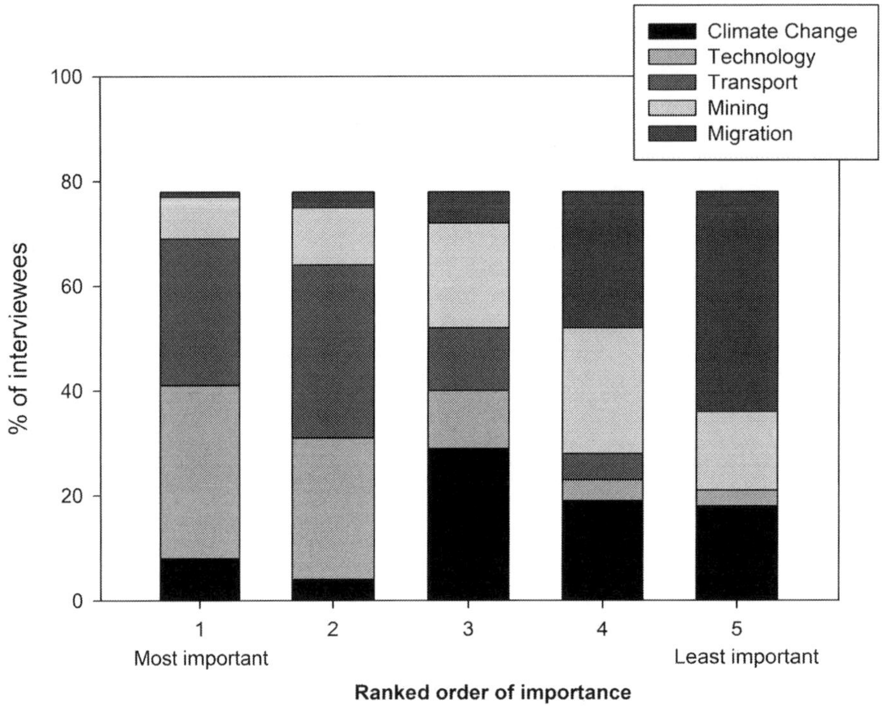

Figure 8.2 Ranking of the different drivers of change according to locally perceived impact on local livelihoods.

While technology and transport were more often ranked under the categories 'most important' (rank 1) and 'very important' (rank 2), migration was the most frequent driver ranked under the category 'least important' (rank 5). Climate change was most often ranked as 'medium important' (rank 3), while mining was in most cases considered as 'less important' (rank 4). Specifically, 42% of the interviewees considered technology as the 'most important' and 35% as 'very important' driver, while 37% of the interviewees ranked transport as 'most important', and 42% as 'very important'. On the other hand, 33% of the interviewees evaluated migration as 'less important' and 54% as 'least important'.

A total of 85% of informants ranked climate change as medium to least important, with a clear tendency toward 'medium important' (37%). Similarly, 76% of the interviewees ranked mining between 'medium important' and 'least important', with the highest frequency under 'less important' (31%) (Figure 8.2).

Perceived positive and negative impacts associated with the five main drivers of local changes

The results show that study participants considered drivers of positive changes as more important than drivers of negative or ambiguous (positive and negative) changes. Technology and transport were overwhelmingly evaluated as drivers of positive changes – by 95% and 96% of the interviewees, respectively (Figure 8.3). In turn, climate change was mostly perceived as a driver of local changes with negative impacts (in 90% of the interviewees), with only 9% of the respondents

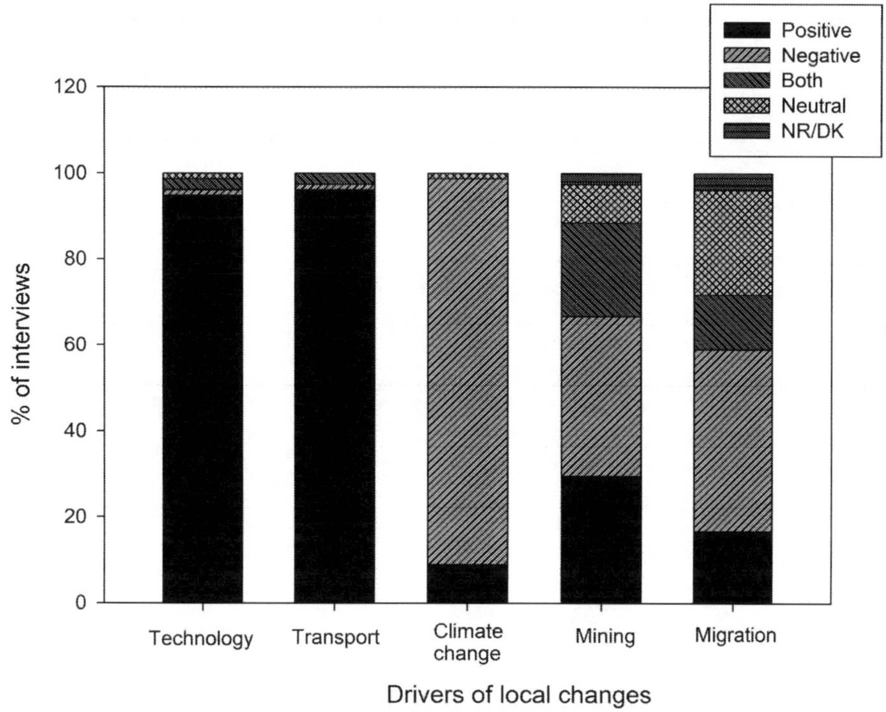

Figure 8.3 Distribution of interviewees' responses regarding the impacts generated by each driver of change in four different directions: positive, negative, both, and neutral; and percentage of non-responders.

perceiving climate change as entirely positive, and 1% of the respondents considering as both, positive and negative (Figure 8.3). Mining was classified more heterogeneously: 37% of the interviewees perceived it as a driver generating negative changes, 29% as a driver of positive changes, 22% as a driver of both negative and positive changes, 9% considered it as neutral, and 3% of the interviewees did not respond or didn't know what to answer (NR/DK). The impacts of migration were considered by 42% of the interviewees as negative, by 24% as neutral, by 17% as positive, and by 13% as both, negative and positive. 4% of the interviewees did not respond or didn't know what to answer (NR/DK).

The different positive and negative impacts reported by interviewees for each driver of local changes are presented in Figure 8.4. When discussing the benefits of technology, interviewees argued that new technologies (e.g., mobile phones, internet) are the best and easiest means of communication (53% of the interviewees) and increase their connectivity with the world, thereby allowing them to meet and make themselves known to other cultures (41%) (Figure 8.4).

The positive impacts of transport included eased mobility (36% of the interviewees), better access to and from the region (29%), better conditions for commerce (23%), and regional development (8%) (Figure 8.4). One interviewee reported exclusively negative impacts referring to the increase in traffic, which disturbs the characteristic peace and tranquility of the region (Figure 8.4).

Among the negative impacts of climate change, interviewees reported a decrease in water availability (69% of the interviewees), harsher living conditions (15%), and other impacts such as higher unpredictability of rainfall or health impacts (4%) (Figure 8.4). The only perceived benefit

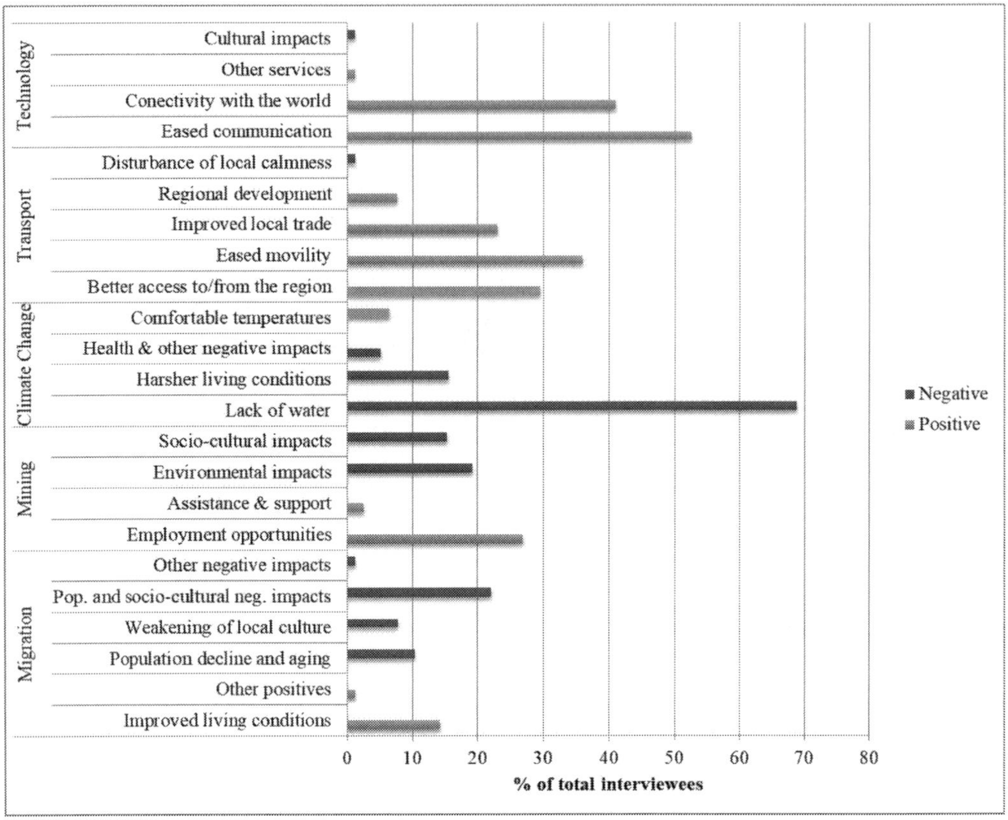

Figure 8.4 Perceived negative and positive impacts of local changes associated with technology, transport, climate change, mining, and migration.

of climate change was that the increase in temperature makes people feel less cold, so they can wear less clothing (6%).

The most heterogeneous responses were provided for the potential impacts and benefits of mining. Interviewees perceived the increased employment opportunities (27% of the interviewees) and assistance (through corporate responsibility programs) (3%) as positive, while environmental (19%) and socio-cultural (15%) impacts were perceived as negative (Figure 8.4).

Responses with respect to migration were also heterogeneous. Some interviewees reported exclusively positive impacts – mainly the improvement of living conditions (14%) and the value of the cultural exchange (1%) –, while others focused on negative impacts, such as population aging and the loss or weakening of local culture (42%) (Figure 8.4).

Discussion

Results show that, although Kolla-Atacameños People and local communities Antofagasta de la Sierra reported more observations of changes in the local environment, livelihoods, and living conditions that are directly driven by climate change, they considered technology and transport as the most important drivers of local changes, while climate change, mining, and migration were considered less important. At the same time, the drivers of changes perceived as most important (i.e., technology and transport) have mostly positive impacts according to interviewees, while less important drivers are associated with rather negative (i.e., climate change) or ambiguous impacts (i.e., mining and migration). This does not only show that the Kolla-Atacameños People and local communities in the study area recognize changes in the living conditions that are directly linked to technology and transport as beneficial, but also that such improvements play a major role and are a priority for them.

The apparent contradiction between the number of reported changes associated with climate change and the lower relative importance people assign to it as a driver of local changes might be due to the fact that parts of the data collection specifically focused on climate change and its impacts on the local environment and livelihoods. In this regard, our results could be understood as a certain discrepancy between the current concerns of Indigenous Peoples and local communities and the scientific agenda. Our results show that Kolla-Atacameños People and local communities in Antofagasta de la Sierra experienced changes and impacts in their local environment, livelihoods, and living conditions from multiple socioeconomic and biophysical drivers. Other studies confirm that climate change, although experienced, is not necessarily perceived as the most impacting local driver, but one among many (e.g., Nyantakyi-Frimpong & Bezner-Kerr, 2015). Hence, our study highlights that climate change impacts are entangled in complex and interacting socioeconomic, cultural, and political processes and dimensions, and need to be addressed as such (Adger et al., 2013; Ensor et al., 2019). Increasingly, climate change research has acknowledged this need and called for more integrated and holistic approaches that frame and understand climate change impacts, vulnerability, and adaptation in a broader social, economic, and political context of multiple drivers, including stressors (e.g., O'Brien et al., 2004; Bennett et al., 2016), resulting in a recently growing number of publications on the topic (Räsänen et al., 2016). Our results reinforce this trend in which a growing number of case studies around the world acknowledge and assess local impacts and adaptation beyond just climate change (e.g., McDowell & Hess, 2012; Porcuna et al. (2024); Reyes-García et al., 2024).

Understanding, challenges, and opportunities derived from multiple drivers, including socioeconomic, political, and environmental transitions, are especially relevant for Indigenous Peoples and local communities who used to live in very isolated and harsh living conditions in mountainous regions and are now confronting unprecedented transformations in the context of globalization (Dhakal & Kattel, 2019). On the one hand, it is recognized that globalization brings some positively evaluated benefits, such as technological advances and new income opportunities in the mining and tourism sector (Izquierdo et al., 2018a). On the other hand, the lower relative importance assigned to climate change is also found in other studies (e.g., Nyantakyi-Frimpong & Bezner-Kerr, 2015). In fact, it could be a co-product of simultaneous trends toward off-farm work and migration that lowers climate change vulnerability by reducing the direct dependency on natural resources (Antwi-Agyei et al., 2014; Cannon, 2014). However, most benefits come along with some trade-offs that should not remain without critical reflection. In this regard, the circumstance that Kolla-Atacameños People and local communities in the study area evaluate changes that improve their lives of higher relevance should not downplay and detract from the fact that they express serious cultural, social, and environmental concerns related to climate change, mining, and migration. Concerns such as cultural and social uprootedness associated with switching from traditional livelihoods to off-farm work and migration expressed by study participants are supported by other studies (Adger et al., 2013). Several studies highlight additional economic and social risks, such as income dependencies on international markets and price fluctuation and a decrease in food sovereignty (Galappaththi & Schlingmann, 2023). Furthermore, a weakening or even loss of Indigenous and local cultures impacts biocultural diversity (Fernández-Llamazares et al., 2020), for which Indigenous and local knowledge is a key component of adaptation strategies. For instance, even though the Argentine Puna ecoregion is experiencing an aridization trend (Morales et al., 2015), a study that analyzed the productivity of 'vegas' (i.e., natural wetlands and key pastures for livestock in the region), indicated that the adoption of management practices traditionally used by Indigenous Peoples and local communities can enhance the productivity, stability, and resilience of these crucial ecosystems (Navarro et al., 2020). Additionally, the current trends toward an expansion of lithium mining activities in the Argentine Puna ecoregion in direct response to a growing global demand for electric vehicles fostered by current climate mitigation policies (Voskoboynik & Andreucci, 2022) are a major concern for both scientists and decision-makers. Scientists point to the socio-environmental impacts (i.e., competition for the limited freshwater resources) (Izquierdo et al 2015, 2018a; Liu & Agusdinata, 2021) or nature conservation issues (Gajardo & Redón, 2019; Marconi et al., 2022); while decision-makers emphasize socioeconomic benefits arising from mining (e.g., local and regional jobs, national royalties) (Ministerio de Desarrollo Productivo, 2021; Díaz Paz et al., 2023). Interestingly, those who are directly affected by the mining activities, that is, Kolla-Atacameños People and local communities in the Argentine Puna region, bring together the views of scientists and decision-makers by referring to both the positive and negative impacts of mining. Importantly, the fact that one-fifth of the study participants report environmental impacts derived from mining, specifically referring to excessive water extraction, while almost 70% expressed concerns about lack of water due to climate change, should be understood as a warning of potential water conflicts between different actors and demands (e.g., mining, livestock, drinking water) (Jerez et al., 2021). To avoid Kolla-Atacameños People and local communities in the Argentine Puna region being left empty-handed and suffering most of the negative consequences, it is important that they are integrated as main actors in decision-making processes surrounding future regional development planning, that should consider potential negative and interacting impacts deriving from climate change and lithium mining (Sovacool, 2021).

The need to bridge different knowledge systems in climate change research and decision-making has been established in political instruments, such as the fifth assessment report (AR5) of the Intergovernmental Panel on Climate Change (IPCC), and scientific articles (e.g., Reyes-García et al., 2016). Nonetheless, how to best realize this integration is still debated (see Salick & Ross, 2009), given the manifold challenges in transferability, integration, and scalability of different systems of knowledge (Reyes-Gracía et al., 2019), and the complexity and potential discrepancies of perceptions at local (Byg & Salick, 2009), regional and global scales, and across different actors (Finnis et al., 2015). These discrepancies do not necessarily mean a problem and should rather be understood as a warning, especially for decision-makers, to be aware of global trends that may produce urgent, critical, and relevant impacts in the future (e.g., Pörtner et al., 2022), without neglecting the priorities and concerns of Indigenous Peoples and local communities. We advocate that the primordial step is to recognize first-hand Indigenous and local knowledge and worldviews and to scale Indigenous Peoples and local communities' primary concerns to the political and research agendas. For this, we should ask these stakeholders about their priorities and urgencies, consider their views from the beginning into the research design, and acknowledge the complexity of multiple drivers that shape and impact the local environment, livelihoods, and living conditions.

Acknowledgments

We thank all the study participants for sharing their thoughts, views, and knowledge with us. We thank Elvira Casagranda, Fernanda Chiappero, Olga Ramos, and Silvina Vazquez for their field assistance and support and we thank Victoria Reyes-García, Reinmar Seidler, and Ramin Soleymani for their valuable reviews and comments on a previous version of this book chapter.

This study has been supported by the National Council of Sciences and Technologies (CONICET) and the National Science and Technology Agency of Argentina through which the fieldwork was financed with the grant PICT2018-04228 (PI: A.E. Izquierdo). Against delivery of the data, the researcher leading to this paper has received funding from the European Research Council under an ERC Consolidator Grant (FP7-771056-LICCI). This work contributes to the 'María de Maeztu' Programme for Units of Excellence of the Spanish Ministry of Science and Innovation (CEX2019-000940-M).

References

Adger, W. N., Barnett, J., Brown, K., Marshall, N., & O'Brien, K. (2013). Cultural dimensions of climate change impacts and adaptation. *Nature Climate Change, 3*(2), 112–117. https://doi.org/10.1038/nclimate1666

Antwi-Agyei, P., Stringer, L.C. & Dougill, A. J. (2014). Livelihood adaptations to climate variability: insights from farming households in Ghana. *Regional Environmental Change, 14*, 1615–1626. https://doi-org.are.uab.cat/10.1007/s10113-014-0597-9

Beniston, M. (2003). Climatic change in mountain regions: a review of possible impacts. *Climate Variability and Change in High Elevation Regions: Past, Present & Future*, 5–31. https://doi.org/10.1007/978-94-015-1252-7_2

Beniston, M. (2005). The risks associated with climatic change in mountain regions. In *Global Change and Mountain Regions. Advances in Global Change Research*, *23*, 511–519. Springer, Dordrecht. https://doi.org/10.1007/978-1-4020-3508-1

Beniston, M., Diaz, H. F., & Bradley, R. S. (1997). Climatic change at high elevation sites: an overview. *Climatic Change*, *36*(3), 233–251.

Bennett, N. J., Blythe, J., Tyler, S., & Ban, N. C. (2016). Communities and change in the anthropocene: understanding social-ecological vulnerability and planning adaptations to multiple interacting exposures. *Regional Environmental Change*, *16*(4), 907–926. https://doi.org/10.1007/s10113-015-0839-5

Burkart, R., Bárbaro, N. O., Sánchez, R. A., & Gómez, D. A. (1995). Eco-regiones de la Argentina (Ecoregions of Argentina). Programa Desarrollo Institucional Ambiental, Buenos Aires, Argentina, 42 pp.

Buytaert, W., Vuille, M., Dewulf, A., Urrutia, R., Karmalkar, A., & Célleri, R. (2010). Uncertainties in climate change projections and regional downscaling in the tropical Andes: implications for water resources management. *Hydrology and Earth System Sciences*, *14*(7), 1247–1258. https://doi.org/10.5194/hess-14-1247-2010

Byg, A., & Salick, J. (2009). Local perspectives on a global phenomenon—climate change in Eastern Tibetan villages. *Global Environmental Change*, *19*(2), 156–166. https://doi.org/10.1016/j.gloenvcha.2009.01.010

Cannon, T. (2014). Rural livelihood diversification and adaptation to climate change. Community-based adaptation to climate change. *Emerging Lessons*, 55–76. https://www.researchgate.net/publication/300232311_4_Rural_livelihood_diversification_and_adaptation_to_climate_change

Carilla, J., Grau, H. R., Paolini, L., & Mariano, M. (2013). Lake fluctuations, plant productivity, and long-term variability in high-elevation tropical Andean ecosystems. *Arctic, Antarctic, and Alpine Research*, *45*(2), 179–189. https://doi.org/10.1657/1938-4246-45.2.179

Chakraborty, R., & Sherpa, P. Y. (2021). From climate adaptation to climate justice: critical reflections on the IPCC and Himalayan climate knowledges. *Climatic Change, 167*(3–4), 49. https://doi.org/10.1007/s10584-021-03158-1

Dhakal, B., & Kattel, R. R. (2019). Effects of global changes on ecosystems services of multiple natural resources in mountain agricultural landscapes. *Science of the Total Environment*, 676, 665–682. https://doi.org/10.1016/j.scitotenv.2019.04.276

Díaz Paz, W. F., Escosteguy, M., Seghezzo, L., Hufty, M., Kruse, E., & Iribarnegaray, M. A. (2023). Lithium mining, water resources, and socio-economic issues in northern Argentina: We are not all in the same boat. *Resources Policy, 81*, 103288. https://doi.org/10.1016/j.resourpol.2022.103288

Dresselhaus, M. S., & Thomas, I. L. (2001). Alternative energy technologies. *Nature*, *414*(6861), 332–337.

Ensor, J. E., Wennström, P., Bhatterai, A., Nightingale, A. J., Eriksen, S., & Sillmann, J. (2019). Asking the right questions in adaptation research and practice: seeing beyond climate impacts in rural Nepal. *Environmental Science & Policy, 94*, 227–236. https://doi.org/10.1016/j.envsci.2019.01.013

Fernández-Llamazares, Á., Benyei, P., Junqueira, A. B., & Reyes-García, V. (2020). Participation in biocultural diversity conservation: insights from five Amazonian examples. In: Baldauf, C. (eds) *Participatory Biodiversity Conservation*. Springer, Cham. https://doi.org/10.1007/978-3-030-41686-7_11

Finnis, J., Sarkar, A., & Stoddart, M. C. (2015). Bridging science and community knowledge? The complicating role of natural variability in perceptions of climate change. *Global Environmental Change*, *32*, 1–10. https://doi.org/10.1016/j.gloenvcha.2014.12.011

Ford, J. D., Cameron, L., Rubis, J., Maillet, M., Nakashima, D., Willox, A. C., & Pearce, T. (2016). Including indigenous knowledge and experience in IPCC assessment reports. *Nature Climate Change, 6*(4), 349–353. https://doi.org/10.1038/nclimate2954

Gajardo, G., & Redón, S. (2019). Andean hypersaline lakes in the Atacama Desert, northern Chile: between lithium exploitation and unique biodiversity conservation. *Conservation Science and Practice, 1*(9). https://doi.org/10.1111/csp2.94

Galappaththi, E. K., & Schlingmann, A. (2023). The sustainability assessment of Indigenous and local knowledge-based climate adaptation responses in agricultural and aquatic food systems. *Current Opinion in Environmental Sustainability, 62*, 101276. https://doi.org/10.1016/j.cosust.2023.101276

Gielen, D. (2021). Critical minerals for the energy transition. *International Renewable Energy Agency: Abu Dhabi, United Arab Emirates*. ISBN 978-92-9260-366-3.

INDEC [Instituto Nacional de Estadısticas y Censos] (2022). Censo nacional de población, hogares y viviendas 2022: resultados provisionales/1a ed. – Ciudad Autónoma de Buenos Aires: Instituto Nacional de Estadística y Censos. https://www.indec.gob.ar/ftp/cuadros/poblacion/cnphv2022_resultados_provisionales.pdf

IPCC. (2022). *Climate Change 2022: Impacts, Adaptation, and Vulnerability. Contribution of Working Group II to the Sixth Assessment Report of the Intergovernmental Panel on Climate Change* (H.-O. Pörtner, D. C. Roberts, M. Tignor, E. S. Poloczanska, Ķ. Mintenbeck, A. Alegría, M. Craig, S. Langsdorf, S. Löschke, V. Möller, A. Okem, & B. Rama, Eds.). Cambridge University Press.

Izquierdo, A. E., Aragon, R. M., Navarro, C. J., & Casagrandra, E. (2018b). Humedales de la Puna: principales proveedores de servicios ecosistémicos de la región. In: Grau, H. R., Babot, J., Izquierdo, A. E., Grau, A., (Eds.) *La Puna Argentina: Naturaleza y Cultura. Serie Conservación de la Naturaleza 24.* Fundacıon Miguel Lillo, Tucumán, pp 96–111. ISSN 0325-9625.

Izquierdo, A. E., & Grau, H. R. (2009). Agriculture adjustment, land-use transition and protected areas in Northwestern Argentina. *Journal of Environmental Management*, *90*(2), 858–865. https://doi.org/10.1016/j.jenvman.2008.02.013

Izquierdo, A. E., Grau, H. R., Carilla, J., & Casagranda, E. (2015). Side effects of green technologies: the potential environmental costs of lithium mining on high elevation Andean wetlands in the context of climate change. *Newsletter Global Land Project*, *12*, 52–56. ISSN 2316-3747.

Izquierdo, A. E., Grau, H. R., Navarro, C. J., Casagranda, E., Castilla, M. C., & Grau, A. (2018a). Highlands in transition: Urbanization, pastoralism, mining, tourism, and wildlife in the Argentinian Puna. *Mountain Research and Development*, *38*(4), 390–400. https://doi.org/10.1659/MRD-JOURNAL-D-17-00075.1

Jerez, B., Garcés, I., & Torres, R. (2021). Lithium extractivism and water injustices in the Salar de Atacama, Chile: the colonial shadow of green electromobility. *Political Geography*, *87*, 102382. https://doi.org/10.1016/j.polgeo.2021.102382

Liu, W., & Agusdinata, D. B. (2021). Dynamics of local impacts in low-carbon transition: agent-based modeling of lithium mining-community-aquifer interactions in Salar de Atacama, Chile. *The Extractive Industries and Society, 8*(3), 100927. https://doi.org/10.1016/j.exis.2021.100927

Marconi, P., Arengo, F., & Clark, A. (2022). The arid Andean plateau waterscapes and the lithium triangle: flamingos as flagships for conservation of high-altitude wetlands under pressure from mining development. *Wetlands Ecology and Management, 30*(4), 827–852. https://doi.org/10.1007/s11273-022-09872-6

Martınez, J. G. (2018). Sociedades prehispanicas de la Puna Argentina: desde el poblamiento temprano hasta los inicios de la produccion pastoril y agrıcola. In: Grau, H. R., Babot, J., Izquierdo, A. E., Grau, A., (Eds.) *La Puna Argentina: Naturaleza y Cultura. Serie Conservación de la Naturaleza 24*. Fundacıon Miguel Lillo, Tucuman, pp 273–294. ISSN 0325–9625.

McDowell, J. Z. & Hess, J. J. (2012). Accessing adaptation: multiple stressors on livelihoods in the Bolivian highlands under a changing climate. *Global Environmental Change, 22*(2), 342–352. https://doi.org/10.1016/J.GLOENVCHA.2011.11.002

Ministerio de Desarrollo Productivo. (2021). Plan Estrategico para el Desarrollo Minero Argentino. Secretaria de Mineria, Buenos Aires. https://www.argentina.gob.ar/sites/default/files/memoria-viva-avance-2021.pdf

Morales, M., Carilla, J., Grau, H. R., & Villalba, R. (2015). Multi-century lake area changes in the Andean high-elevation ecosystems of the Southern Altiplano. *Climate of the Past, 11*, 1821–1855. https://doi.org/10.5194/cpd-11-1821-2015

Morales, M. S., Crispín-DelaCruz, D. B., Álvarez, C., Christie, D. A., Ferrero, M. E., Andreu-Hayles, L., Villalba, R., Guerra, A., Ticse-Otarola, G., Rodríguez-Ramírez, E. C., Loccla-Martínez, R., Sanchez-Ferrer, J. & Requena-Rojas, E. J. (2023). Drought increase since the mid-20th century in the northern South American Altiplano revealed by a 389-year precipitation record. *Climate of the Past, 19*(2), 457–476.

Navarro, C. J. Izquierdo, A. E., Araoz, E., Fogguet, J., & Grau, H. R. (2020). Rewilding of large herbivores communities in high elevation Puna: geographic segregation and no evidence of positive effects on peatlands productivity. *Regional Environmental Change, 20*, 112. https://doi.org/10.1007/s10113-020-01704-8

Nyantakyi-Frimpong, H., & Bezner-Kerr, R. (2015). The relative importance of climate change in the context of multiple stressors in semi-arid Ghana. *Global Environmental Change, 32*, 40–56. https://doi.org/10.1016/j.gloenvcha.2015.03.003

O'Brien, K., Leichenko, R., Kelkar, U., Venema, H., Aandahl, G., Tompkins, H., Javed, A., Bhadwal, S., Barg, S., Nygaard, L., & West, J. (2004). Mapping vulnerability to multiple stressors: climate change and globalization in India. *Global Environmental Change, 14*(4), 303–313. https://doi.org/10.1016/j.gloenvcha.2004.01.001

Olivera, D. E. (2018). Arqueologıa del formativo: los inicios de la agricultura y la ganaderıa. In: Grau, H. R., Babot, J., Izquierdo, A. E., Grau, A., (Eds.) *La Puna Argentina: Naturaleza y Cultura. Serie Conservación de la Naturaleza 24.* Fundacıon Miguel Lillo, Tucumán, pp. 297–318. ISSN 0325-9625.

Porcuna, A., Guillerminet, T., Klappoth, B., & Schlingmann, A. (2024). Agricultural adaptation to multiple stressors in a climate change context. A case study in south-eastern Senegal. In V. Reyes-García, S. Álvarez-Fernández, P. Benyei, L. Calvet-Mir, D. García-del-Amo, A. B. Junqueira, X. Li, V. Porcher, A. Porcuna-Ferrer, A. Schlingmann, & R. Soleymani (Eds.), *Routledge Handbook of Climate Change Impacts on Indigenous Peoples and Local Communities*. Oxfordshire & New York: Routledge. Chp 13, pp 227–243.

Pörtner, H. O., Roberts, D. C., Poloczanska, E. S., Mintenbeck, K., Tignor, M., Alegría, A., Craig, M., Langsdorf, S., Löschke, S., Möller, V., & Okem, A. (2022). IPCC, 2022: summary for policymakers. In: *Climate Change 2022: Impacts, Adaptation, and Vulnerability. Contribution of Working Group II to the Sixth Assessment Report of the Intergovernmental Panel on Climate Change*. Cambridge University Press, Cambridge, UK and New York, NY, USA pp. 3–33.

Pörtner, H. O., Scholes, R. J., Agard, J., Archer, E., Arneth, A., Bai, X., Barnes, D., Burrows, M., Chan, L., Cheung, W. L., Diamond, S., Donatti, C., Duarte, C., Eisenhauer, N., Foden, W., Gasalla, M. A., Handa, C., Hickler, T., Hoegh-Guldberg, O., Ichii, K., Jacob, U., Insarov, G., Kiessling, W., Leadley, P., Leemans, R., Levin, L., Lim, M., Maharaj, S., Managi, S., Marquet, P. A., McElwee, P., Midgley, G., Oberdorff, T., Obura, D., Osman, E., Pandit, R., Pascual, U., Pires, A. P. F., Popp, A., Reyes-García, V., Sankaran, M., Settele, J., Shin, Y. J., Sintayehu, D. W., Smith, P., Steiner, N., Strassburg, B., Sukumar, R., Trisos, C., Val, A.L., Wu, J., Aldrian, E., Parmesan, C., Pichs-Madruga, R., Roberts, D.C., Rogers, A. D., Díaz, S., Fischer, M., Hashimoto, S., Lavorel, S., Wu, N., Ngo, H. T. (2021). *Scientific Outcome of the IPBES-IPCC Co-Sponsored Workshop on Biodiversity and Climate Change*; IPBES Secretariat, Bonn. https://doi.org/10.5281/zenodo.4659158

Prior, T., Wäger, P. A., Stamp, A., Widmer, R., & Giurco, D. (2013). Sustainable governance of scarce metals: the case of lithium. *Science of the Total Environment, 461*, 785–791. https://doi.org/10.1016/j.scitotenv.2013.05.042

Quiroga Mendiola, M., & Cladera, J. (2018). Ganaderıa en la Puna Argentina. In: Grau, H. R., Babot, J., Izquierdo, A. E., Grau, A., (Eds.) *La Puna Argentina: Naturaleza y Cultura. Serie Conservación de la Naturaleza 24.* Fundacıon Miguel Lillo, Tucumán, pp. 387–402. ISSN 0325-9625.

Räsänen, A., Juhola, S., Nygren, A., Käkönen, M., Kallio, M., Monge, A., & Kanninen, M. (2016). Climate change, multiple stressors, and human vulnerability: a systematic review. *Regional Environmental Change, 16*(8), 2291–2302. https://doi.org/10.1007/s10113-016-0974-7

Reid, P., & Vogel, C. (2006). Living and responding to multiple stressors in South Africa—Glimpses from KwaZulu-Natal. *Global Environmental Change, 16*(2), 195–206. https://doi.org/10.1016/j.gloenvcha.2006.01.003

Reyes-García, V., Álvarez-Fernández, S., Benyei, P., Junqueira, A. B., Labeyrie, V., Li, X., Porcher, V., Porcuna-Ferrer, A., Schlingmann, A., & Soleymani, R. (2023). Local indicators of climate change impacts described by indigenous peoples and local communities: study protocol. *PLoS One, 18*(1), e0279847. https://doi.org/10.1371/journal.pone.0279847

Reyes-García, V., Benyei, P., Junqueira, A.B., Huanca, T., Conde, E. (2024). A complex matrix. Reports of environmental change and its drivers by the Tsimane', Bolivian Amazon. In V. Reyes-García, S. Álvarez-Fernández, P. Benyei, L. Calvet-Mir, D. García-del-Amo, A. B. Junqueira, X. Li, V. Porcher, A. Porcuna-Ferrer, A. Schlingmann, & R. Soleymani (Eds.), *Routledge Handbook of Climate Change Impacts on Indigenous Peoples and Local Communities*. Oxfordshire & New York: Routledge. Chp 10, pp 164–179.

Reyes-García, V., Fernández-Llamazares, Á., Guèze, M., Garcés, A., Mallo, M., Vila-Gómez, M., & Vilaseca, M. (2016). Local indicators of climate change: the potential contribution of local knowledge to climate research. *Wiley Interdisciplinary Reviews: Climate Change*, *7*(1), 109–124. https://doi.org/10.1002/wcc.374

Reyes-García, V., García-del-Amo, D., Benyei, P., Fernández-Llamazares, Á., Gravani, K., Junqueira, A. B., Labeyrie, V., Li, X., Matias, D. M., McAlvay, A., Mortyn, P. G., Porcuna-Ferrer, A., Schlingmann, A., & Soleymani-Fard, R. (2019). A collaborative approach to bring insights from local observations of climate change impacts into global climate change research. *Current Opinion in Environmental Sustainability*, *39*, 1–8. https://doi.org/10.1016/j.cosust.2019.04.007

Salick, J., & Ross, N. (2009). Traditional peoples and climate change. *Global Environmental Change*, *19*(2), 137–190. https://doi.org/10.1016/j.gloenvcha.2009.01.004

Scrosati, B., & Garche, J. (2010). Lithium batteries: status, prospects, and future. *Journal of Power Sources*, *195*(9), 2419–2430. https://doi.org/10.1016/j.jpowsour.2009.11.048

Sovacool, B. K. (2021). Who are the victims of low-carbon transitions? Towards a political ecology of climate change mitigation. *Energy Research & Social Science, 73*, 101916. https://doi.org/10.1016/j.erss.2021.101916

Tengö, M., Hill, R., Malmer, P., Raymond, C. M., Spierenburg, M., Danielsen, F., Elmqvist, T., & Folke, C. (2017). Weaving knowledge systems in IPBES, CBD and beyond—lessons learned for sustainability. *Current Opinion in Environmental Sustainability*, *26–27*, 17–25. https://doi.org/10.1016/j.cosust.2016.12.005

Tschakert, P. (2007). Views from the vulnerable: understanding climatic and other stressors in the Sahel. *Global Environmental Change*, *17*(3–4), 381–396. https://doi.org/10.1016/j.gloenvcha.2006.11.008

Urrutia, R., & Vuille, M. (2009). Climate change projections for the tropical Andes using a regional climate model: temperature and precipitation simulations for the end of the 21st century. *Journal of Geophysical Research: Atmospheres*, *114*(D2). https://doi.org/10.1029/2008JD011021

Voskoboynik, D. M., & Andreucci, D. (2022). Greening extractivism: environmental discourses and resource governance in the 'Lithium Triangle'. *Environment and Planning E: Nature and Space, 5*(2), 787–809. https://doi.org/10.1177/25148486211006345

Vuille, M., Francou, B., Wagnon, P., Juen, I., Kaser, G., Mark, B. G., & Bradley, R. S. (2008). Climate change and tropical Andean glaciers: past, present and future. *Earth-Science Reviews*, *89*(3–4), 79–96. https://doi.org/10.1016/j.earscirev.2008.04.002

Wanger, T. C. (2011). The Lithium future—resources, recycling, and the environment. *Conservation Letters*, *4*(3), 202–206. https://doi.org/10.1111/j.1755-263X.2011.00166.x

Yap, M. L. M., & Watene, K. (2019). The sustainable development goals (SDGs) and indigenous peoples: another missed opportunity? *Journal of Human Development and Capabilities, 20*(4), 451–467. https://doi.org/10.1080/19452829.2019.1574725

9

SETTLEMENT, WAY OF LIFE AND WORLDVIEWS

How socio-environmental changes impact and are interpreted by artisanal fishing communities in Portugal

Yorgos Stratoudakis, Patrícia Gonçalves, Daniel Oliveira and Mônica Mesquita

Introduction

The contemporary history of Portugal is tightly linked with fishing. Throughout the Middle Ages and up to the 1830s, small-scale fishing was mainly performed within designated areas, in rivers, lagoons and estuaries, following authorisation of the private, monastic or regal landowner (Coelho, 1995). The activity involved payment of rent and taxes, as well as membership fees in local brotherhoods or *cofrarias*, that reduced fisher's yield and monetary returns (Santos et al., 2012). The liberal revolution in the early 19th century (Costa Pinto and Monteiro, 2020) led to a political and social reconfiguration that increased the importance of fishing for making a living (Garrido, 2018). In the 1830s, the legislative abolition of taxes and religious or social obligations and the passage of fisheries' administration to the state, turned fishing into the most important activity for coastal and riverine communities in Portugal (Garrido, 2018). The importance of fishing was reinforced in 1864 by a regal decree that declared water, coasts and seabed as public property administered by the state. By the end of the 19th century, a large fraction of the active population near the coast were living from fishing (Baldaque da Silva, 1892) or were employed in the sardine (*Sardina pilchardus*) canning industry that was starting to grow in the country (Pulido Valente, 1981).

The western coast of mainland Portugal is exposed to the swell and winds from the northeast Atlantic Ocean. These create highly energetic littoral environments and shape coastal topography. Rivers in northern and central Portugal provide a sedimentary outflow that follows a southwards littoral drift according to the prevailing wave regime (Lira et al., 2016). Long stretches of exposed lowland coastal areas in northern and central Portugal are covered with sandy dunes and its characteristic sparse vegetation (Costa et al., 2000). Sandy dunes could once reach several kilometres wide and become highly mobile under the influence of the wind, particularly in landscapes void of trees. For many centuries most exposed oceanic coasts were probably desert (*terra nullis* in citation of Freitas et al., 2018). Until the end of the 19th century, these stretches had a positive sedimentary budget (accretion) with a net supply of sand piling onto the coast (Freitas et al., 2018).

DOI: 10.4324/9781003356837-12

Such coasts were rather hostile environments for humans (Freitas, 2007), where occasional flooding after heavy rain formed wetlands attractive for mosquitoes, vectors of diseases like malaria (e.g., Palma et al., 2022). Like in most of the European Atlantic coastline, until the 20th century, few people lived along the open coast of mainland Portugal. The main regular human presence was fishers who temporarily settled during the summer months, when storms were less likely and wave height diminished, reducing the risk of capsizing in the surf area (Freitas et al., 2018). Up to the 18th century, most of these fishers created temporary coastal settlements (perishable wooden structures with thatched roofs known as *palheiros* – Freitas et al., 2018) and moved to the interior during the harsher winter months (Garrido, 2018). Accumulated experience from trial-and-error practices that sustained survival in such demanding environments allowed fishing communities to create a wealth of traditional ecological knowledge (TEK) (Freitas et al., 2018). Among other benefits to local communities, TEK allows them to make use of locally available resources and existing technologies to create societal responses to cope with adverse and dynamic environmental conditions (Berkes et al., 2000). This body of knowledge is transmitted across generations through oral history and cultural practices and may be lost when significant sociodemographic changes take place in a community (Freitas et al., 2018; Delgado-Ramírez et al., 2023; Stratoudakis et al., 2023).

Early fishers' settlements in the sandy lowland of the Portuguese coast were mostly linked to the seasonal operation of *Arte-Xávega* (beach seines). Summer weather conditions permitted fishing for sardine, horse mackerel (*Trachurus trachurus*), chub mackerel (*Scomber colias*) and related species in shallow coastal areas with sandy bottoms (Cabral et al., 2003). The *Arte-Xavega* consisted in using an encircling gear towed to the beach using an auxiliary vessel ("*chaveiro*" or "*semi-liua*") and human or animal force to pull the net onshore with long ropes (Martins et al., 2000). Fishing communities using *Arte-Xávega* started to settle permanently along the coast during the late 18th and 19th centuries to provide food for increasing city populations and services for the increasing numbers of beachgoers (Freitas, 2007). Fishing with *Arte-Xávega* occurs at the land-ocean interface and demands a plurality of skills and understandings, simultaneously requiring knowledge on fish species and habitats, understanding of oceanographic forcing in very dynamic environments, and skills to interpret signs in coastal morphology and local economy (Smart Fishing, 2021). Nowadays *Arte-Xávega* is considered Immaterial Cultural Heritage in Portugal (INPCI, 2021) and is still practised in several coastal municipalities, mainly in central Portugal.

The aim of this chapter is to capture and compare the environmental knowledge, practices and adaptations of *Arte-Xavega* over the past century in two neighbouring communities that settled at two different periods of time. The selected communities are less than 10 km away from each other and share a very strong common inheritance related to fishing and, especially, *Arte-Xávega*. In both places, experience, thoughts, emotions and social relationships have been for many generations locally structured by artisanal fishing, from which individuals and communities have been deriving meaning, beliefs, symbols, values and feelings (Chapin and Knapp, 2015). These communities have been locally modifying their socio-environmental surroundings through mind facts and artefacts related to *Arte-Xávega* (D' Ambrosio, 2001). In-depth interviews, directed questionnaires, literature information and participant observations were used to describe a century of experience, observations and interpretations of socio-environmental changes and respective impacts on fishing practice. In the process, sense of place and worldviews based on the local relation with fishing and the sea were described and compared between these two fishing communities on the Atlantic coast of central Portugal, where climatic change is perceived as just one among the drivers of change affecting their daily life and income. Minor, but perceptible, differences emerging in the way of life and worldviews of the two communities are also used to hypothesise on the direction of possible changes in the coming decades.

Material and methods

We worked in two communities founded by *Arte-Xavega* fishers in the late 18th and the early 20th centuries, Costa de Caparica (CdC) and Fonte da Telha (FT), respectively. Both communities are in the Municipality of Almada on the Atlantic coast southwards of the river Tagus' estuary (Figure 9.1). The sites extend along an exposed oceanic stretch of 20 km from the southern tip of the Tagus estuary to the Albufeira lagoon, a semi-enclosed coastal lagoon near Cape Espichel. Landwards of the sandy stretch and its system of dunes (Palma et al., 2022), a Protected Landscape Area that has existed since 1984, classified for its geological, geomorphological and landscape value. The southernmost part of the Protected Landscape has also been a Botanical Reserve since 1971 (Oliveira, 2015).

Although very close to the capital (less than 20 km by road from the centre of Lisbon), the two sites differ in their degree of exposure to urban life. Currently, CdC is a large peri-urban agglomerate of 15,000 permanent inhabitants, administratively considered a city since 2004. During the summer, CdC multiplies several times its population and becomes culturally invaded as a result of beach-going, surfing and tourism by seasonal holiday homeowners from Lisbon and by international visitors. CdC fishers are currently concentrated in a neighbourhood at the southern city limit, close to where they deploy their vessels. This differs from the central location they occupied since the original settlement and up to 50 years ago. The closest fishing port is Trafaria, at the edge of the Tagus estuary (Figure 9.1). FT, although just 6 km southwards of CdC along a continuous

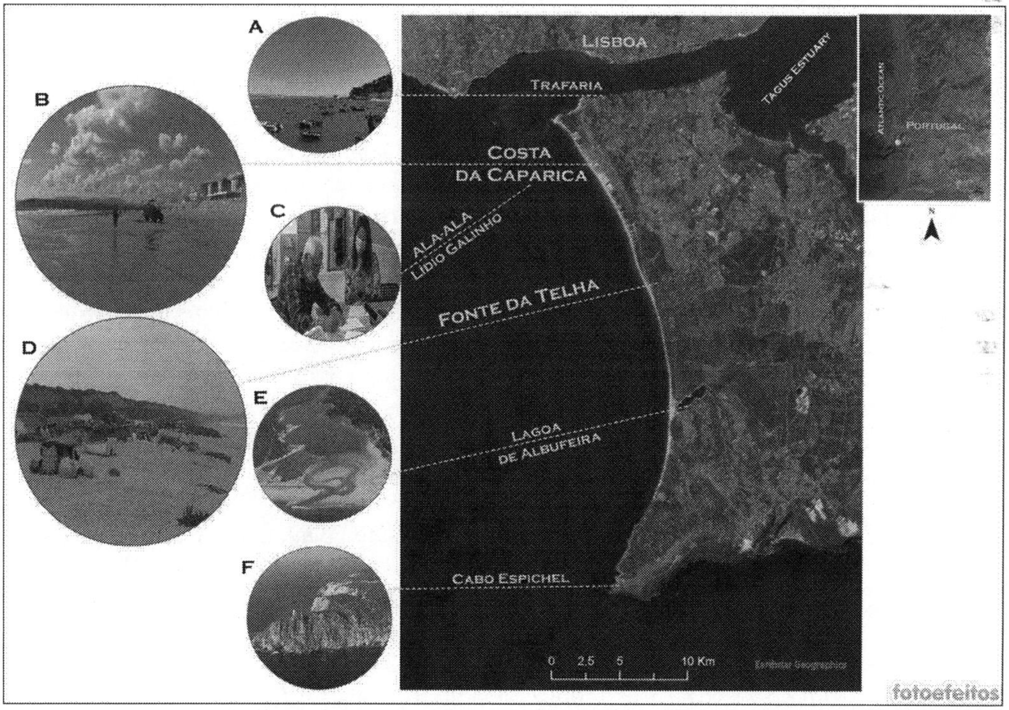

Figure 9.1 Location of study sites in Portugal and related place names. (A) Trafaria, (B) Costa de Caparica, (C) Lídio Galinho (founder and president of fishery association Ala-Ala) signing the Free Prior Informed Consent before the livelihood interview, (D) Fonte da Telha, (E) Lagoa de Albufeira, (F) Cape Espichel.

sandy beach stretch, remains an isolated village with a single road entry that splits along the coast into the fishers' neighbourhood northwards and the touristic neighbourhood southwards. Both sites have a *lota*, a fish auction house where local landings are marketed daily. During the summer, some transactions with tourists also occur on the beach.

Fishing takes place along 10–12 km of sandy beaches (Figure 9.1 – southwards of the groynes and seawall that delimit the touristic area in CdC and northwards of Lagoa de Albufeira) and up to 1.5 km from the coast. Landmarks along this stretch with local toponyms identify the main spots for gear deployment (*lanços*) in a fishing activity without territoriality. The predominant benthic habitat is mobile sediment (sand) with a gentle slope (Freire, 1986) that frequently leads to resuspension due to wave action (fishing area mostly less than 20 m deep). *Arte-Xávega* remains the emblematic fishing activity in both sites, with vessels and gear currently deployed with the help of tractors (Martins et al., 2000). A net of approximately 500 m wide is towed landwards from 1 km offshore by tractors and with the assistance of a crew of about 20 people (*companha,* which can include women and teenagers), of which three to five are in the boat. Activity is seasonal (usually from March–April to October–November) as during rough weather the boats cannot overcome the surf zone. During the summer tourist season, fishing operations are regulated to start in the evening and stop in the morning. Fishers in the two sites also perform other fishing activities, especially with trammel nets operated from the same or from other fishing vessels (Souto, 1998). Vessels are either deployed from the beach or come out from Trafaria. Fishing may also involve gill nets, lines and long-lines, especially within the Tagus estuary, which offers an alternative fishing area under rough oceanic weather. In total, fishing involves 34 small-scale vessels in the two sites (16 in CdC and 18 in FT) and over a hundred fishers in each site, many of them women.

The encounter spaces in an investigation are fundamental elements for the dynamics of the dialogue that sustains it (Mesquita et al., 2016). Previous research carried out in this territory with the fishing communities follows a trajectory of active participation and action, promoting co-construction of knowledge where fishers become researchers of their own practices. These fishing communities have been working with the same research group since 2008, initially on D.A.R. Costa Tr@nsFormArte (a government project of local intervention) and Urban Boundaries (http://fronteirasurbanas.ie.ul.pt/). Ties were further strengthened with the development of the Oceanic Literacy Observatory – OLO (www.olo.blue) in 2016 and the formal integration of some members of the fishing community into it. More recently the Partibridges (https://partibridges.eu/partibridges-portugal/) and Smart Fishing (http://smartfishing.olo.blue/) projects have encouraged the co-construction of a shared interest in the development and sustainability of local artisanal fisheries. In this journey, a bond of belonging to local causes, both fishing and academic, has been created. This facilitated the entry and active participation of fishers in research on climate change impacts following the LICCI protocol (Reyes-García et al., 2023), fostering their interest and willingness to share knowledge and behaviours, strengthen their trust and participatory action. It also allowed fishers to expand their circle of trust by interacting with a new researcher on a new topic – through a critical and slow social pedagogy (Mesquita et al., 2016). Finally, this bond also permitted the inclusion of a critical ethnographic context in the current research (Thomas, 1993), tapping on the past experience of the research team members consolidated during the site contacts for the new research topic (direct and participant observation).

The fisheries LICCI protocol was followed (Miñarro et al., 2020). COVID limitations prolonged the research for almost two years and permanently interrupted the interview with Lídio Galinho, the president of an influential local fishers' association (see acknowledgements). Livelihood interviews took place with the most senior members of the two communities (four men and three women, 64–94 years old). Thematic questions were prepared following a standardised protocol (Miñarro et al., 2020; Reyes-García et al., 2023), but the interviewees were left to provide additional information and memories they considered fit in the conversation. Memories and significant events reported in the interviews were cross-checked with selected literature (Freire, 1986; Rodrigues, 2000; Silva, 2008; Oliveira, 2015) and online sources of information (e.g., historical archives from the Municipality of Almada and Lisboa) to construct the timeline of the study site. An interview regarding species used was transformed into a questionnaire that was responded to by 18 fishers (7 from CdC and 11 from FT, 32–78 years old and a mean fishing experience of 41 years). Livelihood interviews (3) and questionnaires (3) included women respondents from both sites, but gender analysis was not an objective of this study.

Questionnaire respondents had a mean age of 58 years old and came from families that, on average, had already had the three previous generations into *Arte-Xávega* fishing. Some respondents represented the fifth generation, thus also providing information to consolidate the local timeline. Each fisher provided a list of species caught nowadays, in decreasing order of abundance in the catch, and further indicated whether abundance had changed since that person started fishing (five levels response scale: Decreased a lot, Decreased, Unchanged, Increased, or Increased a lot since onset of fishing). Other questions focused on whether species had appeared or disappeared since the onset of fishing, including those with no commercial interest, and whether the number of weekly meals including fish had changed over the past 20 years. The LICCI classification tree was used to specify typologies of change for the atmospheric, physical and life systems (Reyes-García et al., 2023). Informed consent was provided in writing before the interviews. Validation was based on the degree of overlap between the livelihood interviews and the questionnaires. Differences in questionnaire replies between the two sites and between past and present, by site and overall, were tested with Wilcoxon rank sum tests.

Results

The timeline presented in Figure 9.2 spans from the end of the 19th century to the present across four national political phases. It reports temporal marks that have shaped coastal, marine and fisheries operations in Portugal during that period, as well as regional or local development acts and events (the latter reported for CdC but covering the two sites).

During the end of the 19th century, CdC was a large fishing community isolated from the rest of the municipality and from urban centres. Approximately 700 fishers and another 500 people offering land assistance (fish work) were almost exclusively dependent on *Arte-Xávega* and sardine fishing. This period is within the indirect memory span of the oldest fishers interviewed, evoked through comments and memories transmitted from parents or grandparents. During that time, life was harsh and largely dependent on intra-community solidarity, particularly among members of the *companha* of each vessel who tended to live in neighbouring *palheiros*, as well as on the capacity to culturally transmit traditional knowledge on the marine resource system and the fishing practices to guarantee food and survival. Fatal accidents during the artisanal fishing operation occurred frequently and required the solidarity of the families of surviving members of the *companha* to fend for the descendants of the deceased. The *Costa Pinto* fire in the 1880s was the onset of basic improvements in housing in CdC, since, before that moment, *palheiros* did not have even basic infrastructures available. Further improvements were introduced with the *Casa dos Pescadores*, until arriving more recently to the current *Bairro dos Pescadores* with good housing conditions, but peripheral to the city centre.

Events specifically reported for FT in Figure 9.2 demonstrate the delayed process of community settlement, compared to CdC. Contrastingly to CdC, permanent settlement of fishers in FT only started in the 1930s. This is within the direct memory span of some interviewees that moved there as young children within the families of the first settlers. Brick houses, sanitation and

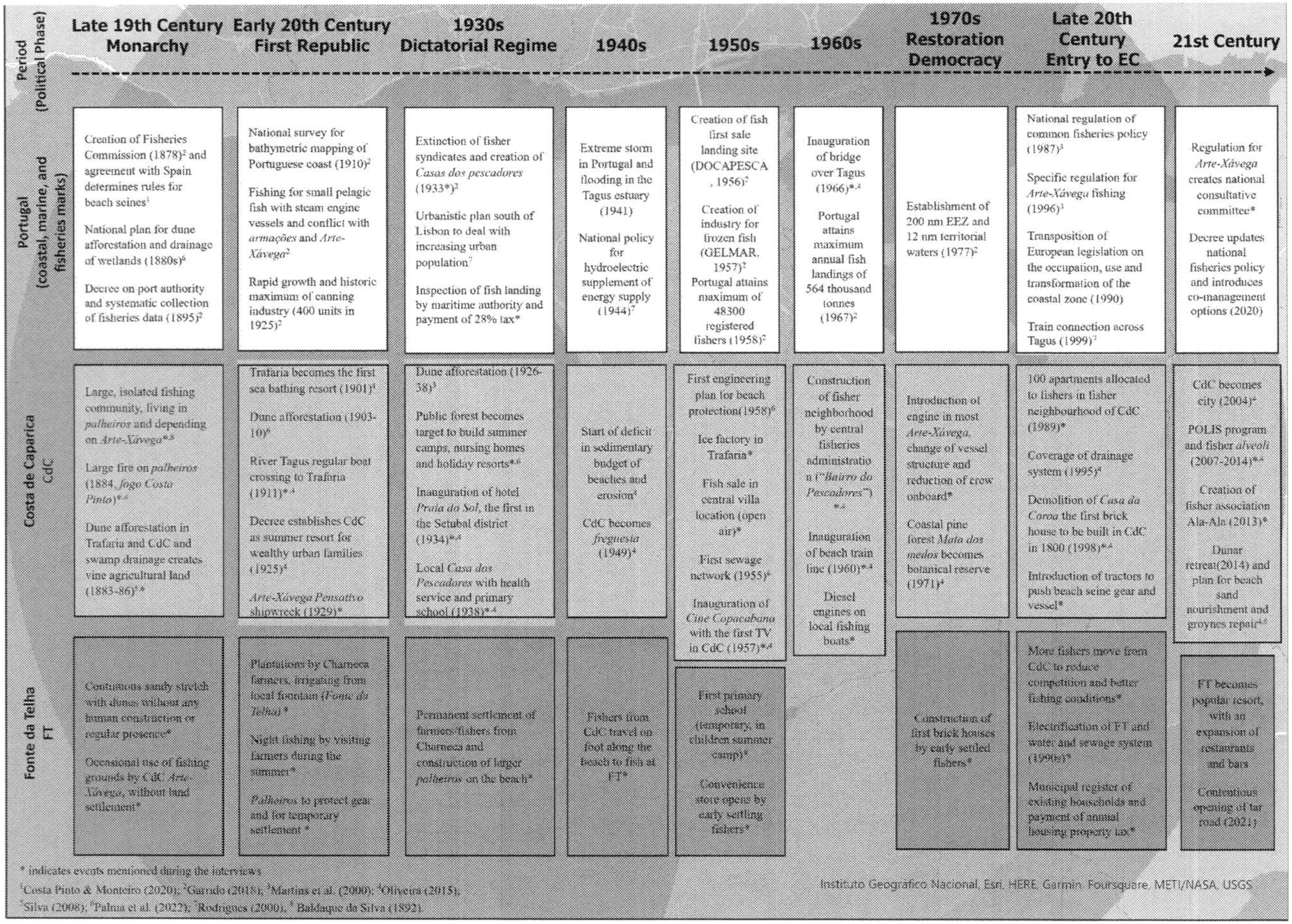

Figure 9.2 Infogram of historical timeline from the late 19th century onwards, based on the span of interviewees' memories and supported by literature review (asterisk indicates events mentioned during the interviews).

electrification only occurred in FT during the 1990s, while the process of property registration is still ongoing. This later settlement and the continuing village isolation from urban settings can explain some of the distinctions in the sense of belonging to the place detected in the two communities (Table 9.1). Despite the close geographical proximity, the common provenance of some community members, and the identical fishing practices, there are detectable differences in the daily life, mind facts and artefacts that currently shape the way of living in the two sites. Some of the fishing materials and practices (including the organisation of storing space for gear next to their home within *quintais*) in FT are still related to the traditional operation at the origin of these communities, while in CdC more modern practices have been adopted. Reduced topological

Table 9.1 Common and distinctive elements in the sense of place attributed to fishers from Costa de Caparica and Fonte da Telha, based on ethnographic observations consolidated with interviews and participant observation

	Costa de Caparica (CdC)	*Fonte da Telha (FT)*
Sense of Place (common element in CdC and FT)	(distinctive element in CdC)	(distinctive element in FT)
Daily life and income (Fishermen, more than fisherwomen, frequent the local *Lota* on a daily basis)	Daily routine not only around fishing and fish work (some, especially in younger generations, also have other jobs) Some fishers do not live in CdC	Daily routine and individual occupations are all around fishing and fish-work, including younger generations All fishers live in FT
Mindfacts (Fishing identity related to *Arte-Xávega*; Relationship with the ocean and its natural elements)	Sense of belonging to the fishing community of CdC is intertwined with a sense of urbanity, where the fisher is inserted and makes part of The fishing community does not reveal unity in its topology, fishers isolate in the modern *Alvéolo*	Sense of belonging to FT is deep and remains inextricably linked to fishing practices, especially *Arte-Xávega* The fishing community reveals unity through its topological distribution, by its feeling of neighbourhood
Artefacts (Fishing arts; Revenue repartition within *companha*; Dynamic and static references to fishing practices; Landmark toponyms as *lanços*)	Some *companhas* need to find fishermen in FT to mend their nets All fishing materials are industrialised Fisher storing places are part of the national coastal rehabilitation project (POLIS). Fishers still denominate the modern structure with its ancestral name – *Alvéolo*. All fishers have an identical space, built by an architect, with reduced space for storage. The houses of fishers are far from this place, sometimes in CdC, sometimes in another municipality.	Fishermen mend their nets locally Some fishing materials are handmade, still produced locally in wood (needles) and cotton (pieces of nets) Fisher storing places are traditional, on sand, between the ocean and the fisher homes. Fishers still denominate the storing places as *Quintais* – as their ancestors. Each *Quintal* occupies the traditional place where each family has constructed them, with a tendency to enlarge them to add new or more artefacts. The *Quintais* vary in size and attend to the space that each fishing family needs.

proximity and diversification of income sources, especially for the younger generations of fishers, are also signs that urbanisation in CdC is modifying daily dynamics and routines away from the all-embracing role that artisanal fishing had on previous CdC generations and still has in FT.

Figure 9.2 also describes the changes that took place in the local *Arte-Xávega* fishing during the past 40 years (reported for CdC but applying to both sites, practically at the same time). Mechanisation, through the use of boat engines and tractors, and technological innovation, through the use of more resistant fishing material and weather predictions, created safer conditions, prolonged oceanic operations and increased fishing effort. In parallel, livelihood interviews indicated that, until the fisher generation before the present, all meals in these two communities were fully dependent on fish, either fresh during the fishing season or salted dry in winter when weather did not permit fishing. In the questionnaires, fishers reported a current median of seven meals per week containing fish. The difference with fish consumption 20 years ago is statistically significant ($W = 47.5$, $p < 0.001$), without significant differences between the two sites. Nowadays, approximately half of the sample (56%) still only consume fish they have captured, while the rest also buy from the local fish market, again without significant differences between the two sites.

The questionnaire replies reported 44 marine taxa as part of the current catch in *Arte-Xávega* belonging to 41 species or genera, of which seven were invertebrates. Three names were colloquial names for juvenile forms of sardine and horse mackerel. More taxa were reported in CdC (median: 9; total: 38) than in FT (median: 5, total: 26), and the difference was highly significant between sites ($W = 8081$, $p < 0.001$), although there was no significant difference in the number of species listed according to fishers´ age, years fishing, or years since settlement. Chub mackerel, horse mackerel and squid (*Loligo* sp.) were mentioned by all fishers, and sardine, Henslow's swimming crab (*pilado*, referring to *Polybius henslowii*), and jellyfish (*alforrecas*, referring to *Catostylus tagii*) were mentioned by all but one. Half or more also mentioned garfish (*Belone belone*) and seabream (*Sparus aurata*). More than a quarter mentioned salema (*Sarpa salpa*), cuttlefish (*Sepia oficinalis*), meagre (*Argyrosomus regius*), seabass (*Dicentrarchus labrax*), weavers (predominantly the lesser weaver *Echiichthys vipera*), bogue (*Boops boops*), annular seabream (*Diplodus annularis*) and other species of the genus *Diplodus* (*sargos*). In terms of current abundance in the catch, there were no significant differences between the two sites: chub mackerel is currently considered the most abundant species, followed by horse mackerel and then sardine. Although mentioned less frequently, salema was ranked as slightly more abundant than the three invertebrate species (swimming crab, jellyfish and squid) that followed.

In relation to climatic changes reported, the most frequent changes observed relate to the 'Marine physical' subsystem and the subsystems 'Air masses' and 'Seasons' of the atmospheric system. Practically, all fishers mentioned either a decrease in the number of windy days (*vendavais*), a decrease in the frequency of storms (*maresias, tempestadas*), or both, in recent years, with a consequent prevalence of milder weather and less agitated sea. A third of fishers from both sites also referred to changes in the seasonality, with the blurring and disappearance of seasons, the rapid (daily) modification of conditions and the prevalence of warmer weather. Several fishers from the two sites also mentioned an increase in sea level and the disappearance of the primary dune system. Some fishers provided more specific or additional information on the topics above. For example, some mentioned a decrease in the intensity of storms, a reduction of the duration of the season with storms (from autumn and winter to just winter), a delay in the period with northerly winds (*nortadas*), a reduction of the sand available to the coastal system, and warmer summer nights in recent times. Interestingly, the above impacted elements were also mentioned in the livelihood interviews, when asked about environmental changes they had experienced. The respondents further mentioned important changes related to a decrease in precipitation, including

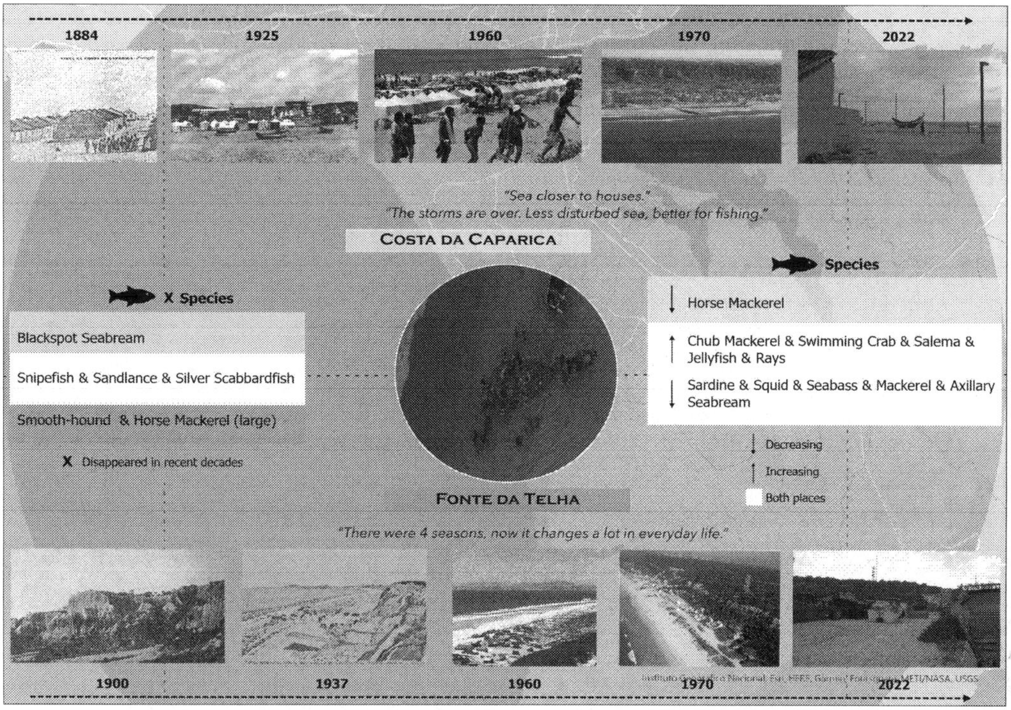

Figure 9.3 Infogram of *Arte-Xávega* fishing and evolution during the study period, both socially (landscape and communities) and biologically (change in species presence and abundance in the catch) in Costa de Caparica and Fonte da Telha.

the reduction of freshwater outflow from the river Tagus, more permanent closure of the Lagoa de Albufeira, and reduction of water runoff to the coast from the cliffs at FT in recent years, but these changes were not mentioned when responding to the questionnaire.

Fishers reported many changes in the marine biological system. All fishers answering the questionnaire mentioned that snipefish (*Macroramphosus* sp.) have disappeared from the catch in recent times and a third of the sample referred the same for the juvenile and/or adult of axillary seabream (*Pagellus acarne*). Three fishers from CdC also mentioned the recent disappearance of juvenile blackspot seabream (*Pagellus bogaraveo*), four fishers from FT mentioned the disappearance of smooth-hound (*Mustelus* sp.), and three mentioned the disappearance of large horse mackerel (*chicharo*). During the interviews, but not during the questionnaires, the disappearance of sandlance (*Ammodytes* sp.) and juveniles of silver scabbardfish (*Lepidopus caudatus*) in recent decades was also mentioned. In relation to species still present in the catch but less abundant in relation to the onset of fishing, questionnaire replies indicated a decrease for sardine, which was found to be highly significant overall ($V = 91$, $p = 0.001$) and significant in each site ($V = 21$, $p = 0.030$ in CdC; $V = 28$; $p = 0.018$ in FT); a decrease for squid, highly significant overall ($V = 111$, $p = 0.003$) and significant in each site ($V = 26.5$, $p = 0.035$ in CdC; $V = 32.5$, $p = 0.040$ in FT); and a decrease for horse mackerel, highly significant overall ($V = 45$, $p = 0.007$), significant in CdC ($V = 21$, $p = 0.032$) but not when considering perceptions of temporal change only in FT ($V = 6$, $p = 0.174$).

A marginally significant decrease in yield was also detected overall for seabass ($V = 15$, $p = 0.048$). A large reduction in yield was also reported for mackerel (*Scomber scombrus*), but the species was reported too few times for the answer to be statistically significant. On the contrary, there was a significant increase in reported current abundance of rays (*Raja* sp.) in the catch ($V = 0$, $p = 0.038$). For the remaining main species currently caught by *Arte-Xávega* (chub mackerel, swimming crab, salema and jellyfish), there was no evidence of significant change based on the interview responses, but some indication of non-significant increase in recent years for the latter two species (Figure 9.3).

When asked in the questionnaire whether the climatic changes reported above have had an impact on fishing, most fishers answered affirmatively (only 22% denied it), but when asked whether this has led to a modification of fishing practices, the answer was predominantly negative (56% reported no change). A reference to climatic impacts on fishing practices was further presented in the livelihood interviews in relation to the past appearance of stranded brown algae on the beach (after southerly winds during winter months) that were collected, dried and sold. In recent decades, these kelp forests near Cape Espichel have disappeared and the activity has stopped.

Discussion

There is no doubt that, from the time the communities were originally settled to the present, artisanal fishing and *Arte-Xávega* have structured the way of living of fishers in CdC and FT. The livelihood interviews and timeline (Figure 9.2) showed that, in both places, fishing and fish work constituted the defining elements of daily family lives and routines and shaped individual and group identities and values. This corresponds to a worldview underpinned by artisanal fishing (Delgado-Ramírez et al., 2023), where TEK contributes to livelihood practices and worldviews across generations (Berkes et al., 2000). However, the present study also revealed that in recent decades, these livelihoods have experienced a change in several dimensions of life (environmental and climatic, social and technological) that have impacted the two communities and shaped individual fishers′ perceptions, interpretations and strategies for adaptation. Some of these impacts and interpretations are common in the two fishing communities and some are distinct, allowing us to hypothesise a diverging direction of possible changes for the two communities in the coming decades.

The technological upgrade of fishing with *Arte-Xávega* since the 1970s occurred in the same way in the two communities (Figure 9.2), with some differences in the evolution of artefacts related to fishing (Table 9.1). On the one hand, construction of better boats with an engine, mechanical support on land, the improvement in the design of the gears and the use of weather information occurred practically simultaneously in the two communities and had the same results. Mechanisation and innovation created safer conditions for sea-going and increased fishing capacity for all fishers, increasing, accordingly, fishing effort and exploitation rate. On the other hand, wider social changes from the 1950s onwards affected the two communities differently. Urbanisation and tourist influx and massification had a much greater impact on the CdC than on the FT fishing community, breaking the topological unity and modifying the sense of belonging in the former but not in the latter (Table 9.1).

The 20th century also brought broader social transformations to the littoral of Portugal (Figure 9.2), including the arrival of new populations and the growth of urban areas (Costa Pinto and Monteiro, 2020). These social changes have also created environmental impacts that were visible to the respondents of the two communities in similar ways. Over the past 50–60 years, fluvial and maritime engineering works in the Tagus river and around its estuary have contributed to a reduction of sedimentary availability and the destruction of the natural defences of coastal systems, leaving them more vulnerable to extreme weather events and erosion (Palma et al., 2022). Several respondents mentioned coastal erosion as clearly visible within their lifetime and a threat for the future of their local community. In the livelihood interviews, less precipitation and construction of dams in the Tagus river were also associated with limited outflow of nutrient-rich brackish waters that attract horse mackerel in more recent decades. Receding freshwater outflow in FT (whose name comes from a natural fountain at the centre of the village) and increasing salinisation of the Tagus estuary (with a concomitant modification in the plant species used in lowland agriculture and increasing residence period for aquatic species that use the estuary seasonally) were also mentioned.

In terms of fishing, changes in the relative abundance of the main target species have been observed over time, without creating major alterations in the fishing practice. More than a century after the original description of Baldaque da Silva (1892), sardine continues to be an important species for *Arte-Xávega*, although in recent decades, it has been substituted by chub mackerel as the most abundant species. Artisanal fishers, used to seasonal and interannual variation in target species abundance, rapidly adapted to the new situations, although markets tended to react more slowly. In the 1990s–2000s, catch in CdC (Martins et al., 2000) and FT (Cabral et al., 2003) was dominated by chub mackerel, with a part being discarded at the time due to limited commercial interest. Nowadays, market circuits on land have adapted to the new reality and all chub mackerel is sold. Fish are mainly exported to Spain for tuna ranching, although there were many complaints on the sale price at *lota*. In relation to wider environmental changes, the main difference in fish species composition compared to the 1990s is the disappearance of snipefish and its substitution by jellyfish, although some differences are also detected between the two communities (Figure 9.3). Large inter-decadal variations in the local abundance of snipefish (Marques et al., 2005), as well as a northwards expansion of chub mackerel in recent decades (Martins et al., 2013) have been reported in Portugal, matching the information provided locally.

Finally, the study also revealed that the above changes contributed towards two prevailing socio-environmental discourses in relation to fishing with *Arte-Xávega* in recent decades. These discourses were not community-specific and both attributed limited importance to climate as a driver of change in local livelihoods. In the first and more prominent discourse, the main changes over time were considered essentially technological, due to the increase in fishing capacity. Under this discourse, environmental and climatic changes were observed but not considered relevant in the formation of daily routines and the provision of income. There was a recognition of cascading environmental impacts on livelihoods due to reduced fish abundance in recent times, but decreasing fish abundance was not associated with changing climatic conditions (in one interview, it was even suggested that milder current weather is mainly a perception resulting from better housing and vessels) or locally-induced overexploitation. *Arte-Xávega* fishers have previously experienced interannual and seasonal variation in pelagic fish abundance, so reduced fish availability and profitability in recent decades were attributed to overexploitation derived mainly from the use of other fishing gears (trawlers) or inefficient function of the local fish market circuits.

In the second discourse, there was a prevalence to point at a socio-environmental interaction as the main driver of change in the local fishing activity. Marine productivity was reported to have been decreasing in recent decades due to reduced storminess and windiness. This, in turn, is considered to reduce the quantity and diversity of available species to fish with *Arte-Xávega*. Under this discourse, climatic drivers of change gained some relevance as they were seen to directly affect the daily routine and income related to fishing by diminishing the food sources for pelagic and semi-pelagic target species. Questionnaire replies corroborated and reinforced the relationship between storminess/windiness and fish productivity, explained through reduced food availability to fish. This could be due to the less-frequent resuspension of organic matter from the sediments during agitation and column mixing, mentioned by several fishers, or due to the diminishing influx of nutrients through upwelling under northerly winds, mentioned in one interview. However, some fishers also indicated that, simultaneously, less adverse weather permitted longer fishing seasons and hauls and less damage on gear, thus increasing fishing effort, catch, and exploitation rate and reducing price for the most abundant species. This narrative re-introduced a social and technological dimension in the main drivers of change, as milder climatic conditions in recent decades were also perceived to facilitate a overexploitation and inefficient market functioning.

Overall, although some socio-environmental changes have a similar impact and interpretation, subtle differences are emerging between these two neighbouring communities that operate in the same ecosystems, under the same rules and using the same fishing gears. Taking a prospective view to coming decades, it seems more likely that FT fishers will maintain the current way of living and practices than CdC fishers. TEK seems likely to continue to play an important role in FT, while in CdC is more likely to emerge the generational gap currently formed in other artisanal fishing communities in Portugal (Stratoudakis et al., 2023). Nuanced differences between apparently similar artisanal fishing communities are starting to be ethnographically registered (Delgado-Ramírez et al., 2023), but further research will need to consolidate concepts and methods to test the above predictions. In particular, it will be interesting to observe how the climatic concerns related to coastal erosion and altering productivity of the coastal pelagic habitat expressed here will be interpreted in the future by two communities under diverging mechanisms for the social transmission of experience and knowledge.

Acknowledgements

We thank all CdC and FT fishers that have provided time and knowledge to participate in the interviews or answer the questionnaire. In particular, we acknowledge Lídio Galinho, founder and president of Ala-Ala until January 2021, to whom we dedicate this work (Lídio contracted COVID two months after the, unfinished, LICCI interview and passed a month later). We also thank Victoria Reyes-García and the LICCI team for the invitation to collaborate in the project and participate in this volume, as well as for constructive reviews of the original manuscript. The fieldwork of the first author was performed during a sabbatical leave granted by the scientific council of IPMA. This study had the support of national funds through Fundação para a Ciência e Tecnologia (FCT), under the strategic project UDIB/04292/2020 granted to MARE and the project LA/P/0069/2020 granted to the Associate Laboratory ARNET.

References

Baldaque da Silva, AA (1892) *Current state of fisheries in Portugal: understanding sea, river and lake fishing across the continent of the kingdom, referring to the year 1886*. National Press, Lisbon (in Portuguese).

Berkes F, Colding J, Folke C (2000) Rediscovery of traditional ecological knowledge as adaptive management. *Ecological Applications* 10:1251–1262.

Cabral H, Duque J, Costa MJ (2003) Discards of the beach seine fishery in the central coast of Portugal. *Fisheries Research* 63:63–71. https://doi.org/10.1016/S0165-7836(03)00004-3

Chapin FS, Knapp CN (2015) Sense of place: a process for identifying and negotiating potentially contested visions of sustainability. *Environmental Science and Policy* 53:38–46. http://dx.doi.org/10.1016/j.envsci.2015.04.012

Coelho MHDC (1995) River fishing in the Portuguese medieval economy and society. *Cadernos Históricos* 6:81–102 (in Portuguese).

Costa JC, Lousã M, Capelo J, Espirito Santo MD, Sevillano JI, Arsénio P (2000) The coastal vegetation of the Portuguese divisory sector: dunes, cliffs and low-scrub communities. *Finisterra* 35:69–93.

Costa Pinto A, Monteiro NG (Coord, 2020) *Contemporary social history: Portugal 1808–2000*. Fundação MAFPRE and Penguin Random House, Lisboa (in Portuguese).

D'Ambrosio U (2001) Inter and intra-cultural dynamics and the quest for human values. Plenary Talk in International Congress of *Universal Values and the Future of Society*, in São Paulo/Brasil, 17–19 September.
Delgado-Ramírez CE, Ota Y, Cisneros-Montemayor AM (2023) Fishing as a livelihood, a way of life, or just a job: the complexity of "fishing communities" in research and policy. *Reviews in Fish Biology and Fisheries* 33:265–280. https://doi.org/10.1007/s11160-022-09721-y
Freire MEF (1986) The littoral plain between Trafaria and Lagoa de Albufeira: study of littoral geomorphology. MSc thesis in Physical and Regional Geography, University of Lisbon.
Freitas JG (2007) The Portuguese coast, images and transformations in the contemporary period: from natural space to humanized territory. *Journal of Integrated Coastal Zone Management* 7:105–115 (in Portuguese).
Freitas JG, Bastos MR, Dias JA (2018) Traditional ecological knowledge as a contribution to climate change mitigation and adaptation: the case of the Portuguese coastal populations. In: Leal Filho W, Manolas E, Azul A, Azeiteiro U, McGhie H (eds) *Handbook of climate change communication: vol. 3. Climate change management*. Springer, Cham. https://doi.org/10.1007/978-3-319-70479-1_16
Garrido A (2018) *Fisheries in Portugal*. Fundação Francisco Manuel dos Santos, Lisboa (in Portuguese).
INPCI – Inventário Nacional do Património Cultural Imaterial (2021). *Ficha de Património Imaterial*. Retrieved from http://www.matrizpci.dgpc.pt/MatrizPCI.Web/InventarioNacional/DetalheFicha/415?dirPesq=0
Lira CP, Silva AN, Taborda R, Andrade CF (2016) Coastline evolution of Portuguese low-lying sandy coast in the past 50 years: an integrated approach. *Earth System Science Data* 8:265–278. https://doi.org/10.5194/essd-8-265-2016
Marques V, Chaves C, Morais A, Cardador F, Stratoudakis Y (2005) Distribution and abundance of snipefish (*Macroramphosus spp.*) off Portugal (1998–2003). *Scientia Marina* 69:563–576.
Martins R, Carneiro M, Rebordão F, Sobral M (2000) Fishing with beach seines. *Relatórios Científicos e Técnicos do IPIMAR*, Lisboa No 48, 32pp. (in Portuguese).
Martins MM, Skagen D, Marques V, Zwolinski J, Silva A (2013) Changes in abundance and spatial distribution of the Atlantic chub mackerel (*Scomber colias*) in the pelagic ecosystem and fisheries off Portugal. *Scientia Marina* 77:551–563. https://doi.org/10.3989/scimar.03861.07B
Mesquita M, Caetano AP, François K (2016) Sociology of space and urban boundaries. Embodiment through communitarian education. In Pirani BM, Smith TS (eds) *Embodiment and cultural differences* (pp. 147–171). Cambridge Scholars Publishing, Cambridge.
Miñarro S, Benyei P, Junqueira AB, Campos-Silva J, Victoria Reyes-García V (2020) LICCI fisheries data collection protocol. Figshare. Book. https://doi.org/10.6084/m9.figshare.11830410
Oliveira MNSC (2015) Natural and anthropic evolution: Trafaria – Cova do Vapor – Costa da Caparica. MSc Thesis in Landscape Architecture. ISA (in Portuguese).
Palma M, Dias JA, Freitas JG (2022) It is not only the sea: a history of human intervention in the beach-dune ecosystem of Costa da Caparica (Portugal). *Journal of Integrated Coastal Zone Management* 21(4):227–247. https://doi.org/10.5894/rgci-n432
Pulido Valente V (1981) The fish canners of Setúbal, 1887–1901. *Análise Social* 17:615–678 (in Portuguese).
Reyes-García V, Álvarez-Fernandez S, Benyei P, García-del-Amo D, Junqueira AB, Labeyrie V, Li X, Porcher V, Porcuna-Ferrer A, Schlingmann A, Soleymani R (2023) Local indicators of climate change impacts described by Indigenous peoples and local communities: study protocol. *PLoS One* 18(1):e0279847. https://doi.org/10.1371/journal.pone.0279847
Rodrigues JS (2000) Infrastructures and urbanization of the South margin: Almada, 19th and 20th centuries. *Análise Social* 35:547–581 (in Portuguese).
Santos M, Seixas S, Aggio R, Hanazaki N, Costa M, Schiavetti A, Dias J, Azeiteiro U (2012) Fisheries as a human activity: artisanal fisheries and sustainability. *Journal of Integrated Coastal Zone Management* 12:405–427 (in Portuguese).
Silva F (2008) Rurality in Almada and Seixal in the 18th and 19th centuries: image, landscape and memory. MSc Thesis in Studies on Heritage, Open University (in Portuguese)
Smart Fishing (2021) Project smart fishing: integrating new technologies for sustainable and safe local fisheries (P01M03 0002P). http://smartfishing.olo.blue (in Portuguese).
Souto H (1998) Artisanal fishing communities in the coast of Portugal – a geographic study. PhD Thesis, New University of Lisbon (in Portuguese).
Stratoudakis Y, Antunes C, Correia C, Belo F, Almeida PR (2023) Riverine communities and management systems for anadromous fisheries in the Iberian Peninsula: global strategy, local realities. *Reviews in Fish Biology and Fisheries* 33:875–892. https://doi.org/10.1007/s11160-022-09742-7
Thomas J (1993) *Doing critical ethnography*. Qualitative Research Methods Series 26. SAGE Publications, London.

10

A COMPLEX MATRIX. REPORTS OF ENVIRONMENTAL CHANGE AND ITS DRIVERS BY THE TSIMANE', BOLIVIAN AMAZON

Victoria Reyes-García, Petra Benyei, André B. Junqueira, Tomás Huanca and Esther Conde

Introduction

The scientific community increasingly emphasises the need to acknowledge the interconnections among climate change and biodiversity decline, as well as their impacts on people's quality of life (e.g., Pörtner et al., 2021). For example, climate change drives geographic range shifts in species, alters phenology and migration patterns, and disrupts key ecological interactions (e.g., García Molinos et al., 2015; Lenoir & Svenning, 2015; Pecl et al., 2017). These changes generate cascade effects that propagate across different components of coupled human–natural systems (Gregr et al., 2020), with climate change impacts aggravating the effects of other drivers of environmental change (Pörtner et al., 2021; Rillig et al., 2019). Despite a growing scientific awareness about the combined impacts of climate and other drivers of change, much remains to be known about how these interactions manifest in local environments and livelihoods. Advancing such understanding is important to promote the design of policies oriented to maximise the co-benefits between climate, biodiversity, and social issues, while minimising co-detrimental outcomes and trade-offs (Pascual et al., 2022). Moreover, given that the additive impacts of climate change and biodiversity decline are likely to be strongly felt in biodiversity-rich areas, and by people who directly depend on nature for their livelihood (Díaz et al., 2019; IPBES, 2019), addressing the issue can help redress social inequalities in climate and biodiversity policies.

Given their holistic nature, Indigenous knowledge (IK) systems are ideally suited to contribute to the understanding of the interlinked nature of climate and other drivers of environmental change, as well as to the understanding of their cascade effects on coupled social-ecological systems, from local to global levels (Brondízio et al., 2021; Junqueira et al., 2021). IK systems have been defined as "the cumulative body of knowledge, practice, and belief, evolving by adaptive processes and handed down through generations by cultural transmission, about the relationship of living beings (including humans) with one another and with their environment" (Berkes et al., 2000, p. 1252). These complex knowledge systems are characterised as being integrated, grounded in territory, diverse, and continuously evolving (Hill et al., 2020). For example, to many Indigenous worldviews and paradigms, the territory plays an active role in creating the world, and the relation between the

DOI: 10.4324/9781003356837-13

mind, the body, and the Earth is the origin of life, promoting language and knowledge (Bawaka Country et al., 2015; Watts, 2013). Moreover, IK systems encompass a relational approach that includes governance, language, institutions, resource use and management practices, social and economic accounting, naming and classification systems, values, spirituality, and worldviews (Hill et al., 2020). Overall, IK systems provide a complex picture of the relations between humans and the environment, a picture that rests on the premise that all aspects of life are interconnected, the world being an integral whole in which knowledge, values, and behaviour are inseparable, often linked through spirituality (e.g., Gram-Hanssen et al., 2021; Posey, 1999; Turner, 1997).

In addition to their intrinsic value, IK systems can complement scientific understandings of the world. Indeed, despite epistemological and ontological differences between IK systems and Western science (Ludwig, 2016; Tengö et al., 2017), many examples exist in which scientists have relied on information derived from IK systems to – for example – monitor the status of nature (e.g., Cartró-Sabaté, 2018), explain causal relations between observations (e.g., Steger et al., 2020), or provide potential explanations to environmental changes and drivers of change (e.g., Klein et al., 2014).

The work presented here builds on the premise that the interpretation of elements derived from IK systems according to the concepts, categories, axioms, methods, and ontologies that characterise science can be relevant for scientists aiming to understand how the synergistic impacts of climate and other drivers of change manifest in local environments and livelihoods. Based on such a premise, we tap into the holistic nature of IK to explore interactions between climate and other drivers of environmental change in the local social-ecological system of the Tsimane', an Indigenous population of the Bolivian Amazon.

The Tsimane' and their social-ecological system

The Tsimane' are one of the largest native Amazonian societies in Bolivia, settled in the Department of Beni (Figure 10.1). The Tsimane' traditionally occupied a territory that extends from the Andean piedmont to the region of Moxos in the northeast, an area mostly covered with open forests and savannas. The climate of the region is equatorial with summer rains from October to April. From May to August, there is a period of reduced precipitation and episodic southern cold winds. The mean annual temperature is 25.8°C and the mean annual rainfall is 1743 mm.

Historical accounts depict the Tsimane' as an elusive and highly mobile society, lacking a hierarchical system of authority (Huanca, 2008). Up until the late 1930s, the Tsimane' maintained a dispersed settlement pattern and a self-sufficient lifestyle and had reduced contact with outsiders. From the 1950s, when the Bolivian government gave Protestant missionaries the responsibility for schooling lowland native Amazonian populations and missionaries' schools were established in their territory, the Tsimane' started to establish long-term settlements near schools. Through their schools and preaching visits to villages, missionaries had profound impacts upon Tsimane' culture and way of life (Huanca, 2008). After the 1960s, the construction of new roads, different waves of government-planned colonisation, the logging boom, and land tenure reforms put the Tsimane' into increasing contact with other segments of the Bolivian society, often through the arrival of non-Tsimane' (Reyes-García, Paneque-Gálvez, Bottazzi, et al., 2014a). While some Tsimane' increased their interactions with non-Tsimane', others moved to farther areas, such as the upstream of the Maniqui and Secure rivers (Reyes-García et al., 2014a, 2014b).

Tsimane' recent history is characterised by population growth, the adoption of new technologies, and increasing integration into the market economy (Godoy et al., 2005; Gurven et al., 2017). The latest estimates put Tsimane' population ca. 14,000 people living in about 125 villages along

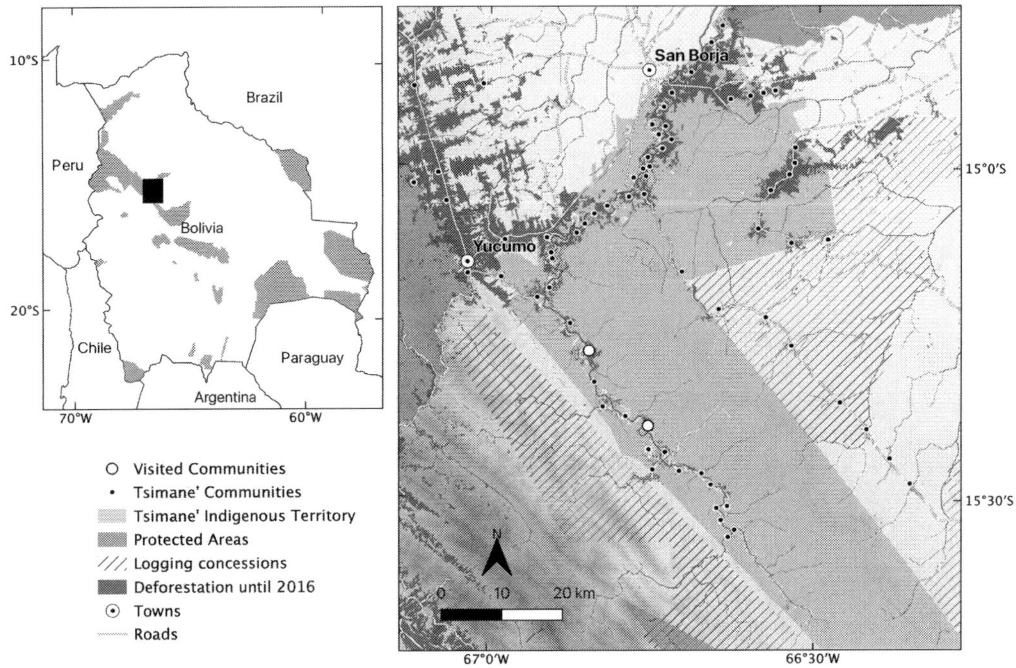

Figure 10.1 Map of the study area.

riverbanks and logging roads (Reyes-García et al., 2014a, 2014b). Nowadays, most Tsimane' continue practising hunting, gathering, fishing, and small-scale shifting agriculture, activities that rely on an extensive knowledge about plants, animals, and landscapes (Díaz-Reviriego et al., 2017; Guèze et al., 2014). Nevertheless, a growing number of Tsimane' households combine these activities with the barter or sale of crops and non-timber forest products, and with occasional wage labour. Some Tsimane', mostly those living close to roads and towns, increasingly depend on market-based economic activities, including cash cropping (particularly rice and plantain), sale of fish and thatch palm (*Geonoma deversa*), and wage labour in logging camps and cattle ranches (Fernández-Llamazares et al., 2016). Many young Tsimane' are moving to agricultural regions where they take paid jobs. These changes have led to changes in dietary patterns and health (e.g., Gurven et al., 2017; Reyes-García et al., 2019), as well as to a general erosion of Tsimane' knowledge system (Reyes-García & Broesch, 2013; Reyes-García, Paneque-Gálvez, Luz, et al., 2014b).

The environment inhabited by the Tsimane' has also experienced important changes. Records from the local weather station reveal that since the 1960s, the area has been subject to a steady increase in temperature, a pronounced decrease in the intensity of precipitation in the rainy season, and an increase in the length of the dry period (Fernández-Llamazares, Garcia et al., 2017b). Analyses of the forest cover between 1986 and 2009 show a general trend towards forest fragmentation, largely due to the clearance of old-growth forests and the subsequent spread of early-growth secondary forests (Figure 10.1; Paneque-Gálvez et al., 2013). A decrease in plant, game, and fish abundance has also been documented, with some reports of local extinctions of species (Guèze et al., 2015; Luz et al., 2015).

To some extent, the social, climatic, and environmental changes affecting the Tsimane' are representative of changes affecting other social-ecological systems across the world, which need to adapt to new conditions brought by multiple drivers of change operating at different scales. This makes the Tsimane' a good case to enrich the scientific understanding of ongoing interactions between climate and other drivers of environmental change in local social-ecological systems.

Methods

Data were collected during October 2019 using a standardised protocol combining group interviews, semi-structured interviews, and a focus group discussion to capture reports of environmental change and its drivers (Reyes-García et al., 2023). Data collection also benefited from authors' long-term (since 1999) and continuous presence (since 2003) in the area (Huanca, 2008; Reyes-García, 2020). Interviews were conducted in Tsimane' language with the collaboration of two experienced Tsimane' research assistants who ensured that responses were adequately interpreted by researchers. Two of the researchers are relatively fluent in Tsimane' language. We obtained the permit to conduct research from the Great Tsimane' Council, the umbrella political organisation of the Tsimane'. In each village, we obtained permission to work from a representative. We also requested free prior and informed consent from each person participating in the research. The research protocol was approved by the Ethics committee of the Universitat Autònoma de Barcelona (CEEAH 4781).

We acknowledge that we are non-Tsimane' scholars and that this positionality affects our interpretations of Tsimane' understanding of environmental change and its drivers. We used a methodological approach that makes an instrumental use of elements derived from IK systems according to concepts and categories common in environmental sciences and that explicitly focus on relational elements embedded in IK systems. Given our positionality and methodological approach, the work presented here should not be understood as a representation of Tsimane' epistemological and ontological grounds, but rather as an attempt to enrich our understanding of ongoing interactions between climate and other drivers of environmental change in a local social-ecological system through the analysis of elements of IK systems.

Sampling

We collected data in two Tsimane' villages along the Maniqui River (Figure 10.1). The selected villages are similar in terms of population size and livelihood activities. They are also representative of other Tsimane' villages in the Maniqui River in terms of environmental and socio-cultural conditions.

We used different sampling strategies for different data collection methods. To select participants for group interviews, we followed purposive sampling, targeting people who had lived in the area for ≥30 years and were locally recognised as knowledgeable. Thirteen people participated in two group interviews. To select participants for semi-structured interviews, we used quota sampling, selecting individuals with different socio-demographic characteristics, who would potentially hold different knowledge and thus contribute to a more enriched picture of environmental change. We interviewed men and women from three different age categories (i.e., <40 years; ≥40 and <60; and ≥60 years). To select people within each quota, in a community meeting, we asked for people willing to participate and selected volunteers that fit our quotas. In total, we interviewed 20 men and 14 women from 16 households. Ten participants were <40 years of age, 11 were ≥40 and <60 years, and 13 were ≥60 years. Finally, we used convenience sampling to select participants for the focus group discussion. We invited all adults living in the second village visited and about 20 people attended, although only eight actively participated in the discussion.

Data collection

As a first step to identify environmental changes perceived by Tsimane', we conducted two group interviews. We started the group interviews by showing participants a seasonal calendar depicting seasonal climatic events and Tsimane' livelihood activities (Figure 10.2). We asked participants to comment on changes regarding the timing of seasonal climatic events and livelihood activities. The information gathered served to adjust the terminology used during semi-structured interviews and focus group discussion.

We then used individual semi-structured interviews to gather Tsimane' observations of environmental changes and potential relations between changes. We started by asking "compared to your youth, what changes in nature have you noticed?" We continued prompting "have you noticed any other change?" Once the informant stopped mentioning changes, we asked more direct questions regarding (a) changes in weather, seasons, temperature, rain, wind, or storms, (b) changes in soils, the river, or streams, (c) changes in wild animals, fish, or plants, and (d) changes in crops or livestock. For each change mentioned, we asked about the driver(s) of the change (i.e., "why do you think this change happened?"). If the answer to this question was another environmental change, we asked again for the driver(s) of the latest mentioned change, until the respondent did not identify any further driver (see Reyes-García et al., 2023) for a full description of the interview protocol).

At the end of fieldwork, we discussed observations that were either contradictory or unclear in a focus group. For instance, one respondent reported that "*Now, small trees (e.g., upuyu, cam) die more frequently than in the past,*" while another respondent reported that "*Trees (e.g., väij, cojma) are not tear by strong winds, as they were in the past.*" We presented these two observations to the

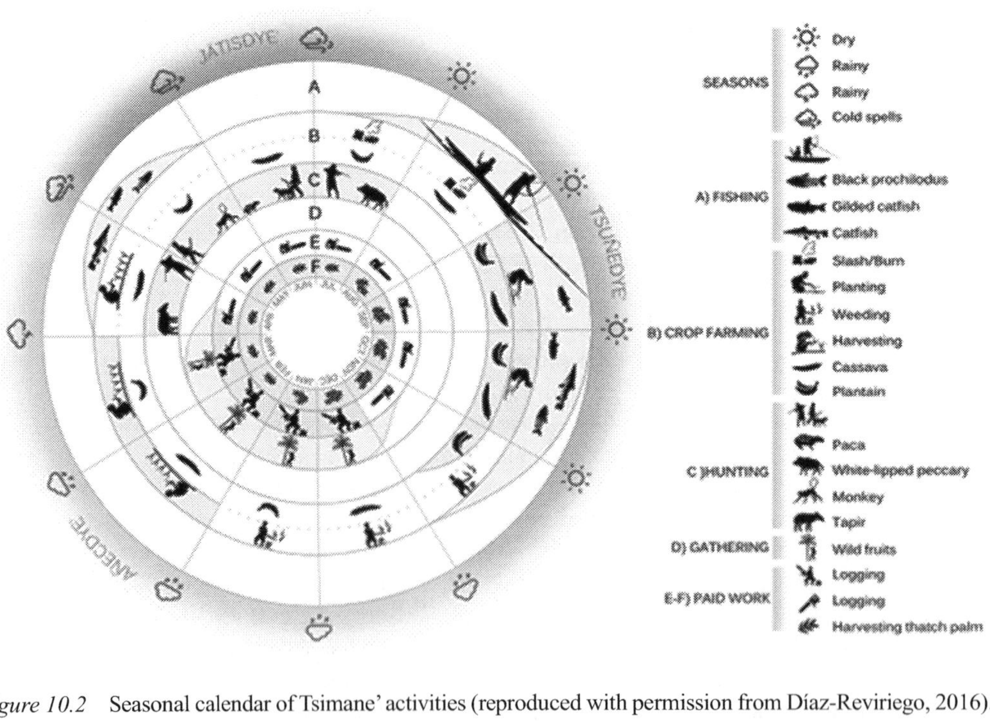

Figure 10.2 Seasonal calendar of Tsimane' activities (reproduced with permission from Díaz-Reviriego, 2016).

group to assess whether there was a consensual perspective on changes in wild plant mortality and on the drivers of such changes. The participatory, collective, and iterative nature of the discussion helped to identify changes and relations that reflect the group's social memory.

Data analysis

For the analysis, we grouped Tsimane' reports of environmental change into categories. We started by grouping *verbatim* reports depicting the same environmental change into indicators. We then organised indicators according to whether they referred to changes in the atmospheric, the physical, or the life systems, further detailing the specific subsystem affected by the change (see Reyes-García et al., 2023). During semi-structured interviews, informants also reported non-environmental changes (e.g., changes in access to education, health or transport, or changes in population). We only include this information when mentioned as a driver of environmental change (see below).

For each indicator of environmental change, we also noted the direction (e.g., increase or decrease) and the overall level of agreement. We considered interviewees 'agreed' in a change (i) when there was agreement in all observations collected in semi-structured interviews or (ii) when participants agreed with the reported change in the focus group discussion. We coded as 'inconclusive' changes mentioned by less than 10% of the sample and not discussed in the focus group. We coded as 'disagreed' changes for which observations with different directions were reported and/or which were not agreed in the focus group discussion.

We then examined responses to our question about drivers of environmental change. To categorise drivers of environmental change, we used the conceptual framework of the Intergovernmental Science-Policy Platform on Biodiversity and Ecosystem Services (Díaz et al., 2019). This framework considers *direct drivers* of environmental change anthropogenic pressures that unequivocally influence biodiversity and ecosystem processes (i.e., climate change, resource exploitation, land-use change, pollution, and invasive species). We considered all responses linking environmental changes to changes in elements of the atmospheric system as 'climate change' driven. The same framework considers *indirect drivers* of environmental change as the underlying causes of direct drivers (i.e., socio-economic and demographic trends, technological innovation, or societal drivers such as cultural change and government policies). Since, according to our protocol, for each change reported, we repeatedly asked about driver(s) of change until the respondent did not identify any further driver, an observation of change might have multiple drivers.

To represent the causal relations explaining environmental change in the Tsimane' social-ecological system, we produced a visual representation (i.e., concept map). The concept map shows indicators and drivers of environmental change. The concept map also shows two types of causal relations linking different nodes: *driving relations*, originating from elements outside the environmental system (black lines) and *cascade effects*, originating from elements of the biophysical systems towards other elements of the biophysical system (grey lines). We use a solid line to represent 'agreed' changes and a dotted line to represent 'inconclusive' changes. Our representation does not include 'disagreed' changes.

Results

Environmental change in Tsimane' social-ecological system

Tsimane' report many observations of environmental change, although we did not find agreement in all of them. The reported observations of change correspond to 57 different indicators in our classification, of which 23 referred to changes in the atmospheric system, 13 to changes in the physical system, and 21 to changes in the life system. From the 57 indicators documented, we found agreement on 26 and we have inconclusive information for 24. For the seven remaining indicators there was no agreement, either on the change reported or on the direction of change.

Changes in elements of the atmospheric system were most frequently reported and most agreed upon. This category included observations of changes in temperature, precipitation, air masses, and seasons. Overall, informants agreed on signalling a temperature increase, both in general (e.g., "*In the past, the weather was colder*" (ID 34, 16/10/2019)) and at specific seasons (e.g., "*Jätisdye* (the cold season) *is less intense*" (ID 3, 07/10/2019)). There was also agreement in a general precipitation decrease, with changes in its temporal distribution (e.g., "*During tsunedye* (the dry season) *there used to be some episodes of rain. Now there are no rains in that season*" (ID 22, 11/10/2019)). Some informants also reported changes in rainfall intensity, frequency of cloudy days, and drought intensity and frequency, although the evidence collected for these observations is inconclusive. We also found agreement in reports of changes in air masses, including changes in the frequency of storms (e.g., "*Now there is less hail falling from the sky*" (ID 22, 11/10/2019)) and changes in wind intensity (e.g., "*Now the wild is less intense. In the past, the wind was so strong that it used to make trees fall down*" (ID 4, 8/10/2019)) and direction (e.g., "*Wind has changed direction. Now it does not come from the south anymore*" (ID 4, 8/10/2019)). Finally, the agreed changes in the length of seasons (e.g., "*Now the dry season is longer*" and "*The jätisdye* (cold season) *is shorter*" (ID 26, 12/10/2019)) seem to be in line with the general perception of a hotter and drier weather.

Tsimane' reports of environmental changes also include reports of changes in the physical system, including observations of changes in water bodies and soils. We found agreement in reports of a reduction in the volume of water bodies, including the river (e.g., "*In the past, the Maniqui river had more water*" (ID 6, 8/10/2019)) and the streams (e.g., "*Now streams get dry faster than they used to*" (ID 1, 7/10/2019)). Tsimane' also reported an increase in the intensity of sedimentation (e.g., "*Now there is more soil in river water*" (ID 31, 15/10/2019)). We also found agreement in reports referring to an increase in freshwater temperature, and to soils being hotter and drier than in the past. There was, however, disagreement in the direction of changes in river floods, with some informants reporting more and some informants reporting fewer river floods than in the past.

Many Tsimane' observations of environmental changes also refer to changes in elements of the life system. Particularly, we found agreement in indicators referring to a decrease in the abundance of fish (e.g., "*Now fish are disappearing*" (ID 6, 8/10/2019)), wild fauna (e.g., "*In the past there were more animals in the river: crocodiles, alligators, turtles*" (ID 4, 8/10/2019)), and wild flora (e.g., "*There is no fine wood to build canoes*" (ID 23, 12/10/2019)). Although some informants also reported changes in plant and animal mortality as well as in plant seasonality and animal behaviour, our information from these indicators is inconclusive. We also found agreement in observations on the decrease in the productivity of crops and the increase in crop mortality rates (e.g., "*Plants 'cook' on the ground. The ground is hot, it rains, and then it is hot again, and this 'cooks' the plants*" (ID 24, 9/10/2019)). Finally, although there were some reports of changes in the frequency of crop pests and animal diseases, we did not find agreement for these observations, as some informants reported an increase in pests due to higher temperatures and some informants reported a decrease due to the use of pesticides.

Drivers of environmental change

When asked about the reasons that explain the observed environmental changes, Tsimane' mentioned many factors that simultaneously and synergistically result in the environmental changes mentioned (Figure 10.3).

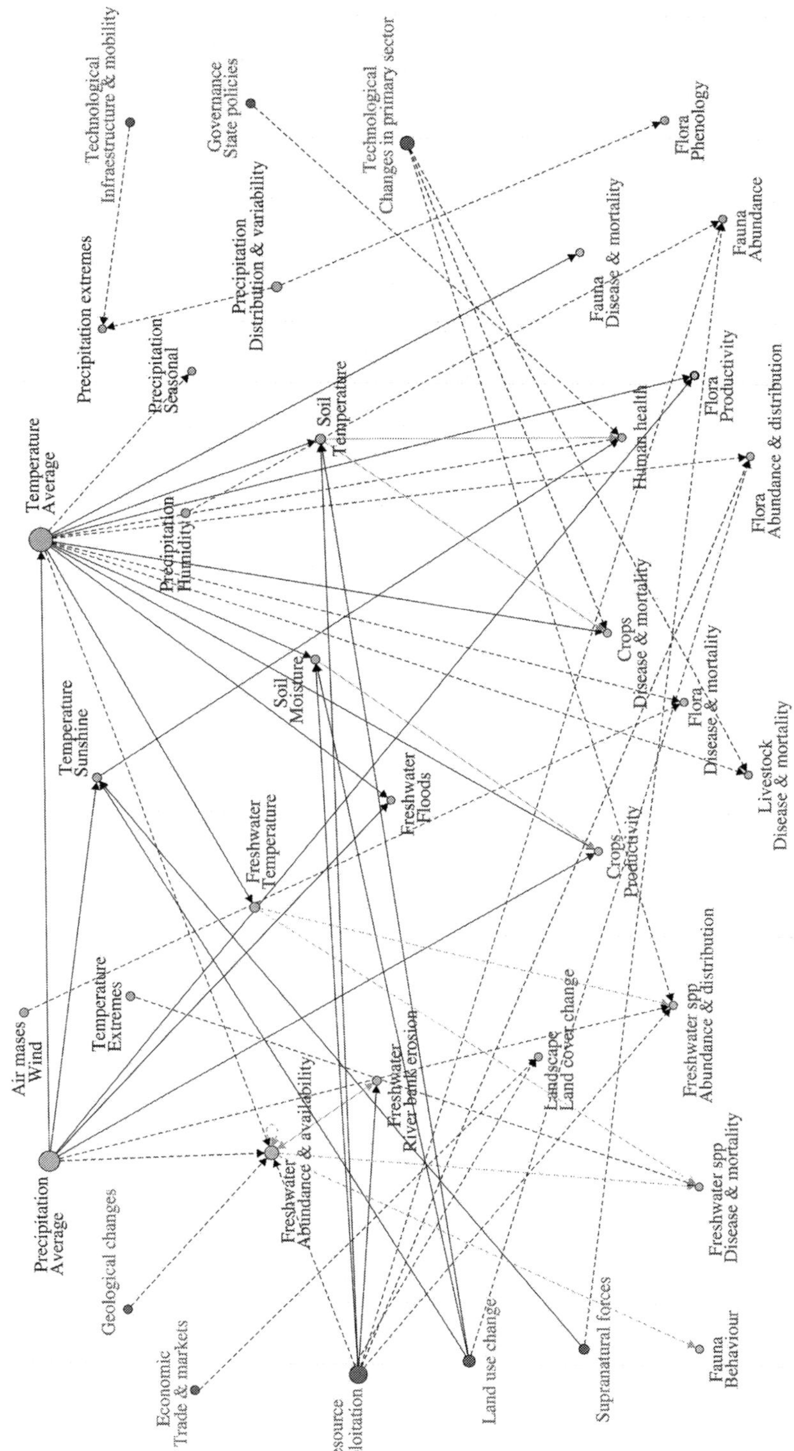

Figure 10.3 Concept map showing documented Tsimane' perceptions of environmental change and their drivers.

According to our classification of Tsimane' responses, changes in elements of the atmospheric system are driven by other changes in elements of the same systems but also by land-use change and supranatural forces. Thus, when asked about the reasons explaining changes in temperature, some Tsimane' referred to changes in rainfall. As one informant reported, "*In the past, the weather was colder than now.* (This is) *because there were more rains in the past.* (ID 3, 7/10/2019)." Also, respondents referred to land-use change as a driver of rainfall changes. Tsimane' explained how the opening of the road linking San Borja to Trinidad (in 1975) resulted in cattle ranching becoming the main economic activity for non-Indigenous peoples (*napo'*) living in the area and that the later opening of the road from Yucumo to Rurrenabaque (1979) facilitated the arrival and establishment of highland colonists (*colla*). These establishments of cattle ranchers and colonists contributed to transform forested areas into pastures and agricultural fields which, according to the Tsimane', has resulted in less rainfall. Finally, some informants also explained changes in elements of the atmospheric system through their worldview. For example, one informant said that changes in temperature and rainfall patterns are a signal that "*Time ages. Just like ourselves. And, at some point, all will die. Time is getting old, that's why things are not producing well anymore*" (ID 11, 10/10/2019).

According to our classification, changes in elements of the physical system (i.e., freshwater and soils) result from the interaction between the cascade effect of changes in elements of the atmospheric system and the pressure of direct drivers (i.e., resource exploitation, land-use change). Thus, changes in soil humidity and temperature were explained through the observed changes in temperature increase and precipitation decrease. The same factors were mentioned to explain the decrease in the river flow and in the decrease of rivers and streams abundance. However, some interviewees argued that the increase in the cultivated area affects both climate and forests, as the loss of trees resulted in less rain and higher temperatures but also in hotter and drier soils. One informant said, "*Our grandparents did small agricultural plots. Now people do large ones. They cut large trees, and that's why the ground gets hotter, because the large trees maintain the moisture in the soil*" (ID 9, 11/10/2019).

Drivers of change in elements of the life system are also multiple and often act in synergistic ways. For example, changes in wild flora and fauna were reported to be driven by changes in elements of the atmospheric and the physical systems, but also by resource exploitation, land use change, technological innovations, and supranatural forces. For example, one informant mentioned that "*Now there is less fish that dies in the river because of the cold weather*" (ID 31, 15/10/2019) and another informant mentioned that "*The trees are fruiting in different times, because it is hotter and there is less rain*" (ID 25, 10/10/2019). Both observations describe how changes in elements of the atmospheric system impact the life system. But informants also mentioned that "*There used to be a lot of fish in the stream. Not anymore, because now there is more drought*" (ID 8, 9/10/2019) and that "*Now wild animals come closer to the village looking for water* (because the streams in the forest are drying due to the decrease in precipitation)" (ID 34, 16/10/2019) thus describing how impacts of changes in elements of the atmospheric system on the life system are mediated through changes in the physical system. Similar relations were described when referring to changes observed in crops. For example, one person mentioned that "*Plants 'cook' on the ground. The ground is hot, it rains, and then it is hot again, and this 'cooks' the plants*" (ID 19, 17/10/2019), a statement that describes a chain of impacts from the atmospheric system to agriculture mediated by changes in soil temperature.

Informants also emphasised that a main driver of change in elements of the life system is increasing resource exploitation, which they relate to wider societal changes such as population growth, the increased presence of loggers and traders, and technological innovations. Thus, when asked about the reason behind the decrease in abundance of wildlife, one informant simply mentioned, "*Now there are many more people. They finish with the fish and animals*" (ID 25, 10/10/2019), highlighting the role of demographic increase in environmental change. In answer to the same question, other people related stories of loggers operating in their territory, first illegally, during the 1970s, and later (i.e., during the 1980s and 1990s) under the auspices of the Great Tsimane' Council. In their accounts, they emphasised the impact logging had on the environment. Tsimane' also provide vivid accounts of the arrival of traders and their interest in natural products, first in pelts and later in mats woven from thatch palm. According to one informant, "*Now there is less 'cajtafa'* [thatch palm]. *The palms die much more because of the heat and because too many people extract them*" (ID 25, 10/10/2019). The presence of loggers and traders was also related to an increase in hunting and fishing pressure and a consequent decrease in wild animal population: "*Many animals have disappeared, because they* [loggers] *hunt more. They kill too many animals*" (ID 15, 07/10/2019). Tsimane' relate these two drivers of change with technological innovations. For example, when talking about the decrease in fish, one informant said "*Now there are less fish because we don't use arrows anymore. We now use hooks, nets, diving, and barbasco* [a fish poisoning technique]" (ID 12, 13/10/2019).

Finally, Tsimane' also mentioned the importance of supranatural forces in explaining the decrease in flora and fauna abundance. For the interviewees, the use of inappropriate, abusive, or culturally disrespectful hunting and fishing techniques awakens the anger of the guardian spirits of nature who punish Tsimane' with animals' disappearance. One informant said that "*People are not hunting respectfully anymore. I found one tapir that was shot, but nobody went to get it. Jäjäbä* [the spirit who guards animals according to Tsimane' cosmology] *gets angry and takes animals away*" (ID 19, 17/10/2019). Another person said that "*Game and fish are disappearing because they go to the other side. The 'muñeco'* [a forest spirit] *takes care of the animals and takes them somewhere else* [to protect them]" (ID 18, 17/10/2019).

Discussion

We explore Tsimane' reports of environmental change and drivers of such change, focusing not only on the description of these elements but also on the relations among them. Our relational analysis presents Tsimane' understanding of environmental change as a network that allows (1) to identify the simultaneous impact of different drivers of environmental change and (2) to detect different processes by which certain changes negatively impact other elements of the system (cascade effects).

Before discussing these results, we start acknowledging two limitations and one important caveat of this work. We acknowledge that our results are limited by our focus on consensual reports and on environmental impacts. During interviews, we found disparity in responses and discrepancy in the arguments, and we were not able to discuss all points in our focus group discussion, for which our data are inconclusive in many points. A larger exploration of data, including disagreements, might provide a better understanding of diverse intra-cultural perspectives about environmental change. We also found changes that go beyond environmental aspects (e.g., problems with river transportation due to rainfall decrease) that we discarded. Therefore, since our analysis focuses on agreed responses and on environmental impacts, it should be considered as a simplification of Tsimane' complex understanding of environmental change. Further research should consider these limitations aiming for a more comprehensive understanding of environmental change.

Moreover, an important caveat of this work refers to the transformation of Tsimane' knowledge through our analysis. Our analytical approach treats reports of environmental change and drivers as separate entities and fits them in a classification system that largely reflects scientific concepts. During interviews, it was evident that the Tsimane' do not necessarily separate concepts in the way we do, for which we are conscious that our procedure entails a transformation of Tsimane' understanding, undoubtedly including misinterpretations derived from the attempts to fit responses into our framework (Nadasdy, 2003). Acknowledging this, we do not claim that our results represent Tsimane' worldview, but rather our own interpretation of Tsimane' understanding of environmental change. Despite these limitations and caveats, we argue that the results presented here enrich our understanding of environmental change by highlighting the importance of the relational approach. We discuss our results through this perspective.

There is now a large body of research discussing the richness of local reports of environmental change and its drivers, which arise from local peoples' intimate connection with the local environment and their forms of knowledge transmission (e.g., Aswani et al., 2018; Klein et al., 2014; Sterling et al., 2017). A main finding of this body of knowledge is that local understandings often weave environmental, social, and cultural elements in explaining environmental change, constituting what Cámara-Leret et al. (2019) call an *Indigenous knowledge network*. Our work is in consonance with this work but extends it using a relational analysis that uncovers two important elements: the simultaneous impact of different drivers of change and the cascade chain of impacts.

The first important finding of this work is that Tsimane' identify the simultaneous impact of different drivers of environmental change. When asked about drivers of environmental change, Tsimane' provided many explanations, some referred to processes that directly affect the environment (i.e., direct drivers such as climate change, resource exploitation, and land use change), whereas other referred to processes that result in environmental change through pressure on the abovementioned drivers (i.e., indirect drivers such as population growth, the arrival of traders and loggers to their lands, or technological innovation that increase resource exploitation or land use change). As scientists are increasingly concerned by the simultaneous and interrelated effect of multiple drivers of change (Pörtner et al., 2021), we argue that the holistic nature of Indigenous knowledge systems sits well with understandings that depart from sectoral approaches to acknowledge complex and synergistic interactions between simultaneous drivers of environmental change. Furthermore, the complex view of the simultaneous drivers of environmental change offers a political view of the environment, a view that emphasises that impacts are not limited to biophysical aspects, but the full social-ecological system. Thus, results from this work show that Tsimane' understand environmental change as resulting from a historical process leading to the appropriation and fragmentation of their lands (see also Reyes-García et al., 2014a, 2014b). In that sense, reacting to environmental change should be considered in the historical and political context in which it occurs.

Moreover, Tsimane' holistic understanding of environmental change incorporates cultural representations of the world that are critical in understanding environmental change, but that are too often underrepresented in science. In that sense, Tsimane' worldviews and spiritual beliefs play a prominent role in their explanation of environmental change (see also (Fernández-Llamazares, Díaz-Reviriego et al., 2017a)). While absent in scientific narratives, explanations based on their worldviews are an essential component in Tsimane' understanding of how and why the world changes, and thus should be acknowledged in any attempt to bridge knowledge systems.

The second, related, finding of this work is that the Tsimane' detect different processes by which certain changes negatively impact other elements of the system (cascade effects). Scientists increasingly recognise the importance of studying how changes in one element of the social-ecological system propagate to other components through cascade effects (Michalak et al., 2017). Moreover, research interest is growing in how the cascade effects of environmental change will affect nature's contributions to people and alter human actions and behaviours, and who will receive the more immediate impacts (Fontúrbel et al., 2018; Michalak et al., 2017). Our work highlights that Indigenous knowledge systems can contribute to trace such processes. For example, the Tsimane' highlight many interactions between elements, such as tracing the impacts of changes in temperature and rainfall into the abundance, mortality, or quality of other elements of their natural and managed landscapes (e.g., fish and crops). Spotting such relations potentially leads to a more complete understanding of environmental change impacts on local social-ecological systems, and thus to the definition of more locally attuned management priorities (Gill & Malamud, 2016).

Before concluding, a last finding merits attention. Our documentation of environmental change and drivers of change reported by Tsimane' did not include important changes documented by researchers working on the area. For example, during interviews, the Tsimane' did not mention pollution as a driver of environmental change, although several mercury-dependent artisanal gold miners operate in the area (Reyes-García et al., 2014a, 2014b) and mercury-dependent artisanal gold mining is the largest source of mercury pollution on Earth, affecting many local communities (Fernández-Llamazares et al., 2020). Similarly, in their account of defaunation, Tsimane' did not refer to a general decrease of fauna, but rather mention that game might have moved somewhere else, for which their account does not seem to capture the large impact of overhunting on regional defaunation (Fernández-Llamazares, Díaz-Reviriego et al., 2017a). These incongruences between scientists' and Tsimane' understanding of environmental change might originate in the perceptual and local nature of Indigenous knowledge systems. Local people are well equipped to detect long-term temporal local trends but may overlook changes happening at global scales or changes that might have impacts not visible to the naked eye. Acknowledging the strengths and weaknesses of the different knowledge systems should encourage further enquiry to obtain a more complete picture of environmental change (e.g., Tengö et al., 2014).

Conclusion

The work presented here provides evidence that, beyond monitoring environmental change, Indigenous knowledge systems can trace its multiple and simultaneous drivers and cascade effects. Findings from this work suggest that local policies to tackle the negative effects of environmental change in the area should equally address simultaneous drivers of change and acknowledge Tsimane' wordlviews of change, as a sectoral approach might not suffice to alleviate the negative impacts of environmental change in local livelihoods. Finally, we argue that, while Indigenous understandings of the complex relations between different elements of the system could contribute to the co-generation of new knowledge, this process requires space for inter-epistemological dialogue.

Acknowledgements

We sincerely thank all Tsimane' who over the years have shared their wisdom and lives with us. The knowledge they shared with us, and that constitutes the basis for the analyses presented here, belong to the communities with whom we have worked. We also thank F. Lesi and M. Pannaccione for data processing, and S. Álvarez-Fernandez for the design of Figure 10.3. Research leading to this chapter has received funding from the European Research Council under an ERC Consolidator Grant (FP7–771056-LICCI). This work contributes to the 'María de Maeztu' Programme for Units of Excellence of the Spanish Ministry of Science and Innovation (CEX2019-000940-M).

References

Aswani, S., Lemahieu, A., & Sauer, W. H. H. (2018). Global trends of local ecological knowledge and future implications. *PLoS One*, *13*(4), e0195440. https://doi.org/10.1371/journal.pone.0195440

Bawaka Country, Wright, S., Suchet-Pearson, S., Lloyd, K., Burarrwanga, L., Ganambarr, R., Ganambarr-Stubbs, M., Ganambarr, B., Maymuru, D., & Sweeney, J. (2015). Co-becoming Bawaka. Towards a relational understanding of place/space. *Progress in Human Geography*, *40*(4), 455–475. https://doi.org/10.1177/0309132515589437

Berkes, F., Colding, J., & Folke, C. (2000). Rediscovery of traditional ecological knowledge as adaptive management. *Ecological Applications*, *10*, 1251–1262. https://doi.org/10.2307/2641280

Brondízio, E. S., Aumeeruddy-Thomas, Y., Bates, P., Carino, J., Fernández-Llamazares, Á., Ferrari, M. F., Galvin, K., Reyes-García, V., McElwee, P., Molnár, Z., Samakov, A., & Shrestha, U. B. (2021). Locally based, regionally manifested, and globally relevant: Indigenous and local knowledge, values, and practices for nature. *Annual Review of Environment and Resources, 46*, 481–509. https://doi.org/10.1146/ANNUREV-ENVIRON-012220-012127

Cámara-Leret, R., Fortuna, M. A., & Bascompte, J. (2019). Indigenous knowledge networks in the face of global change. *Proceedings of the National Academy of Sciences of the United States of America, 116*(20), 9913–9918. https://doi.org/10.1073/pnas.1821843116

Cartró-Sabaté, M. (2018). *Amazo'n'oil: Exposure to oil and lead for Amazonian wildlife*. PhD Dissertation. Universitat Autonoma de Barcelona.

Díaz, S., Settele, J., Brondízio, E. S., Ngo, H. T., Agard, J., Arneth, A., Balvanera, P., Brauman, K. A., Butchart, S. H. M., Chan, K. M. A., Garibaldi, L. A., Ichii, K., Liu, J., Subramanian, S. M., Midgley, G. F., Miloslavich, P., Molnár, Z., Obura, D., Pfaff, A., … Zayas, C. N. (2019). Pervasive human-driven decline of life on Earth points to the need for transformative change. *Science*, *366*(6471), eaax3100. https://doi.org/10.1126/science.aax3100

Díaz-Reviriego, I., Fernández-Llamazares, Á., Howard, P. L., Molina, J. L., & Reyes-García, V. (2017). Fishing in the Amazonian Forest: A gendered social network puzzle. *Society and Natural Resources*, *30*(6). https://doi.org/10.1080/08941920.2016.1257079

Fernández-Llamazares, Á., Díaz-Reviriego, I., Guèze, M., Cabeza, M., Pyhälä, A., & Reyes-García, V. (2016). Local perceptions as a guide for the sustainable management of natural resources: Empirical evidence from a small-scale society in Bolivian Amazonia. *Ecology and Society*, *21*(1). https://doi.org/10.5751/ES-08092-210102

Fernández-Llamazares, Á., Díaz-Reviriego, I., & Reyes-García, V. (2017a). Defaunation through the eyes of the Tsimane'. In *Hunter-gatherers in a changing world* (pp. 77–90). Springer International Publishing. https://doi.org/10.1007/978-3-319-42271-8_5

Fernández-Llamazares, Á., Garcia, R. A., Díaz-Reviriego, I., Cabeza, M., Pyhälä, A., & Reyes-García, V. (2017b). An empirically tested overlap between indigenous and scientific knowledge of a changing climate in Bolivian Amazonia. *Regional Environmental Change*, *17*(6). https://doi.org/10.1007/s10113-017-1125-5

Fernández-Llamazares, Á., Garteigogeascoa, M., Basu, N., Brondizio, E., Cabeza, M., Martínez-Alier, J McElwee, P., & Reyes-García, V. (2020). A state-of-the-art review of indigenous peoples and environmental pollution. *Environmental Health Perspectives* 16(3), 324–341. https://doi.org/10.1002/ieam.4239

Fontúrbel, F. E., Lara, A., Lobos, D., & Little, C. (2018). The cascade impacts of climate change could threaten key ecological interactions. *Ecosphere*, *9*(12), e02485. https://doi.org/10.1002/ecs2.2485

García Molinos, J., Halpern, B. S., Schoeman, D. S., Brown, C. J., Kiessling, W., Moore, P. J., Pandolfi, J. M., Poloczanska, E. S., Richardson, A. J., & Burrows, M. T. (2015). Climate velocity and the future global redistribution of marine biodiversity. *Nature Climate Change*, *6*(1), 83–88. https://doi.org/10.1038/nclimate2769

Gill, J. C., & Malamud, B. D. (2016). Hazard interactions and interaction networks (cascades) within multi-hazard methodologies. *Earth System Dynamics*, *7*(3), 659–679. https://doi.org/10.5194/esd-7-659-2016

Godoy, R., Reyes-García, V., Byron, E., Leonard, W. R., & Vadez, V. (2005). The effect of market economies on the well-being of indigenous peoples and on their use of renewable natural resources. *Annual Review of Anthropology, 34*. https://doi.org/10.1146/annurev.anthro.34.081804.120412

Gram-Hanssen, I., Schafenacker, N., & Bentz, J. (2021). Decolonizing transformations through 'right relations.' *Sustainability Science*, *17*(2), 673–685. https://doi.org/10.1007/S11625-021-00960-9

Gregr, E. J., Christensen, V., Nichol, L., Martone, R. G., Markel, R. W., Watson, J. C., Harley, C. D. G., Pakhomov, E. A., Shurin, J. B., & Chan, K. M. A. (2020). Cascading social-ecological costs and benefits triggered by a recovering keystone predator. *Science*, *368*(6496), 1243–1247. https://doi.org/10.1126/SCIENCE.AAY5342

Guèze, M., Luz, A. C., Paneque-Gálvez, J., Macía, M. J., Orta-Martínez, M., Pino, J., & Reyes-García, V. (2014). Are ecologically important tree species the most useful? A case study from indigenous people in the Bolivian Amazon. *Economic Botany*, *68*(1). https://doi.org/10.1007/s12231-014-9257-8

Guèze, M., Luz, A. C., Paneque-Gálvez, J., Macía, M. J., Orta-Martínez, M., Pino, J., & Reyes-García, V. (2015). Shifts in indigenous culture relate to forest tree diversity: A case study from the Tsimane', Bolivian Amazon. *Biological Conservation*, *186*. https://doi.org/10.1016/j.biocon.2015.03.026

Gurven, M., Stieglitz, J., Trumble, B., Blackwell, A. D., Beheim, B., Davis, H., Hooper, P., & Kaplan, H. (2017). The Tsimane health and life history project: Integrating anthropology and biomedicine. *Evolutionary Anthropology*, *26*(2), 54–73. https://doi.org/10.1002/evan.21515

Hill, R., Adem, Ç., Alangui, W. V., Molnár, Z., Aumeeruddy-Thomas, Y., Bridgewater, P., Tengö, M., Thaman, R., Adou Yao, C. Y., Berkes, F., Carino, J., Carneiro da Cunha, M., Diaw, M. C., Díaz, S., Figueroa, V. E., Fisher, J., Hardison, P., Ichikawa, K., Kariuki, P., … Xue, D. (2020). Working with indigenous, local and scientific knowledge in assessments of nature and nature's linkages with people. *Current Opinion in Environmental Sustainability*, *43*, 8–20. https://doi.org/10.1016/j.cosust.2019.12.006

Huanca, T. (2008). *Tsimane' oral tradition, landscape, and identity in tropical forest*. Imprenta Wagui: La Paz, Bolivia.

IPBES (2019). *Summary for policymakers of the global assessment report on biodiversity and ecosystem services of the intergovernmental science-policy platform on biodiversity and ecosystem services* (S. Díaz, J. Settele, E. S. Brondízio, H. T. Ngo, M. Guèze, J. Agard, A. Arneth, P. Balvanera, K. A. Brauman, S. H. M. Butchart, K. M. A. Chan, L. A. Garibaldi, K. Ichii, J. Liu, S. M. Subramanian, G. F. Midgley, P. Miloslavich, Z. Molnár, D. Obura, … C. N. Zayas, Eds.). IPBES Secretariat.

Junqueira, A. B., Fernández-Llamazares, Á., Torrents-Ticó, M., Haira, P. L., Nasak, J. G., Burgas, D., Fraixedas, S., Cabeza, M., & Reyes-García, V. (2021). Interactions between climate change and infrastructure projects in changing water resources: An ethnobiological perspective from the Daasanach, Kenya. *Journal of Ethnobiology*, *41*(3). https://doi.org/10.2993/0278-0771-41.3.331

Klein, J. A., Hopping, K. A., Yeh, E. T., Nyima, Y., Boone, R. B., & Galvin, K. A. (2014). Unexpected climate impacts on the Tibetan Plateau: Local and scientific knowledge in findings of delayed summer. *Global Environmental Change*, *28*, 141–152. https://doi.org/10.1016/j.gloenvcha.2014.03.007

Lenoir, J., & Svenning, J. C. (2015). Climate-related range shifts – A global multidimensional synthesis and new research directions. *Ecography*, *38*(1), 15–28. https://doi.org/10.1111/ECOG.00967

Ludwig, D. (2016). Overlapping ontologies and Indigenous knowledge. From integration to ontological self-determination. *Studies in History and Philosophy of Science Part A*, *59*, 36–45. https://doi.org/10.1016/J.SHPSA.2016.06.002

Luz, A., Guèze, M., Paneque-Gálvez, J., Pino, J., Macía, M., Orta-Martínez, M., & Reyes-García, V. (2015). How does cultural change affect indigenous peoples' hunting activity? An empirical study among the Tsimane' in the Bolivian Amazon. *Conservation and Society*, *13*(4). https://doi.org/10.4103/0972-4923.179879

Michalak, J. L., Withey, J. C., Lawler, J. J., & Case, M. J. (2017). Future climate vulnerability – evaluating multiple lines of evidence. *Frontiers in Ecology and the Environment, 15*(7), 367–376. https://doi.org/10.1002/fee.1516

Nadasdy, P. (2003). *Hunters and bureaucrats. Power, knowledge, and aboriginal-state relations in the Southwest Yukon*. University of British Columbia Press.

Paneque-Gálvez, J., Mas, J.-F., Guèze, M., Luz, A. C., Macía, M. J., Orta-Martínez, M., Pino, J., & Reyes-García, V. (2013). Land tenure and forest cover change. The case of southwestern Beni, Bolivian Amazon, 1986–2009. *Applied Geography*, *43*. https://doi.org/10.1016/j.apgeog.2013.06.005

Pascual, U., McElwee, P. D., Diamond, S. E., Ngo, H. T., Bai, X., Cheung, W. W. L., Lim, M., Steiner, N., Agard, J., Donatti, C. I., Duarte, C. M., Leemans, R., Managi, S., Pires, A. P. F., Reyes-García, V., Trisos, C., Scholes, R. J., & Pörtner, H. O. (2022). Governing for transformative change across the biodiversity–climate–society nexus. *BioScience*, *72*(7), 684–704. https://doi.org/10.1093/BIOSCI/BIAC031

Pecl, G. T., Araújo, M. B., Bell, J. D., Blanchard, J., Bonebrake, T. C., Chen, I. C., Clark, T. D., Colwell, R. K., Danielsen, F., Evengård, B., Falconi, L., Ferrier, S., Frusher, S., Garcia, R. A., Griffis, R. B., Hobday, A. J., Janion-Scheepers, C., Jarzyna, M. A., Jennings, S., … Williams, S. E. (2017). Biodiversity redistribution under climate change: Impacts on ecosystems and human well-being. *Science*, *355*(6332). https://doi.org/10.1126/SCIENCE.AAI9214/SUPPL_FILE/PECL.SM.PDF

Pörtner, H.-O., Scholes, B., Agard, J., Archer, E., Arneth, A., Bai, X., Barnes, D., Burrows, M., Chan, L., Cheung, W. L. (William), Diamond, S., Donatti, C., Duarte, C., Eisenhauer, N., Foden, W., Gasalla, M. A., Handa, C., Hickler, T., Hoegh-Guldberg, O., … Ngo, H. (2021). *Scientific outcome of the IPBES-IPCC co-sponsored workshop on biodiversity and climate change*. https://doi.org/10.5281/ZENODO.4659159

Posey, D. (1999). *Cultural and spiritual values of biodiversity*. Practical Action Publishing. https://doi.org/10.3362/9781780445434

Reyes-García, V. (2020). Mothering in the field: Participant observation and cultural transmission. In B. Hewlett (Ed.), *The secret life of anthropologists. Lessons from the field* (pp. 182–203). Routledge.

Reyes-García, V., & Broesch, J. (2013). The transmission of ethnobotanical knowledge and skills among Tsimane' in the Bolivian Amazon. In Ellen, R., Johns, and Lycett (eds). *Understanding cultural transmission in anthropology: A critical synthesis* (pp. 191–212). Berghahn: London.

Reyes-García, V., Paneque-Gálvez, J., Bottazzi, P., Luz, A. C., Gueze, M., Macía, M. J., Orta-Martínez, M., & Pacheco, P. (2014a). Indigenous land reconfiguration and fragmented institutions: A historical political ecology of Tsimane' lands (Bolivian Amazon). *Journal of Rural Studies*, *34*, 282–291. https://doi.org/10.1016/j.jrurstud.2014.02.007

Reyes-García, V., Paneque-Gálvez, J., Luz, A., Gueze, M., Macía, M., Orta-Martínez, M., & Pino, J. (2014b). Cultural change and traditional ecological knowledge: An empirical analysis from the Tsimane' in the Bolivian Amazon. *Human Organization*, *73*(2). https://doi.org/10.17730/humo.73.2.31nl363qgr30n017

Reyes-García, V., Powell, B., Díaz-Reviriego, I., Fernández-Llamazares, Á., Gallois, S., & Gueze, M. (2019). Dietary transitions among three contemporary hunter-gatherers across the tropics. *Food Security*, *11*(1), 109–122. https://doi.org/10.1007/s12571-018-0882-4

Reyes-García, V., Álvarez-Fernández, S., Benyei, P., García-Del-Amo, D., Junqueira, A. B., Labeyrie, V., Li, X., Porcher, V., Porcuna-Ferrer, A., Schlingmann, A., & Soleymani, R. (2023). Local indicators of climate change impacts described by indigenous peoples and local communities: Study protocol. *PLoS One*, *18*(1), e0279847. https://doi.org/10.1371/JOURNAL.PONE.0279847

Rillig, M. C., Ryo, M., Lehmann, A., Aguilar-Trigueros, C. A., Buchert, S., Wulf, A., Iwasaki, A., Roy, J., & Yang, G. (2019). The role of multiple global change factors in driving soil functions and microbial biodiversity. *Science*, *366*(6467), 886–890. https://doi.org/10.1126/SCIENCE.AAY2832/SUPPL_FILE/AAY2832_RILLIG_SM.PDF

Steger, C., Nigussie, G., Alonzo, M., Warkineh, B., van den Hoek, J., Fekadu, M., Evangelista, P. H., & Klein, J. A. (2020). Knowledge coproduction improves understanding of environmental change in the Ethiopian highlands. *Ecology and Society*, *25*(2), 1–11. https://doi.org/10.5751/ES-11325-250202

Sterling, E. J., Filardi, C., Toomey, A., Sigouin, A., Betley, E., Gazit, N., Newell, J., Albert, S., Alvira, D., Bergamini, N., Blair, M., Boseto, D., Burrows, K., Bynum, N., Caillon, S., Caselle, J. E., Claudet, J., Cullman, G., Dacks, R., … Jupiter, S. D. (2017). Biocultural approaches to well-being and sustainability indicators across scales. *Nature Ecology & Evolution*, *1*(12), 1798–1806. https://doi.org/10.1038/s41559-017-0349-6

Tengö, M., Brondizio, E. S., Elmqvist, T., Malmer, P., & Spierenburg, M. (2014). Connecting diverse knowledge systems for enhanced ecosystem governance: The multiple evidence base approach. *Ambio*, *43*(5), 579–591. https://doi.org/10.1007/s13280-014-0501-3

Tengö, M., Hill, R., Malmer, P., Raymond, C. M., Spierenburg, M., Danielsen, F., Elmqvist, T., & Folke, C. (2017). Weaving knowledge systems in IPBES, CBD and beyond—lessons learned for sustainability. *Current Opinion in Environmental Sustainability*, *26–27*, 17–25. https://doi.org/10.1016/J.COSUST.2016.12.005

Turner, N. J. (1997). Traditional ecological knowledge. In P. Schoonmaker, B. von Hagen, & E. C. Wolf (Eds.), *The rain forests of home: Profile of a North American bioregion* (pp. 275–298). Island Press.

Watts, V. (2013). Indigenous place-thought and agency amongst humans and non humans (First Woman and Sky Woman Go on a European World Tour!). *Decolonization: Indigeneity, Education & Society*, *2*(1), 20–34. https://jps.library.utoronto.ca/index.php/des/article/view/19145

11

"NOT LIKE IT USED TO BE"

Contending with the altered agricultural calendar in Andean Peru's Colca Valley

Eric Hirsch

Introduction: the premise and the promise

I have found myself with a group of agricultural laborers on a sloped plot of land named Ch'ela, one of Don Gerardo Huaracha's[1] hillside agricultural terrains (or *chakras* in Quechua) in the southern Peruvian village of Yanque, during multiple planting seasons over the past decade. Ch'ela covers five stepwise terraced fields that slouch gently downward toward the Colca River, affording a view of the breathtaking amphitheatric valley and canyon below. The planting season in much of Peru's Andes traditionally begins in August, with some growers seeding fields as early as June. This timing has allowed me to participate in planting quinoa and fava beans on almost every visit. Planting days mean hard work accompanied by sumptuous lunches and generous rest periods where chicha, an alcoholic drink of fermented maize and barley, is shared among the laborers and with the soil (see Figure 11.1).

Agricultural working days, as I experienced them, are occasions for both upbeat conviviality and anxious speculation about the future of growing and selling crops. Laborers would routinely wonder out loud whether the seeds planted in Ch'ela will follow the trend of worsening harvests and plummeting crop prices, which they attribute in part to a destabilized rainy season and in part to the corporate consolidation of Peru's national produce markets. Planting sometimes ends with singing the *Hialeo*, a call-and-response chant that asks the earth (*Pachamama*) to warm and nurture the vulnerable new seeds.

Historically, agriculture here has long been somewhat precarious, given that the highland Colca Valley region's historical climatic baseline means frequent swings between weather extremes upon fertile volcanic soil (Wernke, 2013). Engaging in agricultural work in today's Andes requires the speculative anticipation that fields will not suffer from the newly dangerous weather conditions that villagers widely attribute to climate change, which create extremes that exceed the formerly expected variability in rains, freezes, hailstorms, and droughts. Anthropologists have documented agriculture as a speculative act in diverse political ecologies of uncertainty (Bradley and Stein, 2022; Campbell, 2015; Morris, 2019). Here, when planting days would conclude, my interlocutors

1 All interviewees featured in this article asked to be named. I therefore do not use pseudonyms.

 DOI: 10.4324/9781003356837-14

Figure 11.1 Sharing chicha with the soil during a planting day in Yanque, Peru. Photograph by the author.

wondered whether the labor and the offerings just made to the earth were in vain. This uncertainty was matched in other conversations with growers I was having by the end of the 2010s, where I heard a common refrain: "It's not like it used to be."

This chapter investigates one factor informing that refrain: the Colca Valley region's altered agricultural calendar, whose most prominent change has been a newly unreliable rainy season. While residents never took abundant rains for granted in this dry highland region, the rainy season was once a loosely reliable period of precipitation and warmer weather between November and March. Residents repeatedly emphasized to me that such climate impacts are not harms that came in a vacuum. Rather, they are mediated by structures of environmental injustice rooted in an urban-centric resource capitalism that is an enduring legacy of Peru's colonial past and whose inequalities are intensifying in new ways.

The locally voiced thesis that Colca Valley agriculture is "not like it used to be" exposes the breakdown of a specific promise and the premise that underlies it. The promise is that market development centered on the cash crop will improve villagers' livelihoods rooted in agricultural production. The premise is that the ecosystem where this market-ready agriculture is meant to grow will remain "stable," which in this region means remaining within its usual range of once-expected weather extremes. Recent decades have seen deliberate projects to integrate Colca agriculturalists into regional, national, and global circuits of commerce by way of the cash crop. Non-governmental organizations and state projects trained their focus on developing the *entrepreneur* as the primary figure of progress and Indigenous prosperity (Hirsch, 2022). However, the late 2010s and early 2020s have been an era of revelation. The disappointments of agricultural entrepreneurship are now materializing palpably as plummeting crop prices and declining harvests. I argue that local climate change observations expose major structural shifts at the twilight of market development. Uneven patterns of environmental and economic injustice are essential to understanding the political ecology of climate change in the Colca Valley. Situating climate observations within their irreducibly local historical, political, and cultural contexts, the chapter joins other contributions in this volume in refusing to reduce climate change to an isolated ecological phenomenon to which local communities simply react (see, e.g., Abazeri, 2024; Carmona, 2024).

This chapter locates observations of the altered agricultural calendar within a reverse chronology of broken promises. After contextualizing the study and specifying my methods, I proceed with a discussion of the Colca Valley agricultural calendar and the ways villagers identified and analyzed new deviations from it. Then, I move back in time to locate destabilizing forcings of climate change within the failed long-term promises of entrepreneurship and market development. I conclude by underscoring the importance of connecting emerging research on climate change experiences to localized questions of environmental justice.

Situating the intervention

In his 2010 history of glacier expertise and climate change in the Peruvian Andes, Mark Carey describes previous research on the role of humans in climate change as a series of relatively simple sketches of "broad scenarios that plot polluters against victims of the developed countries' climate-altering emissions" (2010). Carey suggests that it is important to read human experiences of climate change as specific, locally constituted power struggles over basic political questions of justice, authority, expertise, and history. Since Carey's publication, the ethnography of climate change has flourished. Anthropologists and other qualitative social scientists are building a robust archive of local and Indigenous experiences of climate change (for English-language examples, see Callison, 2014; Crate, 2022; Marino, 2015; Schlingmann et al., 2021; Vaughn, 2022).

However, in key policymaking venues, this work is often "overshadowed" by scientific and quantitative approaches (Barnes et al., 2013, 541). The meta-analysis conducted by Victoria Reyes-García and colleagues (2019) indicate that small-scale, locally focused studies of climate experience have seen a productive proliferation, especially over the last decade. But they also suggest that non-western ways of atmospheric knowing remain subjugated forms of expertise when it comes to major policy decisions. For example, Indigenous voices have remained on the fringes of global climate negotiations, even as Indigenous constituencies are now well represented at international climate talks (Muxlow, 2021). Situational inclusion and even participation have all too often failed to translate into significant policy change (Haverkamp, 2022). My interlocutors in Peru made clear that similar structural exclusions are also rampant in local policy arenas, highlighting their feelings of an absent state safety net for agricultural decline, crop failure, and natural disaster impacts. This chapter highlights how agriculturalists in one region of the Andes observe climate change as one linked dimension of an intensifying socioeconomic marginalization of rural voices, expertise, and livelihoods.

I ascertained a grounded sense of the contemporary livelihoods that characterize the routines of many Colca Valley villagers over 11 years of consistent ethnographic visits to the region between 2008 and 2019. As a white male scholar based at a higher education institution in the United States, my positionality is characterized by considerable privilege. I do not profess to tell the full story of how Colca residents understand climate change. Even my invocation of the shorthand "climate change" as a singular concept betrays my epistemological bias. My effort is to offer my own partial interpretations of a small number of testimonies, through limited culturally and disciplinarily constituted intellectual frameworks, in solidarity with a community that has invited my partnership as a researcher and that may or may not incorporate my analyses into their own political, epistemological, and environmental justice projects.

The data I principally draw on for this chapter is an archive of semi-structured interviews ($n = 60$), which I conducted in June and July of 2019, the height of that year's dry season. For the present analysis, I place these interviews in the context of my long-term ties to families in the region and my ongoing engagement with a variety of state and non-governmental sustainable development projects. My interview sampling was partially attained by following the kinship networks of my locally connected host family, and partially made of opportunistic person-on-the-street interviews. I accomplished these interviews with the help of a research team of five undergraduate students from Franklin & Marshall College. Interviews were conducted with villagers residing in the Colca Valley's high, middle, and low elevations to ascertain a vertical diversity of orientations to agricultural work, ecosystem dwelling, and environmental change.[2] Interviews were geared toward ascertaining a diversity of villager voices across categories of class, kin group, gender, generation, and first language.[3] I also targeted influential community members for interviews. These individuals included elders, municipal leaders such as district mayors, and members of influential civic organizations, including elder adult groups and irrigators committees. All interviewees were longtime residents of Colca's rural villages.

The semi-structured interviews I drew on followed the Local Indicators of Climate Change Impacts project protocol for assessing and contextualizing local forms of observing and adapting to climate change (Reyes-García et al., 2023). For this chapter, I coded interviews by examining how climate experiences and testimonies regarding broad environmental marginalization overlapped. Interviews began with a series of questions about daily lives, livelihoods and income, and life stories. This enabled me to discern how villagers located themselves within broader Peruvian contexts of urban–rural spatial inequalities and colonially inflected racial hierarchies (García, 2021; Quijano, 2000), uneven exposure to toxic industry (Gonzalez, 2022; Graeter, 2020), and access to state safety nets (Rasmussen, 2015). Interviews then turned to local observations and experiences of climate change impacts. We asked people to assess the ecological changes they had seen in their home villages since they were born, even if they had left for extended periods, as many have for labor migrations and education. The testimonies I feature below were broadly representative of the themes that spanned my interviews.

Making stable livelihoods in a variegated vertical terrain

The Colca Valley's inhabitants reside in the high southern Andean region, in territory that the Peruvian Ministry of Culture classifies as part of the Quechua Indigenous community. Residents sometimes self-identify as Quechua or Indigenous, but are just as likely to identify themselves as *campesino* (peasant). The long colonial and postcolonial history of subjugated Indigenous status in the Peruvian Andes leads many to disavow being labeled "Indigenous." Whether or not they explicitly invoke Indigenous identifiers (see Salas Carreño, 2020), many Colca residents claim heritage dating to the pre-Inca era. Residents of Yanque, my consistent home village in the middle valley, historically populated by the Collagua ethnic group, suggested that the Incas lifted the Collaguas' amphitheater-style terraced farming and implemented it throughout the Andes.

2 I elaborate the importance of elevation to agricultural diversification and economic status in Hirsch (2018).

3 Most people in the region speak Spanish, the default language of municipal policy, education, and business, and some Quechua. Monolingual Quechua speakers tend to be elderly. When we spoke to primarily Quechua speakers, we drew on my Quechua language instruction and the help of local translators.

The Colca Valley region is a vertical landscape that reaches multiple elevations within a dense three-dimensional space, from the cold highland plains home to dense peat bogs and alpaca herds, down to the soft riverine terrains bathed in lush tangles of bright green vegetation where villagers grow avocados and goldenberries. The steep valley is surrounded by an active volcano, called Sabancaya, and a cluster of mountains whose once-perennial summit snowpack had long been a reliable source of irrigation water. Colca's communities are situated between 2,000 and 6,000 meters above sea level. The high-elevation environment and the dense diversity of distinct elevations for planting and herding mean that this region is home to both considerable environmental challenges and important agricultural innovations and endowments, including terraced fields (see Figure 11.2) and gravity-based irrigation of vast terrains from mountain springs and summit snowpack (Markowitz, 2018; Postigo, 2014, 2019).

This is a region that pairs fertile volcanically enriched soil with unpredictable weather. Weather extremes were a fact of life in the Colca Valley, as archeologists (Wernke, 2013) and villagers (see below) have made clear. Colca agriculturalists draw on historically elaborated ecological interventions to negotiate Colca's long-standing water precarity with sociotechnical irrigation systems creatively constructed around mountain springs (Guillet and Mitchell, 1994). However, recent decades have seen newly devastating drought events that existing infrastructure often cannot accommodate. Consequently, widely grown staple crops throughout the region such as quinoa now need three or more times the amount of water they once demanded. My interviews revealed that by the end of the 2010s, the unpredictability of altered precipitation and dry soil conditions had reached what Adriana Petryna calls a climatic "regime shift," in which local knowledges are forcibly and rapidly rearranged due to changing horizons of possibility for response (2022, 56).

The Colca Valley cradles a tight network of 19 agricultural village districts (populations range from Tapay's 400 inhabitants to Chivay's 6,000). The local market and administrative center is the town of Chivay. The region's villages see a great deal of exchange, interaction, labor migration, and kinship connections with one another and with Arequipa, Peru's second-largest city and the

Figure 11.2 Agricultural labor on terraced fields in Yanque, Peru. Photograph by the author.

department capital, which is about 160 km (three to five hours by bus) from the Colca Valley. Agriculture and pastoralism have long been the core of most villagers' livelihoods across the region. The region's principal cash crops are quinoa, potatoes, fava beans, barley, alfalfa, and the non-food crop straw (*ichu*, traditionally used as a construction material). Tourism, produce markets, hospitality, microfinance, and education and non-governmental organization work have been important additions to the economy, with growth in those sectors gaining momentum in the 1990s. *Campesinos* throughout the valley frequently diversify their incomes with various forms of manual labor, including construction work, migrant labor in mines, and laboring for state-sponsored mega-projects.

The traditional and altered agricultural calendars

Testimonies from my long-form narrative interviews centered on reflections about the altered agricultural calendar. Deviating from recent model-based and large-*n* studies of climate change's impacts on agriculture (see, e.g., Cinner et al., 2022; Lobell et al., 2020; Rosenzweig et al., 2014), the smaller archive of testimonies I draw on to construct an outline of the Colca Valley's altered agricultural calendar engages a more intimate, but inevitably partial, portrayal of one region's experiences of agricultural timing. Table 11.1 represents a schematic overview of calendrical information gleaned from villager testimonies, capturing selected perceptions of how the agricultural calendar has changed in villagers' lifetimes. They parallel other recent scholarly efforts to outline typical Andean agricultural calendars based on interviews with peasant growers (Gurgiser et al., 2016; Vallejo-Rojas, Rivera-Ferre, and Ravera, 2022).

I use the phrase *traditional agricultural calendar* (middle column) to describe the planting and harvesting timeline that Colca Valley villagers articulated to me as having been standard before the acute climatic and ecological instability that has intensified in the most recent decade. That traditional calendar begins in August. It is organized by what villagers label the rainy season (*temporada de lluvias*) lasting from November to March, with the most intense rains taking place in January and February. The rainy season is the fulcrum of Colca Valley agriculture. It is, according to the traditional model, preceded by the planting season (*siembra*) from August to December. The rains were followed by a March to June harvest season (*cosecha*), then a season when fields were left fallow (*barbecho*) during the cold and dry period lasting from June to July. In this time, some families would also begin the *michka*, a round of early planting, in fields they determined to be ready.

My understanding of the *altered agricultural calendar* (right column) is rooted in the testimonies my research assistants and I documented. Colca residents identified new experiences of seasonality, which they highlighted as rainy seasons that were delayed or severely diminished, new levels of frost and intense cold beyond the expected months of June and July, and the unwelcome surprises of erratic out-of-season weather events like hail and snowstorms which once occurred only occasionally during the traditional rainy season. As opposed to how my interlocutors systematically described traditional agricultural timing, the shifts that interviewees noted could sometimes contradict one another and be inconsistent from year to year.

Testimonies demonstrate the intensity and unevenness of the deviations from the traditional agricultural calendar. Julia Quispe Panta is a *campesina* and a village councilperson from Lari, a district that spans middle- and high-elevation settlements. She told us that "sometimes, there are years and years when it rains and there are years that, well, it doesn't rain at its correct time (*a su tiempo*)." The fact that there exists a designated "correct" timing for expected rain indicates that current conditions represent deviations from a known baseline. Eusebia Chavez, a fellow council member in Lari who also works as a teacher in Lari's high school, echoed Julia's observation of rains that are unreliable in timing and intensity. From Eusebia's perspective, climate shifts have been felt

Table 11.1 An outline of Colca Valley's traditional agricultural calendar (middle column) and observed alterations (right column), based on the 2019 testimonies and my broader ethnographic research conducted between 2008 and 2019

Month	*Traditional agricultural calendar* *Expected weather events, with locally associated practices* italicized	*Altered agricultural calendar* *Erratic weather events, with locally associated practices* italicized
August	1 August is a cold (frosts often occur at night, with warmer days) and dry (rain is rare) month. This is the opening month of the agricultural calendar. 2 Some windstorms reach parts of the Colca Valley region. 3 *Rituals* take place across villages that renew the hydrological cycle. These early-August rituals initiate the agricultural calendar. 4 Planting (siembra) begins.	1 Occasional rain, including severe flood-inducing storms in some years. 2 Windstorms last longer than expected. 3 Cold temperatures with higher fluctuations between extremes. 4 *Rituals* take place to renew the hydrological cycle and initiate the agricultural calendar. 5 Main *planting* season begins, but now with *enhanced efforts to calculate risk* from newly extreme heat, cold, and storms.
September	1 Mostly dry, with moderate temperatures (warmer days, night frosts rare). 2 *Planting* month.	1 Dry and moderate, with sudden temperature spikes and troughs and occasional extreme rain and hailstorms. 2 Occasional windstorms. 3 *Planting* continues, along with new *efforts to avoid or plan around extreme weather* by seeding fewer fields, adopting more drought-resilient crops or seed varieties, and utilizing increased fertilizer.
October	1 Mostly dry, with moderate temperatures. 2 *Planting* month.	1 Dry and moderate, with sudden temperature spikes and troughs and occasional extreme storms. 2 *Planting* continues, with above adjustments.
November	1 The rainy season (*temporada de lluvias*) begins. Rains in November tend to indicate a good harvest year to come. 2 Moderate temperatures. 3 *Planting* month. Most planting ends.	1 The rainy season sometimes begins forcefully and overwhelmingly. 2 Rains sometimes fail to appear at all. 3 Moderate temperatures with increased fluctuations. 4 *Planting* month. Most planting ends.
December	1 The first principal month of the rainy season. 2 Most *planting* ends, with field maintenance continuing. 3 Christmas and associated *holidays* offer a reprieve from agricultural labor in much of the month.	1 Rain is sporadic. 2 Moderate temperatures with increased fluctuations. 3 Last planting and field maintenance. 4 Christmas and associated *holidays*.
January	1 Peak precipitation season. Rains are observed almost every day throughout the valley. Snow and hail are also observed. 2 Snow accumulates on mountain peaks. 3 Villages see warmer days and nights.	1 Dramatically decreased or nonexistent rain. 2 Rising number of hot and dry sunny days. 3 *Initiation rituals* for new community leaders. 4 Agricultural maintenance.

(*Continued*)

Table 11.1 (Continued)

Month	*Traditional agricultural calendar* *Expected weather events, with locally associated practices italicized*	*Altered agricultural calendar* *Erratic weather events, with locally associated practices italicized*
	4 *Initiation rituals* for new communal agricultural and hydrological office holders. 5 Minimal agricultural work, with some field maintenance.	
February	1 Peak precipitation season. Rains are observed almost every day throughout the valley. 2 Snow accumulates on mountain peaks. 3 Warmer days and nights. 4 Terrain (chakra) maintenance. 5 Agriculturalists also often engage in small-scale household *maintenance* and construction *projects*, and other *non-field labor*.	1 Dramatically decreased or nonexistent rain. 2 More warm days. 3 The rainy season sometimes starts as late as this month, if it appears at all. 4 Chakra *maintenance*. 5 Other *non-field labor* projects, local obligations, and seasonal employment.
March	1 The rainy season ends; waning precipitation, with some erratic rain and hail storms are expected. 2 The *harvest* season (*cosecha*) begins.	1 Erratic rains that can range from nonexistent or devastatingly sporadic to dangerous heavy rain and surprise intense hailstorms. 2 Increased frost risk. 3 The rainy season sometimes does not begin until this month. 4 *Harvests* begin, if the previous months' weather conditions allow.
April	1 A fertile growth period. The valley is expected to be lush with growing crops and vegetation. 2 *Harvests* take place throughout the month. 3 Holy Week and Easter *holidays* provide a reprieve from agricultural labor.	1 Spotty rain possible. 2 Erratic and occasionally dangerously high levels of precipitation and temperature extremes. 3 Increased frost risk. 4 *Harvests*, newly endangered by frost and erratic weather, are less abundant than expected.
May	1 Colder and drier weather sets in. 2 *Harvests* throughout the month. Harvests end.	1 Colder and drier weather, punctuated by erratic crop-injuring cold snaps and heat waves. 2 *Harvests* are endangered and scarce.
June	1 Colder weather intensifies. 2 Peak frosts in some villages, including Lari. 3 Fallow (*barbecho*) period. *Checking* seeds and fields; routine *maintenance* of agricultural land. 4 Occasional early *planting* (*michka*) occurs.	1 Cold weather intensifies rapidly or unevenly. 2 June is warmer than expected. 3 *Harvests* are endangered and scarce. 4 Occasional early *planting* (*michka*) occurs.
July	1 Coldest and driest month. 2 Fallow (*barbecho*) period. *Checking* seeds and fields. 3 Increased *efforts* to conserve water in fields and at home. 4 Occasional early *planting* (*michka*) occurs.	1 Occasional rain and warmer weather, including severe flood-inducing storms. 2 Erratic frosts. Cold weather is present but can be inconsistent. 3 Occasional early *planting* (*michka*) occurs.

drastically, because, for example, regarding the rains, before they were, how to put it, normal, in their correct moment it rained. Now, instead, suddenly in periods where it's not supposed to, where you shouldn't see that presence, it comes, and sometimes you even see huge rainstorms, which bring landslides.

Eusebia also described a more broadly destabilized seasonal rhythm.

> Look, during this season [early July, the traditional dry season] there shouldn't have been frosts, they should have been done by now…Where you see the last strong frost is the 24th of June. Now it's already July 3rd, and it continues with the same intensity.

Eusebia's calendrical precision demonstrates a scientific habit of rigorous longitudinal observation that scholars have noted in other regions of the Peruvian Andes (Caine, 2021; Orlove, Chiang, and Cane, 2002). Eusebia continued, moving to other anomalies: "Then, for example, last year until the month of September, October, there were these winds, like huge windstorms (*ventarrones*) that before you only ever saw them in the month of August, that was their moment." Turning to the question of snow on surrounding mountain peaks, Eusebia noted that "before, they had, as it's said here, their 'white *poncho*.'[4] With the change now, no, well, you don't see it in the same way."

"It's not like it used to be," Cesaria Maque Flores told us, invoking the refrain in her description of what agriculture has become in the Colca Valley. Cesaria is a part-time vendor in a small store and a part-time agriculturalist from the upper-elevation settlement of Canocota who in 2019 lived mainly in Chivay. Her observations of climate change were diverse, from lamenting the new problem of overheating on stone village roads to sheep that are now plagued with eye mucus. She also lamented the animals in the fields that were forced to burrow underneath one another in desperate search of shade. Cesaria described the need for more irrigation, due to more frequent droughts, by highlighting the plight of alfalfa, a plant known in the region for its ability to grow abundantly with low levels of water:

> you need to be irrigating, because even alfalfa dries with ease. And every fifteen days at least you have to irrigate…you have to always be irrigating, fertilizing, because if you don't do it, if you leave it, your crops will burn with the heat.

High-frequency irrigation and fertilizer use, for Cesaria, were adaptations that the new heat has forced on some families. Cesaria noted the overall breakdown in seasonal rhythms,

> because there are times when it should have rained in the month of December, and it doesn't rain a drop. In January, a tiny bit. And sometimes it only rains in January or February. So, there's no rain. So yes, it's a drastic change.

Like Eusebia, Cesaria also noted a lack of summit snowpack:

> You no longer see snow here above, right? We call [the highland terrain] Patapampa. Before, when I was a child, I saw it covered in white snow. My mother told me, "Don't look, don't

4 The "poncho" invoked here is a reference to the garment traditionally worn for warmth by peasants in the Andean highlands. Its shape over the shoulders is similar to the conical form of snow as it covers a mountain peak.

> look, because you're going to hurt your eyes." Now you look at it, the land is reddish. Snow used to stay, I think. Now there's no snow this season. Every season it used to be totally covered in snow.

Gerardo Huaracha, from Yanque, attributed changes in the weather to "global warming, because before, you almost didn't realize when it was hot. Before there wasn't so much heat, and now there's so much heat, cold, and wind, right?" Here, he emphasizes that today's erratic conditions dwarf the earlier extremes he once observed. Regarding the rains, Gerardo noted:

> Before, the rain began in the month of November. The first days of November the rain began, and then they said that it was going to be a good year. But now, no. It begins sometimes in December. If rain begins in December that's also really good, the little plants are growing, so, in December, it rains, and in January and February, that means good weather. But if it doesn't rain in December, what will the plants do then?

Given these altered rains, he continued, the harvest "has changed. Because before the harvest, right, by March, we were completing the harvest, but now into July, we're just finishing the harvest." Gerardo also commented on the diminished presence of snow.

> Before, Mismi,[5] the whole year it was covered in snow, right? The snow never left, right? [The mountaintops were like this] all the way over to the Sabancaya volcano. But that's no longer the case. All the snow has been erased from the mountaintops. Because of that, the water has also gone down a bit.

Terencia Huaracha Llukra, a *campesina* from Huanca (and a distant cousin of Gerardo's), described parallel struggles even in her lush lower-elevation community. She noted that while the balmy Huanca has not seen water shortages as dire as those in higher-elevation villages, it was still raining less. Agriculturalists in her community were also facing the new problem of frost risk in this usually balmy region. Terencia echoed previous testimonies about diminished snow: "There was snow atop the mountains this year, but only a little bit." She also indicated that before, rains lasted until March, but in recent years, they stopped in February.

Underlying the destabilized agricultural calendar, local histories reveal periodic droughts taking place in the Colca Valley about every five to eight years in the latter half of the 20th century. These histories corroborate publicly available weather data from Peru's weather monitoring service SENAMHI (the National Service for Meteorology and Hydrology of Peru)[6] and the SPEI (Standardized Precipitation-Evapotranspiration Index) Global Drought Monitor service.[7] There exist local memories of the drought years in the early 1980s as particularly intense. However, in both villager testimonies and weather station data, those older droughts pale in comparison to the droughts that have plagued the valley since 2014. These and other testimonies of changes in the agricultural calendar demonstrate a close attunement to altered seasonal rhythms. As the next section illuminates, the practices villagers engaged to adjust their livelihoods to an altered agricultural calendar reveal adaptation to be an uneven and unjust burden.

5 Mismi is the mountain summit and earth-being that supplies the lower part of Yanque with its irrigation water.
6 https://www.senamhi.gob.pe/?&p=lluvia-acumulada
7 https://spei.csic.es/map/maps.html#months=0#month=10#year=2021

Unbeckoned adaptation: adjusting to an altered calendar

The agricultural work I have conducted with Gerardo and members of his network demonstrates the variety of ways they are working to adjust their growing practices as expected weather patterns break down. With less reliable rains, growing becomes less efficient as it is increasingly difficult to coordinate timing and harvesting labor between distinct crops and fields. Agricultural labor has become increasingly uncertain, with an improvised planning that combines knowledge of historical patterns and best estimates based on recent erratic harvest years. Gerardo emphasized that agriculturalists like him are facing the new and burdensome imperative to adjust their livelihoods, usually without any state or regional government support. Similarly, Eusebia summarized her view of local agriculture in Lari as a site of cascading losses:

> Now there is also a lot of migration. Many of the people are leaving from here. And that's because, because, there's a disappointment given the [agricultural] production here, and beyond that, the little terrains are too divided up, so there aren't many opportunities...Because just like that, the costs don't get any cheaper, for example sometimes, potatoes, per kilo sometimes you get from here 20 cents, 25 cents, 30 cents, when the production maybe has been for, well, you've invested more than you're paid.

Eusebia's testimonies moved seamlessly from identifying observed geophysical changes to the era's intensifying socioeconomic marginalization. With her mention of low prices for cash crops, Eusebia captures a sense of disappointment for which observations of decreased harvests and an altered ecology alone would provide an inadequate explanation. Eusebia registered the fact that regional markets for cash crops planted according to a specific calendar were no longer a reliable means of making a living, and that this was also due to the consolidation of large corporate producers within Peru that systematically excluded Colca growers from quinoa markets. Cesaria emphasized, through her own observations of recent changes to the region's agriculture, that climatic changes and sociopolitical changes were inextricably linked. Cesaria made the observation, which I heard frequently from Colca Valley villagers, that local growers tend to use excessive fertilizer. She suggested that these small-scale chemical applications, made with the narrow aim of selling crops at a profit, were a local example of a globally unchecked human tendency to pollute while aiming for short-term gain. Agriculture is changing "significantly," she said:

> it's the climate, and apart from that, unfortunately, it's poor information, or it could be the agriculturalists that have done it, they've put really intense fertilizers and the terrains have turned really poor, and now no longer give good harvests.

In Cesaria's view, fertilizers are a temporary improvement practice that state agents and development projects promised would increase yields, but that ultimately injures fields. This emergency response to the sudden need to integrate fertilizers, Cesaria observed, results in fields that become irreversibly damaged over time.

Terencia noted that "Yes, agriculture has changed, they're now seeding a lot of garlic because at least it sells; there's a market. Potatoes, corn, they no longer plant that much, just a little, because it's not so profitable." To abandon potatoes for garlic in the Andes, the birthplace of the potato, is a striking shift. Describing the food economy, Terencia indicated that "Before, it was healthier, more natural." Under that previous system, she said, "what you produced, you consumed." Now, as she noted in a critique that resonated across interviews from multiple villages, the community was regrettably moving from subsistence farming to "all store-bought goods" ("*pura tienda*"). This move is reflective of the experience of other rural and Indigenous communities that have been unjustly deprived of their food sovereignty due in part to fluctuating markets and prices for local crops (Blue Bird Jernigan et al., 2021).

Terencia echoed Cesaria's and others' concerns about how growing fields were exposed to chemical harms. She addressed the issue of rising pollution when asked about the environmental changes she had observed in her lifetime. "Everyone fumigates, they're now contaminating [the fields] and you can't go back to planting, maybe alfalfa; but even that is being lost." As a related testimony from another anonymous interviewee suggested, when it comes to the land, "People here are really uncaring, and aren't worried for their health or the environment." There was clear intra-community blame for poor harvests and fertilizer and pesticide use that many interviewees entangled with their concerns about newly erratic rains.

The broader context of these diverse testimonies about responses to the changing agricultural economy is a longer local history of marginalization. In the next section, I move into the recent past to understand the development frameworks that by the late 2010s would compound the environmental injustices of the Colca Valley's altered agricultural calendar.

Climate change and the broken promises of market development

Climate justice scholar Kyle Powys Whyte argues that climate change impacts disproportionately harm Indigenous communities around the world, intensifying ongoing structural violences of colonial extraction, exploitation, and abandonment (2016). Interviewees in Colca similarly invoked their experience of climate change as an environmental injustice with roots in the colonial invasion. Eusebia, for example, pinpointed the rise of large mining enterprises and corporate market consolidation at the expense of small rural growers as a continuation of the colonial extractivism that initiated European rule centuries ago. Mercedes Mercado Gonzalez, a beekeeper and entrepreneur from Huanca, described produce markets as spaces of price exploitation by intermediaries that has always been a fact of economic life but is now increasing: "we plant potatoes and corn, and sometimes we go to the market and in the market, they lowball us, or like, we don't make back even what we invested it all goes into the intermediary's hands, leaving us with few profits."

Many testimonies from 2019 wove observations of climate change together with a critique of structural marginalization. This last section contextualizes locally altered agricultural seasonality in deeper colonial forms of extractivism and environmental injustice. I draw on my long-term ethnographic fieldwork and additional scholarly sources to elucidate the history of uneven development that informs many villagers' political readings of the altered agricultural calendar.

The last decade's climate extremes did not touch down upon a blank ecological canvas. Rather, they were layered into the region's recent history of modernizing development projects that worked to inculcate the technocratic promise that markets in agriculture mean widespread prosperity. This promise posited that by commoditizing their crops, agriculturalists could become entrepreneurs that benefit directly from regional market expansion. A commercialized, inclusive market agriculture based on the premise that it would always be possible to grow crops was rooted in the region's mid-20th-century massive infrastructure programs that built out a muscular modernist state vision of techno-scientifically perfected agriculture as a source of ever-increasing improvement. The Majes canal project, at the desert outskirts of the Colca Valley region, was emblematic of this modernist logic. It mobilized irrigation infrastructure to create new agricultural terrains along the coastal desert, setting the stage for expanded cash cropping and, an individualized entrepreneurial logic throughout the region. Projects that approach massive infrastructure as a means of creating the conditions for entrepreneurship have been documented in Peru (Carey, 2010) and elsewhere (Li, 2007; Mitchell, 2002; Scott, 1998).[8]

Building on the new artificially expanded land made available for agriculture, the promise of agricultural market development manifested itself in a series of small-scale initiatives that communicated to Colca Valley villagers that the problems of rural marginality could be solved with the framework of entrepreneurship and the economic technology of the cash crop. Subsistence agriculture dominated most Colca Valley livelihoods until the 1980s. Non-governmental organizations (NGOs) such as Desco (the Center for Studies and Promotion of Development) were instrumental in building out Chivay as a regional market center. Desco was also key to establishing new infrastructures of financial inclusion in Chivay, including banks and microfinance businesses, which worked to commercialize crop growing by providing loans to agriculturalists who framed their work as a business. Desco and other small-scale development projects helped expand local markets for cash cropping beginning in the 1990s. Desco's projects were oriented around themes of commercialization and the effort to transform agriculturalists into self-reliant, growth-ready entrepreneurs who could use their newfound profits to advance themselves. In that time, tourism was also expanding, and with it associated services like high-end restaurants, excursion experiences, and culture-focused museums. Colca Valley tourism pitched an "off-the-beaten-path" appeal to travelers interested in Peruvian cultural and ecotourism beyond the well-traveled attraction of Machu Picchu. By the 2010s, tourism in Colca had expanded massively. Colca is now home to luxury hotels like the Colca Lodge and Las Casitas del Colca.

Despite the expansion of new business in tourism, gastronomy, and hospitality, a combined cash-crop and subsistence agriculture had remained the bedrock of most people's livelihoods in the lower and middle elevation Colca Valley villages, with the addition of pastoralism and livestock care in the high elevation villages. Indeed, in the high-tourism years before the Covid-19 pandemic and Peru's early-2020s political tensions, economic gains meant rising incomes for some agriculturalists and herders. While prosperity from cash-cropping remained deeply uneven, cash crops were a principal means of generating household income. However, by 2019, prices for once-profitable Andean crops were plummeting due to competition with large corporate growing conglomerates on Peru's Pacific coast. For instance, the early 2010s saw a global quinoa boom and a consequent quinoa price spike. News outlets around the world—imagining a one-dimensional, profit-seeking Andean grower—fretted that high quinoa prices would mean the mass replacement of this nutritious local grain for the empty calories of packaged food like pasta as Andean residents moved to selling quinoa instead of consuming it (Blythman, 2013). Certainly, there existed local concerns about food sovereignty, as Terencia's testimony makes clear. Yet a study of Andean quinoa grower households during the early 2010s demonstrates that quinoa's rising price brought prosperity to many growers without forcing them to sacrifice nutrition (International Trade Centre, 2016). My own experience living with a quinoa-selling family corroborates these findings.

However, the late-decade quinoa price *drop* proved devastating for farmers in the Colca Valley and throughout the Andes. In the late 2010s, quinoa prices plummeted from a peak of $4 USD per pound to $0.60 per pound (McDonell, 2018). As Fabiana Li and Claudia Urdanivia note (2020), after quinoa prices in the southern Andes fell from their 2014 peak, farmers began to feel that lower prices did not justify the required agricultural investment. In part due to quinoa's global promotion as a superfood—2013 was deemed the United Nations' "International Year of Quinoa"—corporate actors with more robust access to capital and international markets flocked to quinoa, competing

8 An analysis of water infrastructure projects and climate change can be found in Junqueira et al. (2021).

with small-scale Andean farmers. Quinoa production, alongside the dairy industry and other crops like rice and barley, was increasingly consolidated in larger corporate farms along the Peruvian coast during the 2010s. This created a glut in the market. The encouragement by entrepreneurship-focused Andean development projects to highland farmers to share quinoa with the world led to agricultural expansion followed by agricultural decline due to the absence of market safeguards for small-scale producers (McDonell, 2018).

The corporate capture of what was once a grain targeted for its potential to empower traditional highland farmers had, by the end of the decade, resulted in intensified economic marginalization for many of those farmers. Compounding this marginalization, the altered agricultural calendar undermined the premise of an ecologically stable ground for cash and subsistence crops alike. In 2019, villagers repeatedly told me that cash cropping inevitably meant losing money and was no longer worth the investment. This amounts to the breakdown of the promise of agricultural entrepreneurship: cash cropping in quinoa, potatoes, and other products that Andean farmers once had the reliable ability to sell ceased to be a sure means of generating essential cash income. Many villagers identified their continued agricultural production as "just for us," or exclusively for household consumption, as Yanque resident Edwin Oxa told me. The linked failures of the premise of a consistent agricultural calendar and the promise of entrepreneurial growth reveal how the global heating crisis is inextricable with local crises of structural marginalization and environmental injustice.

Conclusion

The widely offered refrain that agriculture in the Colca Valley is "not like it used to be" entangles observations of new weather unpredictability with a socioeconomic critique of the precipitous fall of the cash crop. The growing scholarly interest in how local communities voice their understandings of climate change demonstrates an important effort in climate science and policy discussions to highlight underrepresented expertise from contexts of marginalized knowledge, where longitudinal insights into ecological disruption are mobilized for forced adaptations. However, as research that highlights local voices on climate change flourishes, it will also be essential for analyses that amplify those voices to foreground the political, historical, economic, and other structural conditions through which locally oriented climate insights are generated.

I repeatedly found Colca Valley growers offering political ecology critiques of climate change, describing their agricultural lives and the adaptations forced upon them without invoking a neat dividing line between the conditions of drastic environmental change and the conditions of uneven market development. Localized observations of the altered agricultural calendar were offered alongside insights about the outdated premise of stable weather and the unfulfilled promise of agricultural entrepreneurship. Testimonies from the Colca Valley transformed my intended study of diverse responses to an atmospheric phenomenon into an analysis of environmental and economic injustice. As research on local indicators of climate change impacts proceeds, an environmental justice framework will be essential to highlighting the distributional politics, capitalism gaps, and deliberate policy decisions that disproportionately expose already-marginalized communities to the uneven burdens of erratic new climatic conditions.

Acknowledgments

I conducted the research for this chapter as a partner of the Local Indicators of Climate Change Impacts (LICCI) project. Thanks to Victoria Reyes-García and Anna Schlingmann for their generous input. Thanks also to my research assistants: Andrea Corilloclla, Nancy Le, Katie McCarthy, Lia

Tavarez, and Matt Turetsky. Fieldwork was supported by the Wenner-Gren Foundation Engaged Anthropology Research Grant and the Department of Earth and Environment, Provosts' Office, and Office of Sponsored Research at Franklin & Marshall College. I would like to express my personal gratitude to Lauren Hock and Anna Hirsch.

References

Abazeri, M. (2024). Strengthening participatory governance to adapt and other sociocultural responses to climate change impacts among farming communities in Kerman, Iran. In V. Reyes-García, S. Álvarez-Fernández, P. Benyei, L. Calvet-Mir, D. García-del-Amo, A. B. Junqueira, X. Li, V. Porcher, A. Porcuna-Ferrer, A. Schlingmann, & R. Soleymani (Eds.), *Routledge Handbook of Climate Change Impacts on Indigenous Peoples and Local Communities*. Oxfordshire & New York: Routledge. Chp 18, pp 303–321.

Barnes, Jessica, Michael Dove, Myanna Lahsen, et al. 2013. "Contribution of Anthropology to the Study of Climate Change." *Nature Climate Change* 3 (6): 541–44.

Blue Bird Jernigan, Valarie, Tara L. Maudrie, Cassandra Jean Nikolaus, et al. 2021. "Food Sovereignty Indicators for Indigenous Community Capacity Building and Health." *Frontiers in Sustainable Food Systems* 5. https://www.frontiersin.org/articles/10.3389/fsufs.2021.704750.

Blythman, Joanna. 2013. "Can Vegans Stomach the Unpalatable Truth about Quinoa?" *The Guardian*, 2013. https://www.theguardian.com/commentisfree/2013/jan/16/vegans-stomach-unpalatable-truth-quinoa.

Bradley, Hannah, and Serena Stein. 2022. "Climate Opportunism and Values of Change on the Arctic Agricultural Frontier." *Economic Anthropology* 9: 207–22.

Caine, Allison. 2021. "'Who Would Watch the Animals?': Gendered Knowledge and Expert Performance Among Andean Pastoralists." *Culture, Agriculture, Food and Environment* 43 (1): 4–13.

Callison, Candis. 2014. *How Climate Change Comes to Matter: The Communal Life of Facts*. Durham, NC: Duke University Press.

Campbell, Jeremy. 2015. *Conjuring Property: Speculation & Environmental Futures in the Brazilian Amazon*. Seattle: University of Washington Press.

Carey, Mark. 2010. *In the Shadow of Melting Glaciers: Climate Change and Andean Society*. Oxford: Oxford University Press.

Carmona, R. (2024) The colonial roots of climate vulnerability. Analising the case of the Mapuche-Pehuenche People in Southern Chile. In V. Reyes-García, S. Álvarez-Fernández, P. Benyei, L. Calvet-Mir, D. García-del-Amo, A. B. Junqueira, X. Li, V. Porcher, A. Porcuna-Ferrer, A. Schlingmann, & R. Soleymani (Eds.), *Routledge Handbook of Climate Change Impacts on Indigenous Peoples and Local Communities*. Oxfordshire & New York: Routledge. Chp 12, pp 196–207.

Cinner, Joshua E., Iain R. Caldwell, Lauric Thiault, et al. 2022. "Potential Impacts of Climate Change on Agriculture and Fisheries Production in 72 Tropical Coastal Communities." *Nature Communications* 13 (1): 3530.

Crate, Susan A. 2022. *Once Upon the Permafrost: Knowing Culture and Climate Change in Siberia*. Tucson: University of Arizona Press.

García, María Elena. 2021. *Gastropolitics and the Specter of Race: Stories of Capital, Culture, and Coloniality in Peru*. Berkeley: University of California Press.

Gonzalez, Adrian. 2022. "What Justice and for Whom? A Political Ecology of Voice Study into 'Senses of Justice' in Peru's Loreto Region." *Environment and Planning E: Nature and Space* 5 (1): 473–504.

Graeter, Stefanie. 2020. "Infrastructural Incorporations: Toxic Storage, Corporate Indemnity, and Ethical Deferral in Peru's Neoextractive Era." *American Anthropologist* 122 (1): 21–36.

Guillet, David, and William P. Mitchell. 1994. "Introduction: High Altitude Irrigation." In *Irrigation at High Altitudes: The Social Organization of Water Control Systems in the Andes*, edited by William P. Mitchell and David Guillet, 1–19. Washington, DC: Society for Latin American Anthropology and the American Anthropological Association.

Gurgiser, Wolfgang, Irmgard Juen, Katrin Singer, et al. 2016. "Comparing Peasants' Perceptions of Precipitation Change with Precipitation Records in the Tropical Callejón de Huaylas, Peru." *Earth System Dynamics* 7: 499–515.

Haverkamp, Jamie. 2022. "The Coloniality of Global Climate Governance." *Hot Spots, Fieldsights* (blog). 2022. https://culanth.org/fieldsights/the-coloniality-of-global-climate-governance.

Hirsch, Eric. 2018. "Remapping the Vertical Archipelago: Mobility, Migration, and the Everyday Labor of Andean Development." *Journal of Latin American and Caribbean Anthropology* 23 (1): 189–208.

———. 2022. *Acts of Growth: Development and the Politics of Abundance in Peru*. Stanford, CA: Stanford University Press.
International Trade Centre. 2016. *Trade in Quinoa: Impact on the Welfare of Peruvian Communities*. Geneva: International Trade Centre.
Junqueira, André Braga, Álvaro Fernández-Llamazares, Miquel Torrents-Ticó, et al. 2021. "Interactions between Climate Change and Infrastructure Projects in Changing Water Resources: An Ethnobiological Perspective from the Daasanach, Kenya." *Journal of Ethnobiology* 41 (3): 331–48.
Li, Fabiana, and Claudia Urdanivia. 2020. *Global Quinoa and Andean Foodways*. https://fabianali.com/global-quinoa-and-andean-foodways/.
Li, Tania Murray. 2007. *The Will to Improve: Governmentality, Development, and the Practice of Politics*. Durham, NC: Duke University Press.
Lobell, David B., George Azzari, Marshall Burke, et al. 2020. "Eyes in the Sky, Boots on the Ground: Assessing Satellite- and Ground-Based Approaches to Crop Yield Measurement and Analysis." *American Journal of Agricultural Economics* 102 (1): 202–19.
Marino, Elizabeth K. 2015. *Fierce Climate, Sacred Ground: An Ethnography of Climate Change in Shishmaref, Alaska*. Fairbanks: University of Alaska Press.
Markowitz, Lisa. 2018. "Experience and Expertise: Confronting Climate Change in the Andes." In *Global Mountain Regions: Conversations toward the Future*, edited by Ann Kingsolver and Sasikumar Balasundaram, 265–271. Bloomington: Indiana University Press.
McDonell, Emma. 2018. "The Quinoa Boom Goes Bust in the Andes." *NACLA Report on the Americas*. https://nacla.org/news/2018/03/12/quinoa-boom-goes-bust-andes.
Mitchell, Timothy. 2002. *Rule of Experts: Egypt, Techno-Politics, Modernity*. Berkeley: University of California Press.
Morris, Meghan. 2019. "Speculative Fields: Property in the Shadow of Post-Conflict Colombia." *Cultural Anthropology* 34 (4): 580–606.
Muxlow, Sheila. 2021. "COP26 Negotiations Close: NDN Collective and Indigenous Climate Action Respond to Outcomes." *Indigenous Climate Action* (blog). November 14, 2021. https://www.indigenousclimate-action.com/entries/cop26-negotiations-close-ndn-collective-and-indigenous-climate-action-respond-to-outcomes.
Orlove, Benjamin S., John C.H. Chiang, and Mark A. Cane. 2002. "Forecasting Andean Rainfall and Crop Yield from the Influence of El Niño on Pleiades Visibility." *Nature* 403: 68–71.
Petryna, Adriana. 2022. *Horizon Work: At the Edges of Knowledge in an Age of Runaway Climate Change*. Princeton, NJ: Princeton University Press.
Postigo, Julio. 2014. "Perception and Resilience of Andean Populations Facing Climate Change." *Journal of Ethnobiology* 34 (3): 383–400.
———. 2019. "Bridging Past and Future to Address Water Stress." *Nature Sustainability* 2: 543–44.
Quijano, Anibal. 2000. "Coloniality of Power, Eurocentrism, and Latin America." *Nepantla* 1 (3): 533–80.
Rasmussen, Matias Borg. 2015. *Andean Waterways: Resource Politics in Highland Peru*. Seattle: University of Washington Press.
Reyes-García, Victoria, Sandra Álvarez-Fernandez, Petra Benyei, et al. 2023. "Local Indicators of Climate Change Impacts Described by Indigenous Peoples and Local Communities: Study Protocol." *PLoS One* 18 (1): e0279847.
Reyes-García, Victoria, D. García-del-Amo, Petra Benyei, et al. 2019. "A Collaborative Approach to Bring Insights from Local Observations of Climate Change Impacts into Global Climate Change Research." *Current Opinion in Environmental Sustainability* 39: 1–8. https://doi.org/10.1016/j.cosust.2019.04.007
Rosenzweig, Cynthia, Joshua Elliott, Delphine Deryng, et al. 2014. "Assessing Agricultural Risks of Climate Change in the 21st Century in a Global Gridded Crop Model Intercomparison." *Proceedings of the National Academy of Sciences* 111 (9): 3268–73.
Salas Carreño, Guillermo. 2020. "Indexicality and the Indigenization of Politics: Dancer–Pilgrims Protesting Mining Concessions in the Andes." *The Journal of Latin American and Caribbean Anthropology*, 25 (1): 7–27.
Schlingmann, Anna, Sonia Graham, Petra Benyei, et al. 2021. "Global Patterns of Adaptation to Climate Change by Indigenous Peoples and Local Communities. A Systematic Review." *Current Opinion in Environmental Sustainability* 51 (August): 55–64.
Scott, James. 1998. *Seeing Like a State: How Certain Schemes to Improve the Human Condition Have Failed*. New Haven, CT: Yale University Press.

Vallejo-Rojas, Virginia, Marta G. Rivera-Ferre, and Federica Ravera. 2022. "The Agri-Food System (Re) Configuration: The Case Study of an Agroecological Network in the Ecuadorian Andes." *Agriculture and Human Values*, 39: 1301–1327.

Vaughn, Sarah E. 2022. *Engineering Vulnerability: In Pursuit of Climate Adaptation*. Durham, NC: Duke University Press.

Wernke, Steven. 2013. *Negotiated Settlements: Andean Communities and Landscapes under Inka and Spanish Colonialism*. Gainesville: University Press of Florida.

Whyte, Kyle Powys. 2016. "Is It Colonial Déjà Vu? Indigenous Peoples and Climate Injustice." In *Humanities for the Environment: Integrating Knowledges, Forging New Constellations of Practice*, edited by Joni Adamson, Michael Davis, and Hsinya Huang, 88–104. London: Earthscan.

12

THE COLONIAL ROOTS OF CLIMATE VULNERABILITY. ANALYSING THE CASE OF THE MAPUCHE-PEHUENCHE PEOPLE IN SOUTHERN CHILE

Rosario Carmona

Introduction

There is consensus on the particular vulnerability of Indigenous Peoples to climate change (IPCC, 2022). However, the meaning of this statement is not free of tensions, as what is understood by climate vulnerability varies depending on who makes the diagnosis (O'Brien et al., 2007). So far, recognition of Indigenous Peoples' vulnerability has reinforced a view that positions them as victims of biophysical events resulting from climate change, such as impacts on their livelihoods (Callison, 2017). As a result, the primary institutional responses to address Indigenous Peoples' vulnerability have aimed at responding to such pressures, in most cases omitting the historical factors that determine vulnerability and affect their capacity to respond (Belfer et al., 2017). Nevertheless, these historical factors, such as colonialism, have been at the heart of historical Indigenous Peoples' complaints and advocacy (Parsons & Nalau, 2016).

In its latest report, the IPCC (2022) recognises that Indigenous Peoples' vulnerability is closely linked to the marginalisation stemming from past and ongoing colonial dynamics. In addition to depriving Indigenous Peoples of their right to self-determination, this marginalisation has excluded them from decision-making processes, and pushed them to inhabit territories with extreme geographical and climatic characteristics (Humphreys Bebbington, 2013). These territories also face disproportionate environmental liabilities from activities that sustain national economies, such as extractivism (Millaleo, 2020). Added to all these pressures is the increasing criminalisation of Indigenous Peoples' demands; the murder of environmental defenders is on the rise, disproportionately affecting Indigenous Peoples, especially those in Latin America (Global Witness, 2021). However, climate policies implemented in Indigenous territories do not address these factors (Parsons & Nalau, 2016). On the contrary, aiming primarily at solving a global problem, these policies do not consider Indigenous Peoples' particular relationships with their territories, the State, and climate change itself (Ulloa, 2017a). Consequently, climate policy has reproduced colonial dynamics that, in addition to producing vulnerability (Whyte, 2021), reinforce the dichotomous illusion that separates humans from the rest of nature, which, according to many Indigenous Peoples, is the root cause of climate change (Redvers et al., 2022).

DOI: 10.4324/9781003356837-15

Indigenous leaders and scholars have highlighted the close relationship between colonialism and climate change (Parsons & Nalau, 2016; Whyte, 2021). Therefore, responding to these problems requires processes of decolonisation that remove the assumptions that underpin our relationship with nature and ourselves (Gram-Hanssen et al., 2021). Decolonisation's first step is to reveal how these assumptions are constructed and sustained (Mohanty, 2008). Concerning Indigenous Peoples, extensive research has addressed the socio-ecological inequality that affects them (Middleton, 2015). In Latin America, multiple studies have analysed the power relations that, since colonial times, determine the administration of the territory and the distribution of environmental liabilities (e.g., Alimonda, 2016; Humphreys Bebbington, 2013). However, a critical review of how national and global dynamics restrict local actors' capacities to cope with climate change's impacts is still incipient (Dietz & Wehrmann, 2016).

The Mapuche People in Chile have faced the multiple social and environmental impacts of forestry extractivism over the last decades (Millaman Reinao, 2008). Although multiple studies set a precedent for understanding how their social vulnerability has been constructed (Aylwin et al., 2013; Klubock, 2014; Mondaca, 2013), the relationship between these processes and climate change remains an under-explored issue and therefore omitted from current climate change measures (Carmona et al., 2021). Against this background, this chapter analyses the social construction of the climate vulnerability of the Pehuenche People – a sub-group of the Mapuche People – in Lonquimay, a rural district in the southern Andes. It investigates the historical and political processes that have mediated their relationship with a territory that has been progressively degraded. Then, it presents the responses that communities and the State are deploying in the face of climate change, considering how different actors understand vulnerability.

Mapuche People and policy. A relationship mediated by extractivism

The Mapuche are an Indigenous People inhabiting south-central Chile and southwestern Argentina. They are composed of diverse sub-groups who share a social, religious and economic structure, and a common language: Mapudungun. Today, the Mapuche People in Chile represent 85% of the Indigenous population and 9% of the national population (INE, 2017). The majority live in the regions of Ñuble, Bío Bío, and La Araucanía, although approximately one-third live in the Metropolitan Region of Santiago. While most Mapuche People live in urban contexts, many maintain strong links with communities in the ancestral territory (Carmona, 2015). Those who live in rural communities practice a subsistence economy based mainly on agriculture and livestock.

The recent history of the Mapuche People illustrates colonial processes based on the exploitation of nature and the subordination of Indigenous Peoples (Alimonda, 2016). The Mapuche were subjected to internal colonisation in the early 1860s, a process that reduced their territory to 5% of the land they occupied in pre-colonial times (Bengoa, 1985). The State presented Mapuche as barbarians, not rooted in the land and its resources, which were considered unexploited (Klubock, 2014).

The usurpation of Mapuche territory by State powers gave way to expanding the agricultural frontier and an incipient timber industry facilitated by forest fires (Otero, 2006). Fire, land accumulation, and inadequate techniques triggered an ecological crisis in the late 19th in the South of Chile, which was addressed through the creation of reserves and reforestation with exotic trees – mainly *Pinus insignis* and *Eucalyptus globulus* – (Klubock, 2014). Since then, plantations have been validated as a technocratic and sustainable response and positioned as a mechanism for soil recovery. During the Pinochet dictatorship, forestry started to be regulated by the market and became the second most important extractive activity in Chile, after mining. Currently, plantations cover 3.6% of the national territory, with the highest concentration (60%) in the regions of Ñuble, Bío Bío, and La Araucanía, home to the highest percentage (42%) of the Mapuche population at the national level (INFOR, 2021).

Like all extractive activities, forestry has generated power asymmetries that have increased marginalisation, poverty, precariousness, and social conflict (Aylwin et al., 2013). Where forestry predominates, unemployment is higher than the national percentage (INE, 2017). Furthermore, logging companies control the territory, do not contribute to local development, and do not mitigate their impacts. Moreover, their management technologies have displaced Indigenous techniques and knowledge (Mondaca, 2013). In environmental terms, extractivism has led to increased degradation, erosion, fires, pollution, and drought, with consequent social, cultural, and economic repercussions. Currently, water scarcity is exacerbated by a legal regime that excludes most local communities from water administration[1] and a megadrought affecting the central-southern part of the country for ten years (Garreaud et al., 2017).

The Mapuche People have consistently denounced these impacts and demanded environmental justice (Schlosberg & Carruthers, 2010). The State has countered their complaints through two main mechanisms: increasing development policies and criminalising the demands. Both are highly controversial. Development policies have not succeeded in reducing poverty; on the contrary, they have often reinforced dependency and favoured the expansion of the neoliberal projects affecting the Mapuche lifestyle (Valdivieso, 2021). The criminalisation of the Mapuche People has progressively militarised communities causing serious rights violations (Bauer, 2018).

One of the policies that have the most significant impact on the territory of the Mapuche People is the Indigenous Peoples Territorial Development Programme (PDTI). Since 2011, the PDTI has aimed at the economic development of communities through an intercultural approach. However, due to its centralised and non-participatory nature, it replicates the procedures of non-Indigenous development policy. Despite its multiple reformulations, its objectives do not meet the communities' needs or adapt to the specific territories (Aninat & Hernando, 2019). By primarily supporting livestock management, the PDTI has led to the homogenisation of agricultural and livestock systems and modelled a relationship with the State based on the delivery of material inputs. Thus, the PDTI has increased Mapuche dependence on state support. It has also impacted social bonds, as it is based on a loan system that prioritises individual work, excludes those who cannot repay, and requires the formation of new directives, overlooking traditional organisation systems. Furthermore, its constraints hamper the work of officials, who lack intercultural competencies and whose roles tend to be misunderstood.

In parallel, the State has increased the criminalisation of minor offences attributed to Mapuche – such as cattle rustling – and the application of emergency laws that suppress the guarantees of due process (Aylwin et al., 2013). As a result, polarisation and radicalisation have escalated. Although police violence is concentrated in an area particularly affected by forestry extractivism, the conflict has increased the stigmatisation of the Mapuche People as a whole.

In this context, the Mapuche People face climate change and its impacts, such as decreasing precipitation and increasing extreme events (ARClim, 2020; Treulen, 2008). However, climate policy does not consider Maoucheparticular vulnerability (Carmona et al., 2021). On the contrary, Indigenous Peoples have been excluded from climate governance (Millaleo, 2020). Most legislation omits or positions them as vulnerable (Carmona et al., 2021). Furthermore, afforestation with exotic species and the reproduction of the exclusionary conservation model have been positioned

1 Chile's Water Code privatises water and separates its ownership from land ownership.

as legitimate mechanisms for mitigation (Millaman Reinao, 2008) – which is inconsistent with evidence showing that increasing plantations and decreasing biodiversity do not translate into carbon storage (Heilmayr et al., 2020).

Material and methods

Lonquimay

Lonquimay is located in an inter-Andean valley in southern Chile (Figure 12.1). As a high mountain territory (1,000 m.a.s.l.), Lonquimay is very sensitive to environmental changes (Marchant, 2011). Projections indicate that the zero isotherm will rise between 300 and 500 metres, increasing watershed runoff and winter flooding of rivers (MMA 2013). This will reduce water storage in the mountains and summer flows, making Lonquimay highly vulnerable to drought (ARClim, 2020). Furthermore, Andean communes in Chile tend to be socially, politically, and economically marginalised (Kronik & Verner, 2010) and have lower levels of quality of life indicators than the rest of the population (Marchant, 2011). Lonquimay is the second poorest commune in La Araucanía, the poorest region in Chile; 64.4% of its population lives in multidimensional poverty (MIDESO, 2017).

About 56% of Lonquimay's population is Mapuche-Pehuenche, living primarily in rural communities (INE, 2017). The main income of these communities comes from state aid, livestock, and the sale of non-timber forest products (e.g., pine nuts and mushrooms).

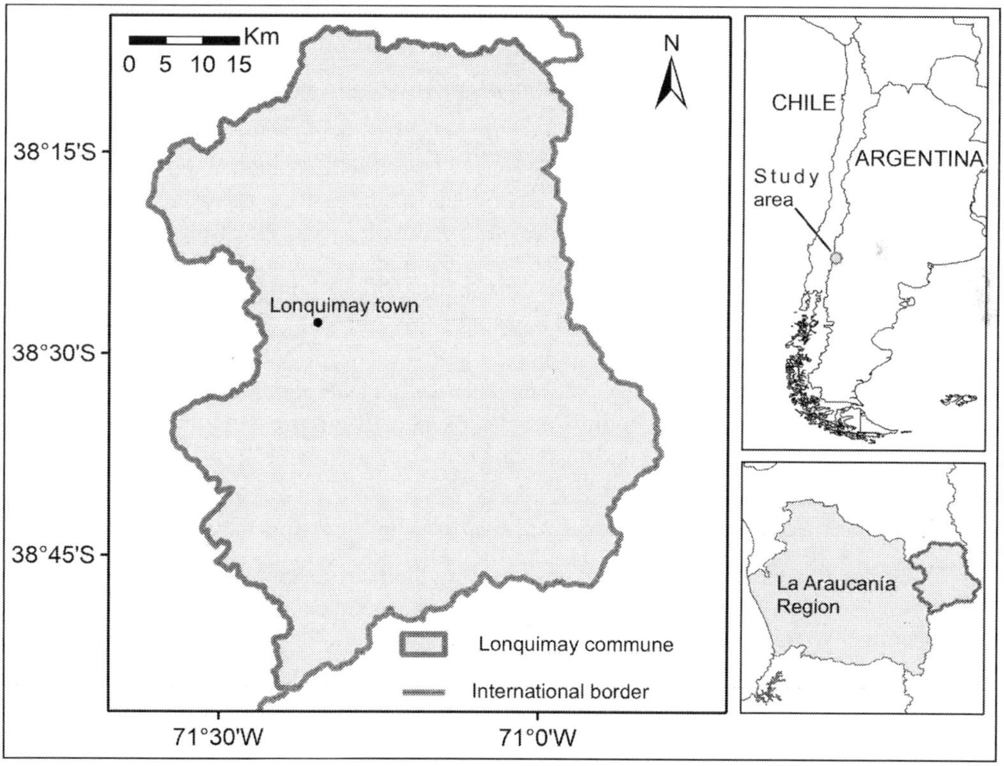

Figure 12.1 Lonquimay. Source: Adrián Fernández (modified) in Carmona 2022.

Data collection

This research draws on ethnographic work conducted between 2017 and 2019. A documental review of literature on the Pehuenche People over the last 30 years was undertaken to investigate the historical and political processes that have mediated the Pehuenche relationship with the territory. Participant observation was conducted in Lonquimay among two Pehuenche communities and during the implementation of three climatic programmes to analyse local responses to climate change. These programmes are EcoAdapt, funded by the European Commission and led by the municipality of Lonquimay, the National Strategy on Climate Change and Vegetation Resources (ENCCRV), funded by the UN-REDD Programme and led by the National Forest Corporation (CONAF), and the Sustainable Mediterranean Communities, funded by the Global Environmental Facility (GEF) and led by the Ministry of the Environment. Local officials' practices and interactions with Pehuenche communities were observed in the households and at community meetings.

In addition to participant observation, 34 semi-structured interviews were conducted. Participants were selected under three criteria: (i) Pehuenche community members whose primary activity is livestock ($n = 12$); (ii) Pehuenche community members who work as municipal officials and are in charge of implementing policies ($n = 8$); and (iii) state officials in charge of climate change programmes implemented in Lonquimay ($n = 14$). The interviews aimed to identify how Pehuenche communities and the State are responding to climate change impacts and how the different actors interact during the implementation of such responses – institutional or local. Furthermore, Pehuenche participants were asked about environmental changes over the last three decades and their effects on socio-ecological systems to better understand the context in which these responses are deployed (Reyes-García et al., 2023).

Data analysis

All information was systematised with the help of Atlas.ti software (Version 8.4.5). I conducted an open coding exercise, which led to the emergence of three codes: causes of environmental degradation, Pehuenche responses to climate change impacts, and state responses to climate change impacts. The relationship between these codes allowed me to access the historical processes and cultural dimensions that construct a scenario of vulnerability to climate change. Specifically, this relationship was analysed based on the two approaches to climate vulnerability proposed by O'Brien et al. (2007): contextual vulnerability and outcome vulnerability. Contextual vulnerability understands climate vulnerability as an outcome of socio-historical processes. It, therefore, assesses the capacity to cope with pressures and changes considering institutional and socio-economic constraints. Outcome vulnerability analyses the negative impacts of climate events, measured through biophysical processes and adaptation options.

Causes of environmental degradation in Lonquimay

The current population of Lonquimay was shaped by the internal colonisation processes of the mid-19th century. Coming from both sides of the Andes Mountains and fleeing from the Chilean and Argentinian armies, several Mapuche families took refuge in its valleys, where they encountered the first Pehuenche inhabitants – who transited between Chile and Argentina. Together, these families survived thanks to the shelter and pine nuts provided by the pehuen (*Araucaria araucana*), a tree they considered sacred. At the beginning of the 20th century, the State granted land titles to some families, but the exclusion of many others led to multiple conflicts (Bengoa, 1992).

The conditions for productive and social development in Lonquimay are minimal. Soil characteristics, mostly of volcanic origin and with steep slopes, and low vegetation cover lead to intense wind erosion, which together with the extreme climate, make agriculture very difficult. Accordingly, the State organised Lonquimay's economy around livestock farming and extraction, first of gold and later of timber. The exploitation of the forests permanently modified the territory; the consequences were irreversible. The stories emerge from the collective memory: "They used to cut araucarias. They came to sell truckloads of pehuen. This was a forest, and now it is empty" (Interview 9, Pehuenche person, 09.01.18). Degradation triggered various conflicts to which the State responded through the sale of timber estates and the creation of forest reserves, which imposed new ways of relating to the forests and limited the extraction of pine nuts (Comisión de Verdad Histórica y Nuevo Trato, 2008).

Land degradation has been reinforced by state policy. Livestock farming was promoted as a status symbol, and today, through the PDTI, the State encourages species and fields homogenisation, with the goal of growing livestock feed. Although most Pehuenche families depend on livestock, this activity is not sustainable, especially cattle. The conditions of the territory and the hard winters make it very difficult for the animals to survive. Moreover, the communities have no processing infrastructure and depend on regional markets for trading, which are very far away. Besides, the mountainous conditions make cow meat tougher and, therefore, less demanded. The participants report that their earnings barely cover their expenses: "We forage for five months because we are not farmers; we have to keep animals ourselves. Then, when we count how much we harvest and spend, we do not earn anything. We exchange money for money" (Interview 7, Pehuenche person, 28.12.18). Nevertheless, participants admit that many families have more animals than their land can support, a pressure currently reinforced by population growth.

To curb soil erosion, the State has implemented reforestation plans with exotic pines, which the Pehuenche consider imposed. "Some professionals explained to the people that we could plant this. The elders believed them, and they planted these trees. But it is harmful! The technicians knew it was harmful. Why did they do it?" (Interview 13, Pehuenche official, 05.12.17). The territory does not allow these trees to grow as fast as they do in the valley and the participants associate them with loss of biodiversity and increased drought. The legal regime, prohibiting communities from drawing water from the rivers that run past their doorsteps, further aggravate the effects of water scarcity. Water scarcity is addressed by the municipality through water tanks. According to the Pehuenche, water from the tanks is insufficient and of poor quality. "There are days when you have water and others when you have to ration it to last a fortnight, so the water is no longer fresh. And we cannot get water from the river, even if we wanted to" (Interview 13, Pehuenche official, 05.12.17).

Despite the difficulties, the Pehuenche maintain that they live in a safe and clean environment. They consider themselves adapted to a climate that, despite being "very harsh", has forged a particular and unique relationship with the territory. Many rely on the territory and its species to heal their ailments; herbal medicine, for example, is in great demand. Trees, especially pehuen, are considered a source of life and sacred.

However, due to external pressures, Pehuenche's knowledge of their land is weakening and, with it, the relationship with nature. Both outsiders and some young Pehuenche no longer express respect for nature. The lack of respect determines that the spirits that take care of nature leave, promoting the illness of the territory. The illness of nature and people affects the physical and spiritual dimensions, reinforcing the illusion of separation that produces ecological damage. "The spirit leaves, and we lose the sense of being Mapuche. Thus, we make many mistakes" (Interview 4, Pehuenche person, 04.01.18).

Local responses to climate change impacts

Pehuenche communities are currently feeling the multiple impacts of climate change, mainly attributed to increased temperature and climate variability. The most critical impact is the decrease in precipitation, especially snowfall, which results in less water storage in the mountains and a drying up of the springs. Households have increased their demand for water from the municipality. Drought is also affecting natural grasslands. To cope with this situation, Pehuenche families, supported by the PDTI, have shifted from growing crops for human consumption, such as wheat and quinoa, to growing livestock feed, such as alfalfa and oat. However, these crops are not very productive in the area due to high pesticide use, land degradation, and climate variability. In addition, the shift leaves families dependent on local markets for food, which generally do not offer nutritious options.

The feed shortage impacts livestock health, which is also affected by other climate impacts. Winter has lengthened, and frosts have increased, weakening calves and burning crops. Livestock mortality and behavioural changes have also increased; females give birth early and do not produce enough milk. In this situation, household's primary strategy is an economic investment: to build sheds and buy substitute milk. However, those who cannot finance these measures must sell their animals, losing, as they call them, the "savings of the Pehuenche".

According to most participants, summer weather has also changed. The mornings and evenings are colder and windier than in the past, making the elders sicker. However, at the same time, the increased intensity of the sun during the day makes it challenging to work in the fields. "The last few years, the heat burns you. You are walking and you get fried" (Interview 4, Pehuenche person, 04.01.18).

Climate change also impacts native flora and fauna. There is a consensus that many *lawenes*[2] have diminished, doing the work of the *lawentuchefe*[3] harder and generating greater dependence on the state health system, which is quite precarious. Native trees are weaker. Many of them break under the weight of the snow, increasing degradation and deforestation. The most affected species is the pehuen. Its phenological changes have opened a national debate on its conservation status,[4] for which Pehuenche fear further restrictions on the traditional use of this species.

To mitigate these impacts and diversify economies, the Pehuenche participate in various training programmes offered by the State. However, according to some Pehuenche leaders and municipal officials, these often reinforce dependency on State assistance.

> After all, we live to pay. It is a vicious circle. The vast majority depend on the various subsidies provided by the State. This is also due to the imposition of a model, of a way of looking at production and the economy, that finally has repercussions on our life.

(Interview 8, Pehuenche person, 26.12.17)

To increase income, many adults migrate to the north for temporary work. However, this strategy impacts the family sphere as the elderly and children are left alone in the fields. Some families, supported by the State, have initiated some touristic activities, but engaging in tourism is not accessible to most families, who do not have the necessary infrastructure and means.

2 Medicinal herb.

3 Herbalist, traditional healer.

4 See www.asemafor.cl/el-cambio-climatico-puede-estar-tras-la-enfermedad-que-afecta-a-la-poblacion-de-araucarias-en-chile/

Despite all these difficulties, the Pehuenche do not feel vulnerable in the face of climate change. They trust their knowledge and are used to facing extreme conditions. However, they do perceive that their capacity to respond has been restricted by state action, as one leader mentions:

> We are not vulnerable. Our rights have been violated. We were reduced, militarily subdued, to stay in the mountains where we do not have the climatic and territorial conditions to generate an autonomous economy. We are not even able to produce food. Furthermore, they violate our rights daily by denying our existence, language, organisation practices, and ways of thinking, by intervening in our processes for their interests.
>
> (Interview 8, Pehuenche person, 26.12.17)

State responses to climate vulnerability

According to non-Indigenous state officials, the United Nations Framework Convention on Climate Change (UNFCCC) defines the criteria to assess vulnerability. Based on these criteria, officials understand vulnerability as a product of the biophysical impacts associated with these indicators on economic development.

Lonquimay's socio-ecological situation has attracted various climate change mitigation and adaptation programmes. The first one was EcoAdapt, which aimed at promoting adaptation through watershed management and was implemented in three communities of Lonquimay. In two of the EcoAdapt experiences, perimeter fences were put in place to protect the water sources from livestock. Not without resistance, these measures also promoted reflection on the sustainability of livestock farming. Specifically, people reflected on the constraints posed by livestock to reforestation, which is critical for recovering water sources.

In the third community, the EcoAdapt project was not implemented due to a lack of coordination. Although dialogues were held to define the objectives, local proposals and needs were disregarded. According to a Pehuenche leader, who is also a municipal official,

> People are asked to solve local problems. Nevertheless, the institution says "you cannot go off-target". So we end up doing what they say. Sure, there was participation, people were consulted. Nevertheless, when people question the project, they say, "No. We have the money, and these are the investment parameters".
>
> (Interview 17, Pehuenche official, 29.12.17)

The second programme was the ENCCRV, which aimed to meet Chile's mitigation targets and has two projects implemented in Lonquimay. The first ENCCRV project directly involved the community of Quinquen in an initiative to capture carbon through reforestation with native species. However, as only one leader was informed, the community perceived the project as imposed and not responding to their immediate pressures. As the international funding was already approved, CONAF had to negotiate with the community. The meetings posed significant intercultural challenges. The dialogues drifted towards long-standing issues, such as the territory's exploitation, and conflicts rooted in state policy surfaced. Although the institution promoted collective work, Pehuenche participants prioritised individual work. The coordination constraints determined that CONAF decided to give up its mitigation targets and focus the project on strengthening local adaptation. During the meetings, it was agreed to recover culturally relevant spaces, generate waste management plans, and build a community nursery. Despite the officials' perception that they could not access Pehuenche's opinion, the Pehuenche said they felt listened to.

The second ENCCRV project implemented in Lonquimay aimed to reforest an area of the China Muerta national reserve, part of which belongs to the Quinquen community and plays a vital role in their livelihoods and spirituality. Quinquen leaders demanded participation in the planning. Nevertheless, CONAF officials had to refuse because the institution was unwilling to reformulate the objectives (MINAGRI et al., 2016). "The institution made the decision without consulting the community" (Interview 5, Pehuenche person, 05.12.17).

The third programme, the GEF project, sought to promote mitigation through sustainable land use. It included three measures in Lonquimay. Two were implemented in Pehuenche communities, the first without significant participation and the second with higher levels of involvement. The former aimed at carbon sequestration through agroforestry. However, the Pehuenche did not want to reduce space for livestock but to increase crops to feed the animals, for which they did not want to participate in the initiative. The lack of coordination led to the institution finally deciding to replicate the practices of the PDTI, that is, financing the growth of oat crops to feed the livestock. The second measure implemented in a Pehuenche community aimed to reforest the forest with native species. However, the involvement of the leaders led to a change of plans, as one leader recalls: "We are tired of having imposed solutions from Santiago. To raise solutions from the territory, that is something important for us" (Interview 8, Pehuenche leader, 26.12.17). The community wanted to recover the forest, but also to strengthen food security and social bonds. To this end, the leaders proposed family gardens. Although the institution did not consent initially, an agreement that combined the construction of the gardens with two nurseries was eventually reached. During implementation, the reorganisation of the plots promoted a profound reflection on the sustainability of livestock farming. Families realised that diversification of the land enabled them to have other sources of income and food. In addition, the space given to local leaders reinforced confidence in local capacities, which led to the creation of a cooperative to trade local products.

Although the ENCCRV and GEF initiatives set a precedent for collaboration between communities and the State, they did not achieve their mitigation goals. The reformulation of objectives led to the delay of resources, generating much disappointment among the Pehuenche. Moreover, in both cases, the nurseries – promoted mainly by the institutions – were semi-abandoned after funding ended. They were not part of the daily and traditional dynamics of the families, so the Pehuenche believed they did not know how to manage them.

Discussion

The case of the Pehuenche People provides an insight into how the vulnerability of Indigenous Peoples to climate change is a socially constructed process. In Lonquimay, climate vulnerability is created through power relations that reinforce marginalisation and colonial dynamics, even those that seek to overcome it.

The situation in Lonquimay, thanks to its socio-ecological conditions, allows us to observe this construction as a dynamic state (O'Brien et al., 2007) and, as other scholars have pointed out in this book (Hirsch, 2024), to understand how climate change is inextricable from the conflicts that stem from structural marginalisation and socio-ecological inequality. The conditions of Lonquimay's territory – unsuitable for agriculture – served as a pretext for the State to promote livestock and extractivism as primary sources of development. The resulting degradation became a justification for constant intervention, reconfiguring the territory and local economies. The social and ecological effects have been disastrous. By reducing socio-ecological inequality to a problem of poverty and encouraging competition for resources, state policy has strengthened dependency and encouraged individualistic practices that are very difficult to reverse.

In this context, the Mapuche-Pehuenche also face the current impacts of climate change. Although they try to respond, many of their responses risk reinforcing the factors that produce their vulnerability (Carmona, 2022). Their identity as herders pushes them to try maintain livestock, even though they recognise that it is neither environmentally nor economically sustainable. However, the Pehuenche do not perceive themselves as vulnerable to climate change; rather, they argue that their capacities have been undermined by impoverishment, precariousness, and marginalisation. Added to the impacts on their territories, they must deal with the assumptions and prejudices emanating from the forestry conflict, which have spread an image of the Mapuche as a conflictive people. Because of this, overcoming climate vulnerability cannot be separated from strengthening their autonomy.

Differently from Pehuenche's view, Chilean climate policy understands vulnerability as a result of biophysical processes and does not consider its institutional and historical causes. Although climate policies in Indigenous Peoples territories are still marginal, they are implemented based on the lessons learned from the development policy for Indigenous Peoples that has failed to resolve territorial conflicts and has strengthened inequality (Aninat & Hernando, 2019), as illustrated by the mitigation project that ended up financing oat crops. In attempting to respond to climate change uncritically, the State does not question the paradigms that have generated the problem – as one that positions certain beings, human and non-human, for the benefit of the few (Redvers et al., 2022). Such acceptance only reduces an issue with environmental and social implications to the technocratic approach (O'Brien, 2012). This perspective generally perpetuates existing dependencies, power asymmetries, and inequalities (Dietz & Wehrmann, 2016), as it allows elites to manipulate the environmental crisis to consolidate their power and economic interests (Bankoff, 1999; Whyte, 2021). This has been the main focus of Chilean climate policy, which, based on international guidelines built on the colonial development paradigm –such as REDD+ as in the case of the ENCCRV – aims primarily at economic growth. This approach prioritises neoliberal measures that victimise those affected and reduce them to objects of policies that dismiss their priorities, worldviews, and knowledge systems (Ulloa, 2017b).

Where technical responses to climate change prevail, the role of state officials is reduced to shaping the capacities of local people to respond according to predefined objectives (O'Brien, 2012). As the nurseries demonstrate, this approach fails to embed itself in local practices. Moreover, in all programmes, communities were presented with prefabricated proposals that neglected local needs and challenges and only attempted to respond to a problem framed globally, which ignores local specificities (Merino, 2018). When participation is encouraged only to respond to demands for international funding, the tensions between the global and local spheres highlight how shifting responsibility for climate change mitigation to communities, while ignoring their needs, replicates the dynamics of coloniality that make Indigenous Peoples vulnerable (Whyte, 2021). Moreover, as the above projects illustrate, institutions miss the opportunity to critically analyse their practices and promote co-construction processes that lead to more just and sustainable responses.

However, the advocacy of the international Indigenous Peoples' movement has positioned the respect for Indigenous Peoples' rights in international discussions, which has been endorsed by the UNFCCC through the Paris Agreement. The interstices of participation that have opened up allow affected communities to challenge the top-down approach. Institutions need to rethink their practices to meet commitments made through international programmes and thus access additional climate finance. Even though the negotiations were challenging, and participation remains weak, the experiences in Lonquimay set a precedent for collaboration at the national level. How the various actors take advantage of the lessons to promote further transformations at the institutional level remains to be seen.

Conclusion

How vulnerability to climate change is understood matters. Different interpretations determine how knowledge is produced and define the institutions, actors, and disciplines that address the problem. Climate change impacts are intertwined with multiple stressors, thus requiring multiple foci of intervention. The processes that have favoured the acceleration of climate change, such as colonisation, exploitation of nature, and capitalism, are the same processes that have pushed Indigenous Peoples into vulnerability.

There is no doubt that we must curb greenhouse gas emissions. However, as long as the inequalities and inequities that have historically disadvantaged Indigenous Peoples are not addressed, any effort to diminish climate change impacts will be unproductive and short-sighted. Analysing the historical and institutional drivers of vulnerability opens up the possibility of reframing climate policy from a more ethical approach that considers social justice and promotes the transformation of institutions.

As the case of Lonquimay illustrates, reducing Indigenous Peoples' vulnerability requires transforming the context in which climate change impacts occur. This implies a review of the activities that have forged the relationship with the territory, reforming the policy tailored to Indigenous Peoples and opening spaces for equal and effective participation. To achieve these objectives, it is necessary to strengthen intercultural competencies and inter-sectoral state coordination, facilitating the effective recognition of Indigenous Peoples' knowledge systems. These efforts must be based on the respect of Indigenous Peoples' rights, ensuring collaboration with communities to meet their goals.

Acknowledgements

Thank you very much to all the participants in this research, especially the Pehuenche families and leaders. Many thanks to Victoria Reyes-García, the LICCI team, and the reviewers of this chapter.

References

Alimonda, H. (2016). Notas sobre la ecología política latinoamericana: Arraigo, herencias, diálogos. *Ecología Política*, *51*, 36–42.

Aninat, I., & Hernando, A. (2019). Mapeando el laberinto de la política pública indígena en Chile. *Estudios Públicos*, *153*, 7–56.

ARClim. (2020). *Atlas teh riesgos climáticos*. https://arclim.mma.gob.cl

Aylwin, J., Yáñez, N., & Sánchez, R. (2013). *Pueblo Mapuche y Recursos Forestales en Chile: Devastación y Conservación en un contexto de Globalización económica*. Observatorio Ciudadano – IWGIA, Copenhagen.

Bankoff, G. (1999). The politics of natural disasters in the Philippines, 1985–95. *The Pacific Review*, *12*(3), 381–420.

Bauer, K. (2018). Not-so-neoliberal governance: Chile's response to Mapuche territorial demands. *Latin American and Caribbean Ethnic Studies*, *13*(3), 214–236. https://doi.org/10.1080/17442222.2018.1457007

Belfer, E., Ford, J. D., & Maillet, M. (2017). Representation of Indigenous peoples in climate change reporting. *Climatic Change*, *145*(1–2), 57–70. https://doi.org/10.1007/s10584-017-2076-z

Bengoa, J. (1985). *Historia del pueblo mapuche* (Siglo XIX y XX). Ediciones Sur, Santiago.

Bengoa, J. (1992). *Quinquén. Cien años de historia pehuenche*. Ediciones Chile América CESOC, Santiago.

Callison, C. (2017). Climate change communication and indigenous publics. In M. Nisbet, S. Ho, E. Markowitz, S. O'Neill, M. Schäfer, & J. Thaker (Eds.), *The Oxford Encyclopedia of Climate Change Communication* (pp. 1–26). Oxford University Press. https://oxfordre.com/view/10.1093/acrefore/9780190228620.001.0001/acrefore-9780190228620-e-411

Carmona, R. (2015). Rukas en la ciudad. Cultura y participación política Mapuche en la región Metropolitana. *Antropologías Del Sur*, *2*(4), 67–87.

Carmona, R. (2022). Resilience requires change: Assessing pehuenche responses to climate change impacts in Southern Chile. *Environmental Justice*. Ahead of Print. https://doi.org/10.1089/env.2021.0044

Carmona, R., Carril, F., & Yon, R. (2021). *Pueblos indígenas y gobernanza del cambio climático en Chile. Algunas consideraciones para su efectiva participación* (p. 19). CIIR. http://www.ciir.cl/ciir/wp-content/uploads/2021/11/policy-paper-cambio-clima%CC%81tico-final.pdf

Comisión de Verdad Histórica y Nuevo Trato. (2008). *Informe de la Comisión de Verdad Histórica y Nuevo Trato con los Pueblos Indígenas*. Gobierno de Chile: Comisión de Verdad Histórica y Nuevo Trato.

Dietz, K., & Wehrmann, D. (2016). *Vulnerability*. InterAmerican Wiki: Terms – Concepts – Critical Perspective. www.uni-bielefeld.de/cias/wiki/v_Vulnerability.html

Garreaud, R., Alvarez-Garreton, C., Barichivich, J., Boisier, J., Christie, D., Galleguillos, M., LeQuesne, C., McPhee, J., & Zambrano-Bigiarini, M. (2017). The 2010–2015 megadrought in central Chile: Impacts on regional hydroclimate and vegetation. *Hydrology and Earth System Sciences*, *21*, 6307–6327.

Global Witness. (2021). *Last line of defence*. Global Witness. https://www.globalwitness.org/en/campaigns/environmental-activists/last-line-defence/

Gram-Hanssen, I., Schafenacker, N., & Bentz, J. (2021). Decolonizing transformations through 'right relations'. *Sustainability Science*. https://doi.org/10.1007/s11625-021-00960-9

Heilmayr, R., Echeverría, C., & Lambin, E. (2020). Impacts of Chilean forest subsidies on forest cover, carbon and biodiversity. *Nature Sustainability*, *3*, 701–709.

Hirsch, E. (2024). "Not like it used to be": contending with the altered agricultural calendar in Andean Peru's Colca Valley. In V. Reyes-García, S. Álvarez-Fernández, P. Benyei, L. Calvet-Mir, D. García-del-Amo, A. B. Junqueira, X. Li, V. Porcher, A. Porcuna-Ferrer, A. Schlingmann, & R. Soleymani (Eds.), *Routledge Handbook of Climate Change Impacts on Indigenous Peoples and Local Communities*. Oxfordshire & New York: Routledge. Chp 11, pp 179–195.

Humphreys Bebbington, D. (2013). Extraction, inequality and indigenous peoples: Insights from Bolivia. *Environmental Science & Policy*, *33*, 438–446.

INE. (2017). *Censo*. Gobierno de Chile: INE.

INFOR. (2021). *Anuario Forestal 2020*. Instituto Forestal. https://www.infor.cl/index.php/destacados-home/732-anuario-forestal-2020-boletin-n-163

IPCC. (2022). Climate Change 2022: Impacts, Adaptation, and Vulnerability. Contribution of Working Group II to the Sixth Assessment Report of the Intergovernmental Panel on Climate Change. In H.-O. Pörtner, D.C. Roberts, M. Tignor, E.S. Poloczanska, K. Mintenbeck, A. Alegría, M. Craig, S. Langsdorf, S. Löschke, V. Möller, A. Okem, & B. Rama (Eds.), Cambridge University Press, Cambridge, and New York.

Klubock, T. M. (2014). *La Frontera. Forests and Ecological Conflict in Chile's Frontier Territory*. Duke University Press, Durham.

Kronik, J., & Verner, D. (2010). *Indigenous Peoples and Climate Change in Latin America and the Caribbean*. The International Bank for Reconstruction and Development/The World Bank, Washington DC.

Marchant, C. (2011). Factores que afectan la sustentabilidad de las comunas de montaña. El caso de la comuna de Lonquimay, Región de la Araucanía, Chile. *Revista de Historia y Geografía*, *25*, 55–73.

Merino, R. (2018). Re-politicizing participation or reframing environmental governance? Beyond indigenous' prior consultation and citizen participation. *World Development*, *111*, 75–83. https://doi.org/10.1016/j.worlddev.2018.06.025

Middleton, B. (2015). Jahát Jatítotòdom: Toward an Indigenous Political Ecology. In R. L. Bryant (Ed.), *The International Handbook of Political Ecology* (pp. 561–576). Edward Elgar Publishing, Cheltenham and Northampton, MA.

MIDESO. (2017). *Encuesta de Caracterización Socioeconómica Nacional CASEN*. Gobierno de Chile: Ministerior de Desarrollo Social.

Millaleo, S. (2020). El Acuerdo de París sobre cambio climático y el proyecto de la ley marco en Chile: Entre el protagonismo y la denegación de los pueblos indígenas. *Anuario de Derechos Humanos*, *Número especial*, 141–153. https://anuariocdh.uchile.cl/index.php/ADH/article/view/60299

Millaman Reinao, R. (2008). The Mapuche and climate change in the Chilean Neoliberal Economic System. *Indigenous Affairs*, *1–2*(8), 66–71.

MINAGRI, CONAF, GEDEFF, & UCCSA. (2016). *Estrategia Nacional de Cambio Climático y Recursos Vegetacionales 2017-2025 Chile*. Ministerio de Agricultura (MINAGRI), Corporación Nacional Forestal

(CONAF) Gerencia de Desarrollo y Fomento Forestal (GEDEFF), Unidad de Cambio Climático y Servicios Ambientales (UCCSA). http://portal.mma.gob.cl/wp-content/doc/ENCCRV-2017-2025-web.pdf

MMA. (2013). *Plan de adaptación al cambio climático del sector silvoagropecuario*. Ministerio de Medio Ambiente, Santiago.

Mohanty, C. (2008). Bajo los ojos de Occidente. Saber académico y discursos coloniale. In S. Mezzadra (Ed.), *Estudios postcoloniales. Ensayos fundamentales* (pp. 69–102). Traficantes de sueños, Madrid.

Mondaca, E. (2013). La re-existencia Mapuche frente al extractivismo forestal en un contexto de neoliberalismo armado. In G. C. Delgado (Ed.), *Ecología política del extractivismo en América Latina: Casos de resistencia y justicia socio-ambiental* (pp. 19–41). CLACSO, Buenos Aires.

O'Brien, K. (2012). Global environmental change II: From adaptation to deliberate transformation. *Progress in Human Geography*, *36*(5), 667–676. https://doi.org/10.1177/0309132511425767

O'Brien, K., Eriksen, S., Nygaard, L. P., & Schjolden, A. (2007). Why different interpretations of vulnerability matter in climate change discourses. *Climate Policy*, *7*(1), 73–88.

Otero, L. (2006). *La huella del fuego. Historia de los bosques nativos. Poblamiento y cambios en el apisaje en el sur de Chile*. Pehuén, Santiago.

Parsons, M., & Nalau, J. (2016). Historical analogies as tools in understanding transformation. *Global Environmental Change*, *38*, 82–96. https://doi.org/10.1016/j.gloenvcha.2016.01.010

Redvers, N., Celidwen, Y., Schultz, C., Horn, O., Githaiga, C., Vera, M., Perdrisat, M., Mad Plume, L., Kobei, D., Kain, M. C., Poelina, A., Rojas, J. N., & Blondin, B. (2022). The determinants of planetary health: An indigenous consensus perspective. *The Lancet Planetary Health*, *6*(2), e156–e163. https://doi.org/10.1016/S2542-5196(21)00354-5

Reyes-García, V., Álvarez-Fernández, S., Benyei, P., García-del-Amo, D., Junqueira, A. B., Labeyrie, V., Li, X., Porcher, V., Porcuna-Ferrer, A., Schlingmann, A., & Soleymani, R. (2023). Local indicators of climate change impacts described by indigenous peoples and local communities: Study protocol. *PLoS One*, *18*(1), e0279847. https://doi.org/10.1371/journal.pone.0279847

Schlosberg, D., & Carruthers, D. (2010). Indigenous struggles, environmental justice, and community capabilities. *Global Environmental Politics*, *10*(4), 12–35.

Treulen, K. (2008). Análisis sobre el impacto de los cambios climáticos en las mujeres mapuche de la región Araucania, Chile. In A. Ulloa, E. Escobar, L. Donato, & P. Escobar (Eds.), *Mujeres indígenas y cambio climático. Perspectivas latinoamericanas* (pp. 71–82). UNAL-Fundación Natura de Colombia-UNODC, Bogota.

Ulloa, A. (2017a). Dinámicas ambientales y extractivas en el siglo XXI: ¿es la época del Antropoceno o del Capitaloceno en Latinoamérica? *Desacatos. Revista de Ciencias Sociales*, *54*, 58–73.

Ulloa, A. (2017b). The geopolitics of carbonized nature and the zero carbon citizen. *South Atlantic Quarterly*, *116*(1), 111–120.

Valdivieso, D. (2021). Confeccionando Territorios. *Cuadernos de antropología social*, *54*. https://doi.org/10.34096/cas.i54.8259

Whyte, K. (2021). Against crisis epistemology. In B. Hokowhitu, A. Moreton-Robinson, L. Tuhiwai-Smith, L. Larkin, & C. Andersen (Eds.), *Routledge Handbook of Critical Indigenous Studies* (pp. 52–64). Routledge, New York.

PART II

Adapting to climate change impacts in the context of global change

INTRODUCTION TO PART II

Rooted wisdom

Indigenous Peoples' and local communities' adaptive responses to climate change

Anna Schlingmann, Laura Calvet-Mir and Xiaoyue Li

As illustrated in the first section of this book, Indigenous Peoples and local communities who rely directly on the environment for their livelihoods are increasingly experiencing climate change and its impacts, often within a broader context of socio-economic, political, and cultural transitions (see also Enso et al., 2019; Nyantakyi-Frimpong & Bezner-Kerr, 2015).

With global greenhouse gas emissions on a continuous upwards trend (Höhne et al., 2021; Liu & Raftery, 2021; Rogelj et al., 2016), climate change impacts are expected to increase in both intensity and damage potential (IPCC, 2021). This reality underscores the imperative of adapting to changing climate conditions as a last resort of social-ecological systems, as emphasized by Pielke et al. (2007). Simultaneously, it raises concerns regarding sustainability, social justice, and the enduring effectiveness of adaptive measures, as discussed by Adger & Barnett (2009). Through adaptation socio-ecological systems adjust "to actual or expected climate and its effects, in order to moderate harm or exploit beneficial opportunities" (IPCC, 2022) and is achieved by reducing vulnerability (Burton et al., 2002; Leary, 1999; Smit & Pilifosova, 2000; Smit et al., 2000) and building adaptive capacity (Adger et al., 2005).

In this section, we provide an introduction to key concepts and frameworks related to adaptation that are crucial for understanding local responses to climate change among Indigenous Peoples and local communities. Following this, we provide an overview of the chapters in the second section of the book, by contextualising these chapters within the introduced framework and highlighting the interrelations among them.

The vulnerability and resilience of Indigenous Peoples and local communities

Vulnerability refers to the likelihood of adverse effects and is determined by a combination of exposure to harm, sensitivity to harm, resilience, and adaptive capacity of systems, institutions, and individuals (Berkes, 2007; IPCC, 2022). In general, social-ecological systems are considered resilient when they possess the ability to absorb and cope with hazardous events, trends, or disturbances, such as natural disasters, by responding or reorganising in ways that maintain their essential function, identity, and structure (Berkes, 2007; IPCC, 2022). The relationship among vulnerability, resilience, and adaptive capacity - all of which are socially constructed - suggests that exposure to climate harm is not the sole determinant of the ensuing damage (Ford et al. 2020).

 DOI: 10.4324/9781003356837-17

It is therefore crucial to critically question approaches that solely focus on the exposure to climate events as the cause for loss and damage without considering the underlying social and political-economic origins of crises (Ribot, 2022). The pre-existing vulnerabilities and precarities on the ground, including issues of undermined sovereignty, power, and social justice rooted in colonial history and exploitative international systems, ultimately define who is most affected by climate hazards (Lahsen & Ribot, 2022). Indigenous Peoples and local communities are often identified as "at-risk" and as the most vulnerable populations (Lahsen & Ribot, 2022). However, evidence suggests that Indigenous Peoples and local communities can be both vulnerable and resilient to climate change, as Indigenous and local knowledge, place attachment, traditional institutions, and collective action contribute to their adaptive capacity, thereby proving to be important tools for adaptation (Ford et al., 2020).

Incremental and transformational adaptation

Discussions on coping, incremental, and transformational adaptation are critical when examining how Indigenous Peoples and local communities are responding to climate change and its impacts. In fact, with water and food security seriously threatened, traditional livelihoods challenged, and lives at risk, these communities have already diversified their adaptation strategies to deal with new climatic conditions around the world (Petzold et al., 2020; Schlingmann et al., 2021; Shaffril et al., 2020). While some of these responses tend to focus on reactive, spontaneous, and rather short-term coping strategies in the face of emergencies, many imply smaller adjustments in the livelihood systems, known as incremental adaptation. However, a few responses refer to transformational, that is, large-scale and system-wide, adaptation (Fedele et al., 2020; Mapfumo et al., 2017; Zant et al., under review). According to the IPCC (2022), coping refers to "the use of available skills, resources and opportunities to address, manage and overcome adverse conditions [...] in the short to medium term". Differently, incremental adaptation is typically more anticipatory and seeks to build stronger resilience by accommodating smaller-scale changes in the system, but without altering its original functions and trajectory (Kates et al., 2012). In the light of exacerbating climate trends, coping and incremental adaptation responses are increasingly being considered insufficient to efficiently deal with climate change, particularly because they fail to address the root causes of climate change impacts and vulnerabilities (Kates et al., 2012; O'Brien, 2012; Ribot, 2014). As a result, there is growing emphasis on the need of transformational adaptation, which involves fundamentally altering the entire system's ecological, social and economic properties and functions, to tackle the root causes of vulnerability. The difference to incremental adaptation lies in the scale and scope of change (Termeer et al., 2017). Transformational adaptation involves changes that are large in magnitude, new to a particular system or location, and have the potential to transform places, shift locations, and bring fundamental changes to a system (Fedele, 2019; Feola, 2015; Kates et al., 2012; O'Brien, 2012; Park et al., 2012; Walker et al., 2004). It is often characterised as a restructuring, path-shifting, innovative, multiscale, system-wide, and persistent process (Fedele, 2019). Transformational adaptation is frequently associated with the goal of achieving higher sustainability (Olsson et al., 2014, 2017).

Adaptation in a sustainable context

The first section of the book exemplifies how Indigenous Peoples and local communities are exposed to multiple drivers that shape their lives and livelihoods. In addition to the impacts of climate change, such impacts include socio-economic transitions arising from technological innovations and globalisation, such as extractivism, (e.g., Carmona, 2024; Izquierdo & Schlingmann, 2024; Stratoudakis et al., 2024). Similar evidence is presented elsewhere (e.g., Bennet et al., 2016; Dhakal & Kattel, 2019). Therefore, communities are simultaneously affected by, and adapting to, new environmental, political, socio-economic, technological, and cultural conditions (e.g., McDowell & Hess, 2012). While adaptive responses can have multiple benefits for several drivers, there is also evidence that responding to one driver may conflict with adapting to other drivers, thus potentially attenuating or limiting overall positive effects (Bennet et al., 2016; Dhakal & Kattel, 2019; Galappaththi & Schlingmann, 2023).

It is therefore important to acknowledge that not all responses to climate change impacts are successful in achieving their intended goals in an effective and sustainable manner. Some responses hardly address underlying vulnerability (Adger et al., 2005), or come with certain trade-offs that can increase or redistribute vulnerability, or even exacerbate the risks of maladaptive outcomes (Atteridge & Remling, 2018; Galappaththi & Schlingmann, 2023; Magnan et al., 2016; Schipper, 2020). Analysing trade-offs and potential negative, unintended side effects is essential to detect and avoid maladaptation. Barnett and O'Neill (2010) describe five types of maladaptation, including actions that result in increasing greenhouse gas emissions, disproportionately burden the most vulnerable, include high opportunity costs, reduce incentives to adapt, and create path dependency. In contrast, Juhola et al. (2016) suggest three forms of maladaptation: measures that rebound vulnerability, shift vulnerability (i.e., transferring the negative effects to others), and erode sustainable development. Due to the many facets of maladaptation, defining successful adaptation requires looking beyond spatial and temporal scales. For example, maladaptation can reinforce existing inequalities by providing benefits for some and negative externalities and spillovers for others (Adger et al., 2005). Moreover, what appears successful today may be unsustainable in the long-term, affecting future generations (Adger et al., 2005; Magnan et al., 2016).

Several authors have emphasised the need to evaluate adaptation to climate change in a sustainability context by connecting climate actions with the Sustainable Development Goals and analysing adaptation success based on environmental, social, cultural, and financial criteria (e.g., Folke et al., 2016; Fuso Nerini et al., 2019). Sustainable adaptation combines principles of environmental integrity with social justice (Eriksen et al., 2011), defining adaptation success under equity and legitimacy aspects (Adger et al., 2005; Singh et al., 2022). Based on a literature review, Owen (2020) identifies five types of indicators of adaptation effectiveness, including reduced risk and vulnerability, enhanced social wellbeing, improved environment, increased economic resources, and strengthened institutions.

The feasibility of local adaptation: understanding barriers and limits

Successful adaptation, however, is not solely dependent on the sustainability of its outcomes, but also depends on the feasibility of adaptation options (Singh et al., 2020). Adaptation feasibility is directly linked to the adaptive capacity of the system in question, which is determined by the persistence of supporting or enabling factors and conditions (i.e., adaptation opportunities), compounding factors and conditions (i.e., adaptation barriers or constraints), and impeding factors and conditions, such as adaptation limits (Klein et al., 2014). While adaptation constraints make adaptation planning and implementation more challenging, less effective and efficient, and more expensive, adaptation limits imply that adaptation is not possible under current conditions, resulting in catastrophic and intolerable climate risks (Dow et al., 2013; Klein et al., 2014; Moser & Ekstrom, 2010). Since 2010, an increasing amount of research has assessed adaptation barriers, with studies on adaptation limits still being rare (Biesbroek et al., 2013; Thomas et al., 2021).

Since climate change adaptation occurs in a broader political context, its feasibility not only depends on bio-physical, including climatic, factors, but is also determined by many socio-economic factors and the interactions between them (Bezner Kerr et al., 2022; Singh et al., 2020; Thomas et al., 2021).

The sustainable livelihood approach defines five categories that determine the adaptive capacity of a system: financial, physical, social, human, and environmental capitals (Morse et al., 2013). Specifically, existing political and economic structures, power dynamics, and inequalities influence the adaptive capacity of a system by providing, hindering, or limiting access to resources, such as secured land (Bezner Kerr et al., 2022). In an evidence-based review on the feasibility of adaptation options across Africa, Williams et al. (2021) detect technological and institutional factors as major barriers to adaptation. In an older case study conducted in Ethiopia, the most common barriers were lack of information, money, and shortage of labour (Deressa et al., 2009). Similarly, a case study in Nepal shows that education, access to credit and extension services, experience with, information on, and belief in climate change impacts, influence adaptation decision-making (Khanal et al., 2018). Other studies support the importance of institutional organisation, stakeholder engagement, and the general support for adaptation by community members (Ford & King, 2015; Naess, 2013). Required resources and costs, for example, financial means, time, and political will are especially high for transformational adaptation (Chung Tiam Fook, 2017; Kates et al., 2012; Pelling et al., 2015; Rickards & Howden, 2012).

Also, the importance of culture, values, and worldviews, including Indigenous and local knowledge and place attachment, for shaping the perception of climate risks, adaptive responses, and outcomes have been the subject of an increasing number of studies (Adger et al., 2013; Clarke et al., 2018; Heyd & Brooks, 2009; McNeeley & Lazrus, 2014; O'Brien, 2009). According to Adger et al. (2009) and Nielsen and Reenberg (2010) ethics and values, knowledge, risk perception, and culture can constitute endogenous social barriers and limits to adaptation within a society; however, since they are socially constructed, they are in many cases mutable and also depend on the adaptation goals. Similarly, Few et al. (2021) highlight that research and policy need to move away from understanding culture as static and in a simplified way as either adaptation enabler or barrier of climate change adaptation.

Therefore, to facilitate and enable future adaptation, especially at the local scale, it is crucial to better understand and overcome existing and contextualised adaptation barriers and limits (Eisenack et al., 2014; Thomas et al., 2021).

Paradigm shift towards bottom-up approaches for climate change adaptation

Within the context of climate justice and in response to questions about who can implement adaptation measures and who benefits most, there is a growing recognition of the need for intersectional approaches. These approaches take into account factors such as gender, class, origin, and ethnicity to understand how they intersect in relation to adaptation choice, feasibility, and success (Kaijser & Kronsell, 2014; Ravera et al., 2016a; Thompson-Hall et al., 2016). Such studies call for more nuanced assessments and perspectives at the local level, both in research and development policies and applied projects (Ravera et al., 2016b).

Similarly, recent studies have called for a paradigm shift in climate change assessment and adaptation, moving from top-down towards bottom-up approaches that promote greater participation, inclusion, sovereignty, and governance by Indigenous Peoples and local communities in decision-making, climate negotiation, and adaptation planning (Conway et al., 2019). Climate change adaptation literature generally distinguishes between autonomous and planned adaptation. Although distinctions are not always clear (Eisenack & Stecker, 2012), autonomous adaptation are generally responses to experienced climate change and its impacts without planning (Füssel, 2007; IPCC, 2021) and initiated by private actors, such as individuals, households, or communities (e.g., Fenton et al., 2017; Mersha & van Laerhoven, 2018), while planned adaptation are intentional intervention strategies that are prepared based on information and implemented by public agents, such as regional and national governments (Füssel, 2007; IPCC, 2021; Smit & Pilifosova, 2001). Typical examples of planned adaptation are countries' National Adaptation Plans (NAP) (Woodruff & Regan, 2019). Recognising that top-down approaches to planned adaptation often fail due to low acceptance by local communities, especially when they are not aligned with Indigenous and local worldviews (Luetz & Nunn, 2020), the concept of community-based adaptation has gained traction in both research and policies (McNamara & Buggy, 2017). Community-based adaptation is a local, community-driven adaptation approach that focuses on empowering and promoting the adaptive capacity of communities. It takes into account the context, culture, knowledge, agency, and preferences of communities as strengths (IPCC, 2022). However, community-based adaptation also comes with challenges and barriers, and requires critical reflection to avoid maladaptation (Ford et al., 2016; Forsyth, 2013; McNamara & Buggy, 2017). Lack of coordination between different stakeholders, culture and tradition, and lack of access to financial resources are particularly pervasive barriers (Piggott-McKellar et al., 2019; Spires et al., 2014). In addition to community-based adaptation, adaptive co-management has been promoted as a promising adaptation option (Galappaththi et al., 2022; Olsson et al., 2004). Co-management involves increasing stakeholder participation through co-producing knowledge and sharing decision-making between government and other stakeholders, such as local communities, the private sector, and non-governmental organisations, thereby strengthening legitimacy (Armitage et al., 2011; Bown et al., 2013; Carlsson & Berkes, 2005; d'Armengol et al., 2018). Adaptive co-management adds the concept of adaptive management, and builds on key elements, such as diversity, resilience, and learning-by-doing (Bown et al., 2013).

In addition to the general need for bottom-up and integrative approaches in climate change adaptation, such as community-based adaptation (Singh et al., 2022) and adaptive co-management (Galappaththi et al., 2022; Olsson et al., 2004), the inclusion of Indigenous Peoples and their knowledge into decision-making and adaptation planning and supporting communities' autonomy are key components for successful, effective, and sustainable adaptation (Eriksen et al., 2011; Klenk et al., 2017; Magni, 2017; Pisor et al., 2022; Yap & Watene, 2019). Many studies have demonstrated the value of Indigenous and local knowledge for successful climate change adaptation (e.g., Leal Filho et al., 2022; Lebel, 2013; Nyong et al., 2007; Rivera-Ferre et al., 2021). However, Indigenous and local knowledge are eroding and weakening in many parts of the world, leading to the loss of these untapped potentials (Fernández-Llamazares et al., 2021).

The aim of the second section of the book is to increase awareness and acknowledgement of valuable knowledge held by Indigenous Peoples and local communities on climate change adaptation by showcasing multiple and diverse case studies from around the globe. With this compilation, we hope to promote the recognition of Indigenous and local knowledge and inspire action in policy, research, and development projects to strengthen the preservation of such knowledge systems and give them a leading role in climate change adaptation.

Insights from the field

The empirical insights on climate change adaptation presented in this book are derived from diverse regions across the globe, including Africa, Asia, Central and South America, North America, and Europe. The case studies cover a wide range of geographic contexts, environmental conditions, and livelihoods. For instance, the case studies involve adaptive responses of farmers from different geographic settings, including in mountainous regions with boreal climate (Tyrol, Austria) and with temperate climate (Central Kenya), in tropical lowlands (Kédougou region, Senegal; Northern Region, Ghana; Ogun State, Nigeria), in dry lowlands (Chiredzi district, Zimbabwe), and in high plains with polar climate (Yunnan, China). The book also includes cases studies of nomadic pastoralists in high plains but dry climate (Inner Mongolia, China), agropastoralists in desert landscapes (Kerman province, Iran), and farming and fishing communities living on the tropical island Viti Levu (Ba Province, Fiji), and hunting and fishing communities settled at the coastal areas in polar climates in the far north of the globe (Nunavut, Canada, and Avannaata, Greenland).

The heterogeneity of case studies and geographic locations is reflected in a high diversity of local adaptation strategies and a broad scope of applied theoretical frameworks. The chapters interlink different approaches discussed above, giving richness and showing the complexity found in the field. For example, several authors discuss the adaptive capacity of Indigenous Peoples and local communities in a multiple-driver framework, also looking at the intersectionality between different socio-economic factors and recommending co-managed adaptation to achieve successful adaptation. This sectional introduction outlines the contribution of the different chapters.

Chapters in this section illustrate how Indigenous Peoples and local communities use their knowledge systems, sometimes in combination with external knowledge systems, to increase their resilience and adaptive capacities, and reduce vulnerabilities in the face of climate change impacts. Several chapters focus on coping and incremental adaptation strategies within their livelihood systems, although certain transformational adaptations, such as switching to alternative livelihoods and migration to other places are also reported. For example, a study by Chakauya et al. (2024) in South-East Zimbabwe describes how farmers use their Indigenous and local knowledge to incrementally adapt to decreasing yields resulting from more frequent droughts, longer dry spells, and higher crop infestations, and to impacts on livestock caused by decreasing grazing pasture. For example, farmers use Indigenous knowledge to select plant species to treat crop pests and livestock diseases, harvest wild fruits as alternative food resources, and apply several techniques of conservation agriculture. Farmers reported using modern methods, too, such as chemical pesticides or modern crop varieties, although they note that these methods are not always effective. Besides incremental adaptations, there were few coping strategies in place, such as the dependence on food aid from donors and governmental assistance, and transformational adaptation, such as switching to non-farming practices, for example selling firewood and making bricks. The authors conclude that a full toolbox of climate adaptation strategies, based on Indigenous, local, and technical knowledge needs to be deployed for farmers to be resilient to climate change.

In a very different context, the study by Fuchs et al. (2024) illustrates how alpine farmers in Tyrol, Austria, respond to climate change impacts mainly by technical and management-related incremental adaptation measures and, to a lesser extent, through transformational adaptation at the community level. The incremental adaptation measures range from changes in the management of water, soils, feed, and pastures, and respond to increasing droughts, decreasing pasture productivity, and shifts of seasons. Transformational adaptation strategies involve abandoning farming and converting to other agricultural production sectors, but also adopting measures to support livestock farming, especially to improve the management of alpine pastures and grassland and strengthen community management and resource sharing. The authors highlight that those transformational responses are not driven solely by climate change but also by other socio-economic changes.

The topic of multiple drivers for adaptation is further deepened in a study conducted by Porcuna-Ferrer et al. (2024) on changes in local agricultural management practices by Bassari farming communities in southeast Senegal and in the study by Jungsberg and Wendt-Lucas (2024) on an Inughuit community in North Greenland. Interestingly, both chapters address the topic of multiple drivers from a slightly different perspective. Porcuna-Ferrer et al. (2024) examine how multiple drivers undermine the sustainability of response outcomes by detecting trade-offs between environmental, social, and economic dimensions. In particular, by switching to new crops and crop varieties, and changing seed and soil management practices, Bassari farmers in Senegal do not only respond to climate change impacts (for example, unpredictable rainfall and shorter rainy seasons, pest manifestations in crops and seeds) but also to socio-economic and cultural stressors, such as higher market integration, external influences of non-governmental organisations and extension services, rural-urban migration, and a decrease in cultural and traditional values, knowledge, and customs. The study shows that being exposed and responding to multiple stressors often imply trade-offs among conflicting interests, thereby undermining the sustainability of the outcomes and increasing the risk of maladaptive outcomes. For example, the study reveals how switching to irrigated horticulture responds to economic drivers, especially the need to cover rising costs of living by increasing market integration and providing additional income opportunities. At the same time, the demand for additional water for irrigation counteracts observed climatic trends towards decreasing and more unreliable rainfall patterns. Similarly, the study by Jungsberg and Wendt-Lucas (2024) shows the trade-offs of certain adaptive measures. For instance, switching to halibut sales and work in the tourism sector strengthen market dependence and make the local population of Qaanaaq more vulnerable to external factors. The same study illustrates how climate change and socio-economic processes interact and reinforce adverse impacts. Specifically, "cheap and fast" housing constructions in the 1960s and 1970s increases the extent of current house damages deriving from thawing permafrost and forcing community members to regularly fix them.

The chapters by Abazeri (2024), Mwangi et al. (2024), Chen (2024), and Chao et al. (2024) shed light on the different aspects of the cultural dimensions of adaptation, emphasising the importance of Indigenous and local knowledge, including tradition, spirituality, context-specific values, and worldviews. These elements are crucial in turning responses to climate change into successful, efficient, and sustainable adaptation. Through their various approaches and focuses, the four chapters reveal the complex, diverse, and interconnected role of culture in the adaptive process. Abazeri's study (2024), based on research in the district of Kerman in Iran, emphasises that using measures based on other societies may be misleading and undermine local priorities, cultural contexts, and social structures. For instance, the study found that local irrigation systems in the region were better suited to the community's needs than modern irrigation systems, which were less efficient and more expensive. Therefore, the research suggests that effective measures for climate adaptation are more likely to emerge at the grassroots level.

Similarly, results from Mwangi et al. (2024) highlight the importance of considering local contexts and cultures when implementing adaptation actions. The study suggests that cultural differences related to ethnicity and corresponding values and social cohesion play a significant role in shaping perceptions of climate change and adaptation strategies. For example, irrigation was found to be more widespread among the Meru than among the Kikuyu smallholder farmers in two mountains in central Kenya. Chen's study (2024) delves into the complex relationship between external factors, namely land-use policy and local knowledge and traditions, and their impact on livelihood diversification among Tibetan people in Shangri-la county. This chapter highlights the importance of understanding the interplay between these factors in the context of climate change adaptation. The study sheds light on some positive outcomes resulting from governmental land-use policies and road constructions, which have led to increased economic opportunities for the local population. However, the research also reveals negative consequences associated with these developments, including the displacement of traditional farming practices and the loss of valuable ecological knowledge. The study emphasises the significance of taking a holistic approach to climate change adaptation that accounts for the complex social and environmental contexts in which it occurs. In contrast, Chao et al. (2024) focus on the crucial role of traditional practices in climate change adaptation, specifically through exploring the Inner Mongolian Ovoo offering ritual's effectiveness in promoting sustainable behaviour and resilience building. The study found that the Ovoo offering ritual can promote sustainable practices, reinforce the Mongolian holistic worldview, bridge social capital, and enhance community's cohesion and solidarity, all of which increase community's resilience capacity. This study aligns with the other chapters in emphasising the importance of Indigenous and local knowledge in climate change adaptation and highlights the significance of spiritual rituals in promoting sustainable behaviour and climate change adaptation. The authors suggest that policymakers and researchers should pay more attention to the role of such traditional practices in promoting sustainable behaviour and resilience building. In line with the other chapters, the study also demonstrates the importance of Indigenous and local knowledge to climate change adaptation discourses and suggests that policymakers and researchers should pay more attention to the role of spiritual rituals in promoting sustainable behaviour and climate change adaptation.

The chapters by Attoh et al. (2024), Ayanlade et al. (2024), and Fuchs et al. (2024) describe how different barriers hinder the feasibility of incremental and transformational adaptation strategies. Specifically, Dagbani Indigenous communities in northern Ghana need to adapt to increased livestock mortality and more frequent crop pests. However, the absence of financial means – the most prominent barrier – and the lack of governmental and non-governmental support hinder poor households from the implementation of adaptation measures, such as the acquisition of new crop varieties and chemicals, or livelihood diversification (Attoh et al., 2024). Similarly, in southwestern Nigeria, where local farmers aim to respond to decreases in cassava yield, a lack of natural, physical, human, social, and financial capitals account for 70% of the reported adaptation barriers. Importantly, in this case study 41.1% of the farmers chose not to respond because they considered they face a low risk of climate change (Ayanlade et al., 2024).

Some authors use a gender or intersectional perspective to understand adaptation barriers in a more nuanced way, by looking at how gender (Abazeri, 2024; Ayanlade et al., 2024) and the interrelation between gender and income (Porcuna-Ferrer et al., 2024) influences response adoption. Ayanlade et al. (2024), working with rural farmers in Nigeria, highlight that all adaptation practices found in the study area are performed by men and women, but a higher percentage of men are involved in all of them. The authors claim to recognize the adaptation capacities of both males and females to understand gender-based adaptation differences since climate change affects men and women differently. Evidence from a case study with Basari people in southeast Senegal shows that women from poor households confront additional adaptation barriers and trade-offs (e.g., lack of access to suitable land and wells, additional workload), and therefore are not able to implement certain adaptive measures, such as irrigated horticulture (Porcuna-Ferrer et al., 2024). Abazeri (2024), whose work focuses on women's organised initiatives, suggests in her chapter that successful adaptation requires strengthening social processes and enabling women's decision-making in grassroots responses.

In the Austrian Alps, rural-urban migration results in underpopulated rural areas, causing a lack of labour forces and leading to abandoned and unmanaged land. Such demographic changes, and especially the lack of labour, hamper transformational adaptive processes, such as livelihood diversification and conversions towards other agricultural production forms (Fuchs et al., 2024). For Fuchs and colleagues (2024), overcoming the barriers to transformational adaptation entails collaboration among farmers since local alliances and knowledge co-production, supported by external competence development, finances, and private sector involvement, is key for reaching successful transformational adaptation.

The five chapters with a focus on adaptation barriers highlight the necessity to implement context-specific adaptation measures that are co-produced between Indigenous People and local communities with other relevant institutions, such as non-governmental organisations or the local government. The authors of these chapters align to the framework of adaptive co-management, claiming for the inclusion of Indigenous and local knowledge into decision-making and adaptation planning, while supporting communities' autonomy.

The chapters from Galappaththi (2024) and Singh et al. (2024) go deeper on the topics of co-production of knowledge and adaptive co-management. They assess how co-produced adaptation between different actors at multiple scales and from different sectors (i.e., Indigenous Peoples and local communities, non-governmental organisations, regional and national governments and the private sector) can foster co-production of knowledge, food and nutrition security, and resource sharing, thereby supporting local adaptation to climate change by strengthening resilience. While Galappaththi (2024) focuses on co-management of fisheries between Inuit, government, and the private sector in the Canadian Arctic, Singh et al. (2024) analyse synergies, complementarities and gaps between autonomous adaptation in an iTaukei community in Fiji and planned adaptation, as specified in the country's National Adaptation Plan. Singh et al. (2024) found that locally implemented adaptation combines autonomous adaptation by *iTaukei* villagers (e.g., the application of ashes and worms as natural pesticides), planned adaptations by the regional and national governments (e.g., evacuation strategy for natural disasters), and co-produced and multiscalar adaptation measures by different actors, including non-governmental organisations (e.g., riverbank protection through reforestation and mangrove rehabilitation). Co-produced and multiscalar adaptation strategies and associated knowledge co-production are promising alternatives to solely autonomous or planned adaptation, but do not come without challenges. To make co-management successful and adaptive, Galappaththi (2024) points to five structural characteristics (i.e., partnerships; vertical and horizontal linkages; sharing of responsibility, authority, and power) and four functional characteristics (i.e., community consultations and knowledge integration for joint decision-making; continuous monitoring, evaluation, and learning). Singh et al. (2024) found that certain (key) adaptive practices and cultural values by *iTaukei* villagers, such as reciprocity and sharing practices, and livelihood diversification towards reduced dependencies on nature, are not sufficiently addressed and represented in Fiji's National Adaptation Plan. To address this gap, the authors present various recommendations and requirements to improve knowledge co-production, such as consulting community representatives, developing sub-national adaptation plans, and fostering adaptive co-management under improved coordination and/through clearly defined roles.

Conclusion

The chapters presented in this section provide only some exemples of the manifold and diverse ways in which Indigenous Peoples and local communities take action and respond to climate change impacts. We have seen how Indigenous Peoples and local communities use and further develop their knowledge systems to increase their adaptive capacity and resilience. For example, they use Indigenous plants to treat crop pests and livestock diseases, diversify their food sources by harvesting wild plants, switch target fish species, and change hunting routes to adopt their traditional livelihoods and practices to new climate conditions. Some also switch to off-farm work to reduce their direct dependency on nature, and thus their sensitivity to climate variability, often at the cost of leaving behind parts of their culture and tradition. Obviously, not all local responses to climate change are feasible, fair, sustainable, and successful. For example, chapters in this section also suggest that not all adaptation strategies are equally benefitting all community members, with women often forming the most disadvantaged group. Also many adaptation strategies come with certain trade-offs between ecological, social, and environmental dimensions, thereby undermining the sustainability of their outcomes. Additionally, Indigenous Peoples and local communities face barriers for incremental and transformational adaptation, such as the lack of financial means and external support, unfavourable demographic changes resulting from rural-urban migration, and time constraints. At the same time, there are promising bottom-up approaches, such as co-production of knowledge and co-management, that bear the potential to overcome certain barriers by equally valuing different knowledge systems, and sharing responsibilities and resources. However, Indigenous Peoples and local communities are still not equally engaged or even having a leading role in adaptation planning and implementation, especially at the national level, for example, in the development, evaluation and improvements of countries' National Adaptation Plans. As a way forward, and based on the insights from the chapters in this part, we conclude that climate change research and decision-making need to address climate change adaptation in a broader context of sustainability, feasibility, and justice. We need a paradigm shift towards greater recognition of the significance of Indigenous and local knowledge, practices, and cultures for climate change adaptation. This would enable Indigenous Peoples and local communities to take a leadership role in climate negotiation, planning, and implementation.

References

Abazeri, M. (2024). Strengthening participatory governance to adapt and other sociocultural responses to climate change impacts among farming communities in Kerman, Iran. In V. Reyes-García, S. Álvarez-Fernández, P. Benyei, L. Calvet-Mir, D. García-del-Amo, A. B. Junqueira, X. Li, V. Porcher, A. Porcuna-Ferrer, A. Schlingmann, & R. Soleymani (Eds.), *Routledge Handbook of Climate Change Impacts on Indigenous Peoples and Local Communities*. Oxfordshire & New York: Routledge. Chp 18, pp 303–321.

Adger, W. N., Arnell, N. W., & Tompkins, E. L. (2005). Successful adaptation to climate change across scales. *Global Environmental Change*, *15*(2), 77–86. https://doi.org/10.1016/j.gloenvcha.2004.12.005

Adger, W. N., & Barnett, J. (2009). Four reasons for concern about adaptation to climate change. *Environment and Planning A*, *41*(12), 2800–2805. https://doi.org/10.1068/a42244

Adger, W. N., Barnett, J., Brown, K., Marshall, N., & O'Brien, K. (2013). Cultural dimensions of climate change impacts and adaptation. *Nature Climate Change*, *3*(2), 112–117. https://doi.org/10.1038/nclimate1666

Adger, W. N., Dessai, S., Goulden, M., Hulme, M., Lorenzoni, I., Nelson, D. R.,… & Wreford, A. (2009). Are there social limits to adaptation to climate change? *Climatic Change*, *93*, 335–354. https://doi.org/10.1007/s10584-008-9520-z

Armitage, D., Berkes, F., Dale, A., Kocho-Schellenberg, E., & Patton, E. (2011). Co-management and the co-production of knowledge: learning to adapt in Canada's Arctic. *Global Environmental Change*, *21*(3), 995–1004. https://doi.org/10.1016/j.gloenvcha.2011.04.006

Atteridge, A., & Remling, E. (2018). Is adaptation reducing vulnerability or redistributing it? *Wiley Interdisciplinary Reviews: Climate Change*, *9*(1), e500. https://doi.org/10.1002/wcc.500

Attoh, E.M.N.A.N., Afriyie, R., Kranjac-Berisavljevic, G., Bessah, E., & Ludwig, F. (2024). Changing terrain: evidence of climate change impacts and adaptive responses of Dagbani Indigenous communities, northern Ghana. In V. Reyes-García, S. Álvarez-Fernández, P. Benyei, L. Calvet-Mir, D. García-del-Amo, A. B. Junqueira, X. Li, V. Porcher, A. Porcuna-Ferrer, A. Schlingmann, & R. Soleymani (Eds.), *Routledge Handbook of Climate Change Impacts on Indigenous Peoples and Local Communities*. Oxfordshire & New York: Routledge. Chp14, pp 244–258.

Ayanlade, A., Oluwatimilehin, I. A., & Ayanlade, O. S. (2024). Climate change impacts on agriculture and barriers to adaptation technologies among rural farmers in Southwestern Nigeria. In V. Reyes-García, S. Álvarez-Fernández, P. Benyei, L. Calvet-Mir, D. García-del-Amo, A. B. Junqueira, X. Li, V. Porcher, A. Porcuna-Ferrer, A. Schlingmann, & R. Soleymani (Eds.), *Routledge Handbook of Climate Change Impacts on Indigenous Peoples and Local Communities*. Oxfordshire & New York: Routledge. Chp 15, pp 259–272.

Barnett, J., & O'Neill, S. (2010). Maladaptation. *Global Environmental Change*, *2*(20), 211–213. https://doi.org/10.1016/j.gloenvcha.2009.11.004

Bennett, N. J., Blythe, J., Tyler, S., & Ban, N. C. (2016). Communities and change in the anthropocene: understanding social-ecological vulnerability and planning adaptations to multiple interacting exposures. *Regional Environmental Change*, *16*, 907–926. https://doi.org/10.1007/s10113-015-0839-5

Berkes, F. (2007). Understanding uncertainty and reducing vulnerability: lessons from resilience thinking. *Natural Hazards*, *41*, 283–295. https://doi.org/10.1007/s11069-006-9036-7

Bezner Kerr, R., Naess, L. O., Allen-O'Neil, B., Totin, E., Nyantakyi-Frimpong, H., Risvoll, C.,… & Eriksen, S. (2022). Interplays between changing biophysical and social dynamics under climate change: implications for limits to sustainable adaptation in food systems. *Global Change Biology*, *28*(11), 3580–3604. https://doi.org/10.1111/gcb.16124

Biesbroek, G. R., Klostermann, J. E., Termeer, C. J., & Kabat, P. (2013). On the nature of barriers to climate change adaptation. *Regional Environmental Change*, *13*, 1119–1129. https://doi.org/10.1007/s10113-013-0421-y

Bown, N. K., Gray, T. S., & Stead, S. M. (2013). Co-management and adaptive co-management: two modes of governance in a Honduran marine protected area. *Marine Policy*, *39*, 128–134. https://doi.org/10.1016/j.marpol.2012.09.005

Burton, I., Huq, S., Lim, B., Pilifosova, O., & Schipper, E. L. (2002). From impacts assessment to adaptation priorities: the shaping of adaptation policy. *Climate Policy*, *2*(2–3), 145–159. https://doi.org/10.3763/cpol.2002.0217

Carlsson, L., & Berkes, F. (2005). Co-management: concepts and methodological implications. *Journal of Environmental Management*, *75*(1), 65–76. https://doi.org/10.1016/j.jenvman.2004.11.008

Carmona, R. (2024). The colonial roots of climate vulnerability. analysing the case of the Mapuche-Pehuenche people in Southern Chile. In V. Reyes-García, S. Álvarez-Fernández, P. Benyei, L. Calvet-Mir, D. García-del-Amo, A. B. Junqueira, X. Li, V. Porcher, A. Porcuna-Ferrer, A. Schlingmann, & R. Soleymani (Eds.), *Routledge Handbook of Climate Change Impacts on Indigenous Peoples and Local Communities*. Oxfordshire & New York: Routledge. Chp 12, pp 196–208.

Chakauya, R., Materechera, S. A., Jiri, O., Chakauya, E., & Machete, M. (2024). Climate change impacts on agriculture, adaptation and resilience: insights from local farmers in South-East Zimbabwe. In V. Reyes-García, S. Álvarez-Fernández, P. Benyei, L. Calvet-Mir, D. García-del-Amo, A. B. Junqueira, X. Li, V. Porcher, A. Porcuna-Ferrer, A. Schlingmann, & R. Soleymani (Eds.), *Routledge Handbook of Climate Change Impacts on Indigenous Peoples and Local Communities*. Oxfordshire & New York: Routledge. Chp 16, pp 273–286.

Chao, O., Li, X., & Reyes-García, V. (2024). Faith, reciprocity, and balance: Inner Mongolian Ovoo offering ritual and its contribution to climate change adaptation. In V. Reyes-García, S. Álvarez-Fernández, P. Benyei, L. Calvet-Mir, D. García-del-Amo, A. B. Junqueira, X. Li, V. Porcher, A. Porcuna-Ferrer, A. Schlingmann, & R. Soleymani (Eds.), *Routledge Handbook of Climate Change Impacts on Indigenous Peoples and Local Communities*. Oxfordshire & New York: Routledge. Chp 21, pp 346–355.

Chen, Z. (2024). Roads, riding stables, and highland barley: livelihood diversification as climate change adaptation among Tibetans in Shangri-La, China. In V. Reyes-García, S. Álvarez-Fernández, P. Benyei, L. Calvet-Mir, D. García-del-Amo, A. B. Junqueira, X. Li, V. Porcher, A. Porcuna-Ferrer, A. Schlingmann, & R. Soleymani (Eds.), Routledge Handbook of Climate Change Impacts on Indigenous Peoples and Local Communities. Oxfordshire & New York: Routledge. Chp 20, pp 335–345.

Chung Tiam Fook, T. (2017). Transformational processes for community-focused adaptation and social change: a synthesis. *Climate and Development*, *9*(1), 5–21. https://doi.org/10.1080/17565529.2015.1086294

Clarke, D., Murphy, C., & Lorenzoni, I. (2018). Place attachment, disruption and transformative adaptation. *Journal of Environmental Psychology*, *55*, 81–89. https://doi.org/10.1016/j.jenvp.2017.12.006

Conway, D., Nicholls, R. J., Brown, S., Tebboth, M. G., Adger, W. N., Ahmad, B.,… & Wester, P. (2019). The need for bottom-up assessments of climate risks and adaptation in climate-sensitive regions. *Nature Climate Change*, *9*(7), 503–511. https://doi.org/10.1038/s41558-019-0502-0

d'Armengol, L., Castillo, M. P., Ruiz-Mallén, I., & Corbera, E. (2018). A systematic review of co-managed small-scale fisheries: social diversity and adaptive management improve outcomes. *Global Environmental Change*, *52*, 212–225. https://doi.org/10.1016/j.gloenvcha.2018.07.009

Deressa, T. T., Hassan, R. M., Ringler, C., Alemu, T., & Yesuf, M. (2009). Determinants of farmers' choice of adaptation methods to climate change in the Nile Basin of Ethiopia. *Global Environmental Change*, *19*(2), 248–255. https://doi.org/10.1016/j.gloenvcha.2009.01.002

Dhakal, B., & Kattel, R. R. (2019). Effects of global changes on ecosystems services of multiple natural resources in mountain agricultural landscapes. *Science of the Total Environment*, *676*, 665–682. https://doi.org/10.1016/j.scitotenv.2019.04.276

Dow, K., Berkhout, F., Preston, B. L., Klein, R. J., Midgley, G., & Shaw, M. R. (2013). Limits to adaptation. *Nature Climate Change, 3*(4), 305–307. https://doi.org/10.1038/nclimate1847

Eisenack, K., Moser, S. C., Hoffmann, E., Klein, R. J., Oberlack, C., Pechan, A.,… & Termeer, C. J. (2014). Explaining and overcoming barriers to climate change adaptation. *Nature Climate Change, 4*(10), 867–872. https://doi.org/10.1038/nclimate2350

Eisenack, K., & Stecker, R. (2012). A framework for analyzing climate change adaptations as actions. *Mitigation and Adaptation Strategies for Global Change*, *17*, 243–260. https://doi.org/10.1007/s11027-011-9323-9

Ensor, J. E., Wennström, P., Bhatterai, A., Nightingale, A. J., Eriksen, S., & Sillmann, J. (2019). Asking the right questions in adaptation research and practice: seeing beyond climate impacts in rural Nepal. *Environmental Science & Policy*, *94*, 227–236. https://doi.org/10.1016/j.envsci.2019.01.013

Eriksen, S., Aldunce, P., Bahinipati, C. S., Martins, R. D. A., Molefe, J. I., Nhemachena, C.,… & Ulsrud, K. (2011). When not every response to climate change is a good one: identifying principles for sustainable adaptation. *Climate and Development*, *3*(1), 7–20. https://doi.org/10.3763/cdev.2010.0060

Fedele, G., Donatti, C. I., Harvey, C. A., Hannah, L., & Hole, D. G. (2019). Transformative adaptation to climate change for sustainable social-ecological systems. *Environmental Science & Policy*, *101*, 116–125. https://doi.org/10.1016/j.envsci.2019.07.001

Fedele, G., Donatti, C. I., Harvey, C. A., Hannah, L., & Hole, D. G. (2020). Limited use of transformative adaptation in response to social-ecological shifts driven by climate change. *Ecology and Society*, *25*(1), 25. https://doi.org/10.5751/ES-11381-250125

Fenton, A., Paavola, J., & Tallontire, A. (2017). Autonomous adaptation to riverine flooding in Satkhira District, Bangladesh: implications for adaptation planning. *Regional Environmental Change*, *17*, 2387–2396. https://doi.org/10.1007/s10113-017-1159-8

Feola, G. (2015). Societal transformation in response to global environmental change: a review of emerging concepts. *Ambio*, *44*(5), 376–390. https://doi.org/10.1007/s13280-014-0582-z

Fernández-Llamazares, Á., Lepofsky, D., Lertzman, K., Armstrong, C. G., Brondizio, E. S., Gavin, M. C.,… & Vaughan, M. B. (2021). Scientists' warning to humanity on threats to indigenous and local knowledge systems. *Journal of Ethnobiology*, *41*(2), 144–169. https://doi.org/10.2993/0278-0771-41.2.144

Few, R., Spear, D., Singh, C., Tebboth, M. G., Davies, J. E., & Thompson-Hall, M. C. (2021). Culture as a mediator of climate change adaptation: neither static nor unidirectional. *Wiley Interdisciplinary Reviews: Climate Change*, *12*(1), e687. https://doi.org/10.1002/wcc.687

Folke, C., Biggs, R., Norström, A. V., Reyers, B., & Rockström, J. (2016). Social-ecological resilience and biosphere-based sustainability science. *Ecology and Society*, *21*(3). http://dx.doi.org/10.5751/ES-08748-210341

Ford, J. D., & King, D. (2015). A framework for examining adaptation readiness. *Mitigation and Adaptation Strategies for Global Change*, *20*, 505–526. https://doi.org/10.1007/s11027-013-9505-8

Ford, J. D., King, N., Galappaththi, E. K., Pearce, T., McDowell, G., & Harper, S. L. (2020). The resilience of Indigenous Peoples to environmental change. *One Earth*, *2*(6), 532–543. https://doi.org/10.1016/j.oneear.2020.05.014

Ford, J. D., Stephenson, E., Cunsolo Willox, A., Edge, V., Farahbakhsh, K., Furgal, C.,… & Sherman, M. (2016). Community-based adaptation research in the Canadian Arctic. *Wiley Interdisciplinary Reviews: Climate Change*, *7*(2), 175–191. https://doi.org/10.1002/wcc.376

Forsyth, T. (2013). Community-based adaptation: a review of past and future challenges. *Wiley Interdisciplinary Reviews: Climate Change*, *4*(5), 439–446. https://doi.org/10.1002/wcc.231

Fuchs, A., Vogel, C. R., & Schunko, C. (2024). Counteracting land abandonment: local adaptation strategies to climate change impacts of alpine farmers in Eastern Tyrol, Austria. In V. Reyes-García, S. Álvarez-Fernández, P. Benyei, L. Calvet-Mir, D. García-del-Amo, A. B. Junqueira, X. Li, V. Porcher, A. Porcuna-Ferrer, A. Schlingmann, & R. Soleymani (Eds.), *Routledge Handbook of Climate Change Impacts on Indigenous Peoples and Local Communities*. Oxfordshire & New York: Routledge. Chp 17, pp 287–302.

Fuso Nerini, F., Sovacool, B., Hughes, N., Cozzi, L., Cosgrave, E., Howells, M.,… & Milligan, B. (2019). Connecting climate action with other Sustainable Development Goals. *Nature Sustainability*, *2*(8), 674–680. https://doi.org/10.1038/s41893-019-0334-y

Füssel, H. M. (2007). Adaptation planning for climate change: concepts, assessment approaches, and key lessons. *Sustainability Science*, *2*, 265–275. https://doi.org/10.1007/s11625-007-0032-y

Galappaththi, E. (2024). How can co-management support adaptation to climate change? The case of co-existing fisheries from Pangnirtung, Nunavut, Canada. In V. Reyes-García, S. Álvarez-Fernández, P. Benyei, L. Calvet-Mir, D. García-del-Amo, A. B. Junqueira, X. Li, V. Porcher, A. Porcuna-Ferrer, A. Schlingmann, & R. Soleymani (Eds.), Routledge Handbook of Climate Change Impacts on Indigenous Peoples and Local Communities. Oxfordshire & New York: Routledge. Chp 23, pp 370–388.

Galappaththi, E. K., Falardeau, M., Harris, L. N., Rocha, J. C., Moore, J. S., & Berkes, F. (2022). Resilience-based steps for adaptive co-management of Arctic small-scale fisheries. *Environmental Research Letters*, *17*(8), 083004. https://doi.org/10.1088/1748-9326/ac7b37

Galappaththi, E. K., & Schlingmann, A. (2023). The sustainability assessment of Indigenous and local knowledge-based climate adaptation responses in agricultural and aquatic food systems. *Current Opinion in Environmental Sustainability, 62*, 101276. https://doi.org/10.1016/j.cosust.2023.101276

Heyd, T., & Brooks, N. (2009). Exploring cultural dimensions of adaptation to climate change. In W. N. Adger (Eds.), *Adapting to Climate Change: Thresholds, Values, Governance* (pp. 269–282). Cambridge University Press. https://doi.org/10.1017/CBO9780511596667.018

Höhne, N., Gidden, M. J., den Elzen, M., Hans, F., Fyson, C., Geiges, A.,… & Rogelj, J. (2021). Wave of net zero emission targets opens window to meeting the Paris agreement. *Nature Climate Change*, *11*(10), 820–822. https://doi.org/10.1038/s41558-021-01142-2

IPCC (2021). Summary for policymakers. In V. Masson-Delmotte et al. (Eds.), *Climate Change 2021: The Physical Science Basis. Contribution of Working Group I to the Sixth Assessment Report of the Intergovernmental Panel on Climate Change* (pp. 3–32). Cambridge University Press. https://doi.org/10.1017/9781009157896.001

IPCC (2022). Annex II: glossary. In H. O. Pörtner et al. (Eds.), *Climate Change 2022: Impacts, Adaptation and Vulnerability. Contribution of Working Group II to the Sixth Assessment Report of the Intergovernmental Panel on Climate Change* (pp. 2897–2930). Cambridge University Press. https://doi.org/10.1017/9781009325844.029

Izquierdo, A., & Schlingmann, A. (2024). Local perceptions of climate change in the context of socioeconomic and political changes in a High Andean community from the Argentine Puna. In V. Reyes-García, S. Álvarez-Fernández, P. Benyei, L. Calvet-Mir, D. García-del-Amo, A. B. Junqueira, X. Li, V. Porcher, A. Porcuna-Ferrer, A. Schlingmann, & R. Soleymani (Eds.), *Routledge Handbook of Climate Change Impacts on Indigenous Peoples and Local Communities*. Oxfordshire & New York: Routledge. Chp 8, pp 136–150.

Juhola, S., Glaas, E., Linnér, B. O., & Neset, T. S. (2016). Redefining maladaptation. *Environmental Science & Policy*, *55*, 135–140. https://doi.org/10.1016/j.envsci.2015.09.014

Jungsberg, L., & Wendt-Lucas N. (2024). Local indicators, adaptation actions, and resilience efforts for a warming climate in North Greenland. In V. Reyes-García, S. Álvarez-Fernández, P. Benyei, L. Calvet-Mir, D. García-del-Amo, A. B. Junqueira, X. Li, V. Porcher, A. Porcuna-Ferrer, A. Schlingmann, & R. Soleymani (Eds.), Routledge Handbook of Climate Change Impacts on Indigenous Peoples and Local Communities. Oxfordshire & New York: Routledge. Chp 22, pp 356–369.

Kaijser, A., & Kronsell, A. (2014). Climate change through the lens of intersectionality. *Environmental Politics*, *23*(3), 417–433. https://doi.org/10.1080/09644016.2013.835203

Kates, R. W., Travis, W. R., & Wilbanks, T. J. (2012). Transformational adaptation when incremental adaptations to climate change are insufficient. *Proceedings of the National Academy of Sciences*, *109*(19), 7156–7161. https://doi.org/10.1073/pnas.1115521109

Khanal, U., Wilson, C., Hoang, V. N., & Lee, B. (2018). Farmers' adaptation to climate change, its determinants and impacts on rice yield in Nepal. *Ecological Economics*, *144*, 139–147. https://doi.org/10.1016/j.ecolecon.2017.08.006

Klein, R. J. T., Midgley, G. F., Preston, B. L., Alam, M., Berkhout, F. G. H., Dow, K., & Shaw, M. R. (2014). Adaptation opportunities, constraints, and limits. In C. B. Field et al. (Eds.), *Climate Change2014: Impacts, Adaptation, and Vulnerability. Part A. Global and Sectoral Aspects. Contribution of Working Group II to the Fifth Assessment Report of the Intergovernmental Panel on Climate Change* (pp. 899–943). Cambridge University Press.

Klenk, N., Fiume, A., Meehan, K., & Gibbes, C. (2017). Local knowledge in climate adaptation research: moving knowledge frameworks from extraction to co-production. *Wiley Interdisciplinary Reviews: Climate Change*, *8*(5), e475. https://doi.org/10.1002/wcc.475

Lahsen, M., & Ribot, J. (2022). Politics of attributing extreme events and disasters to climate change. *Wiley Interdisciplinary Reviews: Climate Change*, *13*(1), e750. https://doi.org/10.1002/wcc.750

Leal Filho, W., Barbir, J., Gwenzi, J., Ayal, D., Simpson, N. P., Adeleke, L.,… & Yaffa, S. (2022). The role of indigenous knowledge in climate change adaptation in Africa. *Environmental Science & Policy*, *136*, 250–260. https://doi.org/10.1016/j.envsci.2022.06.004

Leary, N. A. (1999). A framework for benefit-cost analysis of adaptation to climate change and climate variability. *Mitigation and Adaptation Strategies for Global Change*, *4*, 307–318. https://doi.org/10.1023/A:1009667706027

Lebel, L. (2013). Local knowledge and adaptation to climate change in natural resource-based societies of the Asia-Pacific. *Mitigation and Adaptation Strategies for Global Change*, *18*, 1057–1076. https://doi.org/10.1007/s11027-012-9407-1

Liu, P. R., & Raftery, A. E. (2021). Country-based rate of emissions reductions should increase by 80% beyond nationally determined contributions to meet the 2°C target. *Communications Earth & Environment*, *2*(1), 29. https://doi.org/10.1038/s43247-021-00097-8

Luetz, J. M., & Nunn, P. D. (2020). Climate change adaptation in the Pacific Islands: a review of faith-engaged approaches and opportunities. *Managing Climate Change Adaptation in the Pacific Region*, 293–311. https://doi.org/10.1007/978-3-030-40552-6_15

Magnan, A. K., Schipper, E. L. F., Burkett, M., Bharwani, S., Burton, I., Eriksen, S.,... & Ziervogel, G. (2016). Addressing the risk of maladaptation to climate change. *Wiley Interdisciplinary Reviews: Climate Change*, *7*(5), 646–665. https://doi.org/10.1002/wcc.409

Magni, G. (2017). Indigenous knowledge and implications for the sustainable development agenda. *European Journal of Education*, *52*(4), 437–447. https://doi.org/10.1111/ejed.12238

Mapfumo, P., Onyango, M., Honkponou, S. K., El Mzouri, E. H., Githeko, A., Rabeharisoa, L.,... & Agrawal, A. (2017). Pathways to transformational change in the face of climate impacts: an analytical framework. *Climate and Development*, *9*(5), 439–451. https://doi.org/10.1080/17565529.2015.1040365

McDowell, J. Z., & Hess, J. J. (2012). Accessing adaptation: multiple stressors on livelihoods in the Bolivian highlands under a changing climate. *Global Environmental Change*, *22*(2), 342–352. https://doi.org/10.1016/j.gloenvcha.2011.11.002

McNamara, K. E., & Buggy, L. (2017). Community-based climate change adaptation: a review of academic literature. *Local Environment*, *22*(4), 443–460. https://doi.org/10.1080/13549839.2016.1216954

McNeeley, S. M., & Lazrus, H. (2014). The cultural theory of risk for climate change adaptation. *Weather, Climate, and Society*, *6*(4), 506–519. https://doi.org/10.1175/WCAS-D-13-00027.1

Mersha, A. A., & van Laerhoven, F. (2018). The interplay between planned and autonomous adaptation in response to climate change: insights from rural Ethiopia. *World Development*, *107*, 87–97. https://doi.org/10.1016/j.worlddev.2018.03.001

Morse, S., McNamara, N., Morse, S., & McNamara, N. (2013). The theory behind the sustainable livelihood approach. *Sustainable Livelihood Approach: A Critique of Theory and Practice*, 15–60. https://doi.org/10.1007/978-94-007-6268-8_2

Moser, S. C., & Ekstrom, J. A. (2010). A framework to diagnose barriers to climate change adaptation. *Proceedings of the National Academy of Sciences*, *107*(51), 22026–22031. https://doi.org/10.1073/pnas.1007887107

Mwangi, B., Olaka, L. A., Zafra-Calvo, N., Klein, J. A., & Cuni-Sanchez, A. (2024). The role of culture in climate change adaptation: insights from two mountain regions in Kenya. In V. Reyes-García, S. Álvarez-Fernández, P. Benyei, L. Calvet-Mir, D. García-del-Amo, A. B. Junqueira, X. Li, V. Porcher, A. Porcuna-Ferrer, A. Schlingmann, & R. Soleymani (Eds.), *Routledge Handbook of Climate Change Impacts on Indigenous Peoples and Local Communities*. Oxfordshire & New York: Routledge. Chp 19, pp 322–335.

Naess, L. O. (2013). The role of local knowledge in adaptation to climate change. *Wiley Interdisciplinary Reviews: Climate Change*, *4*(2), 99–106. https://doi.org/10.1002/wcc.204

Nielsen, J. Ø., & Reenberg, A. (2010). Cultural barriers to climate change adaptation: a case study from Northern Burkina Faso. *Global Environmental Change*, *20*(1), 142–152. https://doi.org/10.1016/j.gloenvcha.2009.10.002

Nyantakyi-Frimpong, H., & Bezner-Kerr, R. (2015). The relative importance of climate change in the context of multiple stressors in semi-arid Ghana. *Global Environmental Change*, *32*, 40–56. https://doi.org/10.1016/j.gloenvcha.2015.03.003

Nyong, A., Adesina, F., & Osman Elasha, B. (2007). The value of indigenous knowledge in climate change mitigation and adaptation strategies in the African Sahel. *Mitigation and Adaptation Strategies for Global Change*, *12*, 787–797. https://doi.org/10.1007/s11027-007-9099-0

O'Brien, K. (2012). Global environmental change II: from adaptation to deliberate transformation. *Progress in Human Geography*, *36*(5), 667–676. https://doi.org/10.1177/0309132511425767

O'Brien, K. L. (2009). Do values subjectively define the limits to climate change adaptation. In W. N. Adger (Eds.), *Adapting to Climate Change: Thresholds, Values, Governance* (pp. 164–180). Cambridge University Press. https://doi.org/10.1017/CBO9780511596667.011

Olsson, P., Folke, C., & Berkes, F. (2004). Adaptive comanagement for building resilience in social–ecological systems. *Environmental Management*, *34*, 75–90. https://doi.org/10.1007/s00267-003-0101-7

Olsson, P., Galaz, V., & Boonstra, W. J. (2014). Sustainability transformations: a resilience perspective. *Ecology and Society*, *19*(4). http://dx.doi.org/10.5751/ES-06799-190401

Olsson, P., Moore, M. L., Westley, F. R., & McCarthy, D. D. (2017). The concept of the Anthropocene as a game-changer: a new context for social innovation and transformations to sustainability. *Ecology and Society*, *22*(2). https://doi.org/10.5751/ES-09310-220231

Owen, G. (2020). What makes climate change adaptation effective? A systematic review of the literature. *Global Environmental Change*, *62*, 102071. https://doi.org/10.1016/j.gloenvcha.2020.102071

Park, S. E., Marshall, N. A., Jakku, E., Dowd, A. M., Howden, S. M., Mendham, E., & Fleming, A. (2012). Informing adaptation responses to climate change through theories of transformation. *Global Environmental Change*, *22*(1), 115–126. https://doi.org/10.1016/j.gloenvcha.2011.10.003

Pelling, M., O'Brien, K., & Matyas, D. (2015). Adaptation and transformation. *Climatic Change*, *133*, 113–127. https://doi.org/10.1007/s10584-014-1303-0

Petzold, J., Andrews, N., Ford, J. D., Hedemann, C., & Postigo, J. C. (2020). Indigenous knowledge on climate change adaptation: a global evidence map of academic literature. *Environmental Research Letters*, *15*(11), 113007. https://doi.org/10.1088/1748-9326/abb330

Pielke Jr, R., Prins, G., Rayner, S., & Sarewitz, D. (2007). Lifting the taboo on adaptation. *Nature*, *445*(7128), 597–598. https://doi.org/10.1038/445597a

Piggott-McKellar, A. E., McNamara, K. E., Nunn, P. D., & Watson, J. E. (2019). What are the barriers to successful community-based climate change adaptation? A review of grey literature. *Local Environment*, *24*(4), 374–390. https://doi.org/10.1080/13549839.2019.1580688

Pisor, A. C., Basurto, X., Douglass, K. G., Mach, K. J., Ready, E., Tylianakis, J. M.,… & Jones, J. H. (2022). Effective climate change adaptation means supporting community autonomy. *Nature climate change*, *12*(3), 213–215. https://doi.org/10.1038/s41558-022-01303-x

Porcuna-Ferrer, A., Guillerminet, T., Klappoth, B., & Schlingmann, A. (2024). Agricultural adaptation to multiple stressors in a climate change context. A case study in south-eastern Senegal. In V. Reyes-García, S. Álvarez-Fernández, P. Benyei, L. Calvet-Mir, D. García-del-Amo, A. B. Junqueira, X. Li, V. Porcher, A. Porcuna-Ferrer, A. Schlingmann, & R. Soleymani (Eds.), *Routledge Handbook of Climate Change Impacts on Indigenous Peoples and Local Communities*. Oxfordshire & New York: Routledge. Chp 13, pp 227–243.

Ravera, F., Iniesta-Arandia, I., Martín-López, B., Pascual, U., & Bose, P. (2016a). Gender perspectives in resilience, vulnerability and adaptation to global environmental change. *Ambio*, *45*, 235–247. https://doi.org/10.1007/s13280-016-0842-1

Ravera, F., Martín-López, B., Pascual, U., & Drucker, A. (2016b). The diversity of gendered adaptation strategies to climate change of Indian farmers: a feminist intersectional approach. *Ambio*, *45*, 335–351. https://doi.org/10.1007/s13280-016-0833-2

Ribot, J. (2014). Cause and response: vulnerability and climate in the Anthropocene. *The Journal of Peasant Studies*, *41*(5), 667–705. https://doi.org/10.1080/03066150.2014.894911

Ribot, J. (2022). Violent silence: framing out social causes of climate-related crises. *The Journal of Peasant Studies*, *49*(4), 683–712. https://doi.org/10.1080/03066150.2022.2069016

Rickards, L., & Howden, S. M. (2012). Transformational adaptation: agriculture and climate change. *Crop and Pasture Science*, *63*(3), 240–250. https://doi.org/10.1071/CP11172

Rivera-Ferre, M. G., Di Masso, M., Vara, I., Cuellar, M., López-i-Gelats, F., Bhatta, G. D., & Gallar, D. (2021). Traditional agricultural knowledge in land management: the potential contributions of ethnographic research to climate change adaptation in India, Bangladesh, Nepal, and Pakistan. *Climate and Development*, *13*(7), 644–661. https://doi.org/10.1080/17565529.2020.1848780

Rogelj, J., Den Elzen, M., Höhne, N., Fransen, T., Fekete, H., Winkler, H.,… & Meinshausen, M. (2016). Paris agreement climate proposals need a boost to keep warming well below 2°C. *Nature*, *534*(7609), 631–639. https://doi.org/10.1038/nature18307

Schipper, E. L. F. (2020). Maladaptation: when adaptation to climate change goes very wrong. *One Earth*, *3*(4), 409–414. https://doi.org/10.1016/j.oneear.2020.09.014

Schlingmann, A., Graham, S., Benyei, P., Corbera, E., Sanesteban, I. M., Marelle, A., Soleymani-Fard, R., & Reyes-García, V. (2021). Global patterns of adaptation to climate change by Indigenous peoples and local communities. a systematic review. *Current Opinion in Environmental Sustainability*, *51*, 55–64. https://doi.org/10.1016/j.cosust.2021.03.002

Shaffril, H. A. M., Ahmad, N., Samsuddin, S. F., Samah, A. A., & Hamdan, M. E. (2020). Systematic literature review on adaptation towards climate change impacts among Indigenous people in the Asia Pacific regions. *Journal of Cleaner Production, 258*, 120595. https://doi.org/10.1016/j.jclepro.2020.120595

Singh, C., Ford, J., Ley, D., Bazaz, A., & Revi, A. (2020). Assessing the feasibility of adaptation options: methodological advancements and directions for climate adaptation research and practice. *Climatic Change, 162*, 255–277. https://doi.org/10.1007/s10584-020-02762-x

Singh, C., Iyer, S., New, M. G., Few, R., Kuchimanchi, B., Segnon, A. C., & Morchain, D. (2022). Interrogating 'effectiveness' in climate change adaptation: 11 guiding principles for adaptation research and practice. *Climate and Development, 14*(7), 650–664. https://doi.org/10.1080/17565529.2021.1964937

Singh, P., Schulman, D., Charan, D., Reyes-García, V., Schlingmann, A., Singh, A. A., & Railoa, K. (2024). Climate change adaptation in Fiji: local adaptation strategies to enhance national policy. In V. Reyes-García, S. Álvarez-Fernández, P. Benyei, L. Calvet-Mir, D. García-del-Amo, A. B. Junqueira, X. Li, V. Porcher, A. Porcuna-Ferrer, A. Schlingmann, & R. Soleymani (Eds.), Routledge Handbook of Climate Change Impacts on Indigenous Peoples and Local Communities. Oxfordshire & New York: Routledge. Chp 24, pp 389–408.

Smit, B., Burton, I., Klein, R. J., & Wandel, J. (2000). An anatomy of adaptation to climate change and variability. In S. M. Kane, & G. W. Yohe (Eds.), *Societal Adaptation to Climate Variability and Change* (pp. 223–251). Springer. https://doi.org/10.1007/978-94-017-3010-5_12

Smit, B., & Pilifosova, O. (2001). Adaptation to climate change in the context of sustainable development and equity. In J. J. McCarthy et al. (Eds.), *Climate Change 2001: Impacts, Adaptation, and Vulnerability. Contribution of Working Group II to the Third Assessment Report of the Intergovernmental Panel on Climate Change* (pp. 877–912). Cambridge University Press.

Spires, M., Shackleton, S., & Cundill, G. (2014). Barriers to implementing planned community-based adaptation in developing countries: a systematic literature review. *Climate and Development, 6*(3), 277–287. https://doi.org/10.1080/17565529.2014.886995

Stratoudakis, Y., Gonçalves, P., Oliveira, D., & Mesquita, M. (2024). Settlement, way of life and worldviews: how socio-environmental changes impact and are interpreted by artisanal fishing communities in Portugal. In V. Reyes-García, S. Álvarez-Fernández, P. Benyei, L. Calvet-Mir, D. García-del-Amo, A. B. Junqueira, X. Li, V. Porcher, A. Porcuna-Ferrer, A. Schlingmann, & R. Soleymani (Eds.), *Routledge Handbook of Climate Change Impacts on Indigenous Peoples and Local Communities*. Oxfordshire & New York: Routledge. Chp 9, pp 151–164.

Termeer, C. J., Dewulf, A., & Biesbroek, G. R. (2017). Transformational change: governance interventions for climate change adaptation from a continuous change perspective. *Journal of Environmental Planning and Management, 60*(4), 558–576. https://doi.org/10.1080/09640568.2016.1168288

Thomas, A., Theokritoff, E., Lesnikowski, A., Reckien, D., Jagannathan, K., Cremades, R.,... & Bowen, K. (2021). Global evidence of constraints and limits to human adaptation. *Regional Environmental Change, 21*(3), 1–15. https://doi.org/10.1007/s10113-021-01808-9

Thompson-Hall, M., Carr, E. R., & Pascual, U. (2016). Enhancing and expanding intersectional research for climate change adaptation in agrarian settings. *Ambio, 45*, 373–382. https://doi.org/10.1007/s13280-016-0827-0

Walker, B., Holling, C. S., Carpenter, S. R., & Kinzig, A. (2004). Resilience, adaptability and transformability in social–ecological systems. *Ecology and Society, 9*(2). http://www.ecologyandsociety.org/vol9/iss2/art5

Williams, P. A., Simpson, N. P., Totin, E., North, M. A., & Trisos, C. H. (2021). Feasibility assessment of climate change adaptation options across Africa: an evidence-based review. *Environmental Research Letters, 16*(7), 073004. https://doi.org/10.1088/1748-9326/ac092d

Woodruff, S. C., & Regan, P. (2019). Quality of national adaptation plans and opportunities for improvement. *Mitigation and Adaptation Strategies for Global Change, 24*, 53–71. https://doi.org/10.1007/s11027-018-9794-z

Yap, M. L. M., & Watene, K. (2019). The sustainable development goals (SDGs) and Indigenous Peoples: another missed opportunity? *Journal of Human Development and Capabilities, 20*(4), 451–467. https://doi.org/10.1080/19452829.2019.1574725

Zant, M., Schlingmann, A., Reyes-García, V., & García del Amo, D. (under review). Incremental and transformational adaptation to climate change among IndigenousPeoples and local communities: a global review. *Mitigation and Adaptation Strategies for Global Change* [unpublished manuscript].

13

AGRICULTURAL ADAPTATION TO MULTIPLE STRESSORS IN A CLIMATE CHANGE CONTEXT

A case study in south-eastern Senegal

Anna Porcuna-Ferrer, Théo Guillerminet, Benjamin Klappoth and Anna Schlingmann

Introduction

Smallholder farmers around the world observe inter- and intra-seasonal climate variability and change (Reyes-García et al., 2019; Savo et al., 2016) and use this knowledge to organise their agricultural practices according to local environmental conditions, including climate, soil, and water (Altieri & Nicholls, 2017; Rivera-Ferre et al., 2021). This is also the case for smallholder farmers in West Africa, where episodic droughts have traditionally affected the agricultural rain-fed systems (Nyong et al., 2007), with scientific evidence predicting increasing risks of crop failures due to rising temperatures and changes in the monsoonal rainfall (Carr et al., 2022; Sultan & Gaetani, 2016). At the same time, smallholder farmers also experience other environmental, socio-economic, cultural, and political changes, including demographic changes and rural–urban migration, increasing integration into global markets, the expansion of infrastructure and technology, and off-farm livelihood activities (Ensor et al., 2019). Moreover, this broad range of multi-scalar, multi-temporal, and interacting stressors have differentiated impacts depending on the context-specific vulnerability of individuals and communities (Bennett et al., 2016; O'Brien et al., 2004). Consequently, although research and policy have primarily addressed the impacts of specific stressors, for example, adaptation to climate change or economic development (Bennett et al., 2016), focusing on only one stressor bears the risk of missing unexpected and negative feedbacks and trade-offs produced by other stressors, thereby undermining environmental, social or economic objectives, maintaining or even increasing vulnerability, and potentially leading to maladaptation (Antwi-Agyei et al., 2018; Barnett & O'Neill 2010; Eriksen et al., 2011).

While climate change research considering multiple stressors and their interactions is on the rise (Räsänen et al., 2016), our understanding of context-specific interactions, interferences, and consequences of responses to climatic and other stressors is still limited (Ensor et al., 2019). Importantly, specific responses might have context- and group-specific impacts, for example, if the response favours some, but result in additional efforts, costs, or even negative outcomes for others

 DOI: 10.4324/9781003356837-18

(Segnon et al., 2021). Eriksen et al. (2011) argue that assessing responses from a sustainability perspective entails considering the interdependencies between environmental, social, and economic objectives to understand the vulnerability context in which adaptive responses are framed. Therefore, a thorough multi-stressor and multi-facet analysis that assess environmental, social, and economic implications of adaptation for different social groups is needed to anticipate trade-offs, avoid maladaptation, and foster long-term adaptation.

In response to this need, here we examine smallholder farmers' adaptive responses to multiple stressors in the context of climate change. Our specific objectives are to explore (i) the main changes in local agricultural practices, (ii) the drivers of those changes (also named 'stressors'), and (iii) associated costs, benefits, and trade-offs. In our analysis, we pay attention to the specific implications that changes in agricultural practices have for the different sustainability spheres (i.e., environmental, social, and economic) and social groups (i.e., across gender and wealth). Based on vulnerability and adaptation literature dealing with multiple stressors (e.g., McDowell & Hess, 2012; O'Brien et al., 2004; Tschakert, 2007) and intersectionality (Kaijser & Kronsell, 2014; Ravera et al., 2016a), our discussion enriches the debate on which of the changes in agricultural management practices, beyond bringing climate change resilience, also contribute to social justice, environmental integrity, and economically viable livelihoods (Eriksen et al., 2011).

The Bassari

We conducted research among the Bassari, located in the Kédougou region in south-eastern Senegal (Figure 13.1). The region is characterised by low altitude (approx. 80–380 m a.s.l.), tropical dry or savanna climate, annual mean temperatures around 28°C, and a unimodal rainy season from May to September dominated by the West African monsoon system (ANACIM, 2020; Sultan & Janicot, 2003).

With a wetter climate than the northern regions, south-eastern Senegal was less affected by the Sahel droughts of the 1970s than most Senegalese regions. And although there are important interannual rainfall variations, historical trends show a partial recovery of the lack of precipitation

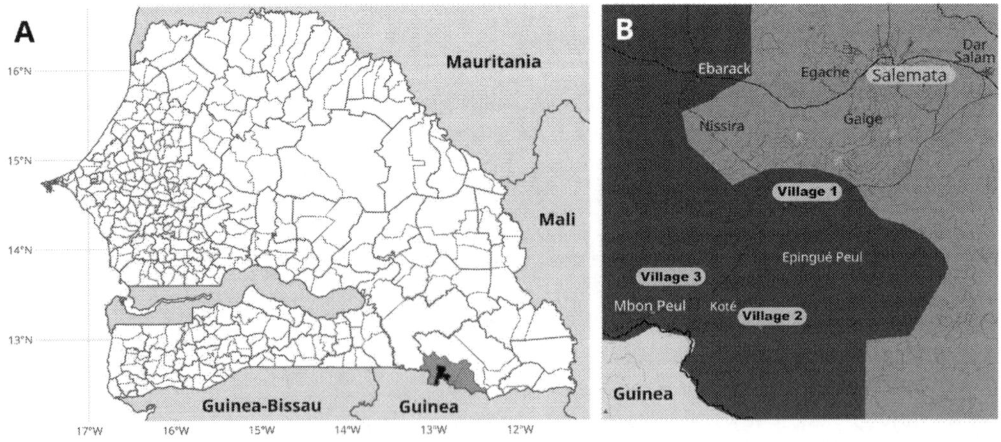

Figure 13.1 Map placing the study area within Senegal. The dark grey and the black polygons represent the department of Salemata and the commune of Ethiolo, respectively.

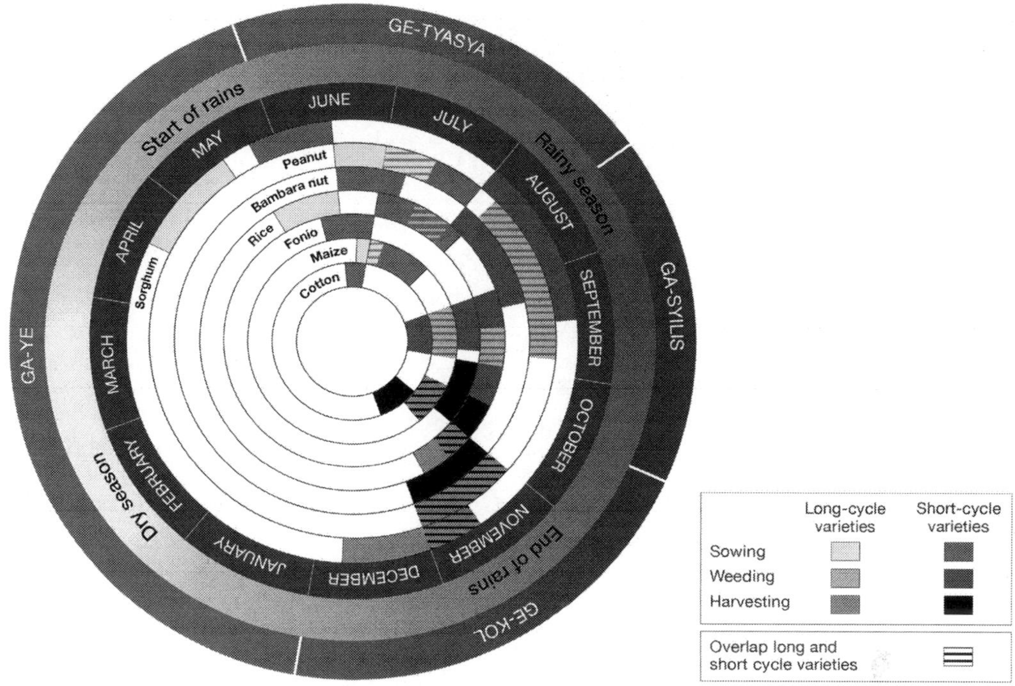

Figure 13.2 Bassari agricultural seasonal calendar.

from the 90s onwards. Nonetheless, future climate predictions for the West African region indicate a trend towards higher temperatures and shorter rainy seasons (Sultan & Gaetani, 2016).

Only 10% of the Bassari region's surface is cultivated and the remaining area preserves large forests and bushlands that are home to many wild flora and fauna. However, population density, currently below 20 inhabitants/km^2, is on the rise resulting in a gradual disappearance of forests and bushlands in favour of cultivated fields (UNESCO, 2011; Porcuna-Ferrer et al., 2023a). The Kédougou area hosts several ethnic minorities, among which we worked with Bassari.

Nowadays, the Bassari mostly rely on small-scale agriculture, which is highly dependent on the use of communal land and shared labour (Nolan, 1986). The importance of agriculture in Bassari life is reflected in the naming of the seasons (Figure 13.2). Agricultural activity is concentrated in the short rainy season. During the dry season, some women practice small-scale horticulture production (e.g., onions, tomatoes, salads, cabbages), while other Bassari engage in off-farm employment, including construction work, wage labour, art craft, tourism, and seasonal migration.

Bassari farming system is organised into four agroecological field types, cultivated with different crop species and managed distinctly (Table 13.1). On average, households cultivate a mean area of 2.1 ha (min: 0.4 ha; max: 9.5 ha). The typical cropping pattern of Bassari smallholders is the rotation of cereals (sorghum, *Sorghum bicolor*; maize, *Zea mays*; fonio, *Digitaria exilis*) and legumes (peanut, *Arachis hypogaea;* Bambara groundnut, *Vigna subterranea*). In the plains, the Bassari also cultivate rice (*Oryza sativa*), which is not subject to crop rotation, and

Table 13.1 Local taxonomy of field types and the main crops and management practices associated

Field type (local name)	*Description*	*Cultivation season*	*Main food crops*	*Crop rotation*	*Fertilisation*	*Irrigation*
'Oxenga'	Soils poor in organic matter (often shallow), mostly on hilly terrains	Rainy season (June–October)	Sorghum, peanut, Bambara groundnut, cotton, fonio (maize)	Yes	Slash-and-burn, fallow system. Few chemical inputs, mostly only in cotton	No irrigation, rain-fed
'Eden'	Soils rich in organic matter with a fine texture,	Rainy season (June–October)	Rice, maize	No	Mostly use of chemical inputs	No irrigation, rain-fed
'Eden'	mostly located in the plains or valleys ('bas-fond') and near the	Rainy season (June–October)	Tubercules: taro, yam	No	Mostly use of biological fertilisation – e.g. tree leaves and branches, and ashes	No irrigation, rain-fed
'Eden'	streams.	Dry season (November/ December–May)	Horticulture: tomato, salad, aubergine, carrot, onion, etc.	No	Use of both, chemical and biological fertiliser (mostly NPK and/or cow dung)	Irrigation, well water
'Enam'	Soils surrounding the households, usually rich in organic matter	All year	Maize, fruit trees, spices, vegetables	No	Fertilised by the livestock kept in the household	No or only punctual irrigation

small-scale horticulture. The main cash crop is cotton (*Gossypium hirsutum*), cultivated as part of a farming contract system (Porcuna-Ferrer et al., 2023b).

Research methods

Data were collected between November 2019 and March 2020, and benefited from the authors' long presence in the community – 16 months in total, between 2019 and 2021 – and the support of four Bassari research assistants. Before starting data collection, we received ethical approval from the Autonomous University of Barcelona (CEEAH 4903) and obtained the permits to conduct the research from the relevant village authorities.

Sampling and data collection were conducted according to the criteria described in the Local Indicator of Climate Change Impact (LICCI) protocol (Reyes-García et al., 2023), summarised below.

Sampling

We worked in three villages in the Bassari country, with 109, 55, and 24 households, respectively. To select participants for semi-structured interviews, we used 'quota sampling' aiming at capturing the diversity of knowledge across gender, age, and wealth. In total, we interviewed 34 men and 13 women from different households, the main social and economic structuring unit in the site. We have a lower sample of women because they were generally less available for interviews; 10 participants were <40 years old, 17 were between 40 and 60 years old, and 20 were >60 years old. Household wealth was defined with the help of the research assistants and based on local conceptions. Research assistants classed households into three groups taking into account herd size, polygamy, cultivated area, and possession of material assets: low resources (~40% of households), intermediate (~50% of households), and wealthy (~10% of households).

Data collection and analysis

To understand how multiple stressors affect the local agricultural system, we first asked interviewees to describe changes in on-farm management practices that they have observed since their youth, including changes in the type of cultivated crop species and varieties, changes in management practices in the different stages of the cultivation cycle – including seed selection and acquisition, land preparation, sowing, cultivation, harvesting –, and changes in post-harvest treatment, storage, and commercialization. For each reported change in local agricultural management practices, we asked the interviewee to explain what had driven such changes.

We additionally explored positive and negative experiences associated with each change in agricultural management practices and recorded the age, gender, and wealth group of each interviewee to assess benefits, costs, and trade-offs for different social groups.

Before presenting our main results, we acknowledge two main limitations of this work. First, our data were not collected to systematically analyse how power relations shape farmers' adaptation options. However, farmers' explanations of the costs, benefits, and trade-offs of different agricultural management practices, together with our understanding of the site dynamics, allowed us to identify that farmers' responses to multiple drivers can simultaneously lead to positive changes for some groups and negative consequences for others. Second, the limited time frame of our research makes it difficult to fully understand or evaluate the long-term outcomes of farmers' strategies. However, results from our empirical work shed light on the complexity of local farmers' adaptation to multiple drivers.

Results

Bassari farmers have adapted their agricultural practices in response to environmental and socio-economic drivers including: (1) climatic (i.e., declining crop yields due to changes in precipitation, temperature, and fog), (2) demographic (i.e., higher land pressure due to population increase and labour shortages due to migration), (3) economic (i.e., increasing monetization of the livelihood system through market integration, which caused higher cash dependency, higher reliance in off-farm activities, and trends towards economic efficiency), (4) social (i.e., increasing influence of NGOs and extension services, and decrease in community's social capital such as communal work and reciprocity), and (5) cultural (i.e., weakening of traditions and cultural norms and changing dietary preferences). These drivers have generated impacts in the local farming system and shaped changes in agricultural management practices, including changes in crop species and varieties, changes in seed management, and changes in soil cultivation practices (see Figures 13.3 and 13.4). Agricultural management changes come along with certain benefits, costs, and trade-offs for different sustainability spheres and social groups, as we next outline.

Adoption of new crop species

The most prominent change in agricultural management practices refers to the adoption of new crop species. Traditional Bassari staple crops were sorghum, fonio, and Bambara groundnut. The adoption of new crops and in particular rice and maize (for subsistence) and cotton and horticulture (for sale) were common changes mentioned by farmers. The introduction and expansion of horticulture (2000s) and cotton (1970s) is relatively recent. Rice (19th century) and maize (1930s)

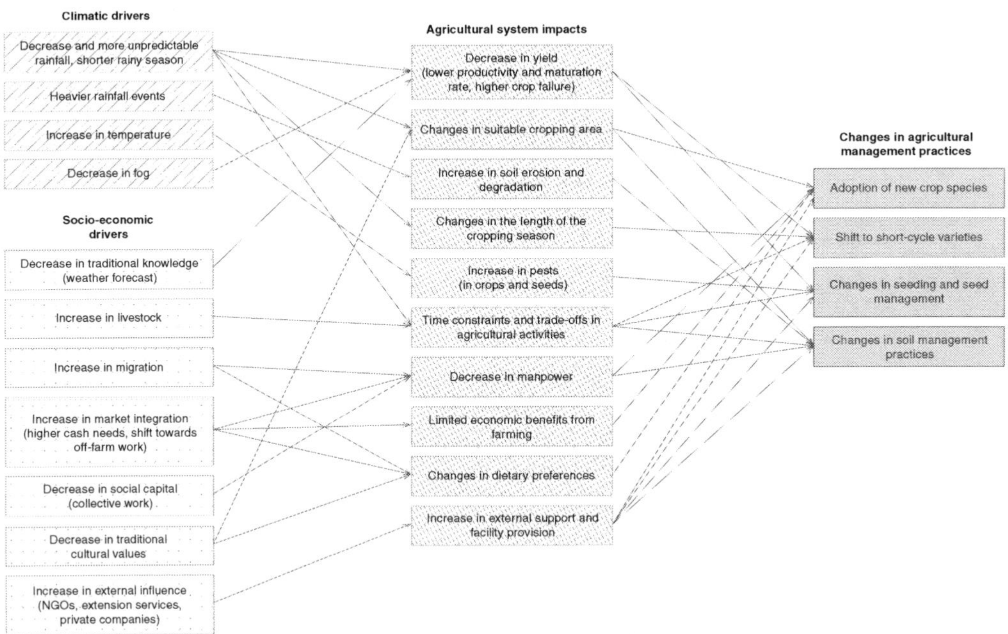

Figure 13.3 Conceptual map showing multiple drivers and impacts on the Bassari agricultural system and resulting changes in agricultural management practices.

A Horticulture has been promoted as an alternative cash crop cultivation
B Many farmers prefer the paddy rice cultivation before traditional crops
C Oxen-ploughing allows more timely and fast cultivation for flat cultivation plots
D The harvest of a cotton and a sorghum field. Farmers often abandon sorghum in favour of the cash crop cotton
E + F Farmers tend to abandon cultivation in the hills (E) to farm in the flat lower areas (F)
G The manufacturing and selling of shea butter has become an important off-farm income alternative

Figure 13.4 Examples of changes in Bassari agricultural practices as a response to different impacts on the agricultural system. Pictures by B. Klappoth, 2019–2020.

have been in the Bassari farming system for a longer time, although their cultivation has experienced a dramatic increase in the last decades. As a direct consequence of adopting new crops, farmers have reduced or even abandoned the cultivation of traditional crops. Due to the higher water and soil fertility requirements of maize, rice, and horticulture, the switch from traditional to new crops also entailed a change in the cultivation location, from hilly areas to the plains (see Porcuna-Ferrer et al., 2024).

Drivers

The adoption of maize and rice involved a mix of economic, social, and cultural reasons. Farmers justified the adoption of maize and rice arguing that these crops provide high yields and require less labour than traditional crops. Among the Bassari, agricultural labour is becoming a limiting factor due to off-farm migration in search for jobs that provide cash income, the decrease in the number of common agricultural working days, and the increase of children schooling (who traditionally were an important help in the fields). Moreover, development projects (in coordination with the Senegalese government) – which intensified their presence in the Bassari area in the recent decades – play a key role in the introduction of new high-yielding rice and maize varieties. Dietary preference for maize and rice have also contributed to the adoption of these crops.

Farmers mentioned mainly economic and social considerations for the adoption of cotton and horticulture. Increased cash needs and the role of external agents – NGOs for the introduction of horticulture and a state enterprise (later privatised) for cotton – were mentioned as the main drivers of cotton and horticulture adoption. Nowadays, cotton is the main source of agricultural income and all interviewed farmers referred to the need to cover the rising costs of living – for example, children's schooling, food, and access to health services –, as well as the rising demands for commercial goods like smartphones or motorbikes. In the discussions with women farmers, they mentioned that horticulture has become increasingly popular because the facilities provided by NGOs – for example, seeds and materials for fencing – make the commercialization of vegetables profitable. None of the farmers interviewed mentioned that their decisions to adopt new crop species reflect a response to climatic stress.

Costs, benefits, and trade-offs

The need to change the planting location was the most frequently mentioned trade-off derived from rice and maize adoption. Given their water requirements, rice and maize cannot be cultivated in the hills, like Bassari traditional staple crops. This shift aggravates land scarcity and promotes unequal land access, also reinforced by population growth and cultural changes in the land tenure system, which has shifted from commonly managed user rights defined by kin groups to private property. The cultivation of maize and rice lowers the value of hilly land and intensifies land pressure in the plains. Bassari recognised that "*[Now] the fields are smaller, and it is difficult to find land in the plains. (…) We are many more people!*" (Anonymous, female farmer, February 2020). In the past, higher-status and wealthier households preferred hilly land. The less desirable plains – characterised by a dense vegetation, soils non-suitable for traditional crops, and protected by cultural taboos – were only used by households settled in the village margins or by ethnic groups who arrived later to the area (e.g., Pular). Nowadays, only those households who traditionally inhabited or cultivated the plains, or very wealthy households who can use their financial and social status to borrow land from others, can cultivate the plains and therefore adopt new and higher-yielding crop species. Interviewees expressed additional concern about the expansion of agriculture into sacred and protected areas like the river basins with dense riparian forests, where, in the past, taboos banned cutting the trees.

Farmers claimed that rice and maize have high water requirements that make them generally more susceptible to dry spells than their traditional crops. However, they also mentioned this drawback to be compensated through the higher water retention capacity of the soils in the plains, which reduces the risk of water stress that occurs in hilly areas. Farmers also referred to the high needs for fertilisers and pesticides as a trade-off for the adoption of rice and maize. According to farmers, the growing demands on land limit the use of traditional crop rotation and fallow periods, which leads to declining soil fertility and increasing soil degradation. Consequently, as fields in the plains are rarely rotated or left fallow and as the new crops have high soil nutrient requirements, regular fertilisation is needed. Some farmers also noticed that the reduction of fallow periods and the lower quality of purchased seeds leads to a greater presence of pests, which requires the application of more pesticides.

For horticulture, farmers reported problems with water scarcity and conflicting interests between household consumption and irrigation needs. This problem was mainly presented by women, who also saw an increase in their work burden due to water scarcity: "*Before sunrise is when there is more water. If you want to irrigate your crops, you need to wake up very early*" (Anonymous, female farmer, March 2020). Horticultural activities also generate disparities, since only some neighbourhoods have access to land suitable for horticulture. As a woman farmer explained: "*I would cultivate tomatoes, onions, and lettuce if I had a place*" (Anonymous, female farmer, March 2020).

Cotton has high requirements for fertilisers and pesticides, representing a form of intensified agriculture that demands more labour to fumigate and fertilise plots, and additional cash to purchase external inputs, thereby creating higher market dependence. Cotton is mostly cultivated by men, who also control the income from cotton sales, which is often used to buy status items (e.g., smartphone, motorbike) and – according to women interviewed – not invested in family expenditures. Hence, the shift towards cotton cultivation had negative implications for women's control over household income and decision-making.

During the 'cotton boom' in the early 2000s, many families almost entirely switched to cotton, reducing the area dedicated to self-consumption. However, in the face of yield failure, low cotton market prices, or unexpected cash needs (e.g., in case of illness), these households became food insecure. Nowadays, most households keep a balance between cotton and subsistence crops. Still, big shares of land and labour are put into cotton cultivation and the proportion devoted to subsistence crops has declined. To compensate for the lack of cultivated subsistence crops, women search for alternative strategies to ensure household food security (e.g., through off-farm work), while at the same time supporting cotton cultivation. In other words, cotton cultivation competes with subsistence crops for land and labour, thereby challenging family food needs and increasing women's work burden.

Shift to short-cycle varieties

The shift from long- to short-cycle varieties was a change commonly mentioned by Bassari farmers and affecting maize, rice, peanut, and fonio, for which short-cycle varieties are available. For sorghum, farmers mentioned adopting a medium-cycle variety. No short- or medium-cycle varieties are available for Bambara groundnut.

Drivers

Farmers reported climatic stressors, scheduling conflicts with other activities (i.e., livestock rearing and off-farm work) and external influences as the main drivers of the shift to short-cycle varieties. Farmers considered traditional long-cycle varieties vulnerable to the changing climate, especially regarding unpredictable rainfall, shorter rainy seasons, and more frequent dry spells. They also explained that in the past, the rainy season lasted from May to October and allowed the cultivation of long-cycle crops, whereas nowadays the rain lasts from mid-June to September and only allows for short-cycle crops and varieties. Respondents also mentioned that in the past communities relied on fog at the end of the rainy season for the final maturation of long-cycle crops, like certain sorghum varieties. Nowadays, however, fog is scarce and too unstable to ensure the final maturation of long-cycle crops.

Farmers also referred to trade-offs between long-cycle varieties and livestock rearing. Free-ranging livestock is kept in communal land during the dry season and at the end of the growing season, when crops are harvested. This practice increases the risk of crop damage by grazing livestock, especially for long-cycle varieties. Farmers also mentioned that the shorter growing cycles fit better with the seasonality of off-farm jobs.

Finally, farmers mentioned that the adoption of short-cycle rice, maize, and peanut varieties was supported and subsidised by NGOs and government development programs, which increases profits by reducing investment costs.

Costs, benefits, and trade-offs

Farmers attributed positive traits to some of the traditional long-cycle varieties that are being lost, which were perceived as more productive, less vulnerable to pests and weeds, more robust to low soil fertility, more durable when stored, and more tasty and nutritious.

Changes in seeding and seed management

Farmers reported three main changes regarding seeding and seed management: (1) changing seed storage techniques by applying chemical pesticides, especially for legume seeds; (2) shifting the planting calendar from the beginning of June to the beginning of July; and (3) re-seeding more often due to the increasing difficulty to find the right moment to seed.

Drivers

Farmers argued that changes in seed storage practices were a response to higher seed degradation from pest infestations due to temperature increase. Pest infestations particularly affected legume crops like peanut and Bambara groundnut. As one farmer explained: "*There are too many insects in the seeds. Before you could store them from one cropping season to the next. Nowadays if you leave it, at the moment of seeding you will just find powder*" (Anonymous, female farmer, January 2020). They acknowledged that the new seed storage practice was initially promoted by NGOs, who used to distribute chemical pesticides for free, although now farmers have to buy them.

Farmers explained the shift in planting schedules as a response to the delayed and shorter rainy season and the higher frequency of dry spells. Farmers described that they cannot seed when the first rain arrives, as they used to do, because no other rains follow, and the germinated seeds dry out: "*Sometimes you organise a big common agricultural working day hoping that the first rains will follow soon, but if the rain does not arrive, you lose everything*" (Anonymous, male farmer, January 2020). At the same time, the shorter rainy season forces farmers to harvest earlier, compared to the past. Seeding later but harvesting earlier makes shifting to short-cycle varieties necessary. Otherwise, crops do not reach maturity before the end of the rainy season, resulting in significant yield losses. As one middle-age male farmer said: "*If you seed too early, you risk that the seeds will dry out in the field; but if you seed too late, crops will not reach maturity*" (Anonymous, male farmer, February 2020).

Unpredictable rainfall, the decreased reliability of traditional indicators of weather forecast, and the loss of weather forecasting knowledge were all considered factors increasing the difficulty to identify the right time to seed.

Costs, benefits, and trade-offs

Additional costs, including risks, were mentioned in relation to changes in seed storage methods. Since NGOs do not freely distribute chemical pesticides anymore, farmers with less economic resources cannot afford buying them and continue to apply traditional forms of seed storage (e.g., mixing seeds with ashes and certain wild plants that keep insects away). Moreover, some farmers complained about the toxic effects of chemical pesticides on human health and refused to use them.

Altogether, there are more costs in accessing and storing seeds, including an increased risk of seed losses during storage and a large dependency on external seed sources (e.g., markets, NGOs). For example, several women farmers mentioned incidents of spurious seeds or of seeds that did not correspond to the announced crop variety. Some farmers also attributed pest infestations to the purchase of seeds from markets or unreliable sources.

Changes in soil management practices

Farmers mentioned two main changes related to soil management practices: the shift from the hand hoe to the oxen-plough and the construction of stone-bands and half-terraces on fields.

Drivers

When asked about the reasons for shifting from the hand hoe to the oxen-plough, most farmers referred first to reduced workload and lower labour demand. The oxen-plough was also related to the increased cultivation in the plains, which allows more efficient ways of ploughing the soil. Several farmers additionally mentioned that owning an oxen-plough was beneficial under current unpredictable rainfall conditions, since the oxen-plough allows quick soil preparation and seeding, thereby offering farmers more flexibility, compared to the hand-hoe. The main reason for farmers to place stone-bands or trunks along the contours of hilly fields, or building half-terraces, was the prevention of soil erosion in response to more frequent flash floods. Both measures were introduced by NGOs and development projects, which made oxen-plough available and trained farmers on the use of stone-bands and terraces on hilly terrains.

Costs, benefits, and trade-offs

Farmers mentioned that the benefits of the oxen-plough are not accessible to everybody. For example, a middle-aged woman explained:

> *If you own an oxen-plough, as soon as the rain comes, you seed. But if you do not own it, then you will have problems, you need to wait for your turn* [the oxen-plough is usually shared in exchange for work, some people also rent it] *and sometimes when they come to plough your field, it is too late.*
>
> (Anonymous, female farmer, February 2020)

Regarding the construction of stone-bands or terraces on hilly fields, interviewed farmers agreed that labour and time constraints were the main factors that hindered the wider implementation of this practice.

Discussion

We structure the discussion around the three main findings: (a) farmers adapt their agricultural management practices in response to multiple stressors; (b) changes in agricultural management practices imply trade-offs among environmental, social, and economic factors; and (c) the distribution of costs and benefits arising from adaptation varies across social groups.

Changes in agricultural management practices are driven by multiple stressors

Changes in farming practices such as changing crop species and varieties and market-based horticulture have been documented and conceptualised as "climate change adaptation strategies" in Senegal (Mertz et al., 2009; Ruggieri et al., 2021) and the world (Schlingmann et al., 2021).

However, the predominant focus on climate change as a main driver of change masks the socio-political root causes of household and individual vulnerability (Ribot, 2014). As our case study shows, farmers do not only switch to shorter-cycle varieties due to the shortening of the cropping season, but also due to labour constraints, higher yields, and dietary changes. In this regard, drivers of change in Bassari agricultural management practices are multifactorial, with socio-economic, political, and cultural stressors being as salient as climate change in guiding farmers' decisions. Other studies have also documented that climate change is not necessarily the main or unique stressor driving livelihood changes in local communities (Nyantakyi-Frimpong & Bezner-Kerr, 2015). Multiple, compounding, interacting, and intertwined stressors that are deeply entangled in the integration of smallholder farmers' communities into globalisation and capitalist economies – such as new economic opportunities and cash needs, population growth, and the weakening of cultural norms and traditions – strongly determine community's and households' vulnerability and are perceived to be equally or even more relevant than climatic stressors (Ensor et al., 2019).

Still, the fact that non-climatic challenges are currently perceived as more significant for local farmers does not downplay the need for adaptation to the mounting threats of climate change. In the light of our findings and in line with McDowell and Hess (2012) and Izquierdo and Schlingmann (2024), we argue for the importance of designing climate change adaptation interventions that allow farmers to better confront multiple stressors according to their own needs and priorities.

Costs and benefits of changes in agricultural management practices result in trade-offs between environmental, social, and economic factors

Our results show that changes in Bassari agricultural management practices entail different costs under the different sustainability spheres, which can have contradicting and sometimes unwanted trade-offs for local communities, leading to maladaptive outcomes and increased vulnerability.

A response that is beneficial with respect to one stressor might be insufficient or ineffective, or even conflict with addressing other stressors (Bennett et al., 2016). For example, a response to new economic conditions does not necessarily increase climate resilience and a response to climatic stressors is not necessarily economically beneficial or viable, as it happens when there is a mismatch between climate-compatible crops and market-driven demand for those crops (O'Brien et al., 2004). Our case study illustrates trade-offs between environmental, social, and economic costs that can lead to maladaptive outcomes. For example, the increase of marketable horticulture and cotton cultivation (short-term economic benefits) contributes to the abandonment of traditional crops, like sorghum, fonio, or Bambara groundnut, which are generally better adapted to local environmental conditions and droughts (environmental costs) (Abrouk et al., 2020; Hadebe et al., 2017; Mayes et al., 2019). Switching to cash-crops can improve households' income (economic benefits), but the additional need for irrigation (i.e., horticulture) can put stress on water resources (environmental costs) and potentially lead to conflicts because of different demands and interests (social costs) (Akinyi et al., 2021; Antwi-Agyei et al., 2018). Similarly, an increasing dependence on research-improved drought-resistant short-cycle seeds of introduced crops (environmental benefits), increases households' vulnerability to market uncertainties and price fluctuations (economic costs) (Galappaththi & Schlingmann, 2023). Introduced crops also tend to be more susceptible to pests, resulting in frequent application of chemical pesticides with risks to human health (social costs), environmental degradation (environmental costs), and additional cash needs (economic costs) (Akinyi et al., 2021; Dhakal & Kattel, 2019). Growing a mixture of traditional crops would allow households to have effective strategies against food insecurity in the face of reduced landholdings, changing rainfall patterns, and decreased soil fertility (environmental and social benefits). However, growing traditional crops implies high labour requirements and low income, since they are not yet integrated into market logics (economic costs) (Galappaththi and Schlingmann, 2023).

In this regard, adaptation to the most impacting stressors in the short term can limit long-term adaptation options and climate resilience when they deliver maladaptive outcomes for one of the different sustainability spheres (McDowell & Hess, 2012; Porcuna-Ferrer et al., under review). Only when farmers' responses reduce harm in all three spheres, they can be considered sustainable adaptation (Eriksen et al., 2011; Wilson, 2014), which emphasises the need for research and policy to design and evaluate climate change adaptation options by considering their trade-offs between environmental, social, and economic spheres in order to avoid lock-ins that could increase future climate vulnerabilities.

Costs and benefits of changes in agricultural management practices are unequally distributed across social groups

Another important finding of our research is that changes in agricultural management practices entail different costs, benefits, and trade-offs for different social groups, with the least vulnerable bearing the benefits and the most vulnerable (i.e., women and poor households) often bearing the costs. Gender or access to financial, physical, and natural capitals define who in the community or within a household can implement and benefit from certain changes in agricultural management practices. This is specifically the case of the usage of oxen ploughing, cotton cultivation, horticulture, and the relocation of fields to the plains, which increased the benefits of some groups and the costs of others. For example, in the study area, the adoption of the oxen-plough mostly benefited a very small number of wealthy households with direct access to an oxen-plough. This is in line with other studies that highlight how unequal access to physical and natural assets affects the scheduling of seasonal agricultural activities of poor households (Roncoli et al., 2001). Regarding cotton cultivation, underlying inequities between men and women in cash access and decision-making influenced their capacity to benefit from cotton expansion, which points to the importance of intra-household power dynamics to access adaptation options – see also, Ravera et al. (2016a). Moreover, our results add empirical evidence to the literature that argues that the implementation of certain adaptation strategies, besides reinforcing pre-existing inequalities, can also entail the renegotiation of local power relations (Ravera et al., 2016b). This is the case for horticulture and the relocation of agricultural fields to the valleys where differential land access between households defined their adaptive capacity and thereby created new inequalities. For example, market-oriented horticulture of water-demanding vegetables enhanced the climatic sensitivity of women with limited access to land with wells, as confirmed by other studies (e.g., Labeyrie et al., 2021).

While the analysis of how the intersection of power and social relations shapes adaptation processes receives increasing scholarly attention (e.g., Kaijser & Kronsell, 2014; Ravera et al., 2016a), it still remains largely unexplored for rural farming communities of the Global South (see Onta & Resurrección 2011 and Carr & Thomson, 2014 for exceptions). Results from our case study reinforce the notion that power-dynamics largely depend on context-specific socio-economic and biophysical characteristics that either catalyse or constraint farmers' adaptation options (Carr & Thompson, 2014; Kaijser & Kronsell, 2014; Thompson-Hall et al., 2016). Our results also offer empirical insights into how power dynamics are not fixed but changing under new conditions as people adapt to change. These findings thus make clear that adaptation is not an homogenous process that equally benefits all community members and highlight the importance of considering how multiple dimensions of social identity interact and jointly influence how farmers differently experience, manage, and benefit from changes in agricultural practices (Ravera et al., 2016b). We argue for the need for grounded power-sensitive approaches as a first step before any intervention to foster climate change adaptation at the local level.

Conclusion

Bassari agricultural system and management practices are impacted and steered by various stressors, from climate change to increased land-scarcity, increasing monetization of the economic system, and changes in social capital and cultural norms. Bassari farmers are responding to those multiple and simultaneously occurring changes by modifying their farming practices. The analysis of changes in agricultural management practices from a 'sustainable adaptation' perspective (*sensu* Eriksen et al., 2011) and using an intersectional approach shows that not every change can be considered a sustainable adaptation, nor equally beneficial for all social groups.

From these results, we derive two important conclusions. First, changes in agricultural management practices have multiple trade-offs derived from different environmental, social, and economic costs: what in the short-term seems a good adaptation option to one stressor can in fact erode access to important assets and therefore, lower farmers' future adaptive capacity to respond to other stressors. An overemphasis of policies on market demand as the only or most important stressor affecting subsistence farmers may deviate attention from other equally important stressors (e.g., climate change), overall resulting in increased vulnerability. Second, our results offer new empirical evidence about the underlying contextualised factors that shape response options and outcomes for different social groups. The changes in agricultural management practices implemented by Bassari farmers led to benefits for some groups, but to costs to others, increasing the burden of the most vulnerable (i.e., women and poor households). We argue that it is important to include power and gender analysis as a first step for any intervention aiming at fostering local adaptation.

Our arguments aim to put the threats posed by climate change on agricultural systems in the context of the multiple cultural, environmental, political, and economic dynamics that intersect with them shaping the life and practices of smallholder farmers of the Global South. Understanding farmers' adaptation processes in their broader context, paying attention to additional risks, unequally distributed costs, and long-term trade-offs allows us to understand climate change impacts within the lived experiences of local communities, and help design adaptation strategies that efficiently reduce vulnerability to climate change and other impacts. Climate change policies should prioritise long-term and multi-beneficial adaptation measures, while avoiding responses that reproduce existing inequalities and undermine farmers' long-term resilience. Adaptation policies should be designed from a holistic intersectional approach that considers multiple-stressors, trade-offs, and power-dynamics.

Acknowledgements

We thank all the Bassari farmers, too numerous to mention, for sharing with us their understanding of the environment and changes in local agricultural systems. We thank Susanne Bonang, Albert Yera Bonang, Jules Yera Bidiar, Pascal Indega Bindia, Dominique Thiaroly Bindia, Joseph Indega Bindia, Henri Sike Bindia, and Théophil Kaly Boubane for all their help and assistance during fieldwork. We thank Victoria Reyes-García, Vanesse Labeyrie, Ndeye Fatou Faye, and Laura Calvet-Mir for invaluable support and shared reflections, pre- and post- fieldwork. We thank Victoria Reyes-García,

Priyatma Singh and Emmanuel M.N.A.N. Attoh for comments on a previous version of the chapter and Eva Porcuna Ferrer for assistance with Figure 13.2. The research leading to these results received funding from the European Research Council under an ERC Consolidator Grant (FP7-771056-LICCI) and contributes to the "María de Maeztu Unit of Excellence" (CEX2019-000940-M).

References

Abrouk, M., Ahmed, H. I., Cubry, P., Šimoníková, D., Cauet, S., Pailles, Y., … & Krattinger, S. G. (2020). Fonio millet genome unlocks African orphan crop diversity for agriculture in a changing climate. *Nature Communications*, *11*(1), 1–13. https://doi.org/10.1038/s41467-020-18329-4

Akinyi, D. P., Karanja Ng'ang'a, S. & Girvetz, E. H. (2021). Trade-offs and synergies of climate change adaptation strategies among smallholder farmers in sub-Saharan Africa: A systematic review. *Regional Sustainability*, *2*(2), 130–143. https://doi.org/10.1016/j.regsus.2021.05.002

Altieri, M. A. & Nicholls, C. I. (2017). The adaptation and mitigation potential of traditional agriculture in a changing climate. *Climatic Change*, *140*(1), 33–45. https://doi.org/10.1007/s10584-013-0909-y

ANACIM. (2020). *National Weather Data, Agence Nationale de l'Aviation Civile et de la Météorologie*. ANACIM. https://www.anacim.sn/

Antwi-Agyei, P., Dougill, A. J., Stringer, L. C. & Codjoe, S. N. A. (2018). Adaptation opportunities and maladaptive outcomes in climate vulnerability hotspots of northern Ghana. *Climate Risk Management*, *19*(November 2017), 83–93. https://doi.org/10.1016/j.crm.2017.11.003

Barnett, J. & O'Neill, S.J. (2010). Maladaptation. *Global Environmental Change, 20*, 211–213. https://doi.org/10.1016/j.gloenvcha.2009.11.004

Bennett, N. J., Blythe, J., Tyler, S. & Ban, N. C. (2016). Communities and change in the anthropocene: Understanding social-ecological vulnerability and planning adaptations to multiple interacting exposures. *Regional Environmental Change*, *16*(4), 907–926. https://doi.org/10.1007/s10113-015-0839-5

Bezner Kerr, R., Naess, L. O., Allen-O'Neil, B., Totin, E., Nyantakyi-Frimpong, H., Risvoll, C., … & Eriksen, S. (2022). Interplays between changing biophysical and social dynamics under climate change: Implications for limits to sustainable adaptation in food systems. *Global Change Biology*, *28*(11), 3580–3604. https://doi.org/10.1111/gcb.16124

Carr, E. R. & Thompson, M. C. (2014). Gender and climate change adaptation in agrarian settings: Current thinking, new directions, and research frontiers. *Geography Compass*, *8*(3), 182–197. https://doi.org/10.1111/gec3.12121

Carr, T. W., Mkuhlani, S., Segnon, A. C., Ali, Z., Zougmoré, R., Dangour, A. D., … & Scheelbeek, P. (2022). Climate change impacts and adaptation strategies for crops in West Africa: A systematic review. *Environmental Research Letters*, *17*(5). https://doi.org/10.1088/1748-9326/ac61c8

Dhakal, B. & Kattel, R. R. (2019). Effects of global changes on ecosystems services of multiple natural resources in mountain agricultural landscapes. *Science of the Total Environment*, *676*, 665–682. https://doi.org/10.1016/j.scitotenv.2019.04.276

Ensor, J. E., Wennström, P., Bhatterai, A., Nightingale, A. J., Eriksen, S. & Sillmann, J. (2019). Asking the right questions in adaptation research and practice: Seeing beyond climate impacts in rural Nepal. *Environmental Science and Policy*, *94*(September 2018), 227–236. https://doi.org/10.1016/j.envsci.2019.01.013

Eriksen, S., Aldunce, P., Bahinipati, C. S., Martins, R. D. A., Molefe, J. I., Nhemachena, C., … & Ulsrud, K. (2011). When not every response to climate change is a good one: Identifying principles for sustainable adaptation. *Climate and Development*, *3*(1), 7–20. https://doi.org/10.3763/cdev.2010.0060

Galappaththi, E. K. & Schlingmann, A. (2023). The sustainability assessment of Indigenous and local knowledge-based climate adaptation responses in agricultural and aquatic food systems. *Current Opinion in Environmental Sustainability, 62*, 101276. https://doi.org/10.1016/j.cosust.2023.101276

Hadebe, S. T., Modi, A. T. & Mabhaudhi, T. (2017). Drought tolerance and water use of cereal crops: A focus on sorghum as a food security crop in Sub-Saharan Africa. *Journal of Agronomy and Crop Science*, *203*(3), 177–191. https://doi.org/10.1111/jac.12191

Izquierdo, A. & Schlingmann, A. (2024). Local perceptions of climate change in the context of socioeconomic and political changes in a High Andean community from the Argentine Puna. In V. Reyes-García, S. Álvarez-Fernández, P. Benyei, L. Calvet-Mir, D. García-del-Amo, A. B. Junqueira, X. Li, V. Porcher, A. Porcuna-Ferrer, A. Schlingmann, & R. Soleymani (Eds.), *Routledge Handbook of Climate Change Impacts on Indigenous Peoples and Local Communities*. Oxfordshire & New York: Routledge. Chp 8, pp 136-150.

Kaijser, A. & Kronsell, A. (2014). Climate change through the lens of intersectionality. *Environmental Politics*, *23*(3), 417–433. https://doi.org/10.1080/09644016.2013.835203

Labeyrie, V., Renard, D., Aumeeruddy-Thomas, Y., Benyei, P., Caillon, S., Calvet-Mir, L., … & Reyes-García, V. (2021). The role of crop diversity in climate change adaptation: Insights from local observations to inform decision making in agriculture. *Current Opinion in Environmental Sustainability*, *51*, 15–23. https://doi.org/10.1016/j.cosust.2021.01.006

Mayes, S., Ho, W. K., Chai, H. H., Song, B., Chang, Y. & Massawe, F. (2019). Bambara groundnut (*Vigna subterranea* (L) Verdc): A climate smart crop for food and nutrition security. In C. Kole (Ed.), *Genomic Designing of Climate-Smart Pulse Crops* (pp. 397–424). https://doi.org/10.1007/978-3-319-96932-9

McDowell, J. Z. & Hess, J. J. (2012). Accessing adaptation: Multiple stressors on livelihoods in the Bolivian highlands under a changing climate. *Global Environmental Change*, *22*(2), 342–352. https://doi.org/10.1016/j.gloenvcha.2011.11.002

Mertz, O., Mbow, C., Reenberg, A. & Diouf, A. (2009). Farmers' perceptions of climate change and agricultural adaptation strategies in rural Sahel. *Environmental Management*, *43*(5), 804–816. https://doi.org/10.1007/s00267-008-9197-0

Nolan, R. W. (1986). *Bassari Migrations: The Quiet Revolution*. https://doi.org/10.1017/CBO9781107415324.004

Nyantakyi-Frimpong, H. & Bezner-Kerr, R. (2015). The relative importance of climate change in the context of multiple stressors in semi-arid Ghana. *Global Environmental Change*, *32*, 40–56. https://doi.org/10.1016/j.gloenvcha.2015.03.003

Nyong, A., Adesina, F. & Osman Elasha, B. (2007). The value of indigenous knowledge in climate change mitigation and adaptation strategies in the African Sahel. *Mitigation and Adaptation Strategies for Global Change*, *12*(5), 787–797. https://doi.org/10.1007/s11027-007-9099-0

O'Brien, K., Leichenko, R., Kelkar, U., Venema, H., Aandahl, G., Tompkins, H., … & West, J. (2004). Mapping vulnerability to multiple stressors: Climate change and globalization in India. *Global Environmental Change*, *14*(4), 303–313. https://doi.org/10.1016/j.gloenvcha.2004.01.001

Onta, N. & Resurreccion, B. P. (2011). The role of gender and caste in climate adaptation strategies in Nepal: Emerging change and persistent inequalities in the far-western region. *Mountain Research and Development*, *31*(4), 351–356. https://doi.org/10.1659/MRD-JOURNAL-D-10-00085.1

Porcuna-Ferrer, A., Calvet-Mir, L., Guillerminet, T., Alvarez-Fernandez, S., Labeyrie, V., Porcuna-Ferrer, E., & Reyes-García, V. (2023a). "So many things have changed": Situated understandings of climate change impacts among the Bassari, south-eastern Senegal. *Environmental Science & Policy*, *148*, 103552. https://doi.org/10.1016/j.envsci.2023.103552

Porcuna-Ferrer, A., Labeyrie, V., Alvarez-Fernandez, S., Calvet-Mir, L., Faye, N. F., Ouadah, S., & Reyes-García, V. (2023b). Crop biocultural traits shape seed networks: Implications for social-ecological resilience in south eastern Senegal. *Agricultural Systems*, *211*, 103750. https://doi.org/10.1016/j.agsy.2023.103750

Porcuna-Ferrer A., Calvet-Mir L., Faye N.F., Klappoth B., Reyes-García V., Labeyrie V. 2024. Drought-tolerant indigenous crop decline in the face of climate change: A political agroecology account from south-eastern Senegal. Journal of Rural Studies.

Räsänen, A., Juhola, S., Nygren, A., Käkönen, M., Kallio, M., Monge, A. & Kanninen, M. (2016). Climate change, multiple stressors and human vulnerability: A systematic review. *Regional Environmental Change*, *16*(8), 2291–2302. https://doi.org/10.1007/s10113-016-0974-7

Ravera, F., Iniesta-Arandia, I., Martín-López, B., Pascual, U. & Bose, P. (2016a). Gender perspectives in resilience, vulnerability and adaptation to global environmental change. *Ambio*, *45*, 235–247. https://doi.org/10.1007/s13280-016-0842-1

Ravera, F., Martín-López, B., Pascual, U. & Drucker, A. (2016b). The diversity of gendered adaptation strategies to climate change of Indian farmers: A feminist intersectional approach. *Ambio*, *45*, 335–351. https://doi.org/10.1007/s13280-016-0833-2

Reyes-García, V., García-del-Amo, D., Benyei, P., Fernández-Llamazares, Á., Gravani, K., Junqueira, A. B., … & Soleymani-Fard, R. (2019, August 1). A collaborative approach to bring insights from local observations of climate change impacts into global climate change research. *Current Opinion in Environmental Sustainability*, *39*, 1–8. https://doi.org/10.1016/j.cosust.2019.04.007

Reyes-García, V., Álvarez-Fernández, S. A., Benyei, P., García-del-Amo, D., Junqueira, A. B., Labeyrie, V., Li, X., Porcher, V., Porcuna-Ferrer, A., Schlingmann, A. & Soleymani, R. (2023). Local indicators of climate change impacts described by indigenous peoples and local communities: Study protocol. *PLoS One*, *18*(1), e0279847. https://doi.org/10.1371/journal.pone.0279847.

Ribot, J. (2014). Cause and response: vulnerability and climate in the Anthropocene. *Journal of Peasant Studies, 41*(5), 667–705. https://doi.org/10.1080/03066150.2014.894911

Rivera-Ferre, M. G., Di Masso, M., Vara, I., Cuellar, M., López-i-Gelats, F., Bhatta, G. D. & Gallar, D. (2021). Traditional agricultural knowledge in land management: The potential contributions of ethnographic research to climate change adaptation in India, Bangladesh, Nepal, and Pakistan. *Climate and Development*, *13*(7), 644–661. https://doi.org/10.1080/17565529.2020.1848780

Roncoli, C., Ingram, K. & Kirshen, P. (2001). The costs and risks of coping with drought: Livelihood impacts and farmers' responses in Burkina Faso. *Climate Research*, *19*(2), 119–132. https://doi.org/10.3354/cr019119

Ruggieri, F., Porcuna-Ferrer, A., Gaudin, A., Faye, N. F., Reyes-García, V. & Labeyrie, V. (2021). Crop diversity management: Sereer smallholders' response to climatic variability in Senegal. *Journal of Ethnobiology*, *41*(3), 389–408. https://doi.org/10.2993/0278-0771-41.3.389

Savo, V., Lepofsky, D., Benner, J. P., Kohfeld, K. E., Bailey, J. & Lertzman, K. (2016). Observations of climate change among subsistence-oriented communities around the world. *Nature Climate Change*, *6*(5), 462–473. https://doi.org/10.1038/nclimate2958

Schlingmann, A., Graham, S., Benyei, P., Corbera, E., Martinez Sanesteban, I., Marelle, A., ... & Reyes-García, V. (2021). Global patterns of adaptation to climate change by indigenous peoples and local communities. A systematic review. *Current Opinion in Environmental Sustainability*, *51*, 55–64. https://doi.org/10.1016/j.cosust.2021.03.002

Segnon, A. C., Totin, E., Zougmoré, R. B., Lokossou, J. C., Thompson-Hall, M., Ofori, B. O., ... & Gordon, C. (2021). Differential household vulnerability to climatic and non-climatic stressors in semi-arid areas of Mali, West Africa. *Climate and Development*, *13*(8), 697–712. https://doi.org/10.1080/17565529.2020.1855097

Sultan, B. & Gaetani, M. (2016). Agriculture in West Africa in the twenty-first century: Climate change and impacts scenarios, and potential for adaptation. *Frontiers in Plant Science*, *7*, 1262. https://doi.org/10.3389/fpls.2016.01262

Sultan, B. & Janicot, S. (2003). The West African monsoon dynamics. Part II: The "preonset" and "onset" of the summer monsoon. *Journal of Climate*, *16*(21), 3407–3427. https://doi.org/10.1175/1520-0442(2003)016<3407:TWAMDP>2.0.CO;2

Thompson-Hall, M., Carr, E. R. & Pascual, U. (2016). Enhancing and expanding intersectional research for climate change adaptation in agrarian settings. *Ambio*, *45*(s3), 373–382. https://doi.org/10.1007/s13280-016-0827-0

Tschakert, P. (2007). Views from the vulnerable: Understanding climatic and other stressors in the Sahel. *Global Environmental Change*, *17*(3–4), 381–396. https://doi.org/10.1016/j.gloenvcha.2006.11.008

UNESCO. (2011). *Pays Bassari Paysages Culturels Bassari, Peul et Bédik Proposition – Proposition d'inscription sur la Liste du patrimoine mondial*. UNESCO. https://whc.unesco.org/fr/list/1407/

Wilson, G. A. (2014). Community resilience: Path dependency, lock-in effects and transitional ruptures. *Journal of Environmental Planning and Management*, *57*(1), 1–26. https://doi.org/10.1080/09640568.2012.741519

14

CHANGING TERRAIN

Evidence of climate change impacts and adaptive responses of Dagbani Indigenous communities, northern Ghana

Emmanuel M.N.A.N. Attoh, Ruddy Afriyie, Gordana Kranjac-Berisavljevic, Enoch Bessah and Fulco Ludwig

Introduction

Climate change remains a global issue creating environmental, social, and economic problems at different levels and scales (Mendelsohn et al., 2006; Popoola et al., 2020). Climate change has severely affected the livelihoods of people, especially those who depend heavily on the natural environment and/or have a low capacity to adapt (Tirivangasi and Nyahunda, 2019). Therefore, efforts are aimed at assisting vulnerable populations to adapt to the changing climate (Pörtner, 2022).

Sub-Saharan Africa is one of the most vulnerable regions, very susceptible to the impacts of climate change due to the overreliance of its economy on climate-sensitive activities and its very low adaptive capacity due to acute poverty levels (AGRA, 2018; Di Falco, 2018). The region highly depends on rainfall to provide water, food, and energy to its inhabitants (Ofori et al., 2021). The prevalence of rain-fed agriculture in most of sub-Saharan Africa makes its food systems highly sensitive to the changing rainfall (Thompson et al., 2010). High poverty rates, conflicts, poor management of natural resources, and low socioeconomic development also increase the region's vulnerability (Maja and Ayano, 2021). Meanwhile, temperatures on the African continent are projected to exceed the global average (with an increased average of 3–4°C). Changes in precipitation are also expected in the 21st century through to the 22nd century (Niang et al., 2014; Thompson et al., 2010).

Changes in the climatic conditions of sub-Saharan Africa have led to many environmental problems, such as an increased infestation of pests and diseases, household asset depletion, rural-urban migration, biodiversity loss, and ecosystem degradation (Abaje and Giwa, 2007; Muli et al., 2018; Theunissen, 2018). Also, climate change has resulted in extreme weather events, such as flooding, droughts, and soil erosion leading to a drastic reduction in crop yields and rising prices of food commodities (Bedeke, 2022; Ibe and Amikuzuno, 2019; Wossen et al., 2018).

The impact of climate change has led to the formulation of measures to help the vulnerable adapt (Bahadur et al., 2013; Berkes et al., 2013). The success of these measures at the local level depends on a deeper understanding of smallholders' perception of climate change impacts (Acquah et al., 2015). An incomplete assessment of climate change impacts that does not consider the conditions of local inhabitants poses a major problem for effective adaptation (Stott and Kettleborough, 2002). Consequently, climate change adaptation and policy formulation processes are

DOI: 10.4324/9781003356837-19

flawed because of unrealistic projections and assumptions made about the conditions of local communities and Indigenous Peoples (Asare-Nuamah and Botchway, 2019).

Indigenous Peoples seldomly adopt transformative adaptation measures that change the fundamental attributes of their systems (O'Brien, 2012; Olsson et al., 2014). Rather, their responses rely on the interaction between people and their collective activities, which are mediated through kinship, friendship, and informal and formal institutions (Goulden et al., 2013; Moutouama et al., 2022). Indigenous Peoples have activities and networks that foster diversification and help reduce the impact of climate variability and change. However, those in Northern Ghana remain the most vulnerable to climate change despite the use of some local adaptation measures. (Armah et al., 2010; Chemura et al., 2020), more so those in Kumbungu (Musah et al., 2013; Nyadzi, 2016). As a result, the goals of this chapter are (1) to recognize and understand changes observed by people of Kumbungu descent in Northern Ghana, (2) to assess which of the observations are caused by the climate change, and (3) to identify local responses to these impacts. It pursues these objectives through the lens of the cultural theory of risk (Douglas and Wildavsky, 1982).

Theoretical framework

This chapter adopted the cultural theory of risk (CTR), which is based on the premise that people's perceptions of changes are influenced by their cultural environments (Dake, 1991; Johnson and Covello, 1987; Stern et al., 1995). A thorough understanding of the cultural influence on risk perception can be attained by using CTR to explain why and how people choose to react to environmental hazards. This can be done by combining CTR and local or traditional environmental knowledge about weather and climate-related threats (McNeeley and Lazrus, 2014).

The development of CTR has demonstrated that people's risk preferences vary over time, in response to various circumstances and experiences (Bellamy and Hulme, 2011; Rayner, 1992; Spickard, 1989). This is consistent with the idea that culture is dynamic and emergent (Wolf, 1982). According to Rayner (1992), specific sorts of social organizations are linked to different kinds of worldviews. Different types of social organizations see and approach dangers in different ways. As a result, conflict over how groups with various institutional cultures organize and approach solutions may emerge (Thompson and Rayner, 1998).

As suggested by CTR, for sustainable adaptation to climate change to occur, all worldviews must be present and able to have a conversation about various perspectives and priorities (McNeeley and Lazrus, 2014). CTR provides a framework for analyzing cultural conflicts, such as those between groups of climate change "believers" and "non-believers," and it reveals how risk perceptions consistent with values and cultures can be identified (McNeeley, 2014). Engaging CTR at the community level enables the understanding of why measures intended to address climate change, such as adaptation and mitigation, are frequently contentious and even divisive within and between varied communities (McNeeley, 2014).

Methodology

Study area

This study was conducted in the Kumbungu district of Northern Ghana (Figure 14.1). The area is located in the tropical zone, characterized by lowlands and covered by savanna grassland (Abdul-Razak and Kruse, 2017). Kumbungu was selected on the basis that instrumental meteorological data was deficient, with little focus on the study of Indigenous indicators for climate change. The

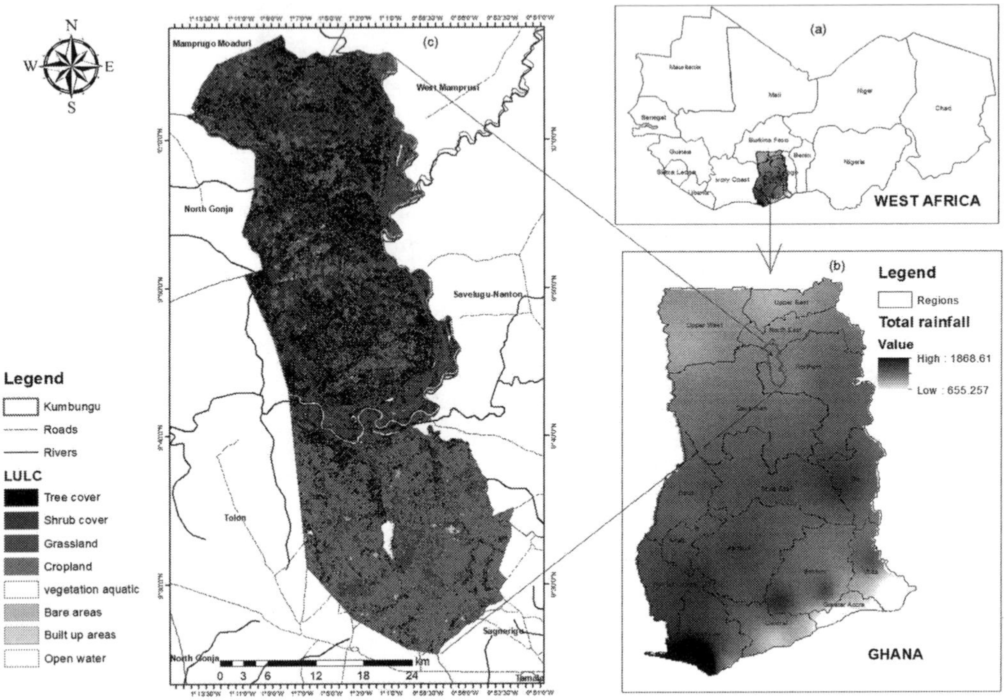

Figure 14.1 Map of the study area. (A) shows Ghana in West Africa. (B) Ghana with annual average rainfall from 1981 to 2010. (C) shows Kumbungu district with land use/cover.

Dagbani ethnic group, estimated at 1.2 million people in Ghana, makes up the local population. Almost all of the households in the district are involved in seasonal small-scale rain-fed agriculture with crops (e.g., sorghum, rice, maize, and millet) and poultry production accounting for about 95% of their total family income (GSS, 2015). Only a few households near the Bontanga Irrigation Project, employ dry-season irrigation (Nyadzi, 2016). Agriculture is negatively affected by the climatic conditions with six to seven months of the dry season (October/November–March/April) and five months of the rainy season (April/May to September/October) (Amikuzino and Donkoh 2012; Nyadzi et al., 2018). The area is also subject to the effects of recurrent floods and droughts (Stanturf et al., 2011; Asante and Amuakwa-Mensah, 2014)

Site selection, sample, and sampling techniques

The villages of Kpasolgu, Kushegu, and Wuba were selected for the study because they have relatively homogenous populations with typical environmental and socio-cultural conditions, and little influence from external forces, such as donor help. The sampling procedure followed those recommended by Reyes-García et al. (2023).

SEMI-STRUCTURED INTERVIEWS

Five key informants were purposely selected for semi-structured interviews to explore and gather site-level data on local livelihoods, local timeline events, and seasonal calendar. Three pilot interviews were conducted first to clarify and contextualize the concepts in local settings and to train translators and research assistants. Also, using the quota sampling technique, the second part of the semi-structured interview was conducted with 18 individuals categorized as young (three men and three women) and old (three men and three women) from each of the three villages. To this sample, we asked questions regarding observable environmental changes attributed to climate change and adaptation/coping measures to respond to these changes (Reyes-García et al., 2023).

FOCUS GROUP DISCUSSIONS

Given that power-play could influence participation, three focus groups were created based on gender and age structure using a convenience sampling technique. The first focus group discussion was for men (mainly household heads) and counted with the participation of 13 people , the second was for women (old and young) with about 10 participants, and the third for the youth only (males) with about 11 participants.

SURVEY

Using information from semi-structured interviews and focus group discussions, a survey with two different parts was conducted. The first part was at the household level with 125 respondents and the second at the individual level with 175 respondents selected using a random sampling technique. The total number of households and individual interviews were uniformly distributed among the three villages. The first version of the survey was pre-tested with ten respondents.

Data collection

Data collection was conducted between November 2019 and June 2021(COVID-19 prolonged the data collection) and the process was divided into four main steps following the protocol developed by Reyes-García et al. (2023). *Step 1:* reconnaissance walks, observations, community meetings and conversations with well-informed locals, literature review, and key informants were used to gather the site and village levels data such as meteorological data, accessibility, infrastructure, seasonal calendar, livelihood and historical timelines of the area. *Step 2:* semi-structured interviews with key informants were used to gather data on observed changes, the direction of changes, the drivers of change, and adaptation techniques used to manage these changes. *Step 3:* The observational changes and adaptation actions identified were further discussed at focus group discussions for validation and elaboration. *Step 4:* A survey at the household and the individual level was organized to gather data on vulnerability, socio-demographics, Local Indicators of Climate Change Impacts (LICCIs), and Local Adaptation of Climate Change Impacts (LACCIs). The data were analyzed using Microsoft excel.

Results

Sample description

About 43% of the respondents opined that their lives have improved over the last 20 years, with the rest of the sample responding to have not seen any changes in their living conditions. Individuals in the study area have received some NGO and governmental sponsorships when undertaking various agriculture projects. Crop production is the primary source of income for all households. Results of the socio-demographics characteristics of individual respondents and household characteristics are shown in Tables 14.1 and 14.2.

Table 14.1 Socio-demographic characteristics of respondents

Individual-level information			
Items	*%*	*Items*	*%*
Gender		**Authority in village**	
Male	57	Yes	20
Female	43	No	80
Age		**Education**	
20–29	5	No formal education	44
30–39	5	Primary school	30
40–49	12	Middle school lever	18
50–59	25	High school graduate	8
60–69	31	**Farm size(ha)**	
70–79	16	<1	1
80–89	6	1–1.9	51
Access to information		2–2.9	47
Almost never	38	3–3.9	17
Almost daily	49	4–4.9	6
Weekly	11	>5	3
Monthly	2		
Sources of information			
Television		**Internet**	
Almost never	61	Almost never	91
Almost daily	29	Almost daily	5
Weekly	7	Weekly	3
Monthly	3	Monthly	1
Radio		**Mobile phone**	
Almost never	0	Almost never	1
Almost daily	95	Almost daily	65
Weekly	5	Weekly	30
Monthly	0	Monthly	4

Authors' construct (2022).

Table 14.2 Household characteristics

Household-level information			
Items	*%*	*Items*	*%*
Monetary debts		**Control over the food they eat**	
Does not owe any money	62	Full control	66
Owes money	38	Some control	34
Number of houses		**Food preference**	
1	1	I mostly eat the food I like, but sometimes have to eat something else	38
2	34	I always eat the food I like	61

(Continued)

Table 14.2 (Continued)

Household-level information			
Items	*%*	*Items*	*%*
3	53	I mostly eat food I do not prefer, but sometimes I could eat the food I like	1
4	11	**Freshwater quality**	
5	1	water quality sometimes was not good	45
Government support		**Water quality is usually good**	54
Yes	53	often not good	1
No	47	**Freshwater quantity**	
NGO support		Sometimes we do not have enough water	45
Yes	70	We always have enough water	54
No	30	often we do not have water	1
Well-being		**Freshwater control**	
Better	43	Full	65
Worse	26	Some	35
Same	31	**Freshwater preference**	
Food quality		I mostly do not have access to my preferred water sources, but sometimes I do	3
I did not always have access to safe and nutritious food	32	I mostly have access to my preferred water sources, but not always	51
I usually had access to safe and nutritious food	68	I always have access to my preferred water sources	46
Food quantity			
We always had enough food	89		
Sometimes we didn't have enough food	11		

Authors' construct (2022).

Observed environmental changes

A large number of survey respondents had observed significant environmental changes and were able to point out the directions and the extent to which these changes are affecting their households (Table 14.3). All the respondents reported an increase in rainfall variability, frequency of heavy rainfall, and decrease in crop maturation time. Almost everyone reported a shortened rainy season, an increased frequency of hot/warm days, wind temperature, and average temperature. More than half of the respondents reported a decrease in the number of windy days. Respondents also mention that these changes significantly affect their households by interfering with their daily lives and affecting farm activities and crop growth. Other observable changes that affect their households at varying degrees include an increase in livestock mortality, an increase in the frequency of crop pests, an increase in wildfire frequency, an increase in wind-induced soil erosion and soil loss, a decrease in the quality of pasture, and a decrease in the intensity/strength of cold waves.

Table 14.3 Observable climate change impacts and their effect on households in the Kumbungu community ($n = 125$)

Observable changes	*Noticed*	*%*	*Direction of change*	*%*	*Effect on household (categories created based on responses)*	*%*
Change in the intensity/ strength of cold waves	Yes	77	Increase	2	Does not affect me at all	89
	No	23	Decrease	98	Affects me a little	1
					Affects me a lot	10
Changes in mean temperature	Yes	73	Increase	98	Does not affect me at all	71
	No	27	Decrease	2	Affects me a little	8
					Affects me a lot	21
Changes in wind-induced soil erosion and soil loss	Yes	78	Increase	100	Does not affect me at all	24
	No	22	Decrease	0	Affects me a little	35
					Affects me a lot	41
Changes in wildfire frequency	Yes	81	Increase	100	Does not affect me at all	8
	No	19	Decrease	0	Affects me a little	23
					Affects me a lot	69
Changes in rainfall variability	Yes	99	Increase	100	Does not affect me at all	1
	No	1	Decrease	0	Affects me a little	9
					Affects me a lot	90
Changes in seasons length /duration/ / disappearance	Yes	100	Shorter	98	Does not affect me at all	1
	No	0	Longer	2	Affects me a little	10
					Affects me a lot	89
Changes in the frequency of heavy rainfall events	Yes	98	Increase	100	Does not affect me at all	4
	No	2	Decrease	0	Affects me a little	7
					Affects me a lot	89
Changes in the frequency of crop 'pests' (insects, birds, larvae, etc.)	Yes	98	Increase	100	Does not affect me at all	1
	No	2	Decrease	0	Affects me a little	5
					Affects me a lot	94
Changes in livestock mortality	Yes	63	Increase	100	Does not affect me at all	44
	No	37	Decrease	0	Affects me a little	24
					Affects me a lot	33
Changes in the number of windy days	Yes	44	Increase	39	Does not affect me at all	58
	No	56	Decrease	61	Affects me a little	16
					Affects me a lot	26
Changes in crop maturation time	Yes	32	Increase	100	Does not affect me at all	53
	No	68	Decrease	0	Affects me a little	29
					Affects me a lot	18
Changes in the abundance or density of wild plant or fungi species		28	Increase	6	Does not affect me at all	55
		72	Decrease	94	Affects me a little	31
					Affects me a lot	14
Changes in the quality of pasture		26	Increase	4	Does not affect me at all	53
		74	Decrease	96	Affects me a little	38
					Affects me a lot	9
Changes in wind temperature	Yes	37	Increase	100	Does not affect me at all	47
	No	63	Decrease	0	Affects me a little	39
					Affects me a lot	14
Changes in the frequency of hot/warm days	Yes	98	Increase	100	Does not affect me at all	18
	No	2	Decrease	0	Affects me a little	20
					Affects me a lot	60

Authors' construct (2022).

Responding to environmental changes: adaptation strategies, enablers, and barriers

In response to the observable climate change impacts listed in Table 14.3, Dagbani people at Kumbungu use some adaptation practices (Table 14.4). The application of chemical fertilizers and changing crop varieties emerged as the most used strategies to battle the impacts of changing climates on agriculture. The abrupt changes in climatic conditions, such as drought, have made farmers adopt harvesting rainwater for irrigation in the dry season and supplement rainfall in the wet season. According to the respondents, the low number of people engaged in building rain harvesting systems is due to the high initial cost involved in constructing the water harvesting system. Farmers in the community also plant more trees to provide shade and fresh air and have adjusted the structure and orientation of buildings to provide ample ventilation during the hot season. The use of sandbags along farmlands and river banks was also adopted to reduce the impacts of flood and erosion. The respondents use chemical insecticides and pesticides to control farm pests, as well as mosquitoes and other harmful household insects. With regards to rainfall variability and unpredictability, they rely on weather and seasonal climate forecast information provided through extension services and or radio stations to estimate the timing and amount of expected rainfall for decisions such as when, what, and how to plant. The forecast information is primarily generated by Ghana Meteorological Agency (GMet) and some private entities like ISOKO. To prevent strong winds from damaging crops/trees, buildings and other farm structures, informants engage in staking crops/trees and fortifying buildings against wind and storm destruction. Few informants engage in off-farm activities such as construction work and selling provisions as an essential adaptation practice to supplement the household income especially when yield is less than anticipated.

New crop varieties have been adopted in the community; the most common being New Rice for Africa (NERICA), which has become the most adopted staple crop in the community. Farmers report that NERICA matures early and withstand long periods of drought, disease, and pest infestations and meet the market demand. Farmers in the area attributed the sudden adoption of the new crop varieties, including NERICA, to the changing rainfall patterns, and temperature changes but also to changes in the market and value chain demands. The same reasons (i.e., changing climatic conditions, changing markets, and value chain) were given as the main drivers for abandoning certain local crop varieties such as foxtail millet, local okra species, and local rice varieties. Also, respondents have abandoned some adaptation measures because of their effectiveness in addressing climate impact, for example, the majority of farmers no longer use climate information from GMet (Table 14.3). In some cases, respondents do not have any adaptation measure to address such impacts, for example, no measures had been put in place to prevent the extinction of some local birds.

From the survey, farmers indicate reasons and enablers that inform the adoption of certain adaptation practices. The high benefits derived from adaptation to reduce climate change impacts, high climate change impacts, high severity/damage potential of climate change and its impacts, and sufficient/high financial capital are some of the driving forces in selecting and implementing an adaptation measure (Table 14.4).

Results also show that certain factors limit the implementation of adaptation strategies in the study area (Table 14.4). A key barrier to adaptation was their inability to afford or develop the right adaptation strategies due to a lack of money. The respondents claim poverty is a major factor as to which adaptation measures to be adopted. For example, modifying building structures requires high financial investments, which are unavailable. Also, results from the survey show that little or no

Table 14.4 Adaptation strategies identified at the Kumbungu community

Adaptation strategies	*Adopted*		*Categorized enablers (out of % of those who adopt)*		*Reasons for adoption*	*Categorized barriers*	
	Item	*%*	*Item*	*%*		*Item*	*%*
Build rain harvesting systems to collect water	Yes	20	(High) benefits from adaptation to reduce climate change impacts	60	• To save water for daily activities during long dry seasons	Lack of financial capital	100
	No	45.6	High climate change impacts	28			
	Not anymore	33.6	High severity/damage potential of climate change and its impacts	51.6			
We plant more trees	Yes	49.6	Belief in the potential for other positive side effects of adaptation/coping measure	3.2	• To provide shade from the hot sun • To provide fresh air	Low benefits from adaptation	0.8
	No	28.8	High climate change impacts	19.4		Time constraint	93.6
	Not anymore	21.6	High severity/damage potential of climate change and its impacts	17.7		Habit/Custom/Preference	5.6
			Other (high) risks (not related to climate change	1.6			
Apply chemical fertilizers	Yes	97.6	(High) benefits from adaptation to reduce climate change impacts	51.6	• To increase the yield of our crops	Lack of financial capital	100
	No	2.4	High climate change impacts	4.1	• To improve the fertility of our soils		
	Not anymore	0	Other (high) risks (not related to climate change	44.3			
Make changes to buildings to make them airy	Yes	36	(High) benefits from adaptation to reduce climate change impacts	35.6	• The weather is hot nowadays	Lack of financial capital	96.8
	No	32	High climate change impacts	44.4	• Reduces heat in the room	Habit/Custom/Preference	1.6
	Not anymore	32	High severity/damage potential of climate change and its impacts	20		Time constraint	1.6
Off-farm activities	Yes	5.6	Other (high) risks (not related to climate change	85.7	• To raise extra income	Lack of financial capital	88.8
	No	38.4	Sufficient/High financial capital	14.3	• To supplement the money gained from agricultural activities	Habit/Custom/Preference	0.8
	Not anymore	56				Low benefits from adaptation	4
						Time constraint	5.6
						Others	0.8

Staking crops/trees and also fortify our buildings	Yes	56 8	(High) benefits from adaptation to reduce climate change impacts	54.9	• To prevent strong winds from pulling down crops/ trees and buildings	Low benefits from adaptation	97.6
	No	15.2	High climate change impacts	26.8		Time constraint	2.4
	Not anymore	28	High severity/damage potential of climate change and its impacts	19.7			
Use insecticides and pesticides chemicals	Yes	86	(High) benefits from adaptation to reduce climate change impacts	72	• Kill insects and pests on the farm • To kill mosquitoes and other harmful organisms in the house	Lack of financial capital	100
	No	6	High climate change impacts	4.7			
	Not anymore	8	High severity/damage potential of climate change and its impacts	1.9			
			Other (high) risks	21.5			
Use sandbags	Yes	59.2	(High) benefits from adaptation to reduce climate change impacts	71.6	• To prevent soil erosion due to rainfall	Low benefits from adaptation	90.4
						Habit/custom/preference	3.2
	No	10.4	High climate change impacts	16.2		Time constraint	3.2
	Not anymore	30.4	High severity/damage potential of climate change and its impacts	10.8		Other trade-offs/ incompatibilities	3.2
			Others	1.4			
Plant short-term season crop varieties	Yes	100	(High) benefits from adaptation to reduce climate change impacts	62.4	• The starting date for many crops has changed and become shorter • Improve yield	Lack of financial capital	100
	No	0	Belief in the potential for other positive side effects of adaptation/coping measure	1.6			
	Not anymore	0	High climate change impacts	19.2			
			High severity/damage potential of climate change and its impacts	11.2			
			Other (high) risks	1.6			
Use weather and climate forecasts for planning	Yes	59.2	(High) benefits from adaptation to reduce climate change impacts	64.9	• To determine the timing and amount of rainfall to take a planting decision	Lack of external support	76
	No	26.4	High climate change impacts	23%		Lack of social capital	0.8
	Not anymore	14.4	High severity/damage potential of climate change and its impacts	12.2			23.2

Authors' construct (2022).

access to weather and climate forecast information impedes farm decision-making. Even individuals who had access to forecast information reported a mismatch between the information provided and farm decision needs given particular local conditions. Some adaptation measures, such as using sandbags to control erosion quickly become ineffective as climate change impacts intensify. Time constraints and habits were some of the limiting factors affecting the implementation of adaptation measures such as planting trees, staking crops and trees, or fortifying buildings.

Discussion

Agriculture employs almost all of the inhabitants in the study area and it is challenged by water scarcity and drought, conditions that are likely to be aggravated due to the observable changing patterns in climatic factors. Informants in the area reported a rising trend in mean temperature, frequency of hot/warm days, and wind temperature, showing a warming climate in the study area. Across Ghana, the rate of warming is most rapid in the northern inland regions than in other parts of the country (Klutse et al., 2020; World Bank, 2010). Rainfall in the study area remains variable over short periods, though an increase in the frequency of heavy events is perceived by farmers. Other studies using farmers' observations and meteorological data have discussed similar trends in Northern Ghana, where Kumbungu is located (Amikuzino and Donkoh 2012; Awuni et al., 2018; Frimpong et al., 2014; Issahaku et al., 2016; Nyadzi, 2016). Almost all of the observable changes identified in Kumbungu have also been found elsewhere in Africa and well documented by previous studies (Agbo, 2013; Farauta et al., 2011; Gebreyesus et al., 2016; Kemausour et al., 2011). In contrast to previous work, our study brings detailed information on the percentage of informants who notice the change, its direction, and its effect on households. The impact of changing temperatures and rainfall will result in a decrease in crop production across sub-Saharan Africa (Barrios et al., 2008; Schlenker and Lobell, 2010).

Studies have reported that changes in climate factors, in particular rainfall and temperature, are crucial to agricultural output (Chepkoech et al., 2018; Nyatuame et al., 2014). The changing pattern in rainfall and temperature in the study area, in addition to a decrease in crop maturation time and shortened length of growing seasons, puts agriculture under threat, as crop failure may increase. As demonstrated by the results of this study, Indigenous People already acknowledge the impact of changing climatic factors on their livelihoods. The negative impact of increasing temperatures and decreasing rainfall has implications for soil and water management and agricultural productivity in general. Also, an increase in temperature, coupled with warm nights and reduced rainfall, affects crops and weed growth, and increases the prevalence of insects, pests, and diseases (Hatfield et al. 2011). Moreover, livestock in the study area are exposed to the direct impacts of high temperatures, affecting their feeding, reproduction, disease prevention, and health in general. Given that farming in the study area is primarily for sustaining family food supply and income (MOFA, 2015), the life and livelihood of the Dagbani people are also under threat.

The type of adaptation practices adopted by the Dagbani suggests the extent of climate impact and the efforts to mitigate them. The unpredictability of the rainfall has increased the demand for weather and seasonal climate information in the study area. However, a major barrier to using climate information was mistrust of the information as the information is not timely provided and does not match needs. Studies have shown that the difficulties to predict the weather and seasonal climate leave serious implications for food production and livelihoods and therefore some farmers combine Indigenous and scientific forecast information to formulate farm decisions. This is also the case in Northern Ghana (Antwi-Agyei et al., 2021; Baffour-Ata et al., 2022; Nyadzi et al., 2018, 2021, 2022; Nyamekye et al., 2021).

The accelerated rate of climate change impact raises new questions regarding the effectiveness of adaptive measures, and what could limit adaptation practices. Results show that a lack of financial resources is a major barrier to implementing appropriate adaptation strategies in the study area. Antwi-Agyei et al. (2015) reported that financial barriers due to a lack of credit facilities are among the most important obstacles hindering the implementation of climate adaptation strategies by farmers in Northern Ghana. The lack of credit facilities as a barrier to adaptation has also been found elsewhere in Africa (Asare-Nuamah et al., 2022; Deressa et al., 2009; Oramah et al., 2022). Therefore we recommend the creation and strengthening of social groups to combine resources, especially financial resources to formulate effective adaptation measures to address climate change impacts

We recognize that while this chapter focuses on identifying climate impact, adaptation, and barriers, it does not examine their multilevel interactions to influence the livelihood of the Dagbani people of Kumbugu. Further research should focus on how barriers influence adaptation and how the adaptation practices will evolve in the future, and the need for transformative adaptation at the local level.

Finally, by applying the cultural theory of risk (CTR), our study provides empirical evidence about local and context-specific impacts of climate change and the factors that influence Indigenous people's adaptation decisions. Moreover, approaching the study from the CTR perspective reveals that the social connections and cultural worldviews of the Dagbani people in Kumbungu, which include their fundamental beliefs about their society and the environment, shape how they perceive the impact of climate change and select the adaptation technique suitable to their local environment.

Conclusion and recommendations

This chapter has provided a nuanced understanding of how climate change poses complex challenges to Dagbani people at Kumbungu in Northern Ghana. Dagbani people have several adaptation strategies to deal with the negative impacts of climate change, but finance remains the major barrier to implementing adaptation strategies. We recommend that social groups should be formed to combine resources to be able to afford and or formulate better adaptation strategies. Households need to be supported with micro-credit schemes. The inclusion of Indigenous people in adaptation policy formulation is highly recommended to offer context, restore trust and increase the useability of these measures. The provision of accurate weather and seasonal climate forecast information that is actionable for climate sensitive decision-making is needed given that most farmers have no confidence in the existing forecast information received. Installing more meteorological stations will serve as a means of obtaining and providing real-time weather and seasonal climate information to the Dagbani people in Kumbugu. Finally, we conclude that through the lens of the cultural theory of risk, empirical evidence has been generated to show that social connections and cultural worldviews of the Indigenous people shape how they perceive context-specific climate change impacts and select suitable adaptation strategies.

Acknowledgments

This study is conducted under the LICCI project. We would like to thank the various families and individuals at Kumbungu who availed themselves during the interviews. The research leading to this paper has received funding from the European Research Council under an ERC Consolidator Grant (FP7-771056-LICCI).

References

Abaje, I. B. & Giwa P. N. (2007). Urban flooding and environmental safety: a case study of Kafanchan Town in Kaduna state. Paper presented at the Golden Jubilee (50th Anniversary) and 49th Annual Conference of the Association of Nigerian Geographers (ANG) Scheduled for 15th–19th October 2007 at the Department of Geography, University of Abuja, Gwagwalada-Abuja.

Abdul-Razak, M., & Kruse, S. (2017). The adaptive capacity of smallholder farmers to climate change in the Northern Region of Ghana. *Climate Risk Management*, 17, 104–122.

Acquah, H. de-G., Nunoo, J., & Darfor, N. K. (2015). Farmers' perceptions and adaptation to climate change: evidence from Ghana. In: Vubo, E. Y. (Ed.), *Environment, Agriculture and Cross-Border Migrations* (pp. 35–52). Dakar, Senegal: CODESRIA.

Agbo, F. U. (2013). Farmers' perception of climate change in Ikwuano local Government area of Abia State, Nigeria. *Journal of Biology, Agriculture and Healthcare*, *3*(6), 36–45.

AGRA, 2018. African Agriculture Status Report 2018. Catalyzing Government Capacity to Drive Agricultural Transformation, Nairobi, Kenya. https://agra.org/wp-content/uploads/2018/10/AASR-2018.pdf

Amikuzino, J., & Donkoh, S. A. (2012). Climate variability and yields of major staple food crops in Northern Ghana. *African Crop Science Journal*, *20*, 349–360.

Antwi-Agyei, P., Amanor, K., Hogarh, J. N., & Dougill, A. J. (2021). Predictors of access to and willingness to pay for climate information services in north-eastern Ghana: a gendered perspective. *Environmental Development*, *37*, 100580.

Antwi-Agyei, P., Dougill, A. J., & Stringer, L. C. (2015). Barriers to climate change adaptation: evidence from northeast Ghana in the context of a systematic literature review. *Climate and Development*, *7*(4), 297–309.

Armah, F. A., Yawson, D. O., Yengoh, G. T., Odoi, J. O., & Afrifa, E. K. (2010). Impact of floods on livelihoods and vulnerability of natural resource dependent communities in Northern Ghana. *Water*, *2*(2), 120–139.

Asare-Nuamah, P., & Botchway, E. (2019). Comparing smallholder farmers' climate change perception with climate data: the case of Adansi North District of Ghana. *Heliyon*, *5*(12), e03065. doi:10.1016/j.heliyon.2019.e03065

Asare-Nuamah, P., Mandaza, M. S., & Amungwa, A. F. (2022). Adaptation strategies and farmer-led agricultural innovations to climate change in Mbire district of Zimbabwe. *International Journal of Rural Management*, *18*(2), 206–231.

Awuni, J. A., Azumah, S. B., & Donkoh, S. A. (2018). Drivers of adoption intensity of improved agricultural technologies among rice farmers: evidence from northern Ghana. *Review of Agricultural and Applied Economics (RAAE)*, *21*(1340-2018-5180), 48–57.

Baffour-Ata, F., Antwi-Agyei, P., Nkiaka, E., Dougill, A. J., Anning, A. K., & Kwakye, S. O. (2022). Climate information services available to farming households in Northern Region, Ghana. *Weather, Climate, and Society*, *14*(2), 467–480.

Bahadur, A. V., Ibrahim, M., & Tanner, T. (2013). Characterizing resilience: Unpacking the concept for tackling climate change and development. *Climate and Development*, *5*(1), 55–65. doi:10.1080/17565529.2012.762334

Barrios, S., Ouattara B., & Strobl E. (2008). The impact of climatic change on agricultural production: is it different for Africa? *Food Policy*, *33*, 287–298.

Bedeke, S. B. (2022). Climate change vulnerability and adaptation of crop producers in sub-Saharan Africa: a review on concepts, approaches and methods. *Environment, Development and Sustainability*, *25*(2), 1017–1051.

Bellamy, R., & Hulme, M. (2011). Beyond the tipping point: understanding perceptions of abrupt climate change and their implications. *Weather, Climate, and Society*, *3*(1), 48–60.

Berkes, F., Colding, J., & Folke, C. (2003). *Navigating Socio-Ecological Systems: Building Resilience for Complexity and Change*. Cambridge: Cambridge University Press.

Chemura, A., Schauberger, B., & Gornott, C. (2020). Impacts of climate change on agro-climatic suitability of major food crops in Ghana. *PLoS One*, *15*(6), e0229881.

Chepkoech, W., Mungai, N. W., Stöber, S., Bett, H. K., & Lotze-Campen, H. (2018). Farmers' perspectives: impact of climate change on African indigenous vegetable production in Kenya. *International Journal of Climate Change Strategies and Management*, *10*(4), 551–579. doi:10.1108/IJCCSM-07-2017-0160

Dake, K. (1991) Orienting dispositions in the perception of risk: an analysis of contemporary worldviews and cultural biases. *Journal of Cross-Cultural Psychology, Special Issue on Risk and Culture*, *22*, 61–82. doi:10.1080/13669870110042598.

Deressa, T. T., Hassan, R. M., Ringler, C., Alemu, T., & Yesuf, M. (2009). Determinants of farmers' choice of adaptation methods to climate change in the Nile Basin of Ethiopia. *Global Environmental Change, 19*(2), 248–255.

Di Falco, S. (2018). Adapting to climate change in sub-Saharan Africa. In: Berck, C., Berck, P., & Di Falco. S., *Agricultural Adaptation to Climate Change in Discussion Paper* 00849. New York: Resources of the Future Press.

Douglas, M. & Wildavsky, A. (1982) *Risk and Culture: An Essay on Selection of Technological and Environmental Dangers*. Berkeley: California University Press.

Farauta, B. K., Egbule, C. L., Idrisa, Y. L., & Agu, V. C. (2011). Farmers' perceptions of climate change and adaptation strategies in northern Nigeria: an empirical assessment. *African Technology Policy Studies Network, Research Paper No. 15*.

Frimpong, K., Oosthuizen, J. D., & Van Etten, E. J. (2014). Recent trends in temperature and relative humidity in Bawku East, Northern Ghana. Journal of Geography and Geology, 6(2), 69–81.DOI:10.5539/jgg.v6n2p69

Gebreyesus, B. T., Mekonen, K., & Tekle, A. (2016). Farmers' perceptions on causes, indicators and determinants of climate change in northern Ethiopia: implication for developing adaptation strategies. *Applied Geography*, *73*, 1–12.

Ghana Statistical Service (GSS). (2015). Ghana demographic and health survey 2014. https://www2.statsghana.gov.gh/docfiles/DHS_Report/Ghana_DHS_2014-KIR-21_May_2015.pdf

Goulden, M. C., Adger, W. N., Allison, E. H., & Conway, D. (2013). Limits to resilience from livelihood diversification and social capital in lake social-ecological systems. *Annals of the Association of American Geographers*, *103*(4), 906–924. doi:10.1080/00045608.2013.765771.

Hatfield, J. L., Boote, K. J., Kimball, B. A., Ziska, L. H., Izaurralde, R. C., Ort, D., … & Wolfe, D. (2011). Climate impacts on agriculture: implications for crop production. *Agronomy Journal*, *103*(2), 351–370.

Ibe, G. O., & Amikuzuno, J. (2019). Climate change in sub-Saharan Africa: a menace to agricultural productivity and ecological protection. *Journal of Applied Sciences and Environmental Management*, *23*(2), 329–335.

Issahaku, A. R., Campion, B. B., & Edziyie, R. (2016). *Rainfall and temperature changes and variability in the Upper East Region of Ghana*. Earth and Space Science, 3(8), 284–294.

Johnson, B. B. & Covello, V. T. (1987). *The Social and Cultural Construction of Risk: Essays on Risk Selection and Perception*. Dordrecht: Reidel Publishing.

Kemausuor F, Dwamena E, Bart-Plange A, Kyei-Baffour N (2011) Farmers' perception of climate change in the Ejura-Sekyedumase district of Ghana. *Asian Res Publishing Network* *6*(10), 26–37.

Klutse, N. A. B., Owusu, K., & Boafo, Y. A. (2020). Projected temperature increases over northern Ghana. *SN Applied Sciences*, *2*(8), 1–14.

Maja, M. M. and Ayano, S. F. (2021). The impact of population growth on natural resources and farmers' capacity to adapt to climate change in low-income countries. *Earth Systems and Environment*, *5*(2), 271–283.

McNeeley, S. M. & Lazrus, H. (2014). The cultural theory of risk for climate change adaptation. *Weather, Climate, and Society*, *6*(4), 506–519. doi:10.1175/WCAS-D-13–00027.1.

Mendelsohn, R., Dinar A., & Williams L. (2006). The distributional impact of climate change on rich and poor countries. *Environment and Development Economics*, *11*, 159–178.

MOFA (Ministry of Food and Agriculture) (2015). Agriculture in Ghana. Facts and figures 2014. Ministry of Food and Agriculture (MoFA). Statistics, Research and Information Directorate (SRID). Accra.

Moutouama, F. T., Tepa-Yotto, G. T., Agboton, C., Gbaguidi, B., Sekabira, H., & Tamò, M. (2022). Farmers' perception of climate change and climate-smart agriculture in Northern Benin, West Africa. *Agronomy*, *12*(6), 1348.

Muli, C., Gerber, N., Sakketa, T., & Mirzabaev, A. (2018). *Ecosystem Tipping Points due to Variable Water Availability and Cascading Effects on Food Security in Sub-Saharan Africa*. ZEF - Working Paper 174, Available at SSRN: https://ssrn.com/abstract=3266355 or http://dx.doi.org/10.2139/ssrn.3266355

Musah, B. A., Mumuni, E., Abayomi, O., & Jibrel, M. B. (2013). Effects of floods on the livelihoods and food security of households in the Tolon/kumbumgu district of the Northern region of Ghana. *American Journal of Research Communication*, *1*(8), 160–171.

Niang, I., Ruppel, O. C., Abdrabo, M. A., Essel, A., Lennard, C., Padgham, J., & Urquhart, P. (2014). Africa. Climate change 2014: impacts, adaptation, and vulnerability. Part B: regional aspects. Contribution of working Group II to the fifth assessment report of the intergovernmental panel on climate change. Climate Change.

Nyadzi, E. (2016). Climate variability since 1970 and farmers' observations in Northern Ghana. *Sustainable Agriculture Research*, *5*, 41–56.

Nyadzi, E., Nyamekye, A. B., Werners, S. E., Biesbroek, R. G., Dewulf, A., Van Slobbe, E., & Ludwig, F. (2018). Diagnosing the potential of hydro-climatic information services to support rice farming in northern Ghana. *NJAS-Wageningen Journal of Life Sciences*, *86*, 51–63.

Nyadzi, E., Werners, S. E., Biesbroek, R., & Ludwig, F. (2021). Techniques and skills of indigenous weather and seasonal climate forecast in Northern Ghana. *Climate and Development*, *13*(6), 551–562.

Nyadzi, E., Werners, S. E., Biesbroek, R., & Ludwig, F. (2022). Towards weather and climate services that integrate indigenous and scientific forecasts to improve forecast reliability and acceptability in Ghana. *Environmental Development*, *42*, 100698.

Nyamekye, A. B., Nyadzi, E., Dewulf, A., Werners, S., Van Slobbe, E., Biesbroek, R. G., … & Ludwig, F. (2021). Forecast probability, lead time and farmer decision-making in rice farming systems in Northern Ghana. *Climate Risk Management*, *31*, 100258.

Nyatuame, M., Owusu-Gyimah, V., & Ampiaw, F. (2014). Statistical analysis of rainfall trend for Volta region in Ghana. *International Journal of Atmospheric Sciences*, *2014*. doi:10.1155/2014/203245

O'Brien, K. (2012). Global environmental change II: From adaptation to deliberate transformation. *Progress in Human Geography*, 36(5), 667–676.

Ofori, S. A., Cobbina, S. J., & Obiri, S. (2021). Climate change, land, water, and food security: perspectives from sub-Saharan Africa. *Frontiers in Sustainable Food Systems*, *5*, 680924.

Olsson, L., Opondo, M., Tschakert, P., Agrawal, A., Eriksen, S., Ma, S., ... & Zakieldeen, S. (2014). Livelihoods and poverty. In *Climate Change 2014 Impacts, Adaptation and Vulnerability: Part A: Global and Sectoral Aspects* (pp. 793-832). Cambridge University Press.

Oramah, C. P., Olsen, O. E., & Arne Pettersen Gould, K. (2022). Assessing the impact of the securitization narrative on climate change adaptation in Nigeria. *Environmental Politics*, 31(6), 1037–1057.

Popoola, K. O., Jerneck, A., & Ajayi, S. A. (2020). Climate variability and rural livelihood security: impacts and implications. *African Handbook of Climate Change Adaptation*, 1–13. https://link.springer.com/referenceworkentry/10.1007/978-3-030-45106-6_200

Pörtner, H. O., Roberts, D. C., Adams, H., Adler, C., Aldunce, P., Ali, E., ... & Ibrahim, Z. Z. (2022). Climate change 2022: impacts, adaptation and vulnerability. *IPCC*.

Rayner (1992). Cultural theory and risk analysis. In: Krimsky, S. and Golding, D. (Eds.), *Social Theories of Risk* (pp. 83–115). Westport, CT: Praeger.

Reyes-García, V., Álvarez-Fernández, S., Benyei, P., García-del-Amo, D., Junqueira, A. B., Labeyrie, V., & Soleymani, R. (2023). Local indicators of climate change impacts described by indigenous peoples and local communities: study protocol. *PLoS One*, *18*(1), e0279847.

Schlenker, W. & Lobell, D. B. (2010). Robust negative impacts of climate change on African agriculture. *Environmental Research Letters*, *5*, 1–8.

Spickard, J V. (1989). A guide to Mary Douglas's three versions of grid/group theory. *Sociological Analysis*, *50*(2), 151–170.

Stern, P. C., Dietz, T., & Guagnano, G. A. (1995). The new ecological paradigm in social psychology context. *Environment and Behavior*, *27*, 723–743.

Stott, P. A. & Kettleborough J. A. (2002). Origins and estimates of uncertainty in predictions of twenty-first-century temperature rise. *Nature*, *416*, 723–726.

Theunissen, A. (2018). Climate change lingo at your fingertips. *FarmBiz*, *4*(4), 42.

Thompson, H. E., Berrang-Ford, L., & Ford, J. D. (2010). Climate change and food security in Sub-Saharan Africa: a systematic literature review. *Sustainability*, *2*, 2719–2733. doi:10.3390/su2082719

Thompson, M. & Rayner, S. (1998). Risk and governance part I: the discourses of climate change. *Government and Opposition*, *35*(2): 139–166.

Tirivangasi, H. M. & Nyahunda, L. (2019). Challenges faced by rural people in mitigating the effects of climate change in the Mazungunye communal lands, Zimbabwe. *Jàmbá: Journal of Disaster Risk Studies*, *11*(1), 1–9.

Wolf, E. (1982). *Europe and the People Without History* (503 pp). Oakland, CA: University of California Press.

World Bank. (2010). *Economics of Adaptation to Climate Change. Ghana Country Study*. Washington, DC: World Bank.

Wossen, T., Berger, T., Haile, M. G., & Troost, C. (2018). Impacts of climate variability and food price volatility on household income and food security of farm households in East and West Africa. *Agricultural Systems*, *163*, 7–15.

15

CLIMATE CHANGE IMPACTS ON AGRICULTURE AND BARRIERS TO ADAPTATION TECHNOLOGIES AMONG RURAL FARMERS IN SOUTHWESTERN NIGERIA

Ayansina Ayanlade, Isaac Ayo Oluwatimilehin and Oluwatoyin Seun Ayanlade

Introduction

Climate change is considered the greatest challenge to agricultural productivity in Africa, with a projected change of about 8% in the average yield of all crops by 2050 (Clay and King, 2019; Collier et al., 2008; Knox et al., 2012; Sultan et al., 2013). The observed changes in rainfall and temperature are expected to pose great challenges to agriculture because rain-fed agriculture provides almost 90% of the staple food in sub-Saharan Africa (SSA) and is the main source of livelihood for about 70% of its total population (Bischiniotis et al., 2018). Agriculture employs about 80% of the labour force in SSA and contributes to about 25% of its GDP (Amani et al., 2021). The Intergovernmental Panel on Climate Change (IPCC, 2014) defines climate change as a change in the state of the climate as indicated by changes in the mean or variability. In recent years, climate change has been reported to make croplands vulnerable to bushfires, pests and pathogens, and to reduce the yields and income expected by farmers, which is likely to deter many from farming in SSA (Trisos et al., 2022). Many factors can affect crop yields, including soil type and moisture, but temperature and rainfall have been shown to be among the main factors affecting crop production, as climate determines water availability for crop growth and production, which directly affect crop yields (Ayanlade et al., 2018; Bang et al., 2019; Cammarano et al., 2019). Rainfall is important for the stability and quality of yields and it can also pose a threat to total yield. For example, reduced or erratic rainfall during cassava germination and tuber development phases can lead to yield losses, just as temperature can favour the growth of weeds and the spread of pests, which in turn leads to yield losses (Oseni and Masarirambi, 2011).

In this chapter, we present a case of climate change impacts on cassava production. Cassava is a crop of the tropical lowlands and is restricted to regions within latitudes 30° north and south of the equator, with higher concentrations near the equator (Onyeneke et al., 2020). It thrives in warm and humid climatic conditions with temperatures between 25 and 29°C and rainfall of about 1,000–15,000 mm, ideally evenly distributed throughout the growing season. However, cassava

DOI: 10.4324/9781003356837-20

has tremendous adaptability to different climatic conditions as it can withstand high temperatures of about 37°C and low temperatures of about 17°C (IITA, 2012). In Nigeria, cassava was traditionally cultivated as a crop, being the base of several Indigenous foods such as gari, eba, fufu, and amala (Eyinla, 2019; Kolawole et al., 2010; Oluwamukomi and Lawal, 2020; Woodley, 2011). Today, cassava is used for animal feed production and processing, as a raw material for starch, and as ethanol for industrial purposes, among others. About two-thirds of Nigerian cassava production is exported to other countries in SSA (Bain et al., 2013; Behnassi et al., 2021; Murumkar et al., 2020; Onyenechere, 2010; Yadav et al., 2019). Since cassava thrives on poorly drained soils and has less need for fertile soils and rainfall than other crops (Descheemaeker et al., 2016), it is grown almost everywhere in Nigeria, although production is concentrated among many smallholder farmers in the central and southern regions, where different varieties of cassava are grown (Diakite et al., 2020; Ekwezuo and Ezeh, 2020).

The local and international demand for cassava and its products has led to an increase in its cultivation in many regions of Nigeria, making it one of the most widely grown crops in different ecological zones in the country (Araro et al., 2020; Gebrechorkos et al., 2020; Kogo et al., 2021). The cultivation of cassava has evolved from traditional production systems to the use of high-yielding varieties and mechanised production and processing. However, agriculture in Nigeria, as in other parts of SSA, is constrained by climate change as current climatic conditions are not close to the optimal conditions for cassava production (Adejuwon, 2004; Ayanlade et al., 2018). Assuming that Nigeria's population is expected to grow to become the third largest country in the world by 2050 (Abubakar and Dano, 2018), the demand for common food crops, including cassava, will increase. There is therefore a need to find ways to cope with climate change impacts on crop production systems (FAO, 1996; Loum and Fogarassy, 2015). While cassava production might have been severely affected by the changing climatic conditions in the country, this is not well documented in the literature. Studies have shown that the relationship between climate change and crop production in Nigeria varies depending on the type of crop, seasonal properties and life cycle of the crop (Niero et al., 2015; Odekunle et al., 2007; Onasanya et al., 2021; Stuch et al., 2021). Generally, climate change impacts have been found to be pronounced on crops outputs, as an increased occurrence of dryness in terms of elongated spell length is capable of reducing yield (Choudhary et al., 2020; Han et al., 2019; Sniderman et al., 2019). The literature, including several chapters of this book, shows that climate change has significant implications for global and regional food production, although the exact extent of these impacts is unknown. This chapter examines the impact of climate change on cassava production in one state in Nigeria. Farmers' reports of climate change impacts, their experiences, and their adaptive capacity were also explored.

Methodology

Research design and data collection

A mixed methods approach was used. The study used historical climate data and survey data. Satellite climate datasets for rainfall and temperature from 1980s to 2020 were used in the analysis. To assess rainfall variability, we considered the number of rain days during various growing seasons and the total rainfall for each year. In this study, a year is divided into two growing seasons: the early growing season from April to June and the late growing season from July to October.

Survey data were collected among farmers in rural communities in freshwater ecological zones situated in Ogun State, Nigeria (Table 15.1). Ogun State is a home of some cassava production in Nigeria (Figure 15.1), with a total land area of approximately 17,000 square kilometres located

Table 15.1 Sampling and sampling locations

State	*Settlements*	*Latitude (°N)*	*Longitude (°E)*	*Samples (%)*
Ogun	Onibode	7.1420	3.4372	30.5
	Kobape	7.0986	3.3925	30.6
	Odeda	7.2328	3.5281	28.8
	Ota	6.7077	3.2560	10.1
	Total			100

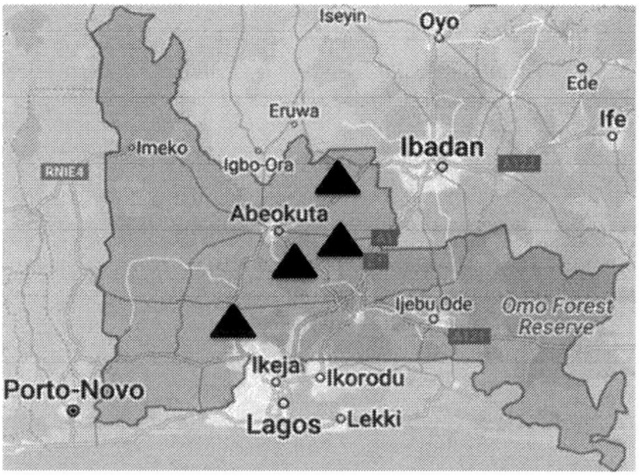

Figure 15.1 Map of the study area showing sample village (black triangle) points.

between latitudes 6012′0″N and 7047′60″N and longitudes 300′0″E and 500′0″ east of the Greenwich Meridian (Solanke, 2013). The area is typical of a tropical climate consisting of two different seasons of wet and dry conditions. The wet season is relatively associated with the dominance of the moist maritime southerly monsoon from the Atlantic Ocean, while the dry season is predicted by the continental northeast trade wind from the Sahara Desert. The average temperature varies according to the month with a distinction of 25.7°C in July and 30.2°C in February. The area belongs to one of the top agrarian states in Nigeria, with abundant cassava crop production (Ayanda et al., 2013; Bamiro et al., 2012; Teeken et al., 2018). The major occupation of the people of Ogun state is agriculture, which employs many people, especially those in rural regions. The state is also noted for its numerous educational institutions, which employ a large number of people. Several other identifiable modern economic activities exist in Ogun state, including insurance and motor companies, petrol stations, light and heavy industries (Solanke, 2015).

The Ogun Agricultural Development Programme (OGADEP) provided cassava yield statistics, which were calculated by dividing the total amount of cassava produced by the size of the farmland. The crop yield data were available only on an annual scale and was obtained by dividing the total production of the year by the total land area from which it was derived. Data on farmers' practices were sourced primarily with the aid of semi-structured interviews (SSI) and focus group discussions (FGD), which were purposely administered to farmers to assess their adaptation options and implementation barriers. The targeted population for the study were farmers 30 years or older. One hundred and twenty copies of a questionnaire were administered to farmers in the study sites, out of which 115 copies were retrieved. Using the purposive sample method, three FGDs with six semi-structured interviews were conducted to assess the farmers' perceptions of climate and experience in different farming communities. Content analysis was performed to assess farmers' cropping experiences. The historical profile of the settlement was collected from the high chief of the settlement, while participatory oral interviews were conducted for each farmer in the settlement. The results of the FGD and SSI were analysed using content analysis, while the data from the questionnaire were analysed using descriptive statistics involving frequencies and percentage error graphs to present the adaptive capacity of farmers and barriers to adaptation. This study followed the research protocol developed by the Local Indicator of Climate Change Impact (LICCI) Research Group regarding SSI, FGD, and survey (Reyes-García et al., 2023). The data from the FGD and SSI were transcribed and analysed using content analysis. Descriptive statistics involving frequencies, and percentage error graphs were used to examine the adaptation methods used by farmers and barriers to adaptation. Gender assessment of impacts and adaptation was carried out, as gender has been identified to play a critical role in the use of adaptive measures of climate change (Obossou et al., In press).

Data analyses

Correlation and multiple regressions were used to show the impact of climate on cassava yield. A line graph was plotted using SigmaPlot 10.0 to show changes in climate parameters, while the standardised anomaly index (SAI) was used to assess the annual variation from the mean. Sunburst diagrams and linear dendrograms were used to illustrate survey results. The SAI has become the most popular precipitation and temperature index based on hypothetical development and usefulness for analysis (Baig et al., 2022; Ghosh and Mujumdar, 2007; Raziei, 2021; Rivera and Penalba, 2014). This measure presents the number of standard deviations from which a climatic record is above or below the climatological average for a station of a particular location. This study employs the use of annual data to calculate the anomaly index as in Oluwatimilehin and Ayanlade (2021). To estimate the parameters, a probability-weighted moment was employed using the classification of the anomaly index shown in Table 15.2. Multiple regression analysis was performed to find the combined impact of climate on each of the crops. With the use of climatic (temperature and rainfall) data, the impacts of climate on crop yield were established with the use of multiple regression.

Survey data were also analysed to assess the gender differences in adaptation and barriers to climate change adaptation. A comprehensive assessment of gender differences was conducted in order to recognise the adaptation capacities of both male and female farmers and to understand gender-based adaptation differences, as climate change affects men and women farmers differently. We establish critical variables that help us in assessing gender differences in climate change

Table 15.2 Anomaly categorisation

Anomaly class	*Categorisation*
2.0+	Extremely wet
1.5–1.99	Very wet
1.0–1.49	Moderately wet
−0.99 to 0.99	Near normal
−1.0 to −1.49	Moderately dry
−1.5 to −1.99	Severely dry
−2 and less	Extremely dry

adaption and adaptation barriers. Utilising quantitative approaches, we determine trends and substantial inequalities between genders, as well as the barriers that prevent women from properly responding to climate change.

Results and discussion

Climate parameters and anomalies

The analysis of meteorological data during the study period revealed high variations in rainfall and temperature both annually and monthly (Figures 15.2 and 15.3). There is a high variation in climatic parameters, as shown in Figure 15.2, as significant annual variations in rainfall were observed for each year in relation to annual total rainfall. Climate variability, as documented in this part of Nigeria, can cause negative departures from the normal weather (Adejuwon, 2005; Bello, 2008). The SAI of annual rainfall and temperature is shown in Figure 15.3. Categorisation around zero is an indication of normal precipitation or temperature, while those markedly above or below zero indicate relatively wet or dry conditions (Figure 15.3). The maximum temperature was above normal in the first and second decades (i.e., SAI > 1), moderate in the third decade, and has increasingly furthered below normal since the fourth decade. The minimum temperature was generally above normal in the first to the middle of the second decade. Rainfall was generally far-off above normal from the early years covering the study period until the 1990s and falls below normal in the consecutive years till 2020.

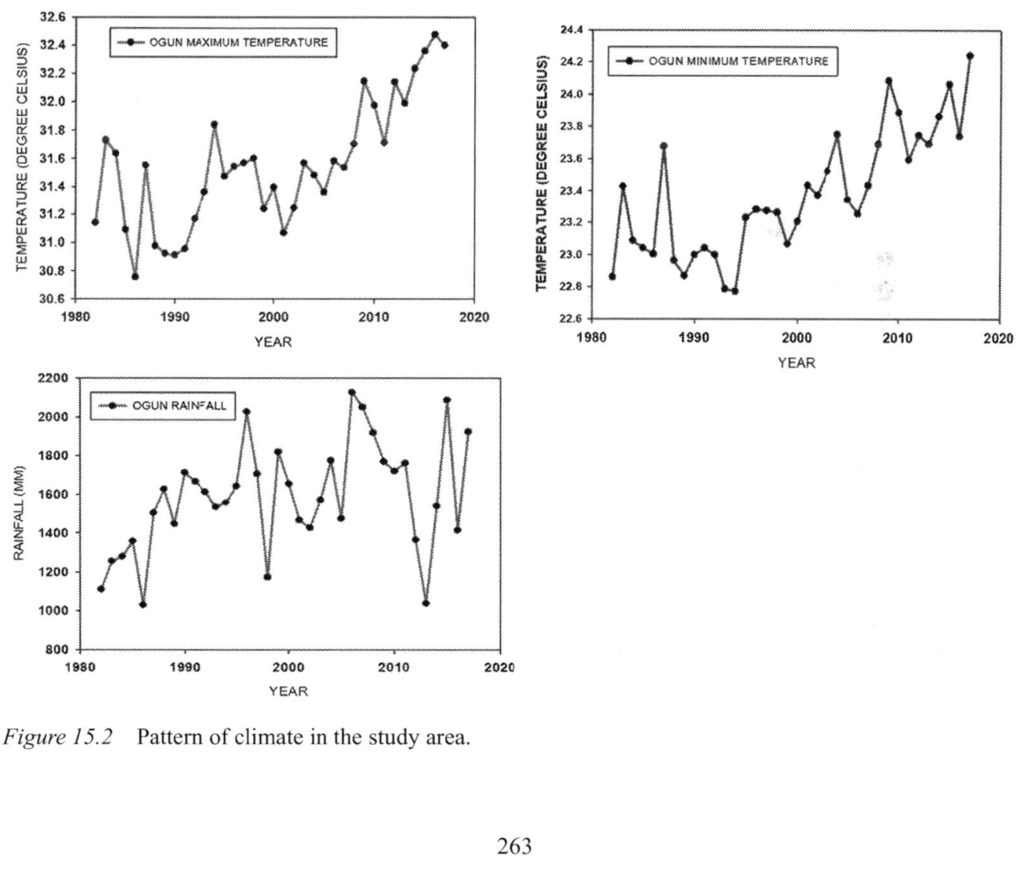

Figure 15.2 Pattern of climate in the study area.

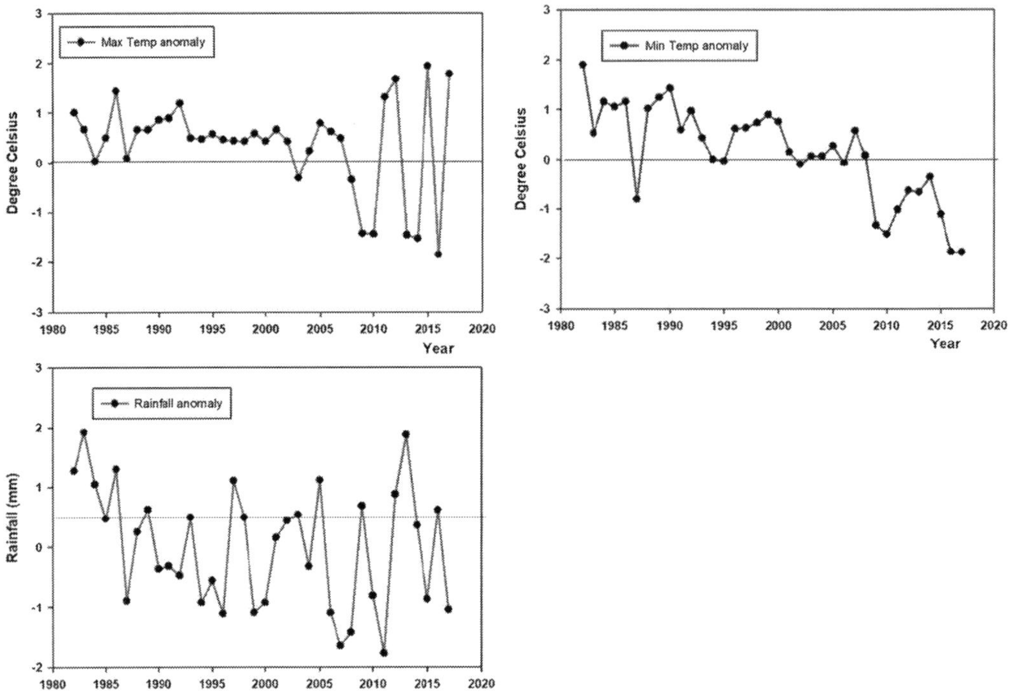

Figure 15.3 Standardised Anomaly Index of climate parameters.

Table 15.3 Correlation between climate and crop yield

Cassava yield	*Rainfall*	*Min temperature*	*Maximum temperature*
Annual	0.84	0.82*	0.87*
Early growing season	0.79*	0.42	0.37
Late growing season	0.75*	0.32	0.57

* Significant at $p < 0.05$, Pearson correlation.

Changes in cassava yield as evidence of climate impacts

The results of the correlation analysis between the annual minimum temperature and crop yields during the growing season are shown in Table 15.3. There is a strong and positive correlation between the yield of cassava and rainfall, and minimum and maximum temperature. However, the relation is only statistically significant for the minimum and maximum temperatures. When combined, rainfall, minimum and maximum temperatures account for over 75% of changes in cassava yield, while the remaining 25% can be explained by other variables, such as edaphic conditions and cassava resistance factors (Table 15.3). Reduction and erratic rainfall both negatively affect the yield of crops. Indeed, the relationship between climate and crop production in Nigeria has been reported to vary depending on the type of crop, seasonal properties and life cycle of the crop under consideration (Ajetomobi, 2016; Odekunle et al., 2007; Olanrewaju et al., 2017).

As shown by the results in Table 15.3, rainfall had a strong relationship with crop yield during both the early and late growing seasons, with $R > 70\%$. The results further showed a very strong relationship between yields and rainfall in both early and late growing seasons (Table 15.3). A strong relationship between rainfall and crop yield is not surprising, as previous studies have reported similar results (Ajetomobi, 2016; Bogale et al., 2022; Odekunle et al., 2007; Olanrewaju et al., 2017). While rainfall variability is capable of influencing maize yield, many of these studies reported that temperature is capable of contributingto changes in yield of approximately 50%; relative humidity can increase yields when it is high throughout the growing season (Ayanlade et al., 2018; Bogale et al., 2022; Odekunle et al., 2007).

Previous work shows that prolonged dry spells combined with high temperatures are likely to have a significant impact on yield (Ayanlade et al., 2018; Bogale et al., 2022; Jägermeyr et al., 2021; Odekunle et al., 2007; Poudel and Kotani, 2013), particularly for cassava. Cassava yields are affected by several factors, though, including the amount of rainfall at the start and end of the growing season, but also fertiliser use, and resistance to pests/mildew, such as nematode moulds (Adiele et al., 2020; De Bauw et al., 2021; Onasanya et al., 2021). With the introduction of climate-smart cassava varieties and the liberalisation of the seed value chain, which enhances the availability of improved seeds to farmers, it has been argued that it will be possible to boost yield even with worsening climatic conditions.

Reported changes and adaptation options

Survey results suggest that most farmers in the sample perceived that the temperature of the hottest month is now warmer than in the past. Many also report that extreme floods are more frequent, while extreme drought and the durations of the rainy and dry seasons have remained the same.

To adapt to those changes, farmers respond in many ways. The adaptation methods used in the area are shown in Figure 15.4. Mixed farming planting of different species or varieties, agricultural

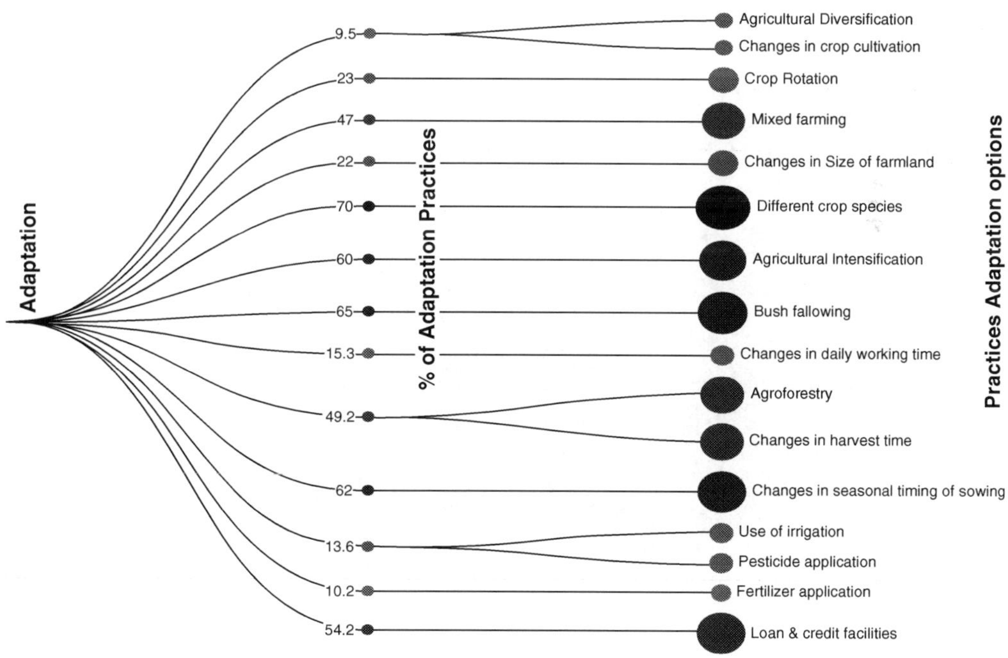

Figure 15.4 Practices adaptation methods for rural farmers.

intensification, bush fallowing, changes in seasonal timing of sowing and harvest, and the use of credit facilities help enhance cassava yields over the study period (Figure 15.4). Some other common adaptations were irrigation, the use of loans and credit facilities, agricultural intensification and agroforestry, among others. Additionally, climate change, together with other societal and ecological changes, has led to farmers abandoning some landrace crops, such as palm, cocoyam and water yam, which are considered to be less productive. Farmers have also adopted new crops, such as watermelon, cucumber and pepper, based on market demand.

It is noticeable from survey results, however, that many farmers can not adopt the desired adaptation options. Barriers to practising adaptation were ranked by farmers, with lack of natural, physical, human, social and financial capital accounting for >70% of the barriers to practising adaptation methods (Figure 15.5). Figure 15.5 shows that 41.1% of people chose not to practise some of the adaptation methods because they considered they faced a low risk of suffering climate change impacts. Many rural farmers in our survey noted that low financial capital is the main

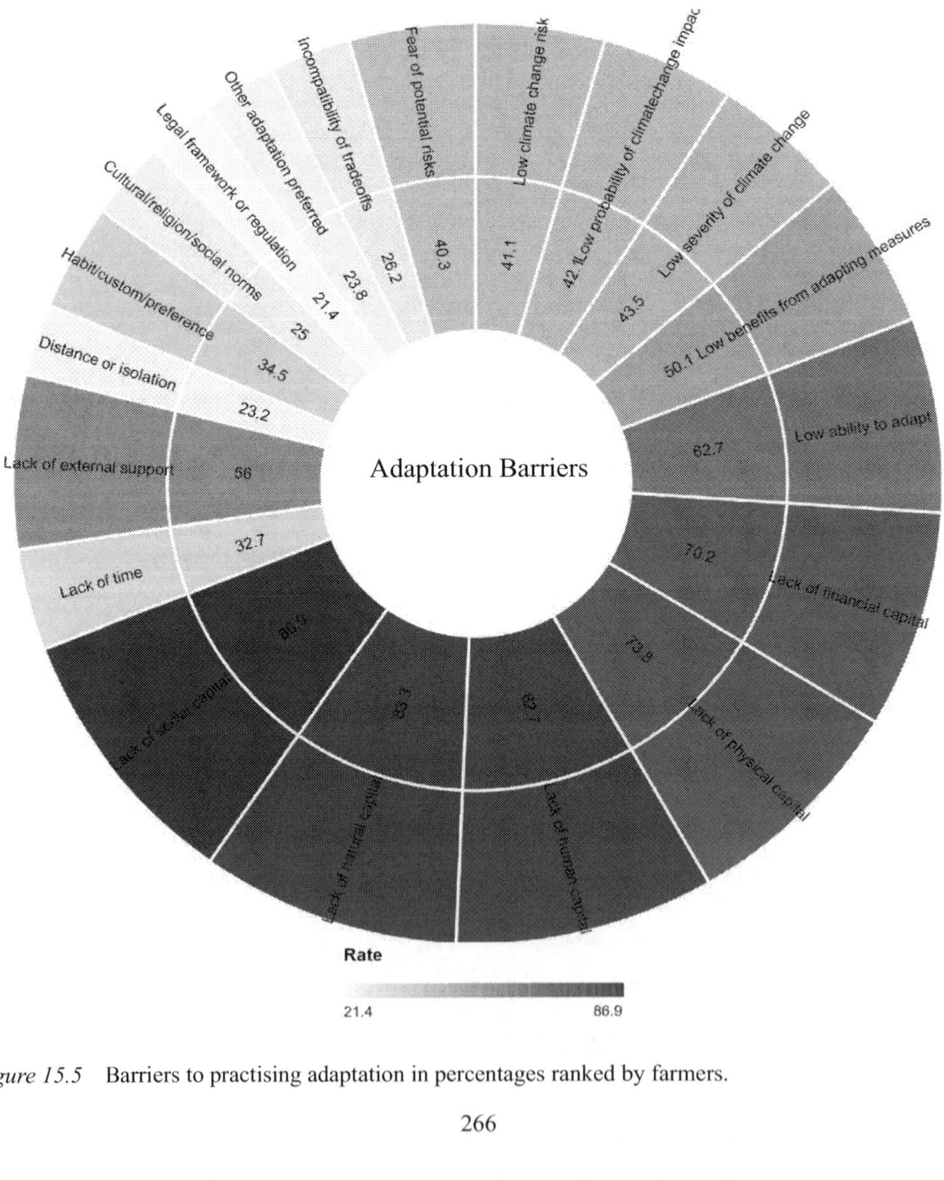

Figure 15.5 Barriers to practising adaptation in percentages ranked by farmers.

barrier to any adaptation method. Lack of financial capital is probably responsible for low adaptation practices in the study area.

Gendered differences in adopting adaptation options

Notably, the present study assesses gender differences in adaptation and barriers to adaptation, as presented in Tables 15.4 and 15.5. Nearly 65% of farmers in the study areas are male. Climate change adaptation strategies are said to vary according to gender, age, farm size, and farming experiences (Obossou et al., In press). Women farmers adopt some adaptation measures, though most of these are largely practised by male farmers. For example, 57.9% of male and 42.1% of female farmers in the study area practice different crop compositions. Similarly, both male and female farmers have nearly equivalent percentages of use for some adaptation methods (Table 15.4), such as irrigation (56.9% male, 43.1% female), fertiliser application (58.5% male; 41.5% female) and pesticide application (58.8% male; 41.2% female). Compared to female farmers, generally >60% of male farmers use most adaptation methods (Table 15.4).

Most female farmers reported not practising specific adaptation methods. It is obvious that men farmers perceive more of these barriers than female farmers, who often perceive less obstacles to adaptation as well as less likelihood and severity of climate change. For female farmers, the lack of financial, natural, social and physical capital are important barriers to adaptation, although some deliberately have not taken any measures or have chosen not to practise adaptation (Table 15.5). Barriers to adaptation practices by rural farmers vary by gender. From Table 15.5, it is noticeable that both males and females have low benefits from adapting measures (55.9% male and 44.1% female farmers). Thus, they have a low ability to adapt (55.3% male and 44.7%

Table 15.4 Adaptation practices by rural farmers based on gender (male 67.1% and female 32.9%)

Adaptation options	Male	Female
Agricultural diversification	61.1	38.9
Changes in crop cultivation	59.3	40.7
Crop rotation	60	40
Mixed farming	57.9	42.1
Changes in size of farmland	61.1	38.9
Different crop composition	57.9	42.1
Agricultural intensification	60.7	39.3
Bush fallowing	57.1	42.9
Changes in daily working time	58	42
Agroforestry	70	30
Changes in seasonal timing of sowing	60	40
Changes in harvest time	73.3	26.7
Use of irrigation	56.9	43.1
Fertiliser application	58.5	41.5
Pesticide application	58.8	41.2
Loan and credit facilities	70.4	29.6
Changes in income	61.2	38.8
Changes in food storage	58.7	41.3
Changes in quantity of food consumed	87.5	12.5
Changes in residence	71.4	28.6

Table 15.5 Barriers to adaptation practices by rural farmers based on gender

Barriers option	Male	Female
Low climate change risk	78.9	21.1
Low probability of climate change impact	80	20
Low severity of climate change	63.2	36.8
Low benefits from adapting measures	55.9	44.1
Low ability to adapt	55.3	44.7
Lack of financial capital	59.5	40.5
Lack of physical capital	56.8	43.2
Lack of human capital	59.1	40.9
Lack of natural capital	61.2	38.8
Lack of social capital	60	40
Lack of time	66.7	33.3
Lack of external support	64.3	35.7
Distance or isolation	60	40
Habit/custom/preference	64.3	35.7
Cultural/religious/social norms	55.6	44.4
Legal framework or regulation	55.6	44.4
Other adaptations with similar purposes preferred	37.5	62.5
Incompatibility of trade-offs	50	50
Fear of potential risks or adverse effects	60	40

female). The major barriers to the use of adaption methods noted by both male and female farmers include lack of financial capital (59.5% male; 40.5% female), physical capital (56.8% male; 43.2% female) and human capital (59.1% male; 40.9% female)(Table 15.5).

Gender has been identified to play a critical role in the use and practices of mitigation and adaptive measures for climate change in many societies (Nyasimi et al., 2018; Pearse, 2017; Prakash et al., 2022). The impact of climate change will be disproportionately harsh for the poorest and most vulnerable people and groups, particularly for women in Africa (Trisos et al., 2022; UNDP, 2012) (Prakash et al., 2022). Women who receive less education tend to be poorer and they are usually excluded from decision-making processes both at the political and household levels. It is therefore important to recognise the adaptation capacities of both males and females to understand gender-based adaptation differences since climate change affects men and women differently (Prakash et al., 2022).

Conclusions

In this study, the impact of climate change on cassava crops was examined and rural farmers' adaptation techniques were studied. It was observed that the rainfall and temperature were marked with high variations, which also correlated strongly and positively with the yield of cassava. A significant number of farmers interviewed reported that temperature has increased and rainfall has declined, affecting their livelihoods. But there are differences in the ability of farmers to adapt, as their adaptive capacities vary significantly. The study recommends the enhancement of adaptation options available to farmers in the region for better yield to unlock new business opportunities for the country through cassava processing.

Other adaptation measures, such as better access to markets, extension and credit services, technology and farm assets are effective and have been ongoing in Africa. These can also be encouraged in the region. Farmers in the study area could also plant crops that require more moisture in the late growing season, while crops that can endure or require higher temperatures should be planted in the early growing season. Additionally, farmers in Nigeria should have access to climatic information, as well as predictions and forecasts for farming seasons. Although the government, international bodies, and local organisations are funding adaptation through loan and credit facilities to farmers, current levels of funding are insuficient. Governments at all levels must support rain-fed agriculture and agricultural research in order to increase crop yields as a result of climate change.

Acknowledgements

The authors wish to express their gratitude to the Ogun State Agricultural Development Programme (OGADEP) for supplying crop yield data for this work.

This work was supported and funded by the Tertiary Education Trust Fund, TETFund NRF 2020 Nigeria (Grant Award – TETF/DR&D-CE/NRF2020/CC/17/VOL.1). A.A. also acknowledges support from a small research grant from the Queen Elizabeth Scholars project on ecological economics, commons governance, and climate justice, with funding from the Canadian Queen Elizabeth II Diamond Jubilee Scholarships Advanced Scholars Program (QES-AS). The research was also partly funded by an ERC Consolidator Grant to Reyes-García (FP7-771056-LICCI).

Conflicts of interest/competing interests

The authors declare that they have no competing interests.

Ethics approval/declarations

This study conforms with the ethics principles of Obafemi Awolowo University. Ethical approval for this study was obtained with standard ethics as the participants were not vulnerable in any way, data was processed in an anonymous procedure, and survey participants had the possibility to skip questions.

References

Abubakar IR and Dano UL. (2018) Socioeconomic challenges and opportunities of urbanization in Nigeria. *Urbanization and Its Impact on Socio-Economic Growth in Developing Regions.* IGI Global, 219–240. Umar Benna and Indo Benna

Adejuwon JO. (2004) Crop yield response to climate variability in the Sudano-Sahelian ecological zones of Nigeria in southwestern Nigeria. *AIACC Report of Workshop for Africa and Indian Ocean Island.* Dakar, Senegal, 15–16. START https://start.org/publication/aiacc-summary-of-the-final-report/

Adejuwon JO. (2005) Food crop production in Nigeria. I. Present effects of climate variability. *Climate Research* 30: 53–60.

Adiele J, Schut AG, Van Den Beuken R, et al. (2020) Towards closing cassava yield gap in West Africa: agronomic efficiency and storage root yield responses to NPK fertilizers. *Field Crops Research* 253: 107820.

Ajetomobi J. (2016) Effects of weather extremes on crop yields in Nigeria. *African Journal of Food, Agriculture, Nutrition and Development* 16: 11168–11184.

Amani A, Sheffield J, Capdevilla A, et al. (2021) Strengthening flood and drought risk management tools for the Lake Chad Basin. *Climate Change and Water Resources in Africa: Perspectives and Solutions Towards an Imminent Water Crisis*, 387–405. Springer

Araro K, Legesse SA and Meshesha DT. (2020) Climate change and variability impacts on rural livelihoods and adaptation strategies in Southern Ethiopia. *Earth Systems and Environment* 4: 15–26.

Ayanda I, Oyeyinka R, Salau S, et al. (2013) Perceived effects of climate change on transhumance pastoralists in Ogun State, Nigeria. *Journal of Agricultural Science* 5: 190.

Ayanlade A, Radeny M, Morton JF, et al. (2018) Rainfall variability and drought characteristics in two agro-climatic zones: an assessment of climate change challenges in Africa. *Science of the Total Environment* 630: 728–737.

Baig MRI, Naikoo MW, Ansari AH, et al. (2022) Spatio-temporal analysis of precipitation pattern and trend using standardized precipitation index and Mann–Kendall test in coastal Andhra Pradesh. *Modeling Earth Systems and Environment* 8: 2733–2752.

Bain LE, Awah PK, Geraldine N, et al. (2013) Malnutrition in Sub–Saharan Africa: burden, causes and prospects. *Pan African Medical Journal* 15: 1–9.

Bamiro OM, Afolabi M and Daramola F. (2012) Enterprise combinations in cassava based food crop farming system in Nigeria: Evidence from Ogun State. *Greener Journal of Agricultural Sciences* 2: 13–20.

Bang S, Bishnoi R, Chauhan AS, et al. (2019) Fuzzy logic based crop yield prediction using temperature and rainfall parameters predicted through ARMA, SARIMA, and ARMAX models. *2019 Twelfth International Conference on Contemporary Computing (IC3).* IEEE, 1–6.

Behnassi M, Rahman MA, D'Silva J, et al. (2021) Enhancing resilience for food and nutrition security within a changing climate. *Emerging Challenges to Food Production and Security in Asia, Middle East, and Africa.* Springer, 1–42. Mohamed Behnassi, Mirza Barjees Baig, Mahjoub El Haiba, Michael R. Reed.

Bello W. (2008) Will capitalism survive climate change? *Chain Reaction* 104: 44.

Bischiniotis K, van den Hurk B, Jongman B, et al. (2018) The influence of antecedent conditions on flood risk in sub-Saharan Africa. *Natural Hazards and Earth System Sciences* 18: 271–285.

Bogale GA, Mengesha M and Hadgu G. (2022) Characterization of local climate and its impact on faba bean (*Vicia faba* L.) yield in Central Ethiopia. *Advances in Meteorology* 2022. 1–11

Cammarano D, Hawes C, Squire G, et al. (2019) Rainfall and temperature impacts on barley (*Hordeum vulgare* L.) yield and malting quality in Scotland. *Field Crops Research* 241: 107559.

Choudhary S, Guha A, Kholova J, et al. (2020) Maize, sorghum, and pearl millet have highly contrasting species strategies to adapt to water stress and climate change-like conditions. *Plant Science* 295: 110297.

Clay N and King B. (2019) Smallholders' uneven capacities to adapt to climate change amid Africa's 'green revolution': case study of Rwanda's crop intensification program. *World Development* 116: 1–14.

Collier P, Conway G and Venables T. (2008) Climate change and Africa. *Oxford Review of Economic Policy* 24: 337–353.

De Bauw P, Birindwa D, Merckx R, et al. (2021) Improved genotypes and fertilizers, not fallow duration, increase cassava yields without compromising arbuscular mycorrhizal fungus richness or diversity. *Mycorrhiza* 31: 483–496.

Descheemaeker K, Oosting SJ, Tui SH-K, et al. (2016) Climate change adaptation and mitigation in smallholder crop–livestock systems in sub-Saharan Africa: a call for integrated impact assessments. *Regional Environmental Change* 16: 2331–2343.

Diakite A, Bayala J and Dembele C. (2020) *End-line study-Capacitating Stakeholders in Using Climate Information for Enhanced Resilience in the Agricultural Sector in West Africa (CaSCIERA-TA).* CGIAR https://cgspace.cgiar.org/handle/10568/111413

Ekwezuo CS and Ezeh CU. (2020) Regional characterisation of meteorological drought and floods over west Africa. *Sustainable Water Resources Management* 6: 1–11.

Eyinla T. (2019) Evaluation of starches digestibility and glycemic index of commonly consumed staple foods from selected varieties of cassava (*Manihot* sp.) and yam (*Dioscorea* sp.). University of Ibadan. IITA https://biblio1.iita.org/handle/20.500.12478/7892

FAO. (1996) Report of the World Food Summit. *World Food Summit.* FAO, Rome.

Gebrechorkos SH, Hülsmann S and Bernhofer C. (2020) Analysis of climate variability and droughts in East Africa using high-resolution climate data products. *Global and Planetary Change* 186: 103130.

Ghosh S and Mujumdar PP. (2007) Nonparametric methods for modeling GCM and scenario uncertainty in drought assessment. *Water Resources Research* 43: 1–19.

Han F, Cook KH and Vizy EK. (2019) Changes in intense rainfall events and dry periods across Africa in the twenty-first century. *Climate Dynamics* 53: 2757–2777.

IITA. (2012) *Growing in Nigeria. Commercial Crop Production Guide Series. Information and Communication Support for Agricultural Growth in Nigeria.* USAID, 1–8. IITA https://biblio.iita.org/documents/U03ManIitaMaizeNothomNodev.pdf-43dafa4ce4033f16250975d3b036c570.pdf

IPCC. (2014) *Climate Change 2014–Impacts, Adaptation and Vulnerability: Regional Aspects.* Cambridge University Press, United Kingdom.

Jägermeyr J, Müller C, Ruane AC, et al. (2021) Climate impacts on global agriculture emerge earlier in new generation of climate and crop models. *Nature Food* 2: 873–885.

Knox J, Hess T, Daccache A, et al. (2012) Climate change impacts on crop productivity in Africa and South Asia. *Environmental Research Letters* 7: 034032.

Kogo BK, Kumar L and Koech R. (2021) Climate change and variability in Kenya: a review of impacts on agriculture and food security. *Environment, Development and Sustainability* 23: 23–43.

Kolawole PO, Agbetoye L and Ogunlowo SA. (2010) Sustaining world food security with improved cassava processing technology: the Nigeria experience. *Sustainability* 2: 3681–3694.

Loum A and Fogarassy C. (2015) The effects of climate change on cereals yield of production and food security in Gambia. *APSTRACT: Applied Studies in Agribusiness and Commerce* 9: 83.

Murumkar A, Durand M, Fernández A, et al. (2020) Trends and spatial patterns of 20th century temperature, rainfall and PET in the semi-arid Logone River basin, Sub-Saharan Africa. *Journal of Arid Environments* 178: 104168.

Niero M, Ingvordsen CH, Jørgensen RB, et al. (2015) How to manage uncertainty in future Life Cycle Assessment (LCA) scenarios addressing the effect of climate change in crop production. *Journal of Cleaner Production* 107: 693–706.

Nyasimi M, Ayanlade A, Mungai C, et al. (2018) Inclusion of gender in Africa's climate change policies and strategies. *Handbook of Climate Change Communication: Vol. 1.* Springer, 171–185. Berlin

Obossou, E., J.M. Chah, I.Q. Anugwa, V. Reyes-García. Gender Dimensions in the Adoption of Climate-Smart Agriculture Technologies in Response to Climate Change Extremes in Benin. Regional Environmental Change, 23, 93 (2023). https://doi.org/10.1007/s10113-023-02085-4

Odekunle T, Orinmoogunje I and Ayanlade A. (2007) Application of GIS to assess rainfall variability impacts on crop yield in Guinean Savanna part of Nigeria. *African Journal of Biotechnology* 6: 2100–2113.

Olanrewaju R, Tilakasir S and Oso C. (2017) *Climate Change and Rice Production: A Case Study in Ekiti State, Nigeria*. Sabaragamuwa University Press, Sri Lanka

Oluwamukomi MO and Lawal OS. (2020) Textural characteristics of Nigerian foods. *Textural Characteristics of World Foods* 361–383. Katsuyoshi Nishinari

Oluwatimilehin IA and Ayanlade A. (2021) Agricultural community-based impact assessment and farmers' perception of climate change in selected Ecological Zones in Nigeria. *Agriculture & Food Security* 10: 3.

Onasanya OO, Hauser S, Necpalova M, et al. (2021) On-farm assessment of cassava root yield response to tillage, plant density, weed control and fertilizer application in southwestern Nigeria. *Field Crops Research* 262: 108038.

Onyenechere E. (2010) Climate change and spatial planning concerns in Nigeria: remedial measures for more effective response. *Journal of Human Ecology* 32: 137–148.

Onyeneke RU, Nwajiuba CU, Tegler B, et al. (2020) Evidence-based policy development: National Adaptation Strategy and Plan of Action on Climate Change for Nigeria (NASPA-CCN). *African Handbook of Climate Change Adaptation* 1: 1–18.

Oseni T and Masarirambi M. (2011) Effect of climate change on maize (*Zea mays*) production and food security in Swaziland. *Change* 2: 3.

Pearse R. (2017) Gender and climate change. *Wiley Interdisciplinary Reviews: Climate Change* 8: e451.

Poudel S and Kotani K. (2013) Climatic impacts on crop yield and its variability in Nepal: do they vary across seasons and altitudes? *Climatic Change* 116: 327–355.

Prakash A, Conde C, Ayanlade A, et al. (2022) *Gender, Climate Justice and Transformative Pathways in IPCC Climate Change 2022: Impacts, Adaptation and Vulnerability*. Contribution of Working Group II to the Sixth Assessment Report of the Intergovernmental Panel on Climate Change [H.-O. Pörtner, D.C. Roberts, M. Tignor, E.S. Poloczanska, K. Mintenbeck, A. Alegría, M. Craig, S. Langsdorf, S. Löschke, V. Möller, A. Okem, B. Rama (eds.)]. Cambridge University Press, Cambridge.

Raziei T. (2021) Revisiting the rainfall anomaly index to serve as a simplified standardized precipitation index. *Journal of Hydrology* 602: 126761.

Reyes-García V, Álvarez-Fernández S, Benyei P, et al. (2023) Local indicators of climate change impacts described by indigenous peoples and local communities: study protocol. *PLoS One* 18: e0279847.

Rivera JA and Penalba OC. (2014) Trends and spatial patterns of drought affected area in Southern South America. *Climate* 2: 264–278.

Sniderman JK, Brown JR, Woodhead JD, et al. (2019) Southern Hemisphere subtropical drying as a transient response to warming. *Nature Climate Change* 9: 232–236.

Solanke M. (2013) Intra-Urban work and school trips characteristics in Ogun state, Nigeria. *International Journal of Development and Sustainability* 2.

Solanke M. (2015) Spatial pattern and organisational structure of intra-urban trips in Ogun State, Nigeria. *Ethiopian Journal of Environmental Studies and Management* 8: 13–27.

Stuch B, Alcamo J and Schaldach R. (2021) Projected climate change impacts on mean and year-to-year variability of yield of key smallholder crops in Sub-Saharan Africa. *Climate and Development* 13: 268–282.

Sultan B, Roudier P, Quirion P, et al. (2013) Assessing climate change impacts on sorghum and millet yields in the Sudanian and Sahelian savannas of West Africa. *Environmental Research Letters* 8: 014040.

Teeken B, Olaosebikan O, Haleegoah J, et al. (2018) Cassava trait preferences of men and women farmers in Nigeria: implications for breeding. *Economic Botany* 72: 263–277.

Trisos CH, Adelekan IO, Totin E, et al. (2022) Africa. In: Climate Change 2022: Impacts, Adaptation, and Vulnerability. Contribution of Working Group II to the Sixth Assessment Report of the Intergovernmental Panel on Climate Change [H.-O. Pörtner, D.C. Roberts, M. Tignor, E.S. Poloczanska, K. Mintenbeck, A. Alegría, M. Craig, S. Langsdorf, S. Löschke, V. Möller, A. Okem, B. Rama (eds.)]. Cambridge University Press.

UNDP. (2012) *Gender and Adaptation*. Available at: https://www.undp.org/content/dam/undp/library/gender/Gender%20and%20Environment/PB2_Africa_Gender-and-Adaptation.pdf.

Woodley E. (2011) Building Nigeria's response to climate change: pilot projects for community-based adaptation in Nigeria. *Experiences of Climate Change Adaptation in Africa.* Springer, 297–315.

Yadav SS, Hegde V, Habibi AB, et al. (2019) Climate change, agriculture and food security. In: Yadav, SS, Redden, RJ, Hatfield, JL, Ebert, AW, Hunter, D., Eds. *Food Security and Climate Change*. Wiley, Hoboken, New Jersey

16

CLIMATE CHANGE IMPACTS ON AGRICULTURE,ADAPTATIONAND RESILIENCE

Insights from local farmers in South-East Zimbabwe

Rumbidzayi Chakauya, Simeon A. Materechera, Obert Jiri, Ereck Chakauya and Machate Machete

Introduction

Zimbabwe is vulnerable to the impacts of climate change, likely to have significant consequences for the country's social, economic and environmental systems. It is estimated that, if climate change is not addressed, 3 billion people will be living in areas considered too hot for human survival in 50 years' time (Xu et al., 2020). The World Health Organisation (WHO) notes that to date 55 million people are affected by drought every year, with women and girl children mostly exposed to severe malnutrition as a result (WHO, 2014). There is no doubt that climate change impacts affect everyone, but they will be particularly impactful for smallholder farmers in marginal climatic conditions, unless adaptation and mitigation solutions are implemented.

Agriculture is the backbone of livelihoods in Zimbabwe's socio-economic development, contributing 14% to the country's Gross Domestic Product (GDP), 40% of foreign currency earnings, 35% to formal employment, and 36% of industrial raw materials (Manzungu, 2014; Nhemachena et al., 2014). Rainfed crops dominate the country's agricultural sector with major crops such as maize, tobacco, sugar cane and cotton. The sector is therefore extremely vulnerable to extreme weather conditions and climate shocks. Shifts in the onset of rains coupled with increasing incidences of flooding complicate human livelihoods in the country (The Government of Zimbabwe and United Nations Development Programme, 2017).

Extreme high temperatures now common in the south-eastern parts of the country have led to significant losses in crop productivity. Literature shows that annual surface temperatures across Zimbabwe have increased by 0.4°C, with the warmest years on record being 1987 (Brazier, 2018). Recurrent droughts are also a common hazard, with evidence of regular occurrence of extreme droughts. Twenty-one mild droughts have been reported between 1951 and 2010 (Nangombe, 2015). These temperatures affect crop yields directly and also indirectly through interaction with other factors such as soil quality, water availability, pest and disease outbreaks. Some projections suggest that crop yields in Zimbabwe could decline by as much as 50% by the end of the century

DOI: 10.4324/9781003356837-21

if no action is taken to adapt to climate change impacts. Already a combination of the reduced crop yields, political and economic challenges have downgraded the status of Zimbabwe from the "breadbasket" of Africa to a "basket case" (Masitera, 2021). The World Food Programme report (2021) estimated that 5.8 million people had insufficient food by October 2022 in Zimbabwe. As such, food security is very fragile and with the country's limited resource mobilisation and assistance from the international community, there is a need for enhancing adaptation and resilience among Zimbabwe's local farmers.

Furthermore, climate change is affecting the availability and quality of water resources in Zimbabwe. Meteorological data suggests that low precipitation, frequent and intense mid-season dry spells during rain seasons have increased during the 2015/2016 season, intensely affecting maize crops in areas like Mt Darwin (Mushore et al., 2017; Brazier, 2018). To date, low rainfall events have been noted in the country, collectively impacting Zimbabwe's agricultural production. Studies have predicted that the region may experience an increase in heavy rainfall, including flash floods and hail storms. To date, there has already been an increase in the incidence of floods in Chiredzi and Muzarabani (Brazier, 2018). In March 2019, cyclone Idai impacted 270,000 people in Manicaland province, leaving infrastructure destroyed, 341 people dead, and at least 17,500 homeless (International Federation of Red Cross Societies, 2019). Intense floods have also affected crop production and other areas of people's well-being, including children's access to food, health, education, clean water and sanitation (Dube, 2017). Put together, extreme climatic events do not just affect agriculture, but also retard the development of rural communities and pose health risks particularly for vulnerable groups.

Climate change is also expected to affect biodiversity and human health. On the one hand, Zimbabwe's unique and diverse ecosystems, including its forests and savannas, are vulnerable to the impacts of climate change. Changes in temperature and precipitation patterns could lead to the loss of biodiversity and ecosystem services, with negative consequences for human well-being and the economy. On the other hand, changes in temperature and rainfall patterns might lead to an increased risk of diseases such as malaria and cholera (Brazier, 2018).

Overall, the impacts of climate change in Zimbabwe are likely to be complex and multifaceted, requiring a range of adaptation and mitigation measures to address. This could include measures to improve agricultural productivity, manage water resources more effectively, and protect biodiversity. Indeed, farmers and the government are increasingly adopting a range of adaptation strategies to address climate-derived challenges. In recent years, the Zimbabwean government has implemented policies to reduce greenhouse gas emissions and to adapt to the impacts of climate change. The National Adaptation Plan (NAP), developed in 2015, outlines the country's strategy for adapting to climate change impacts, including measures like improving water management and irrigation systems, enhancing resilience of agriculture and food systems, and increasing the use of renewable energy. The government of Zimbabwe has also committed to reducing the country's greenhouse gas emissions through the Nationally Determined Contributions (NDCs) under the Paris Agreement. In addition, Zimbabwe has implemented a number of other policies and programs including the development of a National Climate Change Response Strategy, National Climate Policy of 2016 the establishment of a Climate Change Unit within the Ministry of Environment, Water and Climate, and the implementation of the Zimbabwe Environmental Law Association (ZELA) program to support the adoption of clean energy technologies.

Farmers have come up with innovative ways of addressing climate change impacts and improving resilience. Diversifying crops, soil conservation, water conservation and climate-smart agriculture are some of the strategies being used. Among these, diversifying crops is arguably the major adaptation strategy in the toolbox with a variety of crop mixes (Waha et al., 2018). Farmers grow a range of crops with different moisture requirements to increase the chances of a harvest. Since climate change is expected to lead to more frequent and severe droughts in Zimbabwe, farmers are also adopting strategies to conserve water, including the use of irrigation technologies such as drip irrigation, or the use of drought-tolerant crop varieties (Mavhura et al., 2015). As more frequent extreme weather events are expected, farmers are using conservation tillage practices and planting of cover crops to protect soil from erosion. Furthermore, "climate-smart" agriculture practices aimed at increasing productivity and reducing greenhouse gas emissions are being promoted across Zimbabwe. Farmers are using improved crop and livestock management practices. Within this context, this chapter investigates the impacts of climate change in Chiredzi district, Zimbabwe, and the adaptation and coping measures used in households based on local knowledge.

Methodology

Study area

The study was carried out in Chiredzi District situated in South-eastern Zimbabwe (coordinates 21°02′20″S 31°40′40″E) at an elevation of 438 m.a.s.l. (Figure 16.1). The district lies in Natural Region V, a drought prone region where rainfall is minimal (400 mm average annual) and very uncertain (Chikodzi et al., 2013). The main sources of livelihoods in the area are rain-fed agriculture, extensive livestock production, sugarcane production in Hippo Valley and Mkwasine Estates, and game ranching (Unganai et al., 2013). Temperatures are high in summer (plus or minus 39°C), causing evaporation losses of 10 mm–13 mm per day. The climate in Chiredzi is classified as hot and dry (Mupakati & Tanyanyiwa, 2017). A total population of 306,436 lives in Chiredzi, with an estimate of 276,842 people living in the rural areas (Zimstats, 2012). The district houses several protected areas, including Gonarezhou, Malilangwe and Fair Range estate. Vegetation of Chiredzi District is semi-arid, annual grasses with scattered shrubs, mostly mopane (*Colophospermum*) trees cover most of the area. Chiredzi has heavy clay basaltic fertile soils, suitable for both dry land and irrigation agriculture. Crops grown in the district include sorghum, cowpeas, millet, maize, groundnuts and cotton, among others.

Sampling procedure

Data collection followed the protocol outlined by Reyes-García and colleagues (2023). To obtain an overview of observed environmental and climate changes and its related impacts in the district, three villages (Mutapurwa, Njonjonjo and Jekero) were selected with the guidance of the Department of Agricultural Technical and Extension Services (AGRITEX) officers, traditional leadership and assistance of local extension workers. Snowball sampling was used to select 25 participants for semi-structured interviews. The participants sampled consisted of 11 male and 14 female-farmers (crop and livestock production), having lived in Chiredzi for more than 30 years. To validate the data from interviews, two focus groups consisting of nine and ten farmers of mixed gender were sampled using convenience sampling.

Data collection and analysis

The data collection methods employed were in-depth interviews, focus group discussions, and document analysis. Some of the questions which guided the interviews are: "What changes have you noticed in the environment compared to when you were young?," "what caused these

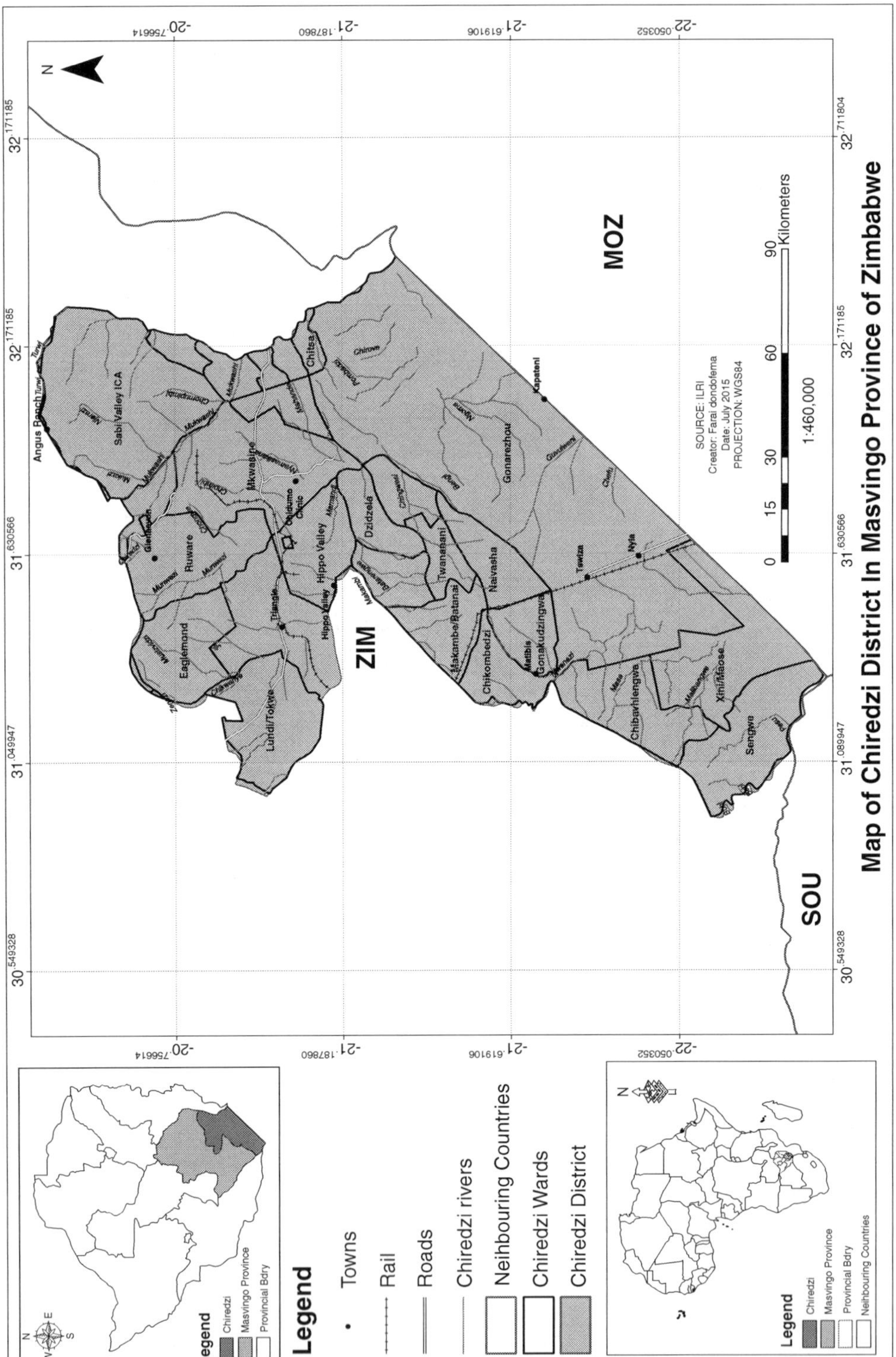

Figure 16.1 The map of Chiredzi District Council, Masvingo, Zimbabwe.

changes?"; and "how are you dealing with the changes noticed?" Two focus group discussions (FGDs) were used to gain in-depth insight and consensus of the information gathered from semi-structured interviews (Reyes-García et al., 2023). A semi-structured interview guide was used to collect data from participants and to moderate the FGDs, whereby the facilitator was leading the discussion and asking probing questions on perceived changes, causes of changes, and adaptation/coping measures, but discussions were allowed to flow freely (Reyes-García et al., 2023). The interviews were audio-recorded and extensive note-taking was done during the sessions. Data collected were subjected to thematic content analysis, which identifies patterns or themes within qualitative data.

Results

General awareness of climate change is high

The results showed varied reports of climate change, causes, and impacts on farmers' livelihoods and their responses. However, this study will only focus on changes in agricultural impacts and responses highlighted by farmers in Chiredzi district. All the interviewees in Chiredzi district reported some changes in the elements of the atmospheric system, which are affecting their agricultural practices. Participants of FGDs corroborated this awareness of climate change impacts.

From the interviews and the FGDs, participants noted that climatic changes in their environment are happening due to natural causes and global warming. Respondents also postulated the importance of local drivers, such as burning bushes. Climate change was also noted by locals as a curse from God due to communal mishaps in the area, such as teenager's abortions and prostitution happening in the community. Abandoning of traditional culture and values also emerged as a cause of change: ancestors are not happy with the abandonment of traditional culture, leading to weather change and variability. Respondents also attributed climate change to human causes such as tree cutting, and cutting of sacred trees and fruit trees such as *nyii* (bird-plum) *and khilluru.*

Farmers' observed climate change impacts on agriculture

Some of the climatic threats affecting agricultural production in Chiredzi observed by subsistence farmers include frequent droughts, flash floods, and long dry spells. Crop failure, less crop yields of maize, sorghum and cowpeas also came out as observations noted by farmers, which they link to the changing climate. Crop pest infestations were postulated by local farmers as a major impact driven by climatic changes affecting agriculture production, both in field and storage. High temperatures in summer and erratic rainfall events were also highlighted as threatening agricultural activities and livelihoods of subsistence farmers in Chiredzi. The impacts of climate change noted by farmers and adaptation or copying measures used are discussed in detail in the following section.

Increased risk of climate-induced crop pests and adaptation measures

The existence of armoured crickets, armyworm and weevils emerged as major climate-related challenges in Chiredzi. Invasion of maize crops by pests such as armyworm, which is very resistant to conventional and locally-derived pest control measures, was reported by all farmers. One respondent noted that

> Before the year 2000, we had access to effective pesticides which helped us to control pests. The appearance of fall armyworms in 2016–2017 got things out of hand, we sprayed more pesticides frequently in the fields but our crops still got destroyed and we suffered huge losses.
>
> (ID A007, 17/12/2019)

The farmers highlighted that low rainfall, increased dry spells, high temperatures and warmer winters influence pest prevalence in Chiredzi district. Weevils were noted to be increasingly affecting cereal crops both in field and in storage, a trend they had never observed in the past. A farmer stated that "It never happened in the past years that grains get affected by weevils whilst in the field before harvesting, but now -with changes in the climate- it's common" (ID A013, 19/12/2019). Moreover, respondents indicated that the weevils are becoming more resistant to both modern pesticides. They do not respond to chemicals anymore when farmers spray.

To treat crop pests, farmers indicated several local strategies which they employ to manage pests, both in the field and in storages (Table 16.1). The farmers used indigenous plant species such as ashes of *mutovhoti (Spirostachys Africana)* and *munondo (Julbernardia globiflora)* trees, which they smear on infected crops. Farmers also cut fresh branches of *mutovhoti* tree, crush or pound them and then dilute them in water and spray the mixture on crops in the field to kill leaf eaters. Smearing of river sand on the crop's meristem to remove armyworms is another short-term strategy used by farmers in the area. Farmers also use smoke residue (*china'i*) to treat weevils affecting crops both in the field and in storage.

To protect stored grains, the smallholder farmers also use *mutovhoti* cut branches and ash, which they spread underneath, in the middle, and on top of their grains in the granaries (*dura*). The tree has a strong scent, which repels weevils. Nevertheless, people interviewed established that some of the medicinal plants they used to treat pests in the past are becoming extinct because of climate change. However, farmers indicated that they are also using affordable modern pesticides such as *Actellic chirindamatura* gold dust, *shumba* super dust grain, and cereal protectant. Due to limited financial resources, some smallholders continue to use traditional pest control mechanisms. Some farmers also mentioned that even though they use modern pesticides to preserve their grain, they are still affected by weevils and the modern crop varieties are more prone to weevils,

Table 16.1 Local strategies for controlling climate-induced crop pests and insects in the fields and in storage in Chiredzi District

Resource	*Local strategies used to treat climate-induced pests and weevils*
River sand	Smear on crop's meristem, particularly maize, to remove armyworm.
Cow dung	Plaster granary (*dura)* walls and floors with cow dung or seal seed bags using cow dung to protect seed from weevils.
Hundi yerukweza (Finger-millet chuff)	Mix grain with chuff in storage bags and seal the bags, to preserve their grain.
Wood ashes	Smear ashes on the bottom and on top of grain to avoid weevil's infection.
Chin'ai (smoke residue)	Hang maize cobs or sorghum in a kitchen hut for preserving seeds to be used in the next ploughing season.
Mutovhoti (Spirostachys Africana)	Cut branches and stem into small pieces and place them underneath, in the middle, on top and on the sides of grain in a bag or granary and seal. Cut fresh branches of mutovhoti tree, crush or pound them, then dilute them in water and spray crops in the field to kill leaf eaters.

getting infected whilst in the field before harvesting. Hence, they resort to use both modern and local strategies to get better results.

Increased climate-induced threats to animal production and adaptation measures

Farmers in the study area reported that, due to climate change and variability, their livestock is more exposed to emerging infectious diseases and deaths particularly in the past three decades. Frequent drought experiences have weakened and made livestock vulnerable. Farmers indicated that livestock diseases such as brucella arbotas, anthrax, bovine tuberculosis, and wounds affect the health and production of their cattle, which also affects their income since they rely on selling livestock to raise income. Tick-borne diseases, which increased, farmers' also relate to climatic changes, cause high cattle mortality in the study area. Locals also indicated that their livestock were more prone to diseases and weakness due to a decrease in grazing pasture in the communities. They indicated that there is less grazing pasture due to less rainfall, persistent droughts, and resettlement of people, who occupy areas previously reserved for grazing. A farmer said that

> More than 30 years ago, we had big portions of land reserved for livestock grazing. My father's household used to have more than 300 cows, but now we cannot even keep more than 20 because there is less grazing pasture due to frequent droughts and climate changes.
>
> (ID A002, 15/12/2019)

Findings from FGD corroborated that farmers were experiencing major losses due to livestock deaths, which they attributed to climate-induced droughts and diseases. One farmer participating in one of the focus groups indicated: "My household's income used to come from selling cattle, but due to droughts and diseases, we no-longer own any cattle. They all perished in 2019 and we now rely on wage labour to survive" (ID A014, 20/12/2019). Loss of livestock due to diseases and frequent droughts was noted by more than 95% of the participants.

Farmers use various local techniques to manage livestock diseases and prevent deaths (Table 16.2). They use medicinal plants like *murunjurunju* or *murumanyama* (*Xeroderris stuhlmannii*) to treat livestock's tapeworms, diarrhoea, and wounds. Farmers crush *muvengahonye* (*Psydrax*

Table 16.2 Local strategies for controlling climate-related livestock diseases

Resource	*Local strategy to treat climate-related livestock diseases*
Murunjurunju (*Xeroderris stuhlmannii*)	Crush leaves and add water to make a drink for treating tapeworms, diarrhoea. Make a paste of crushed leaves and a water to treat septic wounds.
Muvengahonye (*Psydrax livida*)	Crush leaves and make a paste to treat septic wounds on livestock.
Zumbani (Lippia javanica)	Crush leaves and mix with water into a drink for treating tapeworms. Mix *zumbani* and *murumanyama* leaves, mix with water into a drink to stop diarrhoea and worms.
Gavakava (*Aloe* spp)	Infuse crushed leaves with water into a drink to treat respiratory diseases and weakness in chickens, cattle and goats. Apply aloe gel on livestock wounds.
Nhundurwa (Solanum anguivi)	Crush ripe fruit and apply fluid on livestock eyes to treat them.

livida) leaves and make a paste to treat septic wounds on livestock. The knowledge-holders also indicated that they use *zumbani* (*Lippia javanica*) for treating cattle's tapeworms. Subsistence farmers also crush and mix *zumbani* with *murumanyama*, together with water and use the mixture to treat worms and stop diarrhoea in cows. The farmers also use different forms of *gavakava* (*Aloe spp*), where they infuse the leaves in drinking water for livestock diarrhoea, respiratory diseases, general weakness, parasites and wounds. They use this plant for cattle, goats and chicken treatment. However, farmers posited that these medicinal plants are becoming extinct because of climate change and deforestation hence the need for protection and documentation. A farmer noted that "I traditionally use brew sorghum or pearl-millet beer (doro) to treat weak cattle. I make the cattle drink a cup of traditionally brewed beer and the moment it burps, the cattle heals instantly" (ID A009, 17/12/2019). Interestingly, farmers also noted that the cattle breeds nowadays are more prone to diseases compared to their old traditional ones, which were drought and disease resistant.

Local farmers collectively agreed that most farmers 30 years ago owned a lot of livestock (cattle and goats). For example, in the 1980s, it was common for households to own between 200 and 300 cows. This is not common nowadays. In the event of a drought, they exchange livestock with grain for self-consumption and for cash, to pay school fees and other household expenses. Due to climatic and environmental changes, locals say that they no longer keep larger herds of cattle. This helps them avoid large losses since they have inadequate resources to buy supplementary feeds during drought years. To cope and adapt with the aforementioned challenges, some farmers buy livestock feed, while others harvest and reserve crop residue (*mashanga*) for stock feed supplementation. Farmers also use *zhombwe* (*Neorautanenia brachypus*, harms), a shrubby legume with tubers, to feed livestock during droughts.

Frequency of droughts and decreased rainfall amounts affecting crop production

Farmers highlighted that they are experiencing persistent droughts in the district and that local communities mostly rely on food aid. Prominent drought events were in the years 1982–1983, 1992–1993, 1994–1995, 2002–2003, 2015–2016 and 2018–2019.

Changes of rainfall amounts, delay in the timing of rain, and a reduction in the amount of rainfall were also noted. All farmers interviewed perceived the consequences of changes in the rainfall pattern in cropping periods and systems. Farmers also indicated that the amount of rainfall has decreased over the past ten years. Rainfall has become erratic and farmers find it difficult to schedule planting for crops and other agricultural activities. Farmers indicated that the rainfall season now starts as late as December, January and February, not in September or October, as in the past. Farmers noted that "We used to have an accurate rainfall calendar receiving our first rains mid-October, but for the past 20 years we have seen the rains starting in December and even January some years" (IDA011, 13/12/2019; IDA015, 20/12.2019). Farmers indicated that they used to follow a consistent rainfall calendar whereby the first rain, called *odzamashanga,* fell around August, rotting plant residue in the fields. Then, in October, they received *bumharutsva,* a rain helpful for clearing burnt ashes, and in November they received a rain called *nhuruka* to plant seeds. This last rain has now become very inconsistent and mostly absent. Farmers adjust their ploughing times accordingly using their experience and following the rainfall events of that particular year.

To cope with erratic rainfall, farmers implement crop spacing, crop diversification, dry-soil planting, and mixed cropping. Farmers noted that "If we mix crops, some mature and we harvest something, even in drought years" (ID A002, 15/12/2019; ID A023, 21/12/2019). The farmers also indicated that they practise conservation agriculture whereby they dig holes in the field, and add manure in each hole before the onset of first rains. The holes help to conserve moisture and nutrients for crops like maize and sorghum. A farmer noted that

> Conservation agriculture is being encouraged by local extension workers, and farmers are doing it at a larger scale, but it's a practice we have been using for ages and we learned this from our elders. We have however been using it at a lower scale, but now due to its promotion by the government we use it at a larger scale.
>
> (ID A009, 17/12/2019)

Farmers also indicated that they no longer practice dry soil planting (a practice commonly used in Chiredzi in the past whereby farmers seeded their crops in dry soil three weeks before the first rains). Nowadays, they only put seeds in the ground after the rain falls. Three farmers noted that "With the current rainfall patterns, when we dry-soil plant, some seasons we receive less rainfall and the seeds rot in the soil" (ID A017, 20/12/2019).

Farmers also stated that during years of good harvest, they protectively store their grain, (e.g., rapoko, sorghum, pearl millet) in big sealed bags for use in years of less or no harvest. They also highlighted that some families survived on wild fruits such as sosoti (*Schotia brachypetala*), manzviru (*Vangueria infausta*) and makwakwa (*Strychnos spinose*). In addition, during drought periods, heavily affected families would survive on *chikezakeza* (*rapoko* grass), a weed with edible grains. In drought seasons, m*agwangwata* (bulrush) spongy rhizomes found in the river banks are pounded and eaten in the form of porridge.

In terms of crop production, farmers now resort to the use of short-season varieties of sorghum and pearl-millet and grow crops like red finger millet which were not commonly grown in the district. In addition, crops like sesame, which had been abandoned for decades, are now commonly being grown in the area. Subsistence farmers also actively participate in irrigation schemes growing beans, vegetables and maize mostly for sale, but also for consumption. Three farmers highlighted that they had shifted focus to non-agricultural activities like firewood selling and brick-making because of continuous drought events in the district. One farmer indicated that

> The time, effort, and resources we put in crop production no longer benefit us because of droughts. We now put more effort in brick –making for sale, and the money we raise, we buy food and other basic household needs.
>
> (ID A024, 22/12/2019)

Farmers noted that nowadays they rely mostly on food aid from donors and government assistance programmes for survival in years of drought, but in the absence of aid, they go back to their tried and tested traditional practices.

Rising temperatures affecting livelihoods and crop production and copying measures

All respondents noted that the temperatures in Chiredzi have become warmer. Inasmuch as the low-veld region is generally hot, the climate has become hotter in the past two decades. Farmers indicated that their district is too hot and they experience sunshine intensity. One farmer reiterated that "It feels like the sun has been lowered down. The heat is too much" (ID A007, 17/12/2019). It was indicated that due to the hot temperatures, the ground is very dry. Another consequence of the high temperatures is that when it rains, flash floods occur. Farmers also highlighted that the air feels very hot and dry to such an extent that in 2019, birds fall dead off the sky due to heat. It was noted that "We have never experienced such heat where birds just drop dead from the sky" (ID A013, 19/12/2019). Another farmer also highlighted that

> Some days in summer the heat is too much. Even the air we breathe is too hot; to the extent that we feel like we will suffocate. Whether we stay indoors, or sit in the shade outdoors, there is no difference.
>
> (ID A005, 16/12/2019)

Crops like maize and groundnuts as well as livestock pasture were indicated to be drying often because of intense temperatures. One respondent noted that "Often the moisture in the soil dries up maize seeds quickly and the seeds rot, so we have to plough again and it's costly for us" (ID A002, 15/12/2019). The respondents also indicated that 20 years ago, they used to have long sorghum varieties of crops like *Mutode*. In terms of crop selection, to adapt to climate-induced droughts and long dry spells, local farmers indicated that they are opting for short-season sorghum varieties like *macia* and *chihumani*, which mature within three months.

To adapt to the rising temperatures, farmers are planting fruit trees around their yards for shade. Farmers also mentioned that they have changed their working times in the fields. Now, they start around four early morning and break at nine, to reduce heat effects. They also wear protective clothing like sun hats when working in the fields or walking in the sun to reduce sunburns.

Discussion

Climate change is largely affecting many subsistence farmers. The World Food Programme estimates that 30% of the SADC region will be vulnerable to climate change by 2050 if adaptation measures are not put in place (WFP, 2021). Farmers in the study area confirmed this by indicating climate changes such as temperature changes, changing rainfall patterns, and droughts, and their impacts including increasing pests, livestock deaths and diseases, and increasing food insecurity.

Local farmers indicated an increase in both existing and new pests affecting crop production in Chiredzi. Farmers linked climatic conditions, like hot temperatures, high rates of erratic rainfall and dry conditions in the district, to the high prevalence of pests, both established and emerging. Mafongoya et al. (2019) note that climate and weather patterns are key for the "distribution, development and population dynamics of pests and insects". A surge of pests affecting crop yields has also been noted globally, with worst cases noted in Africa. The presence of armyworms has been confirmed in Zimbabwe, where the pest affected 130,000 hectares of maize in the 2016/2017 season. Armyworms are spreading both in the southern African region and West Africa. Mwadzingeni and Mugandani (2022) confirmed that fall armyworms, tomato leaf miners and cotton mealy bugs are new pests befalling crop production in Zimbabwe due to climate change. Mafongoya et al. (2019) also corroborates that pests are causing at least 40% of crop losses worldwide and will continue to cause more losses unless adaptation measures are implemented.

Smallholder farmers are struggling to control the aforementioned and other emerging pests because of a lack of funds to buy pesticides. Though some farmers can afford modern pesticides, using them alone no longer suffices. Due to pest increases, farmers alternate the use of traditional and affordable conventional pest control measures. A study done in Rushinga established similar findings, as it found that farmers increase the volume and frequency of application of pesticides as a response to climate change impacts (Zinyemba et al., 2021). Farmers with no easy access to pesticides continue to use traditional means, which they have used for decades. Farmers also use cultural practices like crop rotation and intercropping to control pests. There is therefore a need for the government to support pest management strategies, which are informed by both conventional and traditional knowledge systems, and are affordable, safe and sustainable among rural farming communities.

Low crop yields came out as another of the major impacts of climate change in Chiredzi district. Decreasing crop yields is not only a challenge for the study area, but all semi-arid regions are vulnerable to change. For example, the WFP (2021) notes deficits in cereal production ranging between 0.1 and 8.9 million metric tonnes in SADC. Similarly, Nhamo et al. (2019) note a reduction in cereal production and negative impacts on high-value crops caused by rising temperatures and changes in rainfall patterns across Sub-Saharan Africa. Furthermore, yields of non-cereal crops, such as beans, are estimated to decrease by the mid-21st century with banana and plantain production most likely to decrease in western and eastern Africa (Pereira, 2017). Farmers linked increasing temperatures and less rainfall to reduced crop and livestock production in Chiredzi. Schlenker and Lobell (2010) also note the vulnerability of maize production in southern Africa, with specific reference to Zimbabwe and South Africa, predicting 30% yield losses. Considering this estimation, together with the impacts of crop yield losses predicted by IPCC, there is a need for intervention solutions to assist farmers to adapt to climatic impacts. Although the yield of some crops for example, cassava and peanut yields, is expected to increase with climate change by 2030s (Pereira, 2017), it is evident that negative impacts outweigh the positive impact of climate change on crop production. This means that adaptation solutions are critical to help subsistence farmers survive and become food secure under climate change.

Farmers' perceptions of warmer or increasing temperatures in the district are consistent with Davis-Reddy and Vincent (2017), whose findings portray an increase in temperatures in Southern Africa. The World Bank (2021) predicts the prominence of extreme temperature events, with an increase in days exceeding 35°C in the period 2040–2059. Farmers also observed an increased frequency and intensity of droughts. This is in line with observations by SADC, which declared drought a regional disaster with over 40 million people, including Zimbabwe (SADC, 2016) The Zimbabwe Vulnerability Assessment Committee (ZIMVAC) 2015's rapid assessment report showed that in 2015, 2.8 million people were food insecure in the country due to climate-induced drought impacts.

With respect to adapting to climate change impacts, farmers indicated the use of various coping strategies. In that sense, farmers in the area have developed and perfected reliable local knowledge-based coping strategies, such as planting short-season, drought-tolerant small grain varieties for sorghum and millet. Farmers also practise conservative agriculture, a practice also noted byMakwara (2010). Farmers also practise water harvesting and irrigation, though the irrigation fields are often not enough for all members of a community (Nangombe, 2015). This is consistent with the observations across other African countries like South Africa, Malawi, and Uganda where farmers use irrigation schemes to adapt to the effects of climate change. Crop diversification, dry-soil planting, mixed cropping, varying harvesting dates, and crop spacing to improve resilience to drought stress practices implemented by farmers in Chiredzi are consistent with findings by Mtambanengwe et al. (2012). In this study, we observed clear evidence of the use of traditional knowledge, which is the major source of knowledge for local communities and followed by government extension services. However, the government is increasing access to extension services, training farmers on climate-smart solutions. Other modern tools in the box include crop modelling and climate forecasting. These are yet to make an impact on the community as they depend on the strength of information sharing in the community.

The use of traditional knowledge by farmers in Chiredzi dovetails with findings in literature stating that most rural communities in Africa have relied on local knowledge in their agricultural activities. This knowledge is used to determine timing of agricultural activities, type of seeds to grow, managing pests and livestock diseases as well as storing grain and other livelihood activities (Ebhuoma & Simatele, 2017; Karki et al., 2017; Mafongoya & Ajayi, 2017). Rankoana (2017) indicates that traditional knowledge has been used over centuries in crop production by farmers without access to external inputs, capital as well as modern scientific know-how. Furthermore, traditional knowledge also plays a critical role in enhancing rural communities' ability to acquire food resources for their families and their livestock. Traditional knowledge is also imperative in community agriculture because it enables farmers to adapt their farming activities to seasonal changes and to plan and cope amid changing climatic events, as also shown by Schipper et al. (2014). Jiri et al. (2015) note that rural farmers in Zimbabwe have relied on the use of traditional weather forecasts to choose crops and plan planting dates. Although traditional knowledge remains poorly embedded in climate adaptation policies, it remains an integral part of rural agricultural production.

Conclusion and recommendations

The study indicates that smallholder farmers in Chiredzi are aware that the climate is changing and they are not spared from experiencing climate change impacts. The impacts noted include frequent droughts, increasing temperatures, livestock diseases and deaths, reduced crop production, and changing rainfall patterns. Farmers are implementing both conventional and traditional knowledge-based adaptive measures to cope with the changing climate, but they remain exposed and vulnerable to threats. Whether with access to conventional adaptation measures or not, traditional knowledge remains valuable, embedded in rural farming practices and still in use in the Chiredzi district to adapt to climate change impacts. Therefore, governments, local leadership and traditional knowledge holders should work together in developing context-specific climate solutions borrowing from both conventional and context-specific traditional knowledge practices. We conclude that as the food production systems are evolving and countries are moving into knowledge-based economies, all knowledge systems at our disposal be deployed for farmers to improve climate resilience based on knowledge. We recommend further studies on the impact of climate change on animal husbandry and interventions needed based on grassroot and inclusive innovation.

Acknowledgements

We acknowledge the Chiredzi Rural District community for participating in the study. This research was partly funded by an ERC Consolidator Grant FP7-771056-LICCI.

References

Brazier, A. (2018). *Climate Change in Zimbabwe: A Guide for Planners and Decision-Makers* (Alwyn Francis (ed.); 2nd ed., Issue January). Harare, Konrad-Adenauer-Stiftung.

Chikodzi, D., Murwendo, T., & Simba, F. M. (2013). Climate Change and Variability in Southeast Zimbabwe: Scenarios and Societal Opportunities. Vol.2 No.3A, September 2013, 36–46.

Davis-Reddy, C. L., & Vincent, K. (2017). *Climate Risk and Vulnerability: A Handbook for Southern Africa.* CSIR.

Dube, E. (2017). *Southern African Development Community's Preparedness to Deal with Terrorism and Extremist: The Case of Zimbabwe 2008–2016*. 196.4.80.91. http://196.4.80.91/jspui/bitstream/handle/10646/3316/Dube_Southern_African_development_community's_preparedness_to_deal_with_terrorism_and_extremism_the_case_of_Zimbabwe_2008-2016.pdf?sequence=1

Ebhuoma, E., & Simatele, D. (2017). Defying the Odds: Climate Variability, Asset Adaptation and Food Security Nexus in the Delta State of Nigeria. *International Journal of Disaster Risk Reduction*. https://www.sciencedirect.com/science/article/pii/S2212420916302126

International Federation of Red Cross Societies (2019). Zimbabwe: Tropical Cyclone Idai, Final Report.

Jiri, O., Mafongoya, P. L., & Chivenge, P. (2015). Indigenous Knowledge Systems, Seasonal 'Quality' and Climate Change Adaptation in Zimbabwe. *Climate Research*. http://www.int-res.com/abstracts/cr/v66/n2/p103–111/

Karki, M., Pokhrel, P., & Adhikari, J. R. (2017). A 10 Climate change integrating indigenous and local knowledge into adaptation policies and practices a case study from Nepal. *Shifting Cultivation Policies: Balancing Environmental and Social Sustainability*. Edited by Malcolm Cairns. Wallingford: CABI.

Mafongoya, P. L., & Ajayi, O. O. C. (2017). *Indigenous Knowledge Systems and Climate Change Management in Africa*. Ajayi, O. C., & Mafongoya, P. L. (2017). Indigenous knowledge systems and climate change management in Africa. CTA.

Mafongoya, P., Gubba, A., Moodley, V., Chapoto, D., Kisten, L., & Phophi, M. (2019). Climate change and rapidly evolving pests and diseases in Southern Africa. New frontiers in natural resources management in Africa, 41–57.

Makwara, E. C. (2010). Sustainable and profitable farming through conservation agriculture in Zimbabwe: Prospects, opportunities and constraints. *Journal of sustainable development in Africa*, 12(8), 180–190.

Manzungu, E. (2014). *A Comprehensive Scoping and Assessment of Climate Smart Agriculture Policies in Zimbabwe*. Report. FANRPAN. April.

Masitera, E. (2021). Philosophical Perspectives on Land Reform in Southern Africa. In *Philosophical Perspectives on Land Reform in Southern Africa*. https://doi.org/10.1007/978-3-030-49705-7

Mavhura, E., Manatsa, D., & Mushore, T. (2015). Adaptation to Drought in Arid and Semi-Arid Environments: Case of the Zambezi Valley, Zimbabwe. *Jàmbá: Journal of Disaster Risk*. https://journals.co.za/content/jemba/7/1/EJC168492

Mtambanengwe, F., Mapfumo, P., Chikowo, R., (2012). Climate Change and Variability: Smallholder Farming Communities in Zimbabwe Portray a Varied Understanding. *African Crop Science*. https://www.ajol.info/index.php/acsj/article/view/81755

Mupakati, T., & Tanyanyiwa, V. I. (2017). United Nations Development Programme, Strengthening National Capacity for Climate Change Programme, Ministry of Environment, Water and Climate, Zimbabwe. *Jàmbá – Journal of Disaster Risk Studies*, *9*(1), 1–10. https://doi.org/10.4102/jamba

Mushore, T., Manatsa, D., Pedzisai, E., Muzenda-mudavanhu, C., Mushore, W., & Kudzotsa, I. (2017). *Investigating the Implications of Meteorological Indicators of Seasonal Rainfall Performance on Maize Yield in a Rain-Fed Agricultural System: Case Study of Mt. Darwin District in Zimbabwe*. 1167–1173. https://doi.org/10.1007/s00704-016-1838-2

Mwadzingeni, L., Mugandani, R., & Mafongoya, P. (2022). *Risks of climate change on future water supply in smallholder irrigation schemes in Zimbabwe*. Water, 14(11), 1682.

Nangombe, S. S. (2015). Drought Conditions and Management Strategies in Zimbabwe. *Proceedings of the Regional Workshops on*…. http://www.ais.unwater.org/ais/pluginfile.php/601/mod_page/content/21/Zimbabwe.pdf

Nhamo, L., African, S., Matchaya, G., & Mabhaudhi, T. (2019). *Cereal Production Trends under Climate Change: Impacts and Adaptation Strategies in Southern Africa Cereal Production Trends under Climate Change: Impacts and Adaptation Strategies in Southern Africa*. February. https://doi.org/10.3390/agriculture9020030

Nhemachena, C., Mano, R., Mudombi, S., (2014). Climate Change Adaptation for Rural Communities Dependent on Agriculture and Tourism in Marginal Farming Areas of the Hwange District, Zimbabwe. *African Journal of*…. http://www.academicjournals.org/journal/AJAR/article-full-text-pdf/05190DE45750

Pereira, L. (2017). *Oxford Research Encyclopedia of Environmental Science Climate Change Impacts on Agriculture across Africa* (Issue April). https://doi.org/10.1093/acrefore/9780199389414.013.292

Rankoana, S. A. (2017). *The use of indigenous knowledge in subsistence farming: Implications for sustainable agricultural production in dikgale community in Limpopo Province, South Africa. In Toward a sustainable agriculture*: Farming practices and water use (Vol. 7). MDPI, Basel.

Reyes-García, V., Álvarez-Fernandez, S., Benyei, P., García-del-Amo, D., Junqueira, A.B., Labeyrie, V., Li, X., Porcher, V., Porcuna-Ferrer, A., Schlingmann, A., Soleymani, R. (2023). Local Indicators of Climate Change Impacts Described by Indigenous Peoples and Local Communities: Study Protocol. *PLoS One*, *18*((1): e0279847), 1–15. https://doi.org/10.1371/journal.pone.0279847

Schipper, E. L. F., Ayers, J., Reid, H., Huq, S., & Rahman, A. (2014). *Community-Based Adaptation to Climate Change: Scaling It Up*. books.google.com. https://books.google.com/books?hl=en&lr=&id=zw-pAgAAQBAJ&oi=fnd&pg=PP1&dq=Adaptation+policy+AND+Zimbabwe&ots=6ixV6NyKxC&sig=rFhFcVrU5GJXN-d7IGXe2tA_4gA

Schlenker, W., & Lobell, D. B. (2010). Robust negative impacts of climate change on African agriculture. *Environmental Research Letters*. http://iopscience.iop.org/article/10.1088/1748-9326/5/1/014010/meta

The Government of Zimbabwe and United Nations Development Programme (2017). *Zimbabwe Human Development Report 2017*. https://hdr.undp.org/system/files/documents/undpzw2017zhdrfullpdf.pdf

The World Bank (2021). *Climate Risk Country Profile*. https://climateknowledgeportal.worldbank.org/sites/default/files/2021-05/14956-WB_Zimbabwe Country Profile-WEB %281%29.pdf

The World Food Programme (2021). *Climate Change in Southern Africa* (Issue June). https://executiveboard.wfp.org/document_download/WFP-0000129015

Zimbabwe Vulnerability Assessment Committe. Zimbabwe Vulnerability Assessment Committee (ZimVAC) 2016 Rural Livelihoods Assessment, Food and Nutrition Council; Zimbabwe Vulnerability Assessment Committee: Harare, Zimbabwe, 2016

Unganai, L. S., Troni, J., Manatsa, D., (2013). Tailoring seasonal climate forecasts for climate risk management in rainfed farming systems of southeast Zimbabwe. *Climate and …*. http://rsa.tandfonline.com/doi/abs/10.1080/17565529.2013.801823

Waha, K., Van Wijk, M. T., Fritz, S., See, L., Thornton, P. K., Wichern, J., & Herrero, M. (2018). Agricultural Diversification as an Important Strategy for Achieving Food Security in Africa. December 2017, 3390–3400. https://doi.org/10.1111/gcb.14158

World Health Organization (2014). *WHO Guidance to Protect Health from Climate Change through Health Adaptation Planning*. apps.who.int. http://apps.who.int/iris/bitstream/10665/137383/1/9789241508001_eng.pdf

Xu, C., Kohler, T. A., Lenton, T. M., Svenning, J., & Scheffer, M. (2020). *Future of the Human Climate Niche*. https://doi.org/10.1073/pnas.1910114117

Zimstats (2012). Zimbabwe Population. *World Population Review*, 1–152. http://worldpopulationreview.com/countries/Zimbabwe/

Zinyemba, C., Archer, E., & Rother, H. A. (2021). Climate Change, Pesticides and Health: Considering the Risks and Opportunities of Adaptation for Zimbabwean Smallholder Cotton Growers. *International Journal of Environmental Research and Public Health*, *18*(1), 1–11. https://doi.org/10.3390/ijerph18010121

17

COUNTERACTING LAND ABANDONMENT

Local adaptation strategies to climate change impacts of alpine farmers in Eastern Tyrol, Austria

Anna Fuchs, Christian R. Vogl and Christoph Schunko

Introduction

Mountain regions are multifunctional landscapes that provide a wide range of material, non-material, and regulatory contributions to people at local to global scales (EEA, 2010; Schirpke et al., 2019). Yet, mountain regions are fragile ecosystems that are severely affected by the ongoing and intensifying climatic changes around the globe (Adler & Wester, 2021). Already small changes in temperature and precipitation lead to drastic impacts, such as shifting spatial boundaries, that is, the tree- and snow-lines, negative effects on water cycles, and increased likelihoods of natural hazards (Adler & Wester, 2021). In the remaining 21st century, further increases in temperature, changes in seasonal cycles of precipitation and radiation, as well as changes in weather extremes are anticipated, as for example, in the European Alps (Gobiet et al., 2014).

Ongoing and future expected climate change impacts in mountain regions have wide-reaching adverse effects on local people who depend on these ecosystems, such as farmers. Mountain farmers are directly affected by the spread of invasive species, pests and diseases, changes in vegetation periods and the phenology of plant species, and increases in weather extremes, like droughts and floods (Adler & Wester, 2021). All these impacts lead to a deterioration of local livelihoods through a decline in agricultural and livestock productivity (Adler & Wester, 2021). This is problematic because farming in mountain areas has a diversity of vital functions, including the maintenance of natural resources such as soil, water, and biodiversity, the protection of human settlements against natural hazards, the provision of high-quality food stuff, and maintaining cultural landscapes. Only active land management through extensive agriculture can continue to secure many of these functions (Dax, 2001).

To secure the functions provided by mountain farming, it is key that mountain farmers successfully adapt to climate change impacts. Adaptation strategies may include a wide range of social (e.g., behavioural, educational), institutional (e.g., economics, laws), and physical and structural activities (e.g., technology, ecosystem-based) (Owen, 2020). Adaptation strategies may be implemented autonomously, by individuals or organisations, or collectively- These strategies may vary from incremental adaptations, that initially adapt existing activities and have limited scopes and depths, to transformative adaptations, that are based on fundamentally altered paradigms and wide scopes and depths (Termeer et al., 2017). For example, incremental adaptations are technical and

DOI: 10.4324/9781003356837-22

managerial adaptations of agricultural production systems, whereas transformative adaptations include changes in types of agricultural production and relocation of cultivated fields to locations with more suitable climatic conditions (Vij et al., 2021). In mountain regions, incremental adaptations tend to be more frequent than transformative adaptations (Adler & Wester, 2021). In European mountain regions, for example, about two-thirds of the adaptations reported are incremental (Vij et al., 2021).

Adaptation strategies can be top-down or bottom-up (Grüneis et al., 2018), whereby in either case, the involvement of local people and their knowledge is essential for adaptation planning success (Bezner Kerr et al., 2022), as local people have considerable knowledge about local climatic changes and their impacts on the natural elements that support their livelihood activities (Reyes-García et al., 2019). Research shows that local people hold knowledge about the most pressing drivers of change and develop their own adaptation strategies to mitigate the effects of climate change impacts (Petzold et al., 2020; Schlingmann et al., 2021). The co-production of new knowledge based on different knowledge systems leads to the most applicable and effective results for adaptation (McDowell et al., 2021). Whereas a considerable number of studies on local adaptations to climate change impacts were completed in Asian and African countries, few studies have been conducted in Europe so far (Schlingmann et al., 2021).

Here, we contribute to filling this geographical gap by conducting research in Eastern Tyrol, Austria. We aimed to answer: Which climatic changes do local people in mountain regions observe? Which climate change impacts on mountain farming do local people report? How do local people perceive the incremental and transformative adaptations of mountain farming practices in reaction to climate change impacts?

Methods

Field site

Research took place in Eastern Tyrol (synonymous with the political district of Lienz), federal state of Tyrol, Austria. Having a surface of 2,020 km^2 and 48,800 inhabitants, Eastern Tyrol is part of the Eastern Alps and characterised by a mountainous landscape with an altitude range from 668 m.a.s.l up to 3,797 m.a.s.l. (Figure 17.1). The agricultural land of the district (360 km^2) consists of alpine pastures (224 km^2), mainly used for cattle grazing in summer, valley bottom grassland (125 km^2), used for cattle grazing and hay-making as winter fodder, and arable farming and orchards (12 km^2) (AMA, 2022). Only 9% of the district's surface is a permanent settlement area (WKO Tirol, 2021), whereas 30% is a nature area worthy of protection and located within the Hohe Tauern National Park (Land Tirol, 2022).

Until the mid-20th century, the main income of Eastern Tyrolean inhabitants came from agriculture and small businesses, but from the 1960s onwards, the tourism and industrial sectors became important economic branches. In the year 2019, 7% of Eastern Tyrolean employees worked in agriculture and forestry, while 61% of employees worked in the services sector (including tourism) and 32% in the industrial and commercial sectors (including construction). The number of agricultural and forestry holdings dropped from 3,080 holdings in 1960 to 2,545 holdings in 2010, whereas in 2010 only 26% of all holdings were full-time farms, 56% part-time farms and 18% owned by legal entities or groupings (Land Tirol, 2022).

Within Eastern Tyrol, the villages Kals am Großglockner, Virgen, and Prägraten am Großvenediger, all located in the proximity of 20 km to each other, were selected because of their socio-ecological similarity and the posibility to access to village mayors via the regional management

Figure 17.1 The mountainous landscape of Eastern Tyrol is used for alpine pastures, grassland, and forestry (view to the Virgen valley). Photo by C.R. Vogl (2017).

agency, our local counterpart. Each village has a population of between 1,100 and 2,100 inhabitants and is situated between 1,194 and 1,335 m.a.s.l. The three communities were highly nature-depended and self-sufficient through farming until the 1960s. Mixed farming systems, including crop cultivation and livestock production relying on many different animal species, were common. Today, the cultivation of crops is largely abandoned, and former fields are used as grassland for haymaking and pastures for livestock farming. Cattle farming dominates the agricultural sector. Homegardens and fruit trees are tended for self-consumption and in Virgen some farms are engaged in fruit growing. However, only a few farmers still make their living exclusively from farming. Most farms are run as part-time businesses.

Sampling

To select participants for semi-structured interviewing and focus group discusions, we used snowball sampling (Bernard, 2011). We selected respondents who had lived in the region for 25 years and had in-depth first-hand experiences in nature-related livelihood activities. Mayors and municipal employees of the villages selected were the door openers for the first contact with respondents. After each interview, each respondent was asked if they knew further persons fulfilling our sampling criteria. Overall, 18 respondents participated in the semi-structured interviews (Kals: six respondents, Virgen: five, Prägraten: seven). Thirteen respondents were farmers, six worked in forestry, and ten in other nature-dependent sectors, such as mountaineering and hunting. We interviewed ten men and eight women. Two respondents were under the age of 40, 11 between 40 and 65, and five older than 65. Subsequently, three focus groups, one in each village, were conducted (Kals: four respondents, Virgen: three, Prägraten: three). Of the ten focus group participants, three respondents were farmers, three others worked in forestry, and four more in other nature-dependent sectors. Nine men and one woman participated in focus group discussions. One respondent was under the age of 40, eight between 40 and 65, and one older than 65. One respondent participated in both, semi-structured interviews and focus groups, so altogether 27 different respondents were involved.

Data collection

The first author collected data in two phases. The first phase took place in July and August 2020, when data were collected using semi-structured interviews (Bernard, 2011). The interview guide was structured along the research protocol of the local indicators of climate change impacts (LICCI) project, where indicators of change are classified according to whether they affect elements of the climatic, physical, and life systems (Reyes-García et al., 2023). The first author inquired (1) changes respondents had observed compared to when they were young adults, (2) time since when the changes were observed, (3) the perceived drivers of changes, (4) actual and potential future adaptation measures, and (5) barriers to adaptation. In this chapter, we only present the perceived changes in the climatic system, their impacts, and adaptations in mountain farming. When changes in mountain farming are reported to be driven jointly by climatic and non-climatic factors, we present all of these drivers. Impacts on mountain farming that are solely driven by non-climatic factors are omitted.

In the second phase, which took place between January and March 2021, the changes reported in phase one were validated in focus group discussions (Reyes-García et al., 2023), whereas we did not validate the reported adaptations. Only perceptions of changes that were mentioned by at least three respondents, but with contradictory information (e.g., opposite trends) or incomplete (e.g., unclear drivers,) were discussed. During the focus groups, each report of change that was pre-selected for discussion was first presented. Then the respondents were asked (1) whether they had noticed the change, (2) since when they had noticed the change, (3) what they thought the driver of the change was. Reported changes that had not yet been reported in the semi-structured interviews, were added to our list. Altogether, between 7 and 13 reports of change were validated in each of the three focus groups. This chapter only presents those on which a general consensus was reached and to which no one has objected.

Semi-structured interviews and focus groups were audio-recorded and transcribed for data analysis. Each semi-structured interview lasted between 45 and 90 minutes and each focus group about two hours. The semi-structured interviews were held in respondents' houses. The focus groups were held online, because group meetings in person were not possible at the time of data collection due to the COVID pandemic. All interviews and focus groups were conducted in German.

Data analysis

Data from semi-structured interviews and focus groups were coded and summarised with qualitative content analysis (Mayring, 2014). The local indicators of climate change framework (Reyes-García et al., 2023) served as a coding frame to classify reported changes. The codes were applied on the level of subsystems (e.g., temperature, precipitation, wind, etc). The drivers of change and adaptation measures were coded inductively, whereas transformative adaptations were understood as fundamental, system-wide reorganizations, and thus distinguished from incremental adaptations (Brondízio et al., 2019). Finally, a descriptive summary of each reported change, the drivers of change, and adaptations was made according to the steps of paraphrasing, generalisation, and reduction (Mayring, 2014). All transcripts were coded in Atlas.ti (atlasti.com) and further analysed in a spreadsheet software.

Results

Local reports of climatic changes

Temperature

Among all climatic changes, respondents reported temperature changes, including changes in mean and seasonal temperatures, temperature extremes, and sunshine intensities, most frequently. A common observation was that the mean temperature has been increasing within the last years. For ten of 18 respondents, an increase in the mean temperature is especially apparent in winter. Respondents reported that, over the last years, winters became milder and that the usually long cold periods, which could reach up to minus 20°C, are now reduced to only a few days, which are also less cold. Moreover, cold periods are reported to be nowadays regularly interrupted by warmer phases with degrees above zero. Five respondents reported that the months of February, March, and April tend to be warmer than in the past and that the season of spring thus starts earlier. Yet, it was often mentioned that cold spells still occur during these months and that the temperatures may drop again in later spring. Seven respondents also mentioned explicitly that the mean temperatures in summer have increased in comparison to the past. A common observation was that it has become normal to have several days with up to 30°C in summer, while this was an exception in the past. The summers of the years 2017–2019 were reported as especially hot. Some respondents also emphasised that days with especially high temperatures have become more frequent. Five respondents also reported that the mean temperature in autumn has increased. Moreover, it was mentioned that the season of autumn is prolonged. Seven respondents reported that the temperature decreases less during summer nights, which were also described as more humid than in the past. In Prägraten, for example, it was noted that the temperature can still reach 20°C at 10 pm, which was not common in the past. Three respondents reported increasingly rapid temperature fluctuations in summer, but also in winter. They described this as very fast changes between high and low temperatures instead of more gradual changes. Three respondents reported that the sunshine intensity has increased over the last few years.

Precipitation

The second most frequently mentioned changes were changes in precipitation. These included changes in seasonal precipitation, precipitation extremes, precipitation predictability, and occurrence of droughts. Ten respondents reported summer rainfall events as heavier than in the past. They described them as occurring suddenly and locally, releasing a high amount of precipitation in a short period of time. Heavy rainfall events were also described as occurring more frequently now than before. Two respondents stated that in addition to the rainfall intensity, the temporal distribution of rainfall has changed. According to these informants, in the past rains used to be lighter and more constant, whereas nowadays, there are longer periods without rain followed by heavy precipitation in a short time. This matches with the reports of eight respondents that now it is more difficult to predict rain or snow. Overall, the weather was reported as less stable and changing faster. A common observation was also that it was drier in the last few years. Many people stated that this was especially the case since the year 2015, yet others had the impression that a dryer weather is already a longer-lasting trend. Dry spells and periods of drought were mainly observed to occur in spring and summer.

Wind

Respondents also reported climatic changes related to wind, although there were disagreements in such perceptions. Five respondents reported an increase in wind strength. These statements were often related to a heavy windstorm in the year 2018, with very high wind speeds of more than 200 km/h, which greatly impacted the region. This windstorm was a one-time extreme event and few respondents reported that wind strength or frequency have changed noticeably, other than in that event. Moreover, three respondents reported an increase in the number of windy days and 13 respondents reported that storms have become more intense, including electrical storms, snowstorms, and windstorms.

Local reports of climate change impacts on mountain farming

Pastures

Nine respondents reported a rise of the tree line and seven reported the encroachment of shrubs on alpine pastures. They perceived that the increase in mean temperature and the prolongation of the vegetation period contributed to these changes, although the main driver was reported to be a reduction of the management intensity of alpine pastures. Alpine pastures that are less or no longer grazed, mown, and cleared of woody vegetation by cutting and burning were reported to be completely overgrown within a few decades by dwarf shrubs, including black alder (*Alnus glutniosa*), alpine roses (*Rhododendron ferruginosum*; *Rhododendron hirsutum*), dwarf juniper (*Juniperus communis alpina*), and trees, including larch (*Larix decidua*) and cembra pine (*Pinus cembra*) (Figure 17.2). Respondents explained that reduced land management is due to the abandonment of agriculture. In former times, at least ten persons per household helped to manage the pastures, whereas nowadays, this is rarely done because households have become smaller and employees more expensive. Moreover, in the past the animal species that grazed on the alpine pastures were more diverse, including goats, sheep, cows and horses, which were perceived to have more efficiently kept the pastures free from woody plants. Moreover, animals used to be herded and guided to specific spots to keep them clean from shrubs and saplings, while nowadays animals roam freely in fenced paddocks. Several respondents reported that the increase in mean temperatures also contributed to the fact that trees have started to grow in areas that were not forested before, including areas above the earlier tree line. Shrub encroachment replaces a high diversity of herbaceous plant species with few species of shrubs and is thereby reported to contribute to the loss of plant diversity in mountain pastures. Moreover, two respondents reported that the alpine pastures are affected by an increase in landslides, which are driven by the more frequent and intense heavy rainfall events.

Another change mentioned by five respondents is that the abundance of certain weeds has increased, which again they attribute to simultaneous changes in management and higher mean temperatures. For example, the increasing abundance of European white hellebore (*Veratrum album*), which is poisonous and presents a danger for grazing animals, was associated with a decrease in weed management of alpine pastures, while the increasing abundance of species of the genus *Adenostyles* (*A. alliariae* and *A. alpine*) was attributed to higher temperatures. In grasslands at the bottom of the valley, more nitrophilous and deep-rooting plant species, for example, dock (*Rumex* spp.), common hogweed (*Heracleum sphondylium*) and caraway (*Carum carvi*), are perceived to be growing more abundantly. Respondents reasoned that these changes came with changes in valley bottom grassland management, including more frequent mowing, grazing with higher stocking

Figure 17.2 Alpine pastures are increasingly overgrown by shrubs and trees due to climatic changes and declines in alpine pasture grazing. Photo by A. Fuchs (2020).

densities, less manual weeding, use of heavy machinery, and increased fertilisation. However, some respondents also reported a climatic influence, saying that deep-rooting plants may dominate as they are better adapted to the higher temperatures and drier conditions caused by climate change.

Seven respondents reported that, both on the alpine pastures and on valley bottom grasslands, the vegetation starts to grow earlier, and four respondents reported that crop species, mainly fruit trees, tend to sprout and flower earlier. In both cases, these trends were attributed to the earlier onset and increased mean temperature in spring. People stated that there were years when the grass started to grow two to three weeks earlier than in former times and that this is the case even if there is a lot of snow in winter. According to informants, this is so because the snow now melts faster due to higher temperatures and more intense sunshine.

Although the vegetation period was reported to start earlier, seven respondents observed that the total pasture productivity has decreased over the last years. Many respondents recounted that in recent years, the hay yields of at least one cut were lower than in the past. This partly resulted in fodder shortage for livestock. Farmers reported that the increasing drought in spring and summer, in combination with the higher mean temperatures and the more frequent heat waves in summer, hampered vegetation growth. People explained that, in the past, the constant and even rainfall kept the soil moist. Nowadays, with the heavy rainfall events that bring a lot of precipitation in a short time, the soil is not capable of taking up all the water. Thus, a great part runs off without serving as a resource for vegetation growth.

Moreover, six respondents reported that the June beetles (generic name for *Phyllopertha horticola, Amphimallon solstitiale,* among other beetle species) increasingly damage pasture. The larvae of the June beetle overwinter in the soil, feed on the roots of the vegetation, and damage the turf. Moreover, the adult June beetles feed on the leaves of fruit trees and on their ripe and unripe fruits. Respondents reported that pastures infestation is more intense for those pastures suffering from drought and heat. The increase in June beetles was also attributed to the fact that temperatures rarely fall below 0°C in winter, so the ground freezes less often. This makes it easier for the larvae to overwinter. The respondents argued that, since the June beetles prefer a warm climate, the higher mean temperatures also facilitate their reproduction. Some respondents reported that the common cockchafer (*Melolontha melolontha*) is also more abundant now, having similar negative impacts on the grassland. However, other respondents disagreed with this, saying that common cockchafers only occur at altitudes lower than the grassland in their village.

Local responses to climate change impacts

Incremental adaptations

The incremental adaptation measures reported range from changes in the management of water, soils, feed, and pastures to changes in livestock management and mainly respond to increasing droughts, decreasing pasture productivity, and seasonal shifts. Respondents reported that, in response to droughts and resulting hay yield losses, grassland irrigation has increased. Mainly simple, self-made irrigation systems (e.g., old fire pump with water cannon; hose in stream) taking surface waters for irrigation have been applied. Although many respondents highlighted the positive effects of irrigation on hay yields, numerous barriers were also related to this adaptation measure. Several respondents thought that the water from mountain streams and springs might be too cold to effectively foster vegetational growth and that it might be required to prewarm the water before irrigation. Respondents also questioned whether the composition of glacial water, especially the high amount of glacial sediment, was suitable for grassland vegetation. Technical aspects also seemed challenging, as effective irrigation requires high water pressure and a sophisticated water supply system to reach all grassland. Such technical equipment demands financial investment and administrative efforts, which were perceived as unlikely to be worthwhile for many part-time farmers.

Another response to increasing drought and resulting hay yield losses was the improvement of soil quality by building up humus. Some respondents stated that they do so by fertilising with fresh and composted manure. Another adaptive measure mentioned to prevent damages and yield losses due to increased drought was to keep the vegetation cover in grassland closed through regular re-sowing. The larvae of the June beetles are thought to be contained by ploughing up grassland and re-sowing infested areas. Yet, respondents remarked that ploughing-up was not viable on steep grassland and that re-seeding provided more fodder for the larvae. A common adaptation measure to the drought-induced hay yield losses was the purchase of hay from community-external sources. Most farmers interviewed mentioned that they did not purchase more hay than needed to get through the winter. Yet, respondents reported that some farmers import larger amounts of hay, because they keep more cattle than their land can feed or to ease their work processes. Respondents mentioned that barriers to this adaptation measure were the difficulty of planning whether and how much additional hay needs to be bought, the difficulty of buying hay produced nearby, and the lower quality of imported hay compared to hay harvested on-farm.

To cope with the unpredictability of rainfall, farmers now do more silage production than hay-making. In contrast to hay-making, silage can be produced during shorter dry periods, that are not sufficient to completely dry the mown grass, and thus helps to decouple fodder production from the increasingly unstable local weather conditions. Yet, not all respondents consider this measure as sustainable in the long term, due to the lower fodder quality of silage and the use of plastic foil for the wrapped silage bales, which generates waste. One respondent mentioned that the installation of hay ventilation plants is also a response to the unpredictability of rainfall.

Another adaptation measure to sustain a good hay harvest is the reintroduction of mowing steep mountain meadows, which used to be mown under difficult and dangerous conditions, in order to obtain additional, high-quality forage. Yet, again the lack of workforce on farms and the lack of time are barriers for this practice to spread. Respondents also stated that accessibility to alpine pastures needs to be improved and that more roads are required to facilitate hay transport.

Some respondents reported that more farmers are re-introducing locally adapted cattle breeds, like *Tiroler Grauvieh* and *Original Pinzgauer Rind*, replacing previously common and highly productive cattle breeds, like *Holstein Friesians.* Locally adapted breeds have a lower fodder demand, which is more compatible with the decreasing pasture productivity caused by increasing droughts. Moreover, some respondents thought about reducing the number of cattle to be able to provide them with their reduced farm-grown fodder. However, this response has been hardly implemented so far because the price for livestock is especially low in the periods of fodder shortage.

Finally, the respondents reported that the earlier onset of spring and the related earlier start of the vegetative growth have made it possible to let ruminants start grazing earlier in the year. This shift can be up to three weeks earlier than 40–50 years ago for alpine pastures. Similarly, some farmers stated that hay in grassland may be mown up to two weeks earlier than in the past. On favourable grassland, it has become possible to cut three, instead of two, times per season, which was unimaginable 40–50 years ago. Moreover, some farmers stated that during some extreme heat days, as some that had occurred in the last years, they started to keep the farm animals in the stable during the day and release them on the pasture only during the night, to reduce the heat stress for the animals.

Transformative adaptations

The reported transformative adaptation measures range from the abandonment of farming and conversion to other agricultural production sectors, to comprehensive proposals for measures to support livestock farming, especially to improve the management of alpine pastures and grassland and strengthen community management and resource sharing. Reported repsonses respond´to the impacts of drought and decreasing pasture productivity, but they also respond to non-climatic socio-economic changes, especially the decreasing availability of workforce.

Most drastically, interviewees emphasised that more and more farmers are ceasing farming for different reasons. Farms are lacking successors, agriculture no longer seems to be profitable, and young people prefer to work in less nature-dependent sectors, with fixed working hours, holidays, and secured income. Most farmers are part-time farmers anyway. This trend is not primarily driven by climatic changes, yet climate change impacts are perceived as intensifying the situation. Increasing problems with natural hazards, droughts, yield losses, pest infestations, and shrub encroachment, which are projected to amplify in the future, are an additional motivator to become financially independent from farming. Some respondents noted though, that the implementation of home office due to the COVID-19 pandemic enhanced the compatibility of off-farm and farm work because of the reduced need to commute.

Instead of stopping farming, some farmers have shifted to different agricultural production sectors, for example, crop cultivation or producing hay for sale. Respondents thought that it has become easier to cultivate cereals, due to the increasing mean temperatures and the longer vegetation period, and some local farmers have restarted cereal cultivation, which was more common in the past. Yet, respondents were sceptical, whether there will be a long-term shift to cereal cultivation because the required effort was thought to be higher than the financial profit. Barriers for cereal cultivation include the already very high workload for most part-time farmers due to farming and off-farm activities, the general lack of workforce in agriculture, the need for investment in specialised machinery for cereal cultivation, and the steepness of the area, which makes mechanised cereal cultivation difficult. Increasing cultivation also means that mountain meadows and alpine pastures would be less maintained, leading to less hay production.

For several respondents, abandoning livestock farming is a step in the wrong direction, and they argue that instead comprehensive measures to improve the management of alpine pastures would help enhance the local adaptive capacity to climate change impacts. Letting farm animals graze on the alpine pastures was considered the best way to protect them from heat stress caused by the increasing temperatures and intense heat waves in summer. Furthermore, grazing limits the progressive shrub encroachments and rising treelines, and maintains the available grazing area. Yet, most respondents reported that the management of alpine pastures is progressively decreasing. Efficient management of alpine pastures would require a sufficiently high stocking density of grazing animals, manual cleaning, for example cutting and burning dwarf shrubs, and the maintenance of roads and paths. However, while the amount of work to be done to maintain pastures is high, the number of people working per farm and the stocking density of animals on alpine pastures are decreasing.

Moreover, re-establishing the formerly common *feldfluren*, which are trees, hedges and dry-stone walls that separate individual plots of grassland, was considered important to maintain grassland's productive capacity. Some respondents mentioned that the *feldfluren* protected grassland against the stronger wind, which, in combination with the higher temperature, dries the grassland. Moreover, respondents reported that *feldfluren* improve soil fertility, create favourable microclimates, store humidity after rainfall, and provide shade for grazing animals. Despite these advantages, respondents stated that the *feldfluren* are impractical for the work with agricultural machinery. In addition, trees and hedges need to be maintained to limit their growth. *Feldfluren* still characterise the cultural landscape of Virgen today, but in the past, they also existed in Prägraten and Kals, where they were partly removed to facilitate work with machinery. Yet, in these two communities, interviewees stated that the awareness of the ecological and protective benefits of *feldfluren* is leading to their restoration. Respondents also reported that community ownership of alpine pastures or community management by farming communities, cooperatives, or associations would help increase the efficiency of alpine pasture management. This is the case because community ownership would offer the possibility of sharing work and costs of activities such as cleaning pastures, constructing stables, and herding. Moreover, respondents argued that the maximum livestock density of alpine pastures should be exhausted to keep them free from shrub encroachment.

Community-based sharing of machinery and tools, as well as sharing workforce and helping each other out were also considered good opportunities to reinforce land management practices which strengthen the adaptive capacity of local communities. Again, these responses are reported to be hampered by the lack of workforce and time. Many interviewees had the impression that co-operation and sharing have declined since the 1970s. Most people either have their own machinery or employ agricultural contractors from outside the communities. A barrier to community-based sharing of machinery and tools is that they are needed at similar times of the year. Good organisation and agreements are required, but not always easily achieved. Moreover, involving external voluntary help, for example of students, local clubs for civil protection, or other young people who are concerned about nature conservation was also seen as a chance to cope with the increasing damage on alpine pastures.

Discussion

In this study, we aimed to understand local people's reports of (i) climatic changes, (ii) their impacts on mountain farming, and (iii) adaptations to climate change impacts in Eastern Tyrol. We found (i) that respondents perceive a range of climate change impacts on mountain farming, with most impacts being perceived as adverse, creating new challenges or exacerbating long-standing ones, and (ii) that farmers identified a range of responses to climate change impacts and associated stressors, with incremental adaptations being widely implemented while transformative adaptations in support of mountain farming lack implementation (Figure 17.3).

Most climate change impacts are reported to be due to rises in mean temperature and to less but heavier precipitation events, which leads to less water being stored in the soils and more frequent droughts. These changes provide new socio-economic and ecological challenges but also to exacerbate adverse developments which have been affecting the area for decades. For example, the advancing shrub encroachment in alpine pastures is a longstanding concern, which was initially caused by socio-economic changes, including low profitability of farming, declines of alpine pasture grazing, and lack of labour in farming. However, shrub encroachment is now aggravated by increasing mean temperatures and resulting increases in vegetation cover. In contrast, the increasing hay harvest losses due to increasing temperatures and drought and the increased reproduction and spread into higher altitudes of pests, such as the June beetles, were reported to be recent impacts providing new challenges. This finding suggests that, while the continued shift of the labour force away from mountain farming has in itself been a longstanding and hardly stoppable trend in Eastern Tyrol – as in many other regions across Europe (Schuh, 2020) –, climatic changes and their impacts make it even more difficult to counter the trend.

Whereas adverse climate change impacts are, by far, the most widespread impacts in Eastern Tyrol, as in mountain regions throughout the globe (Adler & Wester, 2021; Babai, 2024; Mattalia et al., 2024), climate change impacts are also perceived to offer opportunities. Especially, the extension of the vegetation period was mentioned as a positive impact as it allows earlier and more frequent mowing of hay, earlier start of livestock grazing, earlier start of the use of alpine pastures, and an increased potential for cereal cultivation. While these opportunities may not compensate for adverse impacts, they can still offset some difficulties and offer new farming

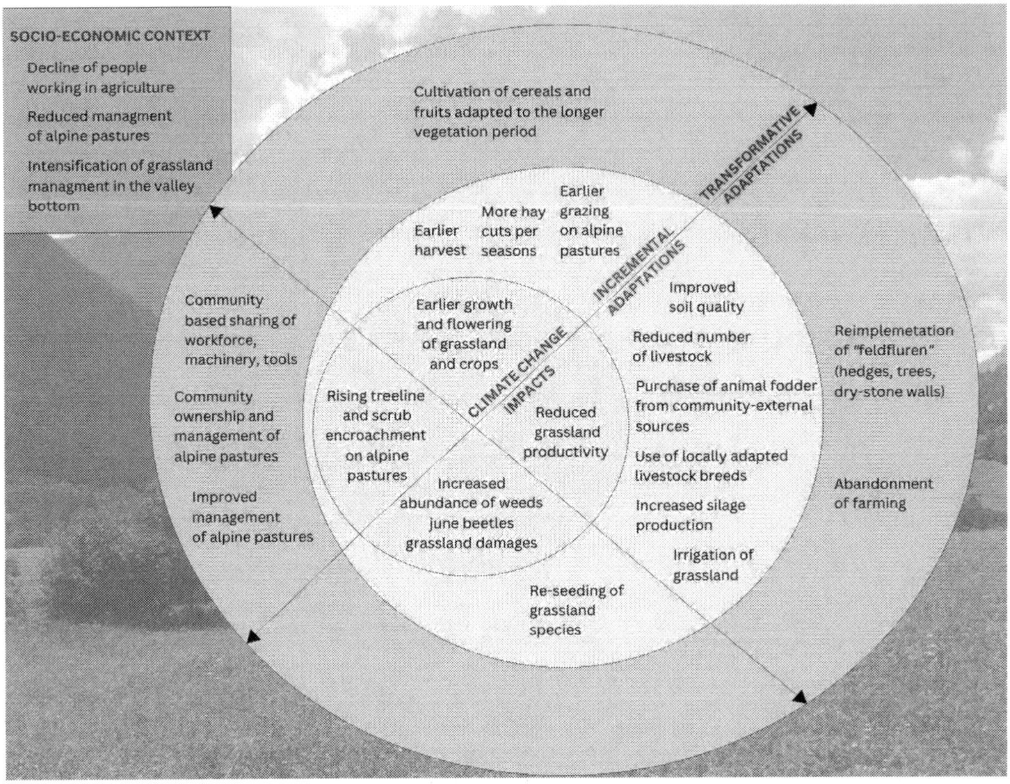

Figure 17.3 Overview of climate change impacts and local adaptations in mountain farming in Eastern Tyrol.

opportunities. Research in other areas has also shown how local people perceive new opportunities with the changed environment. For example, in pine forests in Spain, mushroom abundance is predicted to increase in upcoming decades (Karavani et al., 2018), and sugarcane yields are projected to increase in Southern China (Ruan et al., 2018). In a case study conducted in western Amazonia, as much as about 20% of climate change impacts related to fishing and 15% of climate change impacts related to manioc cultivation were perceived as positive rather than neutral or negative, although for other livelihood activities, like rubber production and hunting, climate change impacts were perceived purely as negative (Estevo et al., 2023). Understanding the differentiated climate change impacts on diverse livelihood activities is important to seize opportunities in a changing environment.

Moreover, we found that Eastern Tyrolean farmers report a range of incremental to transformative responses to climate change impacts oriented to maintain the productivity of their livestock farming systems. The incremental adaptations reported tend to be technical and managerial, quickly implemented by individual farmers. These include earlier start of grazing and mowing, irrigating grassland, fertilising with manure and switching to locally adapted cattle breeds. Due to their low complexity, these adaptations are already in place, at least for some farmers. In contrast, the transformative adaptations reported tend to be ideas that are to be implemented at community level, or beyond, rather than by individuals and that currently are only partially implemented or not implemented at all due to their high complexity. Ideas for transformative adaptations include large-scale technical and managerial adaptations, for which collaboration and mutual support is needed, for example re-establishing *feldfluren* and cultivating cereals, but also social adaptations, such as community-based management of alpine pastures and sharing of machinery and labour force. An exception is the abandonment of agriculture, which is a transformative adaptation based on individual decisions. However, while being a response for individual farmers, the abandonment of agriculture does not contribute to supporting and maintaining agriculture and the benefits it provides. In other words, collective decisions and cooperation are required for positive transformative adaptations that give a future to livestock farming and agriculture in the region.

Our findings dovetail with findings from other mountain areas around the globe, where most adaptations implemented modify existing technical and management-related practices to cope with pressing stressors and do not provide more comprehensive, structural responses to the compound of climate change impacts and socio-economic challenges (Adler & Wester, 2021; Berrang-Ford et al., 2021; McDowell et al., 2021; Vij et al., 2021). Moreover, social factors are common barriers to transformative adaptations (Bezner Kerr et al., 2022). Positive transformative adaptations are difficult to implement because of their complexity, but incremental adaptations may also become transformative when taken together or expanded (Termeer et al., 2017). In our study area, the first steps on the way towards the implementation of transformative adaptations could be the reintroduction of locally adapted cattle breeds, the improvement of soil quality of grassland, increased silage production, and mowing steep mountain meadows. When taken together, these responses may result in considerably better adapted, and more resilient, farming systems. Moreover, the beginning recovery of *feldfluren* or cultivation of cereals may set examples that, if properly supported, may be adopted on a wider scale and thus become transformative not only for individual farmers but for the whole area.

However, the same socio-economic trends that have challenged mountain farming in Eastern Tyrol since decades, especially the diminishing labour force and increasing land abandonment, hinder positive transformative adaptations too. Transformative adaptations are considered to be work intensive and require a constant level or even increase of labour force, knowledge, and skills. Land abandonment by farmers is a widespread and persistent problem in Europe. Remote and marginal areas, such as mountain areas, where biophysical conditions are difficult for agricultural production, are among the most affected areas (Ustaoglu & Collier, 2018). For example, abandonment of mountain pastoralism is the major driver of landscape change in the Italian Alps (Cocca et al., 2012). Tyrol, where the current level of mountain farming is only maintained by significant subsidies and additional farm-income from tourism (Schuh, 2020; Terres et al., 2015), is considered a hotspot of land abandonment in Europe (Schuh, 2020). Overcoming the barriers to transformative adaptation must therefore take these socio-economic conditions into consideration. The strengthening of collaborations among farmers, as suggested by respondents, should be further supported in this respect, since local collaborations and knowledge co-production, supported by external competence development, finances, and private sector involvement, is key for reaching positive transformative adaptations (Adler & Wester, 2021; McDowell et al., 2021).

Conclusion

In this study, we found that Eastern Tyrolese report a range of adverse climatic change impacts on mountain farming. In response to those, farmers are (partly) implementing technical and management-related incremental adaptations and have ideas for transformative adaptations at the community level that will support and maintain agriculture. Implementation of these ideas is necessary to safeguard the multiple functions of mountain farming in the face of intensifying climate change impacts. However, implementing positive transformative adaptations is challenging as a range of local stakeholders need to be involved and deep and long-standing socio-economic challenges to the region, such as labour force decline and land abandonment, need to be addressed. Local farmers, who are among those most affected by climate change impacts, can and must actively contribute to this change. As the impacts of climate change are currently progressing faster than the implementation of adaptation strategies, local actors should act quickly to counteract the expected dramatic impacts of climate change.

Acknowledgements

First of all, we wish to thank all respondents for sharing their valuable time, knowledge, and experiences. Our sincere appreciation goes to the team of the Regionsmanagement Osttirol who helped to identify the study areas in Eastern Tyrol. Also, we want to thank the mayors of Kals, Prägraten, and Virgen and our contact persons – Erika Rogl, Anton Steiner, Dietmar Ruggenthaler, and Albin Mariacher for welcoming us into their communities and their continuous support during the field research.

References

Adler, C., & Wester, P. (2021). *IPCC WGII Sixth Assessment Report—Cross-Chapter Paper 5: Mountains.*

AMA. (2022). *AMA Flächenauswertung—Bezirk Lienz.* https://flaechenauswertung.services.ama.at

Babai, D. (2024). *'The weather is more erratic; it changes faster...' local perceptions of climate change in the Eastern Carpathians, Romania.* In V. Reyes-García, S. Alvarez-Fernandez, P. Benyei, L. Calvet-Mir, D. García-del-Amo, A. B. Junqueira, X. Li, V. Porcher, A. Porcuna-Ferrer, A. Schlingman, & R. Soleymani (Eds.), Routledge handbook of climate change impacts on indigenous peoples and local communities. Chapter 4. Routledge, Oxon and New York. pp 74–94

Bernard, H. R. (2011). *Research methods in anthropology: Qualitative and quantitative approaches* (5th edition). Lanham: Rowman & Littlefield.

Berrang-Ford, L., Siders, A. R., Lesnikowski, A., Fischer, A. P., Callaghan, M. W., Haddaway, N. R., Mach, K. J., Araos, M., Shah, M. A. R., Wannewitz, M., Doshi, D., Leiter, T., Matavel, C., Musah-Surugu, J. I., Wong-Parodi, G., Antwi-Agyei, P., Ajibade, I., Chauhan, N., Kakenmaster, W., … Abu, T. Z. (2021). A systematic global stocktake of evidence on human adaptation to climate change. *Nature Climate Change, 11*(11), Article 11. https://doi.org/10.1038/s41558-021-01170-y

Bezner Kerr, R., Naess, L. O., Allen-O'Neil, B., Totin, E., Nyantakyi-Frimpong, H., Risvoll, C., Rivera Ferre, M. G., López-i-Gelats, F., & Eriksen, S. (2022). Interplays between changing biophysical and social dynamics under climate change: Implications for limits to sustainable adaptation in food systems. *Global Change Biology, 28*(11), 3580–3604. https://doi.org/10.1111/gcb.16124

Brondízio, E., Settele, J., Díaz, S. M., & Ngo, H. (Eds.). (2019). *The global assessment report on biodiversity and ecosystem services.* IPBES Secretariat. https://doi.org/10.5281/zenodo.3831673

Cocca, G., Sturaro, E., Gallo, L., & Ramanzin, M. (2012). Is the abandonment of traditional livestock farming systems the main driver of mountain landscape change in Alpine areas? *Land Use Policy, 29*(4), 878–886. https://doi.org/10.1016/j.landusepol.2012.01.005

Dax, T. (2001). Endogenous development in Austria's mountain regions. *Mountain Research and Development, 21*(3), 231–235. https://doi.org/10.1659/0276-4741(2001)021[0231:EDIAMR]2.0.CO;2

EEA. (2010). *Europe's ecological backbone: Recognising the true value of our mountains* [Publication]. European Environment Agency. https://www.eea.europa.eu/publications/europes-ecological-backbone

Estevo, M. O., Junqueira, A. B., Reyes-García, V., & Campos-Silva, J. V. (2023). Understanding multidirectional climate change impacts on local livelihoods through the lens of local ecological knowledge: A study in Western Amazonia. *Society & Natural Resources, 36*(3), 232–249. https://doi.org/10.1080/08941920.2022.2153294

Gobiet, A., Kotlarski, S., Beniston, M., Heinrich, G., Rajczak, J., & Stoffel, M. (2014). 21st century climate change in the European Alps—A review. *Science of the Total Environment, 493*, 1138–1151. https://doi.org/10.1016/j.scitotenv.2013.07.050

Grüneis, H., Penker, M., Höferl, K.-M., Schermer, M., & Scherhaufer, P. (2018). Why do we not pick the low-hanging fruit? Governing adaptation to climate change and resilience in Tyrolean mountain agriculture. *Land Use Policy*, *79*, 386–396. https://doi.org/10.1016/j.landusepol.2018.08.025

Karavani, A., De Cáceres, M., Martínez de Aragón, J., Bonet, J. A., & de-Miguel, S. (2018). Effect of climatic and soil moisture conditions on mushroom productivity and related ecosystem services in Mediterranean pine stands facing climate change. *Agricultural and Forest Meteorology*, *248*, 432–440. https://doi.org/10.1016/j.agrformet.2017.10.024

Land Tirol. (2022). *Regionsprofil Lienz*. https://www.tirol.gv.at/fileadmin/themen/statistik-budget/statistik/downloads/Regionsprofile/Stat_profile/bezirke/Lienz.pdf

Mattalia, G., Stryamets, N., Toscano Rivalta, V., & Reyes-García, V. (2024). Correspondence between local and scientific knowledge of climate change: The case of Hutsuls, Northern Romanian Carpathians. In V. Reyes-García, S. Alvarez-Fernandez, P. Benyei, L. Calvet-Mir, D. García-del-Amo, A. B. Junqueira, X. Li, V. Porcher, A. Porcuna-Ferrer, A. Schlingman, & R. Soleymani (Eds.), *Routledge handbook of climate change impacts on indigenous peoples and local communities*. Chapter 1. Routledge, Oxon and New York. pp 30–44

Mayring, P. (2014). *Qualitative content analysis: Theoretical foundation, basic procedures and software solution*. https://nbn-resolving.org/urn:nbn:de:0168-ssoar-395173

McDowell, G., Stevens, M., Lesnikowski, A., Huggel, C., Harden, A., DiBella, J., Morecroft, M., Kumar, P., Joe, E. T., Bhatt, I. D., & the Global Adaptation Mapping Initiative (2021). Closing the adaptation gap in mountains. *Mountain Research and Development*, *41*(3), A1. https://doi.org/10.1659/MRD-JOURNAL-D-21-00033.1

Owen, G. (2020). What makes climate change adaptation effective? A systematic review of the literature. *Global Environmental Change*, *62*, 102071. https://doi.org/10.1016/j.gloenvcha.2020.102071

Petzold, J., Andrews, N., Ford, J. D., Hedemann, C., & Postigo, J. C. (2020). Indigenous knowledge on climate change adaptation: A global evidence map of academic literature. *Environmental Research Letters*, *15*(11), 113007. https://doi.org/10.1088/1748-9326/abb330

Reyes-García, V., Álvarez-Fernández, S., Benyei, P., García-del-Amo, D., Junqueira, A. B., Labeyrie, V., Li, X., Porcher, V., Porcuna-Ferrer, A., Schlingmann, A., & Soleymani, R. (2023). Local indicators of climate change impacts described by indigenous peoples and local communities: Study protocol. *PLoS One*, *18*(1), e0279847. https://doi.org/10.1371/journal.pone.0279847

Reyes-García, V., García-del-Amo, D., Benyei, P., Fernández-Llamazares, Á., Gravani, K., Junqueira, A. B., Labeyrie, V., Li, X., Matias, D. M., McAlvay, A., Mortyn, P. G., Porcuna-Ferrer, A., Schlingmann, A., & Soleymani-Fard, R. (2019). A collaborative approach to bring insights from local observations of climate change impacts into global climate change research. *Current Opinion in Environmental Sustainability*, *39*, 1–8. https://doi.org/10.1016/j.cosust.2019.04.007

Ruan, H., Feng, P., Wang, B., Xing, H., O'Leary, G. J., Huang, Z., Guo, H., & Liu, D. L. (2018). Future climate change projects positive impacts on sugarcane productivity in southern China. *European Journal of Agronomy*, *96*, 108–119. https://doi.org/10.1016/j.eja.2018.03.007

Schirpke, U., Tappeiner, U., & Tasser, E. (2019). A transnational perspective of global and regional ecosystem service flows from and to mountain regions. *Scientific Reports*, *9*(1), Article 1. https://doi.org/10.1038/s41598-019-43229-z

Schlingmann, A., Graham, S., Benyei, P., Corbera, E., Martinez Sanesteban, I., Marelle, A., Soleymani-Fard, R., & Reyes-García, V. (2021). Global patterns of adaptation to climate change by Indigenous Peoples and local communities. A systematic review. *Current Opinion in Environmental Sustainability*, *51*, 55–64. https://doi.org/10.1016/j.cosust.2021.03.002

Schuh, B. (2020). *Research for AGRI Committee—The challenge of land abandonment after 2020 and options for mitigating measures* (p. 98). European Parliament, Policy Department for Structural and Cohesion Policies. https://www.europarl.europa.eu/RegData/etudes/STUD/2020/652238/IPOL_STU(2020)652238_EN.pdf

Termeer, C. J. A. M., Dewulf, A., & Biesbroek, G. R. (2017). Transformational change: Governance interventions for climate change adaptation from a continuous change perspective. *Journal of Environmental Planning and Management*, *60*(4), 558–576. https://doi.org/10.1080/09640568.2016.1168288

Terres, J.-M., Scacchiafichi, L. N., Wania, A., Ambar, M., Anguiano, E., Buckwell, A., Coppola, A., Gocht, A., Källström, H. N., Pointereau, P., Strijker, D., Visek, L., Vranken, L., & Zobena, A. (2015). Farmland abandonment in Europe: Identification of drivers and indicators, and development of a composite indicator of risk. *Land Use Policy*, *49*, 20–34. https://doi.org/10.1016/j.landusepol.2015.06.009

Ustaoglu, E., & Collier, M. J. (2018). Farmland abandonment in Europe: An overview of drivers, consequences, and assessment of the sustainability implications. *Environmental Reviews*, *26*(4), 396–416. https://doi.org/10.1139/er-2018-0001

Vij, S., Biesbroek, R., Adler, C., & Muccione, V. (2021). Climate change adaptation in European mountain systems: A systematic mapping of academic research. *Mountain Research and Development*, *41*(1), A1–A6. Scopus. https://doi.org/10.1659/MRD-JOURNAL-D-20-00033.1

WKO Tirol. (2021). *Bezirk Lienz—Zahlen und Fakten zur Wirtschaft 2021/22* (p. 44). Wirtschaftskammer Tirol. https://www.wko.at/service/t/bezirksstellen/BIZ-Lienz-2021_22-End-oS.pdf

18

STRENGTHENING PARTICIPATORY GOVERNANCE TO ADAPT AND OTHER SOCIOCULTURAL RESPONSES TO CLIMATE CHANGE IMPACTS AMONG FARMING COMMUNITIES IN KERMAN, IRAN

Mariam Abazeri

Introduction

Human-induced climate change has resulted in widespread impacts on the livelihoods and cultures of people around the world (IPCC, 2022). Vulnerability, or the propensity to be adversely affected by climate change, is especially high among people productively dependent on their environments, such as smallholder farmers and agropastoralists (Misra, 2017; IPCC, 2022). Climate change can increase the vulnerability of agrarian communities by affecting agricultural yields and income as well as household well-being, water and food security, and social relationships (Alam et al., 2017). However, impacts can vary substantially among and within regions and are difficult to measure through quantitative methods such as large-scale weather records and modeling techniques (Reyes-Garcia et al., 2019). Increasingly, scientists call for more grounded studies with local communities in geographically diverse cropping systems (Rosenzweig & Neofotis, 2013). This chapter draws on a mixture of grounded qualitative, quantitative, and participative visual data to understand how sociocultural elements that create 'place' influence the ways people respond to climate change impacts.

Local impacts and adaptations to climate change can vary substantially according to dynamic and evolving social and cultural interactions within and between communities (Weber, 2010; Adger et al., 2013; Kim & Wolinsky-Nahmias, 2014). Social dimensions, in the way they are examined here, refer to aspects of life such as livelihood, demographics, and community relations, while cultural dimensions refer to values, practices, beliefs, narratives, rituals, and 'collective understandings that enable social life' (Appadurai, 1996; Winthrop, 2014; Saunders et al., 2020; Strand et al., 2022). Culture is often intimately connected to place or physical spaces given meaning by people (Basso, 1996; Escobar, 2001). Climate change can thus impact the relationships and meanings people craft toward place, changing cultures and communities often in ways that people consider negative or undesirable within a given cultural frame of reference (Kirsch, 2001).

Analyses of risks, adaptations, and implementation methods often overlook the social and cultural dimensions of contemporary ecological challenges, prioritizing instead quantitative economic and ecological impacts over more qualitative ethnographic data (Satterfield et al., 2013;

 DOI: 10.4324/9781003356837-23

Berkes, 2015; Gee et al., 2017; Muhl & Sowman, 2020; Strand et al., 2022). Consequently, many contemporary responses to climate change fail to address cultural dimensions of climate impacts and can even exacerbate and put at risk important elements of social life (Adger et al., 2013). Some of the consequences of these changes are ambiguous and difficult to predict, while others are more tangible. For example, the displacement of people due to climate change impacts such as drought or sea-level rise and adaptation strategies like forced or incentivized migration have a clear impact on culture with often no effective replacement or compensation for lost sites of cultural importance (Crate, 2008; Garrett, 2010). More insight into how culture interacts with climate-related impacts and adaptation strategies can transform our understanding of social responses to climate change and improve future policies and interventions as well as advance social justice, sociocultural diversity, and the well-being of vulnerable communities (Adger et al., 2011; Saunders et al., 2020).

Accordingly, here I examine some of the ways social and cultural elements, especially livelihoods, social infrastructure, and narratives of place, drive and influence responses to reported environmental changes. The novelty of this study is in the use of qualitative, quantitative, and participatory visual methods to consider the historic and social conditions that shape climate action in a remote rural community in Southern Iran. This approach enables us to understand how people use ecological knowledge alongside new and existing practices, institutions, and experiences to adapt to evolving conditions and needs. In the following sections, I explain the approach I used to collect and analyze data, outline the impacts and adaptation measures residents report, and analyze the socio-cultural elements that influence and inform climate-related observations and actions.

Background

The agricultural sector plays a pivotal role in providing food and welfare in Iran, accounting for about 20% of employment and providing 90% of domestic food demands (Mesgaran et al., 2016). For smallholder farmers, the majority of whom already cultivate under extreme temperatures and rain conditions, climatic changes are expected to exert even more tension on agricultural yields, livelihoods, and well-being (Hashemi et al., 2015). Despite the importance and major role of the agricultural sector in Iran, grounded studies of climate change impacts and adaptive measures are limited (e.g., Mesgaran et al., 2017; Karimi et al., 2018; Azadi et al., 2019; Mirzaei et al., 2019).

Throughout much of Iran, villages have provided the social boundary as well as the production unit of peasant life and labor. The historic mode and method of production was communal, commonly described as a *boneh* (Katouzian, 2009). This system owes its origins to the fact that water is the scarcest agricultural resource in the region. The climatic aridity and scarcity of water encouraged cooperative farming with specialized roles to govern and distribute agricultural resources among households, enabling villages to operate as an independent unit of residence and livelihood. Following the land reform law of 1962, many functions and relationships within the *boneh* system were eliminated without a replacement. Many new landowning farmers lacked specialized skills and the weakening of agricultural cooperation brought about and intensified socioeconomic disparities in rural areas (Lob, 2020). Instead, the land tenancy reforms encouraged the development of industrial farm corporations, which turned many peasant properties into paper shares of agri-businesses, thereby concentrating land ownership for absentee capitalist farmers, and increasing rural unemployment and migration to urban areas for smallholder farmers (Katouzian, 1981; Ashraf, 1991).

A critical aspect that had made this region habitable for centuries is a cooperative water governing system known in Persian as a *qanat* (Beaumont et al., 1989; Labbaf Khaneiki, 2019). Through a gently sloping subterranean network of channels, groundwater is extracted from aquifers and transported to farmlands and residences for agricultural and domestic use (Jomehpour, 2009). The development and maintenance of this infrastructure not only depends on physical aspects of the natural environment, such as soil type, size and location of aquifers, or amount of rainfall, but also social aspects of its users, such as organization of labor roles and responsibilities, division and accountability of water access and rights, and oversight of channels and other related infrastructure (Honari, 1989; Wessels, 2008). In recent decades, this system has been increasingly replaced by modern methods of water extraction, such as semi-deep and deep pumped wells, which while offering private and unregulated access to water resources, compromises the quality of groundwater and sustainability of other extraction methods (Ehlers & Saidi, 1989; Semsar Yazdi & Labbaf Khaneiki, 2019).

State interventions through land and water management have also significantly altered the characteristics of freshwater resources in the country through a set of techniques and policies related to storage, redistribution, and consumption (Sadegh et al., 2010; Frans et al., 2013; Mirchi et al., 2014; Mehran et al., 2017). A continued increase in irrigated agricultural production, despite a reduction in precipitation, has only been possible through extensive reservoir building, interbasin water transferring, overusing of groundwater, and increased fertilizer use (Gohari et al., 2013; Madani, 2014; Madani et al., 2016; Mirzaei et al., 2019). These interventions have affected regional hydrologic cycles and catalyzed the growth of 'anthropogenic droughts' (AghaKouchak et al., 2015) that reflect increasing human-induced water stress intensified by unsustainable water use relative to available water supply. Increased pressure on surface and groundwater resources, as well as recent droughts, have resulted in the drying of lakes, rivers, springs, and water channels, declining groundwater level and well discharges, soil erosion, saltwater intrusion, desertification, and biodiversity loss (Madani, 2014; AghaKouchak et al., 2015; Ashraf et al., 2017; Mirzaei et al., 2019; Maghrebi et al., 2020). Coupled with global climatic changes, rural Iranian communities productively dependent on groundwater face mounting vulnerability and risk of food insecurity, unemployment, migration, conflict, and socioeconomic pressure in Iran (Maghrebi et al., 2020).

Methods

Field site

This study takes place in the district of Takab, a cluster of villages located along the Western periphery of the Lut Desert in the province of Kerman, Iran (Figure 18.1). Takab is home to a population of about 1,700 households spread across 30 villages whose livelihoods have been historically based on agropastoralism. Customary crops farmed include grains, fodder, dates and other fruit, herbs, and vegetables, and livestock raised include sheep, chicken, and cattle. This study was conducted in three villages in Takab: Shafiabad, Deh Seyf, and Valiabad, consisting of about 140 households in total. Smallholder farmers in this region cultivate in one of the hottest and driest climates in the world. A recent study recorded the hottest land surface temperature on Earth of 80.8°C in the nearby Lut Desert (Zhao et al., 2021). Precipitation averages are often under 75 mm per year and precipitation is absent about 300–350 days a year (Ghorbani, 2013). Endemic vegetation is well adapted to salinity and aridity. Halophytic plants such as *Tamarix* trees, many of which form phytogenic mounds known locally as *nebkas*, are found along the desert plains in areas of notable ambient salinity and relatively shallow underground water tables (Akhani & Mucina, 2015).

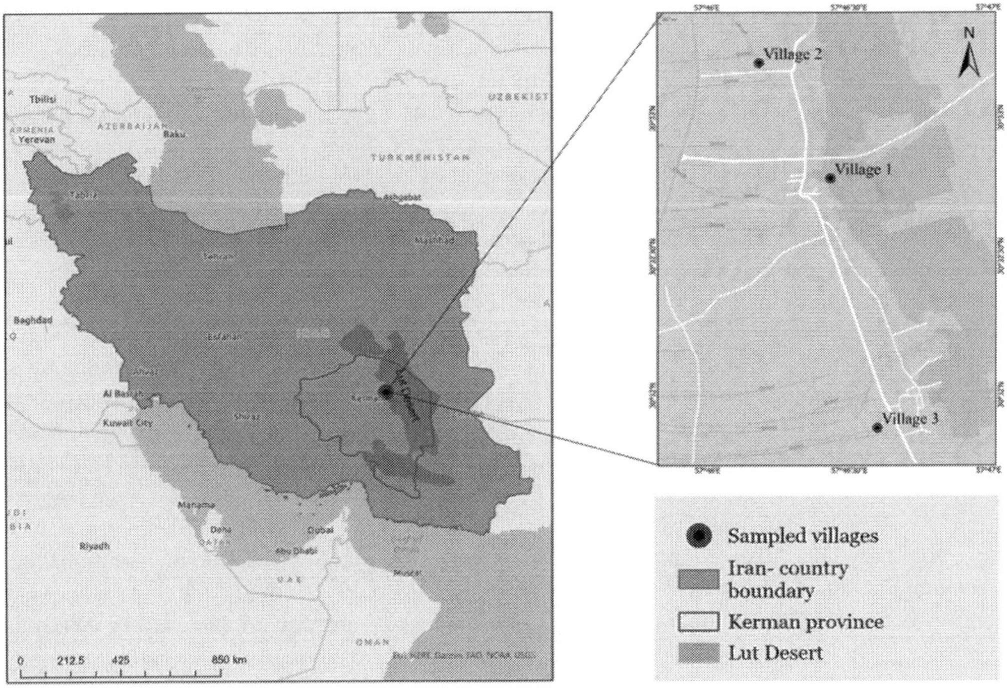

Figure 18.1 Map of field sites.

Sampling and data collection

This work was carried out alongside a collaborative filmmaking project in which residents were involved in and that addressed interrelated social and environmental issues (Abazeri, 2022). It is also part of the Local Indicators of Climate Change Impacts (LICCI) project, which explores the use of Indigenous and local knowledge systems in climate research and policy (Reyes-Garcia et al., 2023). Primary data on climate change impacts and adaptation measures were collected through qualitative, quantitative, and participatory visual methods, which included in-depth semi-structured interviews, focus group discussions, meteorological data, and video footage shot during the filmmaking project. Through semi-structured interviews, I sought to understand residents' perceptions of local climate change impacts and responses to these changes. Focus group discussions helped substantiate data from the interviews, while the weather station data served as a reference for local reports. Video footage contextualized some of the impacts and adaptation measures residents discussed throughout the film intervention and served to further highlight local narratives and perspectives on environmental challenges and adaptive responses. Before starting, I followed the ethical guidelines of the LICCI protocol as described in Reyes-Garcia et al. (2023), including securing ongoing free, prior and informed consent with village leaders and participants, and obtaining institutional review board clearance from the University of Miami. The interviews and focus group discussions were audio recorded with permission from interviewees and video footage was referenced with permission from participants of the video project.

Twenty semi-structured interviews were carried out using quota sampling across three categories of interest: age, gender, and livelihood. To select people within each quota, I relied on snowball sampling. The interviewees consisted of ten women and ten men ranging from 22 to 76 years of age. Participants were all engaged in crop cultivation and animal husbandry, with some men also active in waged labor and tourism services and some women in handcraft production, tourism services, and plant foraging. Residents engaged in different activities according to the time of year, demand for labor, and physical abilities. In the semi-structured interviews, I asked informants to describe observed changes in their environment using a benchmark (e.g., since you were young) and a timeline of the district's local history created with village elders in previous interviews. Participants were encouraged to think of changes in atmospheric, physical, and life systems related to their livelihoods and well-being including health, husbandry, and cultivation practices. Each described observation of change was followed by questions on its potential drivers, types of responses, timing of implementation.

Two focus group discussions were held in Takab to validate the observed changes collected during the semi-structured interviews (Reyes-Garcia et al., 2023). For both discussions, I used convenience sampling aimed at capturing social diversity, including livelihood activities, expertise (i.e., preference given to elders and local experts), age, and gender. Groups included men and women ranging from 29 to 64 years of age who worked in crop cultivation, animal husbandry, and handcrafting. To compare trends of long-term climatic changes, meteorological data were also obtained from a weather station located about 20 kilometers south of the district at an elevation of 482 meters. The available weather data from this station had a limited time series beginning in 2003 with records of monthly max, min, and mean temperatures (°C) and R24 precipitation (mm) parameters.

The video footage I reference was taken from 2014 to 2018 by six young local women living in Takab, ages initially ranging from 19 to 33, and a documentary filmmaker from Tehran, as part of a participatory video project (PV). The PV group used video cameras and the filmmaking process to discuss and address gendered struggles (Pink, 2007; Strand et al., 2022; Milne et al., 2012; White, 2003). Their footage shows the women discussing at length amongst themselves and with other residents' concerns about their social and physical environment, specifically women's livelihoods and water issues, and documents the practices they use to raise consciousness and organize local action. It is important to note these topics emerged through a co-generative and dialogic process and were not imposed by external facilitators. The collection consists of over 1.2 Terabytes of High-Definition video shot by nine people, eight of whom are local women from Takab, over a span of four years, roughly amounting to over 600 hours of footage.

Data analysis

Data on observed environmental changes collected during interviews and focus groups were recorded and organized according to a classification system that differentiates between impacts on atmospheric, physical, and life systems (Reyes-Garcia et al., 2020). Each responsive strategy described by residents was grouped into a three-level classification system that differentiates between response *sectors* (i.e., livelihood activities), *domains* (i.e., activities' timing, location, products, resource input, capacity building, livelihood systems), and *types* (i.e., quantitative, or qualitative changes) (Schlingmann et al., 2021). Video footage was organized by date, location, subject, and setting. A random sample of footage was selected (n = 100 hours) and through thematic analysis, footage was coded into categories of social response to environmental changes until themes began to surface revealing narratives, motivations, and experiences that informed interventions (Kiger & Varpio, 2020). I continued this process, reviewing, revising, and consolidating overlapping categories until broad sociocultural trends emerged.

Results

Local reports of environmental changes

Residents in Takab have observed several changes in atmospheric, physical, and life systems. Through interviews, focus groups, and PV footage, 44 indicators of environmental change were identified. The indicators have been categorized into elements of climatic, physical, and life systems according to the Reyes-Garcia et al. (2020) classification system (Table 18.1). Changes in elements of the climatic system were noted across temperature, precipitation, seasonal, and air

Table 18.1 Perceived indicators of environmental changes ($n = 20$)

System	*Sub-system*	*Impacted element*	*Impact indicator*	*Direction*
Climatic	Temperature	Mean temperature	Change in frequency of cold days	Decrease
		Mean temperature	Change in frequency of hot days	Increase
		Mean temperature	Change in temperature during the night	Increase
		Mean temperature	Change in temperature during the day	Increase
		Temperature extremes	Change in frequency of heat waves	Increase
		Temperature extremes	Change in length/duration of heat waves	Increase
		Temperature extremes	Change in intensity/strength of heat waves	Increase
		Seasonal temperature	Change in mean temperature in given season	Increase
		Seasonal temperature	Change in intensity/strength of extreme hot season	Increase
		Sunshine	Change in sunshine intensity	Increase
	Precipitation	Mean precipitation	Change in mean rainfall	Decrease
		Mean precipitation	Change in number of rainy days	Decrease
		Precipitation distribution	Change in length/duration of rainfall events	Decrease
		Seasonal precipitation	Change in amount of rainfall in given season	Decrease
		Seasonal precipitation	Change in intensity/strength of rainfall in given season	Decrease
		Drought	Change in frequency of drought events	Increase
			Change in length of drought	Increase
		Clouds and fog	Change in color of clouds	Lighter
			Change in cloud size/thickness	Decrease
			Change in frequency of cloudy days	Decrease
		Moisture, dew, and frost	Change in amount of dew	Decrease
			Change in intensity of frost	Decrease
			Change in air moisture/humidity	Decrease
			Change in frequency of frost days	Decrease
	Seasons	Duration and timing	Change in length/duration of seasons	Shift
			Change in timing (onset or end) of seasons	Shift
	Air masses	Wind	Change in wind temperature	Increase

(*Continued*)

Table 18.1 (Continued)

System	*Sub-system*	*Impacted element*	*Impact indicator*	*Direction*
		Storm	Change in frequency of lightning and thunder	Decrease
Physical	Freshwater	Freshwater availability	Change in freshwater availability	Decrease
		Underground water	Change in phreatic level	Decrease
	Terrestrial	Soil moisture	Change in soil moisture/humidity	Decrease
		Soil erosion	Change in wind-induced soil erosion and soil loss	Increase
		Soil degradation	Change in soil desertification	Increase
	Cryosphere	Snowfall and snow cover	Change in extent of permanent snow	Decrease
Life	Terrestrial wild fauna	Terrestrial fauna abundance	Change in abundance of terrestrial fauna	Increase
		Terrestrial fauna composition	Change in species composition of terrestrial fauna	Shift
	Terrestrial wild flora	Wild flora abundance	Change in abundance of wild plant species	Increase
		Wild flora species composition	Change in distribution of a given vegetation type	Shift
	Cultivated plants	Cultivated plant productivity and quality	Change in crop productivity/yield	Decrease
			Change in cultivated species' fruit size	Decrease
		Disease/pest/mortality of crops	Change in frequency of crop pests	Increase
			Change in crop mortality rate	Increase
		Phenology and reproduction	Change in crop-suitable cultivation areas	Decrease
	Livestock	Livestock productivity and quality	Change in livestock productivity (e.g., milk, meat, wool)	Decrease

mass subsystems. Nearly all participants observed higher temperatures throughout a daily and yearly cycle, less rainfall and atmospheric moisture throughout the year, and especially during the rainier winter season, and a shift in the onset and duration of seasons. Perceptions of changes in the physical system included changes in the freshwater, terrestrial, and cryosphere subsystems. All participants observed a decrease in the amount of freshwater available, the amount of moisture in the soil, and the extent of snow coverage over nearby mountains. The meteorological data, while limited in the duration of observed parameters, reflects the same trends showing a general increase in temperature and a decrease in precipitation (Figures 18.2 and 18.3).

Regarding changes in the life system, all participants reported an increase in the abundance of insects and wild flora by their households and fields, a decrease in the productivity and quality of cultivated plants and livestock, an increase in the frequency of crop pests, and a decrease in suitable land for cultivation. The interrelated effects of climate and other drivers of change were repeatedly highlighted by participants. For example, changes in phreatic levels were associated with atmospheric changes in precipitation as well anthropogenic changes in water extraction, storage, and irrigation.

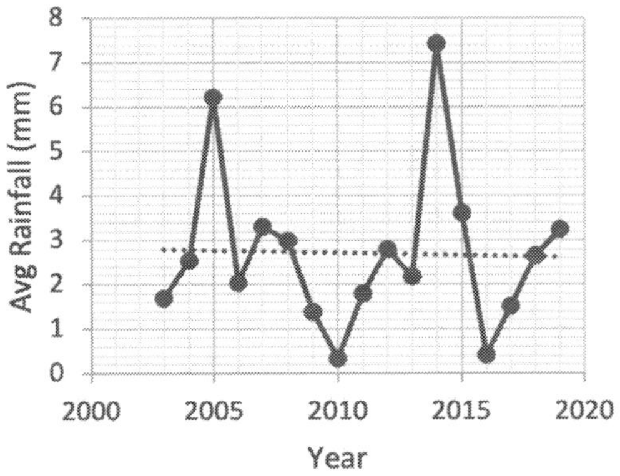

Figure 18.2 Avg. yearly rainfall from 2003 to 2019.

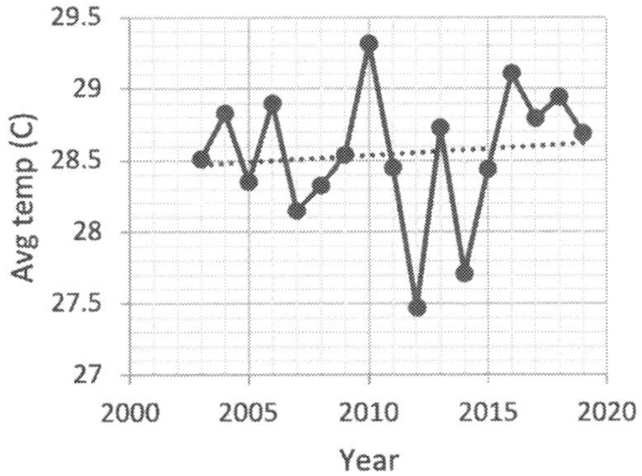

Figure 18.3 Avg. yearly temperature from 2003 to 2019.

Source: Shahdad weather station, Kerman, Iran.

Social responses to changes in environment

In response to environmental impacts, Takab residents reported adopting changes to the ways they cultivate plants, raise animals, supplement their agricultural income, organize their daily household and social activities, and govern their resources. These changes are oriented to increase their adaptive capacity and have been categorized into different social domains (Table 18.2) according to the classification system proposed by Schlingmann et al. (2021) using data collected through interviews, focus groups, and video footage.

Residents reported important changes in crop cultivation and animal husbandry as a result of less available water and warmer temperatures. Farmers described a decrease in crop and animal diversity and a shift from the cultivation of various grains, herbs, fruits, and vegetables to a monoculture of garlic. Several residents also reported a change in the quantity, breed, and species of animals reared, most notably from large herds of local sheep varieties to few cattle, describing a decrease and, in many cases, a discontinuation of activities related to agropastoralism. Residents explained that limited arable land for fodder along with other economic factors affected the market value of local animal husbandry, resulting in a general abandonment of herding lifestyles and livelihoods in the past decade.

Table 18.2 Adaptation indicators to perceived environmental changes ($n = 20$)

Sector	*Domain*	*Class*	*Adaptation indicator*	*Sub-system drivers*
Crop and tree cultivation	Livelihood Organization	Livelihood (sub)system	Change in livelihood: crop cultivation	Temperature; Precipitation; Freshwater; Cultivated plants
			Change in livelihood: agropastoralism	Temperature; Precipitation; Freshwater; Cultivated plants; Livestock
		Size, structure, & composition	Change in size of the area under food/ and or fodder crop cultivation	Precipitation; Freshwater
			Change in crop species composition (including fodder crops)	Temperature; Precipitation; Freshwater; Terrestrial; Cultivated plants
			Change in tree species composition	Temperature; Precipitation; Freshwater; Cultivated plants
		Timing	Change in timing and duration of seasonal activities	Temperature
		Infrastructure, construction, & technology	Change in the use/ construction of irrigation channels	Precipitation; Freshwater
			Change in the use of other technical irrigation infrastructure	Precipitation; Freshwater
			Change in the use of other technologies (items/tools, material) related to plant cultivation (e.g., tractors)	Precipitation; Freshwater; Terrestrial
		Mobility & Location	Change in location of and distance to water sources for irrigation	Precipitation; Freshwater
			Change in permanent/ long-term close-by horizontal relocation of cultivation fields (e.g., within the community/district)	Precipitation; Freshwater
	Livelihood related activities	Weeding & pesticide demand	Change in the use of chemical pesticide on plant cultivations	Temperature; Precipitation; Cultivated plants

(*Continued*)

Table 18.2 (Continued)

Sector	*Domain*	*Class*	*Adaptation indicator*	*Sub-system drivers*
		Fertilizer demand	Change in the use of chemical fertilizer on plant cultivations	Precipitation; Freshwater; Terrestrial; Cultivated plants
		Water demand	Change in the quantity of groundwater used for irrigation	Precipitation; Freshwater
Livestock rearing/ Animal husbandry	Livelihood organization	Livelihood (sub)system	Change in livestock activity: livestock keeping	Precipitation; Freshwater
			Change in livestock activity: mainly stall-feeding system	Precipitation; Freshwater
		Size, structure, & composition	Change in quantity of livestock	Precipitation; Freshwater
			Change in size of grazing land	Precipitation; Freshwater
			Change in composition of livestock species	Precipitation; Freshwater
			Change in composition of livestock breeds	Precipitation; Freshwater
		Timing	Change in timing and duration of seasonal pastoral activities	Seasons
		Mobility & location	Change in temporal/ seasonal close-by vertical relocation of livestock (change in elevation)	Precipitation; Freshwater
	Livelihood related activities	Feed demand	Change in quantity of feed from grazing	Precipitation; Freshwater
			Change in cultivating fodder or using by-products as fodder	Precipitation; Freshwater
			Change in composition of extra feed for livestock	Precipitation; Freshwater
Household capacities	Household organization	Financial capital	Change in income through market interactions/selling livelihood-related products	Precipitation; Freshwater; Cultivated plants; Livestock

(*Continued*)

Table 18.2 (Continued)

Sector	*Domain*	*Class*	*Adaptation indicator*	*Sub-system drivers*
			Change in income from selling prepared food, handcraft/artisan/artwork/medicine	Precipitation; Freshwater; Cultivated plants; Livestock
			Change in income from other work (in non/low natural dependent sector)	Precipitation; Freshwater; Cultivated plants; Livestock
			Change in loans, credits, and insurance	Precipitation; Freshwater
			Change in receiving remittance or other financial support	Precipitation; Freshwater
		Natural capital	Change in land and water access due to ownership change	Precipitation; Freshwater
			Change in other means to increase or protect natural capital	Precipitation; Freshwater
		Human capital	Change in education, knowledge, awareness raising, and practical skills	Precipitation; Freshwater
			Change in gender roles in livelihood and household activities	Precipitation; Freshwater; Livestock
		Social capital	Change in relationships with family members/relatives/neighbors	Precipitation; Freshwater
			Collective/community-based sharing/pooling/gifting/non-monetary exchanges	Precipitation; Freshwater
			Changes in participation in community-based/common/collective actions/support	Precipitation; Freshwater
			Change in receiving regular free external support/assistance/material supply and aid, excluding financial support	Precipitation; Freshwater

(*Continued*)

Table 18.2 (Continued)

Sector	*Domain*	*Class*	*Adaptation indicator*	*Sub-system drivers*
		Timing	Change in timing and duration of daily household activities	Temperature
		Mobility & location	Change in permanent/ long-term faraway relocation of residence (e.g., outside the community/district; migration to urban area)	Precipitation; Freshwater
			Change in temporal/ seasonal faraway relocation of residence (e.g., outside the community/district; to urban area)	Precipitation; Freshwater
	Household related activities	Food demand	Change in diet/food composition	Precipitation; Freshwater
		Other household resource demand	Change in quantity of other household resource use (other than water and food)	Temperature
		External resource dependency	Change in buying food and water, including food crops	Precipitation; Freshwater
			Change in buying other products and resources related to livelihood (e.g., feed)	Precipitation; Freshwater
			Changes in buying other means and resources (e.g., petroleum)	Precipitation; Freshwater; Terrestrial; Cultivated plants

Ironically, while garlic requires more water than other previously cultivated crops, residents have responded to decreases in the water supply by using less land and modifying land and water tenure practices related to garlic cultivation. Water rights in Takab are based on shares of hours of *qanat* water use within a period, for example, a household may have a four-hour share of water within a seven-day irrigation cycle. These cycles can range anywhere from six to 21 days, depending on cropping patterns. Though most hours are inherited, shares can be traded and sold at the start of the agricultural season, with prices based on the quality and supply of groundwater of corresponding *qanats*. An overall decrease in the amount of water per hour extracted through the *qanat* system has resulted in some merging of land and water shares among residents to increase the total amount of irrigated land available. It has also driven increased reliance on pumped wells. Whereas most households still largely organize water distribution through hours of access to collectively governed *qanat* waters, an increase in pumped wells highlights the growing privatization of water resource extraction and its effects on the *qanat* system. Some residents regard pumped wells as a less laborious and more autonomous measure for irrigation. However, they also lead to a drawdown of the water table, forming a cone of depression around the well that disrupts the direction of groundwater flow, moving it toward the well and away from the *qanat* system[1] (Semsar Yazdi & Labbaf Khaneiki, 2019). Several residents complained that along with lower groundwater levels, farmers were experiencing more soil salinity. Over-pumping allows for saline water below or surrounding an aquifer to creep into the freshwater zone, reducing the aquifer's water quality.

Nonetheless, *qanat*s remain a central locus of agrarian life and culture in Takab. As a result, the *qanat* system has been one of the foci of intervention of the participatory video project. Since 2012, a group of women have participated in several interventions that address livelihood and water concerns. Their interventions include training and hosting workshops on local tapestry-making techniques and skill development, organizing handcrafters into a labor cooperative, cutting out brokers from the management and distribution process of handcrafts, creating a communal fund that supports the *qanat* system of water governance, and participating in the decision-making structure of *qanat* governance. Through their communal fund, which the cooperative finances through a percentage of handcraft profits, the group provides additional resources to help maintain and repair the extensive underground network of water channels.

Though residents have consolidated agricultural activities, they are progressively relying on other sources of income to compensate during years of drought and water shortage. Households increasingly rely on tourism services, marketing their homes as guesthouses, as well as providing guided excursions and meals to tourists visiting the Lut Desert. Public and private actors have also invested in the restoration of historic sites and the development of hospitality facilities around the area to help drive more tourism. Despite these efforts, high unemployment and urban-rural inequalities are among the most pressing factors that drive rural migration to urban areas.

To cope with rising temperatures, residents have also cited changing the frequency, onset, and duration of intraday behavior and events, like outdoor leisure and working activities. In general, they reported more time indoors and a heavier reliance on air conditioning units and other motorized cooling devices than before. These adaptations increase residents' energy dependency, especially as previous technologies and practices used to cool buildings, people, and perishable items are replaced by motorized ones, further reducing opportunities of socializing during the day, and delaying many outdoor activities to nighttime, which also has important social and health implications (Fan et al., 2023).

Geocultural narratives: impacts on notions of place and self

Different ways of place-making can produce and shape the materiality of an area. The forms in which Takab households make meaning of place combine geographic elements with the imaginative, leading to a development of mythologies, political orders, and relationships that link notions

1 The lowered water levels can also limit vegetation access to groundwater, leading to increased desertification and wind erosion, as well as drain the porous layers of an aquifer, resulting in the compaction of soil and diminishing the capacity of the sedimentary layers to store groundwater. The drawdown of groundwaters and drying of *qanats* has left areas in the region uninhabitable and forced entire villages to migrate elsewhere (Semsar Yazdi & Labbaf Khaneiki 2019; Foltz 2002; Hoogesteger and Vincent 2006; Wessels 2008).

of place and self (Kia, 2020). Consequently, changes in the climatic, physical, and life systems in Takab affect the narratives and relationships people build with the place they call home. These narratives impact the way people think about environmental changes and the measures they take in response.

Results from this study indicate a range of strategies people have taken to cope and adapt to climatic changes, many of which are grounded in social norms, beliefs, and customs residents carry out and hold about themselves and their place in the world. This is most notable in the social organization of water resources, where both individual and collective forms of governance are on display, showing the benefits and limitations of each arrangement. Modern techniques of water extraction highlight a general shift in agrarian life that has taken place since the Shah's reforms toward more independent and privatized methods of access and distribution of common resources. Yet, the vulnerability that cultivators face due to climatic, environmental, and market changes has also incentivized a resurgence of collective modes of social organization and governance advanced through the PV group's cooperative interventions throughout the district. Footage of their interventions documents the process through which the PV group and later other residents make sense of the relation between social hierarchies and climate change vulnerabilities. Within this process, they confront narratives about proper behavior for women through discussions of handcraft working conditions. The women intervene by creating a cooperative that manages the production and sale of their work, organizing handcrafters around the region to work and sell as a collective institution. The women then use a portion of their profits to form a communal fund to lead qanat restoration. In this process, the PV group not only challenges the exploitative conditions of gendered labor but also inserts more women into public roles of decision-making while highlighting the benefits of collective action.

While women have circulated narratives of cooperative governance and legacies of sustainable water practices through their videos, the oral circulation of folklore, poems, and adages that communicates knowledge about local environmental phenomena has notably decreased between generations (see also Garteizgogeascoa et al., 2020). For example, in Takab seasons are divided into intervals of 40 days, an auspicious period in Islam that symbolizes spiritual maturity—a "big forty" signifies a 40-day period and a "small forty," 20 days. Each "forty" is characterized by specific cultivation and husbandry activities and weather patterns. For example, the poem *Ahman va Bahman* correlates the appearance and location of certain constellations with the end of the "small forty" of winter, which brings with it a variety of atmospheric conditions like cloudy days, nightly rain, and strong winds, which residents say affect the behavior of domesticated animals and set up suitable conditions to begin cultivating crops. The poem explains the variability in weather conditions as the taunting between brothers *Ahman* and *Bahman*. Barley and other grains are among the first crops cultivated and harvested, which are timed from when these conditions first appear. Indigenous and local traditions that forecast weather patterns, often including astronomical, atmospheric, and botanical indicators, are an important element of a broader tapestry of folk knowledge that informs and supports how communities survive and respond to their environments (Orlove et al., 2002; Bardsley & Wiseman, 2012). Many young residents surveyed in Takab, however, were largely unaware of adages and folklore relaying local environmental knowledge. When asked about this gap, residents often mentioned changes in lifestyles, livelihoods, and climate that render much of this wisdom irrelevant or obsolete.

Shifts in the composition of crops and animals have also not only changed diets and agricultural practices in the district but also changed the historic associations and circulated ideas that relate human physical health with the land. In Persianate mythology, elements like air, land, and water affect the humors of a porous body, supporting the notion of individual and collective selves in harmony with their environment (Kia, 2020). Elders frequently correlated changes in climate and land, especially drought, with a sense of deteriorating social well-being, discussing the ways in which adaptive responses have made life more comfortable, but compromised the overall physical health of the land and people connected to it. For example, date palm trees have a historic presence in Takab as an endemic crop well suited for the arid climate and are widely regarded as sacred symbols of prosperity and fertility. The leaves and fibers have customarily been used for handcrafts and the fruits have been an essential ingredient in much of the region's cuisine, a staple of Persian medicine, and a supplement to animal fodder. Decreases in the cultivation of date palms in this area, due to decreases in water resources and increases in soil erosion, which lower the quality of its fruit and thus the income derived from its sales, not only change the profile of the local economy but affect the diets, medicines, and social interdependencies created through the various applications of date palm products with residents. In recent years, the handcraft cooperative, named after a local palm frond weaving technique, has tried to educate younger residents and visitors about the plant's many applications in an attempt to revive its relevance in everyday life beyond its currently symbolic state. Despite their efforts, date palm trees are no longer widely cultivated in Takab, revealing changes in the way residents use and depend on endemic flora.

Many of the changes listed above result from an interplay of climatic, economic, and social factors (Fuchs et al., 2024; Reyes-Garcia et al., 2024). Residents tend to make changes to mitigate immediate impacts based on their financial capacity (Race et al., 2016; Attoh et al., 2024; Ayanlade et al., 2024), though these changes may not necessarily result in adaptation (Maru et al., 2014). In Takab, some residents can afford to invest in pumped wells or pay for more hours and access to a better *qanat,* while some cannot. Additionally, measures that address one area of impact (e.g., shifting to garlic cultivation) may not necessarily be a long-term response to multi-driver changes (e.g., decreasing water supply). Similar to findings noted by Fuchs et al. (2024), more transformative adaptations resulted from collective action initiated by the women of the PV group. As women organized to assert more decision-making power in Takab, their interventions enabled collective responses to socio-economic and environmental challenges in order to better sustain the livelihoods of residents in their community.

Effective adaptation needs to balance feasibility, affordability, and community engagement (Adger et al., 2005; Dahle, 2007) with the diverse experiences that create local knowledge to nourish transformative measures beyond short-term gains. Public investment in sustainable adaptive strategies for agricultural communities remains important, but there also needs to be parallel investments in the social processes and infrastructure that enable more decision-making power for women and other marginalized people in a community along with more blending and exchange of local ecological knowledge and climate science to inform contextualized and socioculturally situated responses. Such thinking could be the catalyst for transformative adaptation that enables remote desert communities, like the one in Takab, to revisit and restore sustainable and energy-efficient infrastructure, diversify livelihood sources, incentivize more resilient and endemic crops and food systems, and reinforce collective and inclusive decision-making institutions.

Conclusion

In this work, I have examined how Takab residents recognize and respond to changes in climate through various adaptive measures. These approaches reveal dynamic cultural and social elements embedded in the local communities that are otherwise overlooked in broad-based assessments of a community's adaptive capacity. For this reason, using measures based on other societies may be misleading and undermine local priorities, cultural contexts, and social structures, leading to misinformed suggestions and investments (Cameron, 2012; Race et al., 2016). Excessive emphasis on exogenous practices and assets may fail to recognize the social processes and local infrastructure that supports sustainability and a situated sense of self in place. Indeed, it may well be the social processes addressing place-based struggles that the livelihoods of residents in their community transformative adaptation measures.

Acknowledgments

I am grateful to all the participants in Takab who generously shared their time, energy, history, and personal stories. I also want to thank Hamed Zolfaghari for assisting in data collection and Victoria Reyes-Garcia, peer reviewers from the LICCI team, Kenny Broad, and Stephanie Seidel for their helpful feedback on this chapter.

References

Abazeri, M. (2022). A chronology of seeing: Subverting gender norms through a long-term participatory video project with rural women in Iran. *Visual Anthropology Review* 12278, 2–21.

Adger, W.N., Arnell, N.W., & Tompkins, E.L. (2005). Successful adaptation to climate change across scales. *Global Environmental Change* 15, 77–86.

Adger, W.N., Barnett J., Brown K., Marshall N., & O'Brien, K. (2013). Cultural dimensions of climate change impacts and adaptation. *Nature Climate Change* 3, 112–117.

Adger, W.N., Brown, K., Nelson, D.R., Berkes, F., Eakin, H., Folke, C., Galvin, K., Gunderson, L., Goulden, M., O'Brien, K., Ruitenbeek, J., & Tompkins, E.L. (2011). Resilience implications of policy responses to climate change. *WIREs Climate Change* 2, 757–766.

AghaKouchak, A., Feldman, D., Hoerling, M., Huxman, T., & Lund, J. (2015). Water and climate: Recognize anthropogenic drought. *Nature: Opinions*, 524, 409–411.

Akhani, H., & Mucina, L. (2015). The Tamaricetea archeuthoidis: A new class for the continental riparian thickets of the Middle East, Central Asia and the subarid regions of the Lower Volga valley. *Lazaroa* 36. https://doi.org/10.5209/rev_LAZA.2015.v36.50200.

Alam, M.M., Siwar, C., Talib, B.A., & Wahid, A.N.M. (2017). Climatic changes and vulnerability of household food accessibility: A study on Malaysian east coast economic region. *International Journal of Climate Change Strategies and Management* 9(3), 387–401.

Appadurai, A. (1996). *Modernity at Large: Cultural Dimensions of Globalization*. University of Minnesota Press.

Ashraf, A. (1991). State and agrarian relations before and after the Iranian Revolution, 1960–1990. In J. Waterbury and F. Kazemi (Eds.), *Peasants and Politics in the Modern Middle East* (pp. 277–311). Florida University Press.

Ashraf, B., AghaKouchak, A., Alizadeh, A., et al. (2017). Quantifying anthropogenic stress on groundwater resources. *Nature: Scientific Reports* 7, 12910.

Attoh, E.M.N.A.N., Afriyie, R., Kranjac-Berisavljevic, G., Bessah, E., & Ludwig, F. (2024). Changing terrain: Evidence of climate change impacts and adaptive responses of Dagbani Indigenous communities in Northern Ghana. In V. Reyes-Garcia, et al. (Eds.) *Routledge Handbook of Climate Change Impacts on Indigenous Peoples and Local Communities*. Oxfordshire & New York: Routledge. Chp 14, pp 244–259.

Ayanlade, A., Oluwatimilehin, I.A., & Ayanlade, O.S. (2024). Climate change impacts on agriculture and barriers to adaptation technologies options among rural farmers in Southwestern Nigeria. In V. Reyes-Garcia, et al. (Eds.) *Routledge Handbook of Climate Change Impacts on Indigenous Peoples and Local Communities*. Chapter 15. Routledge, Oxfordshire & New York: Routledge. Chp 15, pp 259–272.

Azadi, Y., Yazdanpanah, M., & Mahmoudi, H. (2019). Understanding smallholder farmers' adaptation behaviors through climate change beliefs, risk perception, trust, and psychological distance: Evidence from wheat growers in Iran. *Journal of Environmental Management* 250, 109456.

Bardsley, D.K., & Wiseman, N.D. (2012). Climate change vulnerability and social development for remote indigenous communities of South Australia. *Global Environmental Change* 22, 713–723.

Basso, K. (1996). Wisdom sits in places: Notes on a western Apache landscape. In K. Basso & S. Feld (Eds.) *Senses of Place* (pp. 53–90). School of American Research Press.

Beaumont, P., Bonine, M., & Mc Lachlan, K. (Eds.). (1989). *Qanat, Kariz and Khattara: Traditional Water Systems in the Middle East and North Africa*. London: Middle East and North African Studies Press.

Berkes, F. (2015). *Coasts for People: Interdisciplinary Approaches to Coastal and Marine Resource Management*. New York: Routledge.

Cameron, E. (2012). Securing indigenous politics: A critique of the vulnerability and adaptation approach to the human dimensions of climate change in the Canadian Arctic. *Global Environmental Change* 22, 103–114.

Crate, S. A. (2008). Gone the bull of winter? Grappling with the cultural implications of and anthropology's role(s) in global climate change. *Current Anthropology* 49, 569–595.

Dahle, K. (2007). When do transformative initiatives really transform? A typology of different paths for transition to a sustainable society. *Futures* 39, 487–504.

Ehlers, E., & Saidi, A. (1989). Qanats and pumped wells: The case of Assad'abad, Hamadan. In P. Beaumont, K. Borine, and K. McLachlan (Eds.), *Qanat, Kariz, and Khattara* (pp. 89–112). London: Middle East and North African Studies Press.

Escobar, A. (2001). Culture sits in places: Reflections on globalism and subaltern strategies of localization. *Political Geography* 20, 139–179.

Fan, Y., Wang, J., Obradovich, N., & Zheng, S. (2023). Intraday adaptation to extreme temperatures in outdoor activity. *Nature: Scientific Reports* 13, 473.

Foltz, R.C. (2002). Iran's water crisis: Cultural, political, and ethical dimensions. *Journal of Agricultural and Environmental Ethics 15*(4), 357–380.

Frans, C., Istanbulluoglu, E., Mishra, V., Munoz-Arriola, F., & Lettenmaier, D.P. (2013). Are climatic or land cover changes the dominant cause of runoff trends in the Upper Mississippi River Basin? *Geophysical Research Letters* 40(6), 1104–1110.

Fuchs, A., Vogl, C.R., & Schunko, C. (2024). Counteracting land abandonment: Local adaptation strategies to climate change impacts of alpine farmers in Eastern Tyrol, Austria.In V. Reyes-García, S. Álvarez-Fernández, P. Benyei, L. Calvet-Mir, D. García-del-Amo, A. B. Junqueira, X. Li, V. Porcher, A. Porcuna-Ferrer, A. Schlingmann, & R. Soleymani (Eds.), *Routledge Handbook of Climate Change Impacts on Indigenous Peoples and Local Communities*. Oxfordshire & New York: Routledge. Chp 17, pp 287–302.

Garrett, B. L. (2010). Drowned memories: The submerged places of the Winnemem Wintu. *Archaeologies* 6(2), 346–371.

Garteizgogeascoa, M., Garcia-del-Amo, D., & Reyes-Garcia, V. (2020). Using proverbs to study local perceptions of climate change: A case study in Sierra Nevada (Spain). *Regional Environmental Change* 20(59), 1–12.

Gee, K., Kannen, A., Adlam, R., Brooks, C., Chapman, M., Cormier, R., Fischer, C., Fletcher, S., Gubbins, M., Shucksmith, R., & Shellock, R. (2017). Identifying culturally significant areas for marine spatial planning. *Ocean & Coastal Management* 136(1), 139–147.

Ghorbani, M. (2013). *The Economic Geology of Iran: Mineral Deposits and Natural Resources*. Springer.

Gohari, A., Eslamian, S., Abedi-Koupaei, J., Massah Bavani, A., Wang, D., & Madani, K. (2013). Climate change impacts on crop production in Iran's Zayandeh-Rud River Basin. *Science of the Total Environment* 442, 405–419.

Hashemi, S.M., Bagheri, A., & Marshall, N. (2015). Toward sustainable adaptation to future climate change: Insights from vulnerability and resilience approaches analyzing agrarian system of Iran. *Environmental Development and Sustainability* 19, 1–25.

Honari, M. (1989). Qanats and Human Ecosystems in Iran. In P. Beaumont, M. Bonine, & K. Mc Lachlan (Eds.), *Qanat, Kariz and Khattara. Traditional Water Systems in the Middle East and North Africa* (pp. 61–87). London: Middle East and North African Studies Press.

Hoogesteger, J., & Vincent, L. (2006). *Institutional Change and Farmer Participation in the Abshar Irrigation System, Iran in Proceedings*. Amsterdam: CERES.

IPCC (2022). Summary for policymakers. In H.-O. Pörtner, D.C. Roberts, M. Tignor, E.S. Poloczanska, K. Mintenbeck, A. Alegría, M. Craig, S. Langsdorf, S. Löschke, V. Möller, A. Okem, B. Rama (Eds.), *Climate Change 2022: Impacts, Adaptation and Vulnerability. Contribution of Working Group II to the Sixth Assessment Report of the Intergovernmental Panel on Climate Change* (pp. 3–33). Cambridge: Cambridge University Press.

Jomehpour, M. (2009). Qanat irrigation systems as important and ingenious agricultural heritage: Case study of the qanats of Kashan, Iran. *International Journal of Environmental Studies* 66, 297–315.

Karimi, V., Karami, E., & Keshavarz, M. (2018). Climate change and agriculture: impacts and adaptive responses in Iran. *Journal of Integrative Agriculture* 17(1), 1–15.

Katouzian, H. (1981). *The Political Economy of Modern Iran: Despotism and Pseudo-Modernism, 1926–1979*. New York: New York University Press.

Katouzian, H. (2009). *The Persians: Ancient, Medieval, and Modern Iran*. New Haven: Yale University Press.

Kia, M. (2020). *Persianate Selves: Memories of Place and Origin before Nationalism*. Stanford: Stanford University Press.

Kiger, M.E., & Varpio, L. (2020). Thematic analysis of qualitative data: AMEE Guide No. 131. *Medical Teacher*. https://doi.org/10.1080/0142159X.2020.1755030.

Kim, S.Y., & Wolinsky-Nahmias, Y. (2014). Cross-national public opinion on climate change: The effects of affluence and vulnerability. *Global Environmental Politics* 14(1), 79–106.

Kirsch, S. (2001). Lost worlds: Environmental disaster, 'culture loss,' and the law. *Current Anthropology* 42, 167–198.

Labbaf Khaneiki, M. (2019). Qanat and territorial cooperation in Iran. *Water History* 11, 185–206.

Lob, E. (2020). *Iran's Reconstruction Jihad: Rural Development and Regime Consolidation after 1979*. Cambridge: Cambridge University Press.

Madani, K. (2014). Water management in Iran: What is causing the looming crisis? *Journal of Environmental Studies and Sciences* 4(4), 315–328.

Madani, K., AghaKouchak, A., & Mirchi, A. (2016). Iran's socio-economic drought: Challenges of a water-bankrupt nation. *Iranian Studies* 49(6), 997–1016.

Maghrebi, M., Noori, R., Bhattarai, R., Mundher Yaseen, Z., Tang, Q., Al-Ansari, N., Danandeh Mehr A., Karbassi A., Omidvar, J., Farnoush, H., Torabi Haghighi, A., Klove, B., & Madani, K. (2020). Iran's agriculture in the anthropocene. *Earth's Future* 8, e2020EF001547.

Maru, Y.T., Stafford Smith, D.M., Pinho, P.F., & Dube O.P. (2014). A linked vulnerability and resilience framework for adaptation pathways in remote disadvantaged communities. *Global Environmental Change* 28, 337–350.

Mehran, A., AghaKouchak, A., Nakhjiri, N., Stewardson, M.J., Peel, M.C., Phillips, T.J., Wada, Y., & Ravalico J.K. (2017). Compounding impacts of human-induced water stress and climate change on water availability. *Nature: Scientific Reports* 7, 6282.

Mesgaran, M.B., Madani, K., Hashemi, H., & Azadi, P. (2016). Evaluation of land and precipitation for agriculture in Iran, Working Paper 2. *Stanford Iran 2040 Project*. Stanford University.

Mesgaran, M.B., Madani, K., Hashemi, H., & Azadi, P. (2017). Iran's land suitability for agriculture. *Nature: Scientific Reports* 7, 7670.

Milne, E.P., Mitchell, C., & de Lange, N. (2012). *Handbook of Participatory Video*. Berkeley, CA: Altamira Press.

Mirchi, A., Watkins Jr., D.W., Huckins, C.J., Madani, K., & Hjorth P. (2014). Water resources management in a homogenizing world: Averting the growth and underinvestment trajectory. *Water Resource Research* 50(9), 7515–7526.

Mirzaei, A., Saghafian, B., Mirchi, A., & Madani, K. (2019). The groundwater-energy-food nexus in Iran's agricultural sector: Implications for water security. *Water* 11(9), 1835. https://doi.org/10.3390/w11091835

Misra, M. (2017). Smallholder agriculture and climate change adaptation in Bangladesh: Questioning the technological optimism. *Climate and Development 9*(4), 337–347.

Mitchell, T. (2000). *Questions of Modernity*. Minneapolis: University of Minnesota Press.

Muhl, E.-K., & Sowman, M. (2020). Rights, resources, rezoning, and the challenges of governance in South Africa's oldest marine protected area. *Conservation & Society* 18(4), 366–377.

Orlove, B., Chiang, J.C.H., & Cane, M.A. (2002). Ethnoclimatology in the Andes: A cross-disciplinary study uncovers a scientific basis for the scheme Andean potato farmers traditionally use to predict the coming rains. *American Scientist* 90(5), 428–435.

Pink, S. (2007). *Visual Interventions: Applied Visual Anthropology*. Oxford: Berghahn Books.

Race, D., Mathew, S., Campbell, M., & Hampton, K. (2016). Understanding climate adaptation investments for communities living in desert Australia: Experiences of indigenous communities. *Climatic Change* 139, 461–475.

Reyes-Garcia V., Alvarez-Fernandez S., Benyei P., Garcia-del-Amo D., Junqueira A.B., Labeyrie V., et al. (2023). Local indicators of climate change impacts described by indigenous peoples and local communities: Study protocol. *PLoS One* 18(1), e0279847.

Reyes-García, V., Benyei, P., Junqueira, A.B., Huanca, T., Conde, E. (2024). A complex matrix. Reports of environmental change and its drivers by the Tsimane', Bolivian Amazon. In V. Reyes-García, S. Álvarez-Fernández, P. Benyei, L. Calvet-Mir, D. García-del-Amo, A. B. Junqueira, X. Li, V. Porcher, A. Porcuna-Ferrer, A. Schlingmann, & R. Soleymani (Eds.), *Routledge Handbook of Climate Change Impacts on Indigenous Peoples and Local Communities*. Oxfordshire & New York: Routledge. Chp 10, pp 164–179.

Reyes-Garcia, V., Garcia-del-Amo, D., Benyei, P., Fernández-Llamazares, A., Gravani, K., Junqueira, A.B., Labeyrie, V., Li X., Matias, D. M.S., McAlvay, A., Mortyn, P.G., Porcuna-Ferrer, A., Schlingmann, A., & Soleymani-Fard, R. (2019). A collaborative approach to bring insights from local observations of climate change impacts into global climate change research. *Environmental Sustainability* 39, 1–8.

Reyes-Garcia, V., Garcia-del-Amo, D., Benyei, P., Junqueira, A.B., Labeyrie, V., Li, X., Porcuna-Ferrer A., Schlingmann, A., Soleymani-Fard, R., & Porcher, V. (2020). Local indicators of climate change impacts. Data collection protocol. *Figshare* 124. http://doi.org/10.6084/m9.figshare.11513511.v3.

Rosenzweig, C., & Neofotis, P. (2013). Detection and attribution of anthropogenic climate change impacts. *WIREs Climate Change* 4, 121–150.

Sadegh, M., Mahjouri, N., & Kerachian, R. (2010). Optimal inter-basin water allocation using crisp and fuzzy Shapley games. *Water Resources Management* 24(10), 2291–2310.

Satterfield, T., Gregory, R., Klain, S., Roberts, M., & Chan, K.M. (2013). Culture, intangibles and metrics in environmental management. *Journal of Environmental Management* 117, 103–114.

Saunders, F., Gilek, M., Ikanuniece, A., Tafon, R. V., Gee, K., & Zaucha, J. (2020). Theorizing social sustainability and justice in marine spatial planning: Democracy, diversity, and equity. *Sustainability* 12(2560), 1–18.

Schlingmann, A., Graham, S., Benyei, P., Corbera, E., Martinez Sanesteban, I., Marelle, A., Soleymani-Fard, R., & Reyes-Garcia, V. (2021). Global patterns of adaptation to climate change by Indigenous Peoples and local communities: A systematic review. *Current Opinion in Environmental Sustainability* 51, 55–64.

Semsar Yazdi, A.A., & Labbaf Khaneiki, M. (2019). *Veins of the Desert: A Review on Qanat/Falaj/Karez*. Aalborg: River Publishers.

Strand, M., Rivers, N., & Snow, B. (2022). Reimagining ocean stewardship: Arts-based methods to 'hear' and 'see' indigenous and local knowledge in ocean management. *Frontiers in Marine Science* 9, 886632.

Weber, E. U. (2010). What shapes perceptions of climate change? *WIREs Climate Change* 1(3), 332–342.

Wessels, J. (2008). *To Cooperate or Not to Cooperate: Collective Action for Rehabilitation of Traditional Water Tunnel Systems (Qanats) in Syria*. University of Amsterdam Press.

White, S. (2003). *Participatory Video: Images That Transform and Empower*. Sage Publications.

Winthrop, R. H. (2014). The strange case of cultural services: Limits of the ecosystem services paradigm. *Ecological Economics* 108, 208–214.

Zhao, Y., Norouzi, H., Azarderakhsh, M., & AghaKouchak, A. (2021). Global patterns of hottest, coldest, and extreme diurnal variability on Earth. *BAMS* 102(9), 1672–1681.

19

THE ROLE OF CULTURE IN CLIMATE CHANGE ADAPTATION

Insights from two mountain regions in Kenya

Ben Mwangi, Lydia A. Olaka, Noelia Zafra-Calvo, Julia A. Klein and Aida Cuni-Sanchez

Introduction

There is increasing recognition that cultural dimensions shape climate change adaptation (Adger et al., 2013; Thomas et al., 2019; Few et al., 2020). Culture, understood as shared traits across social groups expressed in worldviews, values, knowledges, norms, taboos, customs, and behaviours (Spencer-Oatey, 2012), affects not only the nature and tolerability of risks and the faith in information sources, but also which actions should be prioritised, the acceptability of responses, and expectations for who can or should take action (Few et al., 2020). For example, research on place attachment suggests that place-related identities and meanings that are incorporated into policy and planning processes (by formal or traditional authorities, which include trust, transparency, and inclusion) facilitate effective adaptation actions (Baxter et al., 2012). Social networks, such as the linkages and interactions among different actors, also play a key role in the local adaptation process to climate change; and under specific cultural characteristics, social interactions could be employed as an effective tool to improve farm households' adaptation to climate change (Dapilah et al., 2019).

As recently highlighted, culture is inherently dynamic, including in what is considered a culturally appropriate adaptation at a given place and point in time (Few et al., 2020). For example, in Senegal Sereer farmers have changed preferences for crop diversity to adapt to climate variations (Ruggieri et al., 2021), while in northern Kenya, Borana pastoralists have changed preferences for livestock mix to adapt to climate variations and new market opportunities in camel milk trade (Few et al., 2020). Notably, few examples of the effects of culture on climate change adaptation are available from mountain regions, even if these are known to be not just climatically but also culturally complex socio-ecological systems (Klein et al., 2019).

The mountains of central Kenya offer a unique setting for exploring the effects of culture on climate change adaptation. First, different ethnic groups inhabit these mountains. Second, climate change impacts are likely to be already noticed by local farmers. For example, increased temperatures, changes in rainfall amount and distribution, decreased streamflow, and reduced crop yields in the past 20–30 years have been reported by local farmers in mountains both north (e.g., Mt Marsabit in Kenya, Cuni-Sanchez et al., 2018) and south (e.g., Mt Kilimanjaro, Uluguru

DOI: 10.4324/9781003356837-24

et al., 2010; Francis & Kraybill, 2013; Velempini et al., 2016; Kassian et al., 2017; Schumacher, 2018; Kangazi et al., 2021). Like in these other mountains (see Kaganzi et al., 2021 for a review), smallholder farmers are likely to be implementing several adaptation strategies. Third, farmers' changes in crop choice in central Kenya are likely to be negatively interacting with climate change impacts, as Leclerc et al. (2014) suggested. Although maize is known to be more susceptible to drought than traditional crops such as sorghum or pearl millet, there have been increasing rates of maize adoption in Kenya since 1965 (Ouma et al., 2002), and maize became the dominant crop through the 1990s (Smale & Jayne, 2003). In fact, Bernard (1972) already noted the remarkably rapid diffusion of maize as a consequence of new roads and markets, even in areas where the crop cannot successfully be grown. Among the Kikuyu farmers, the increasing popularity of maize occurred even earlier, in the beginning of 1950s (Middelton, 1953). Apart from maize, smallholder farmers in Mount Kenya are increasingly cultivating bananas as cash crop, especially in areas negatively affected by declining incomes from traditional cash crops such as coffee (Wambugu & Kiome, 2001).

This chapter, focused on two ethnically distinct (Meru and Kikuyu) smallholder farmer communities living in two nearby mountains of Central Kenya, aims at: (1) identifying the changes in climate and their impacts on the biophysical system in the past 30 years as observed by these farmers; (2) determining which strategies they are using to adapt to these impacts; and (3) reflecting on how cultural differences related to ethnicity affect farmers' reports of both climate change impacts and adaptation responses.

Methodology

Study areas

We focused on the smallholder farmer communities living in two mountain regions in central Kenya: Mount Kenya and the Aberdare Range (Figure 19.1). Mount Kenya (5,198 m) is an extinct volcano and the Aberdare Range (4,001 m) is also of volcanic and tectonic origin. The region has a bimodal rainfall regime with long rains (*Mvua refu* in Swahili) falling between March to May and short rains (*Mvua fupi*) from October to December. On the southern-eastern slopes of Mt Kenya annual rainfall is 2,300 mm (KFS, 2019), while in the Aberdare Range annual rainfall is 2,200 mm (Nathan, 2020).

The slopes of Mt Kenya have a high population density: 318 people per km^2 (Meru CIDP, 2018), Meru being a dominant ethnic group in the area. The Meru or Amîîrú people are a farmer ethnic group of Bantu origin whose ancestral lands are the eastern slopes of Mt Kenya. There are about 2 million Meru in Kenya (KNBS, 2019). The Meru cultivate both subsistence (e.g., maize, beans, Irish potatoes) and cash crops (e.g., Arabica coffee, tea, bananas, avocados, macadamia, and khat *Catha edulis*). Most Meru also raise animals, typically cows and goats.

The slopes of the Aberdare range also have a high population density: 342 people per km^2 (Nyeri CIDP, 2018). There are around 8 million Kikuyu in Kenya (17.1% of the Kenyan population) (Samantha et al., 2021). The Kikuyu also cultivate both subsistence (e.g., maize, beans) and cash crops (e.g., Arabica coffee, tea, bananas). Most Kikuyu farmers also raise animals, typically cows, sheep, and goats. While for the Meru green bananas are often consumed as staple, this is not the case for most Kikuyu families, for which their main staple is *ugali* (made of maize). Access to urban markets is similar for both areas.

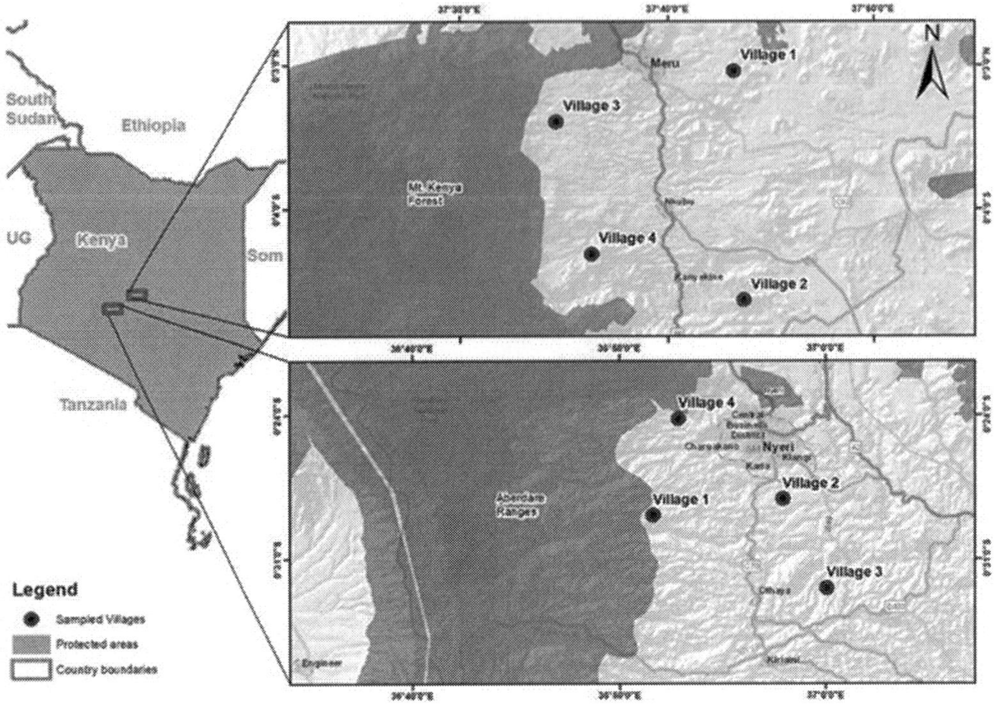

Figure 19.1 Study areas with the location of the villages sampled.

Data collection and analysis

We first conducted exploratory focus group discussions (FGDs) with four to five elders in four villages on each mountain (Figure 19.1). Results from FGDs were used to adapt a common semi-structured questionnaire (which included questions on climate change impacts and adaptation responses) to each study context. During the FGDs, we also gathered information on change agents (defined as the person or organisation that had initiated a given climate change adaptation strategy). Then, in each mountain, we administered semi-structured questionnaires to 150 randomly selected household heads using a purposive sample (50% male and 50% female). Questionnaires addressed household characteristics and asset ownership, observed climatic changes and impacts on the biophysical domain in the past 30 years, strategies used to adapt to observed changes and impacts, and information on weather forecasts and climate change literacy (see Supplementary Material). Climate change literacy was defined as a combination of climate change awareness (having heard of the concept of climate change) and the knowledge and acceptance of its anthropogenic cause. Climate change literacy is recognised with 'high confidence' as contributing towards climate action (IPCC, 2019). The methodological approach and the questionnaire used follows Reyes-Garcia et al. (2023).

The exploratory FGDs and the interviews were conducted mostly in Swahili, with some clarifications made in English, Kikuyu, or Meru, and were facilitated by the first author between April and August 2021. All study participants (FGDs and interviews) participated on a voluntary basis and were first informed that the aim of the study was to better understand the climatic changes and impacts observed in these mountains and their climate change adaptation strategies. Due to the predominance of agriculture-based livelihoods, historical sedentary settlements, and culture, throughout the paper, we refer to our study respondents as 'farmers', but we acknowledge that they might pursue multiple livelihood strategies. Descriptive statistics were used for data analysis. We report responses as frequencies of Meru (n = 150) or Kikuyu (n = 150) farmers reporting each climate change impact or adaptation strategy mentioned.

Results

The socio-economic characteristics of the Meru and Kikuyu farmers interviewed were relatively similar. Among the Meru, households had three adults on average, 99% of the households interviewed relied on farming for their livelihoods, 44% farmed coffee as cash crop, and average farm size was 1 hectare. All participants owned the house they resided in, all had a radio. Only 3% of the respondents had not completed primary school. Among the Kikuyu, households also had three adults on average, 90% of the households interviewed relied on farming for their livelihoods, 50% farmed coffee as cash crop, and average farm size was 2.4 hectares. All participants owned the house they resided in, all had a radio, and only 2% of the respondents had not completed primary school.

Climatic changes and impacts

Most Meru and Kikuyu respondents (>70%) reported that in the past 30 years, they had observed increased temperatures during dry and rainy seasons and important changes in rainfall amount and distribution, including late onset of the long rains, more dry spells during the long rains and more showers during the dry season (Figure 19.2). Most respondents also highlighted that now there were fewer foggy days and fewer hails storms than in the past. Most Meru respondents (>70%) also reported more extreme droughts, less frost, and increased wind, but fewer Kikuyu respondents (<50%) reported such changes.

Most Meru and Kikuyu respondents (>70%) also reported reduced streamflow during the rainy seasons, and most Meru respondents also reported increased soil erosion (Figure 19.3), which they mostly related to observed changes in rainfall amount and distribution, although some Meru

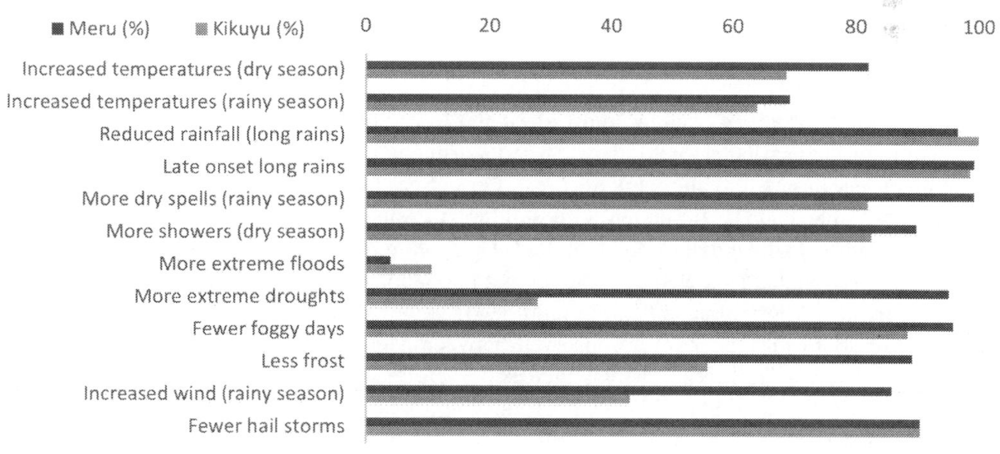

Figure 19.2 Observed changes in climate in terms of the percentage of respondents that pertain to any of the ethnical groups (Meru n=150, Kikuyu n=150)

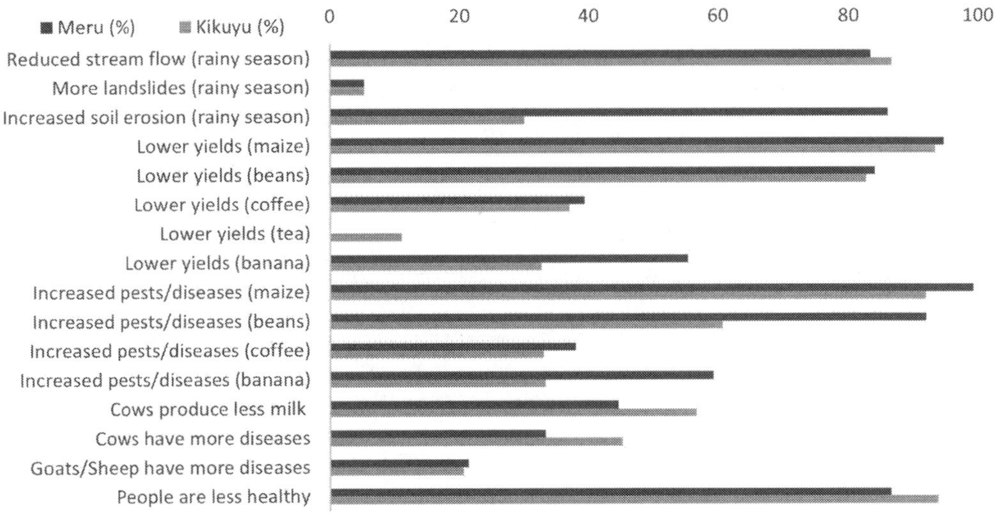

Figure 19.3 Observed impacts in the biophysical domain regarding the percentage of respondents that pertain to any of the ethnical groups (Meru n = 150, Kikuyu n = 150)

farmers also mentioned that increased local deforestation around some villages might also be driving increased soil erosion. Most Meru and Kikuyu respondents (>70%) reported decreased yields of maize and beans, an increase in pests and diseases for these crops (particularly Fall armyworm for maize and aphids in beans), and that people was overall less healthy. Respondents related all these impacts to increased temperatures and observed changes in rainfall amount and distribution. The Kikuyu specifically said that they had observed an increase in respiratory diseases, which they specifically linked to colder nights during the dry season – although they said that overall day temperatures had increased. Some Meru and Kikuyu respondents (20–40%) also reported reduced yields for coffee and banana, increased pests for these crops (mostly coffee borer amongst the Meru, coffee berry disease amongst the Kikuyu), reduced milk production in cows, and an increase in livestock diseases (Figure 19.3).

Climate change adaptation strategies

In response to the observed changes, most Meru and Kikuyu respondents (>70%) have increased the use of improved varieties (particularly early maturing maize), inputs (e.g., fertiliser, pesticide), and soil conservation techniques (Figure 19.4). For coffee, increased use of improved coffee varieties, inputs, and shade (planting more trees in farms to shade coffee plants) were also reported. Over 60% of Meru and Kikuyu respondents had also diversified their livelihoods by starting to rear animals and/or producing vegetables. Diversifying into labour and selling timber or firewood were reported only by a few respondents (see Figure 19.4).

Some important differences were observed between Meru and Kikuyu. Irrigation was more widespread amongst the Meru, whereas increasing veterinary care and supplementary feed and sowing seeds earlier or replanting seeds (if they die) was more widespread amongst Kikuyu. Main soil conservation techniques also differed across groups: while terracing was more common among the Meru (52%), establishing perennial grass cover was most common among the Kikuyu (70%). Meru farmers accessed irrigation through a ‘water users association’, which has

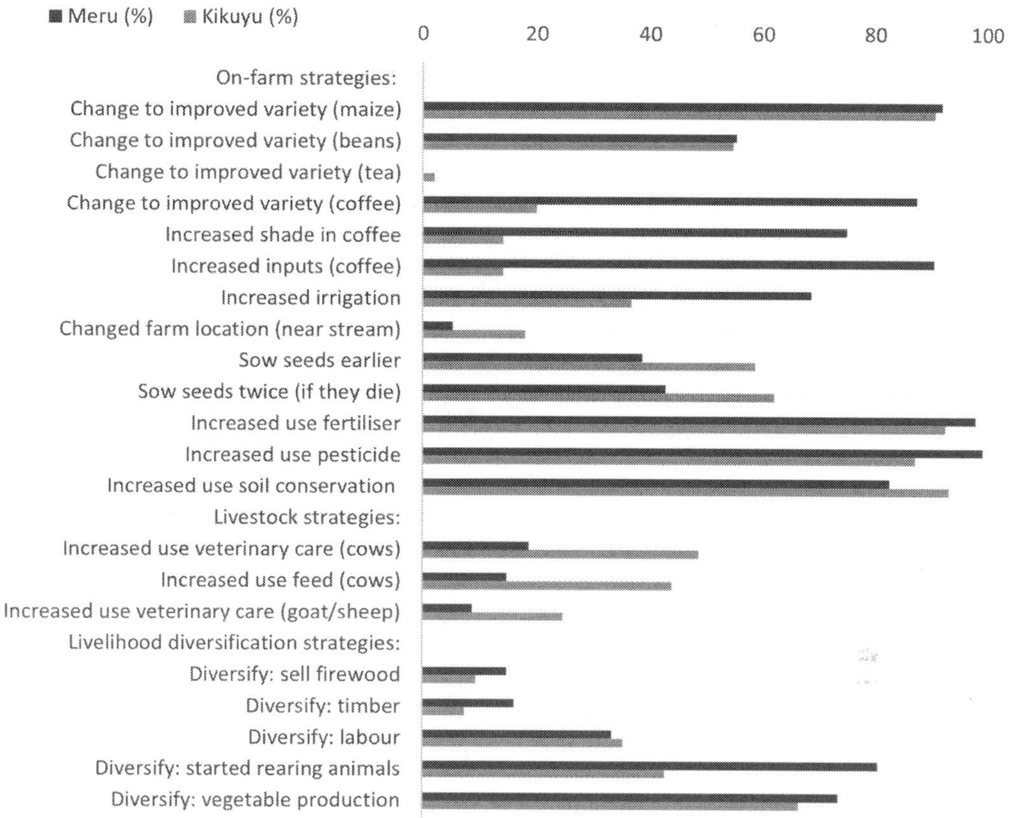

Figure 19.4 Adaptation strategies used regarding percent of respondents that pertain to any of the ethnical groups (Meru n = 150, Kikuyu n = 150)

a relatively low membership fee (mostly used for pipe maintenance). Such a system did not exist among the Kikuyu in the Aberdare Range.

Agents of change and climate change literacy

Amongst the Meru, most climate change adaptation strategies – except two explained below – had been initiated by local farmers without support from external agents of change, and often following advice from rich farmers who have better access to information, technology and inputs, and who are keen to advise their fellow farmers, for example, by giving them some improved seeds 'to try out', here referred to as 'top farmers'. Not all rich farmers are top farmers though. Top farmers created space for mutual learning – increasing trust and farmers' confidence. Such top farmers were not found in the Kikuyu villages studied. In four villages studied, small non-governmental organisations (NGOs) were found to be acting as agents of change promoting some climate change adaptation strategies: (i) in one Meru village, an NGO supported improved coffee varieties and coffee certification (to achieve higher prices), (ii) in another Meru village, another NGO promoted improved soil conservation, (iii) in one Kikuyu village, an NGO promoted improved maize seeds, and (iv) in another Kikuyu village, another NGO promoted diversification into vegetable/fruit farming.

The percentage of climate change literate respondents was similar across groups: 82% amongst the Meru and 81% amongst the Kikuyu. Interestingly, though, most Meru and Kikuyu respondents (82% and 68%, respectively) used their personal experience or information from elders, relatives, and neighbours to decide when to sow seeds. When inquired, respondents said that they believed that weather forecasts presented on the radio were not accurate or were given at 'short-notice'.

Discussion

Our study approach has some limitations. First, we only studied four villages per ethnic group. In our study area, biophysical or socio-economic differences (e.g., market access) across villages of one ethnic group are rather limited. It is possible that by including more villages, different adaptation strategies could be identified, as the presence of 'external' change agents such as NGOs is an important driver of farmers' adoption of some adaptation strategies, as we discuss below. Also, although we interviewed female respondents, most female respondents were married, which – in the study context – could be limiting their participation in decision-making regarding adaptation options. Future work should also investigate adaptation options for female-headed households. We should acknowledge that we focused on climatic changes as the main challenge to farmers' livelihoods. However, other drivers of change are likely to be acting synergistically with climatic change, constraining adaptation options, including changes in market access, agricultural policies (e.g., Chen, 2024; Nkurunziza et al., 2023), and trends in food preferences in nearby urban areas.

Climatic changes and impacts

Overall, both Meru and Kikuyu respondents reported increased temperatures, changes in rainfall quantity and distribution, reduced fog, frost, and hailstorms, with few differences across ethnic groups studied. The climatic changes reported in this study are in line with changes reported in previous studies on climate change perceptions in Kenya's mountains (see Table 19.1), except for frost, for which an increase was reported from Taita Hills (Motaroki et al., 2021). Most previous studies (see overview in Table 19.1) did not investigate changes in fog, frost, wind, or heavy rainfall- probably because the methods used did not consider such climatic variables. Recent work on smallholder farmer perceptions in Tanzania and DR Congo also highlighted the importance of considering multiple climate variables when studying climate change impacts in mountain regions, if we are to really understand climatic changes and their impacts at the local scale (Batumike et al., 2021; Kaganzi et al., 2021; Amani et al., 2022).

The perceived changes in rainfall reported by Meru farmers for the past 30 years dovetail with research based on meteorological data in Mt Kenya. For example, a small increase in mean annual temperature and a small decrease in annual rainfall (period 1980–2017) were reported by Karienye and Kamiri (2020), while a negative trend for interannual variation of seasonal rainfall (period 1961–2006) was reported by Leclerc et al. (2014). We could not find similar studies conducted in the Aberdare Range to compare. Numerous studies have reported an agreement between meteorological data and climate perceptions (e.g., see Savo et al., 2016 for a review), including for fog (Cuni-Sanchez et al., 2018). For variables rarely recorded in climatic stations (e.g., fog), local knowledge could be used to fill in the knowledge gap and could help drive future lines of research.

Regarding impacts in the biophysical domain, all impacts mentioned by our study participants have been reported by other studies in Kenya's mountains (Table 19.1) or in Tanzania's mountains (e.g., see Kaganzi et al., 2021). Some differences were observed between ethnic groups, which might be related to differences in local biophysical conditions. For example, an increase in soil

Table 19.1 Climatic changes and impacts reported by other studies on farmers' perceptions in Kenya's mountain regions

	This study	*Aberdare Range*	*Mt Kenya*	*Central Kenya*	*Kakamega*	*Mau*	*Mt Elgon*	*Taita Hills*
Increased temperatures	x	x	x	x	x		x	x
Late onset rainy season	x				x			x
Reduced rainfall during rainy seasons	x	x	x	x	x	x		x
Increased rainfall during rainy seasons			x				x	
Increased drought			x			x		x
Reduced fog	x							
Increased flood			x		x			
Increased wind	x						x	x
Increased frost								x
Increased heavy rainfall					x			
Reduced crop yields	x	x	x	x	x	x		x
Reduced coffee yields				x				
Reduced maize yields				x	x			x
Reduced banana yields			x					
Increased crop pests and diseases	x	x	x	x	x	x		x
Reduced livestock health	x				x	x		
Reduced human health	x							

Jairo and Korir (2019); Karienye and Kamiri (2020); Asayehegn Gebreeyesus et al. (2017); Saalu et al. (2019); Chepkoech et al. (2018); Kong'ani et al. (2018); Nyberg et al. (2020); Motaroki et al. (2021).

erosion – cited by more Meru than Kikuyu participants – could be linked to stepper slopes or local deforestation acting synergistically with heavier rains in Mt Kenya compared to the Aberdare Range.

In our study, Meru and Kikuyu respondents related reduced maize yields to moisture and heat stress, while reduced bean yields were related to increased pests (aphids, bean weevils, and cutworms), which is in agreement with farmers' reports in Taita Hills (Motakori et al., 2021). Only two previous studies report reduced yields for banana, including in Mt Kilimanjaro (Kaganzi et al., 2021). Regarding livestock, Meru and Kikuyu study participants linked reduced cattle health to reduced fodder availability due to increasing droughts, as did respondents in northern Kenya (Cuni-Sanchez et al., 2018).

In this study, an increase in respiratory diseases was reported in the Aberdare Range, where Kikuyu farmers linked the impact to colder nights during the dry season – although they said that overall day temperatures had increased. This is different from other studies on mountains in East Africa, where increasing temperatures have influenced the prevalence of water and vector-borne diseases such as cholera, typhoid, dysentery, malaria, amoebic diseases as is the case in West Kenya highlands (Wandiga et al., 2010) or Mt Kilimanjaro (Kaganzi et al., 2021). Future work should address the climate change impacts in human health in detail. As highlighted by other authors (e.g., Junqueira et al., 2021), local perceptions can help identify climate change impacts that may be overlooked by scientists, government, or development agencies.

Climate change adaptation strategies and culture

Most Meru and Kikuyu respondents adapted to climate change impacts by modifying farming practices and, to a certain extent, diversifying livelihoods. All strategies cited by Meru and Kikuyu participants have been mentioned in previous studies on Kenyan mountainous regions (see Table 19.2). Beyond Kenya, increased use of improved varieties and fertilisers has been reported in Jimma Mountains in Ethiopia, central Uganda, Mt Elgon, and Mt Kilimanjaro (Tiyo et al., 2015; Mulinde et al., 2019; Eshetu et al., 2021; Kaganzi et al., 2021) and increased use of irrigation has been reported for the Rwenzori Mountains in Uganda (Zizinga et al., 2017). Contrary to other studies (e.g., central

Table 19.2 Climate change adaptation strategies reported by other studies on farmers' perceptions in Kenya's mountain regions. (few) refers to few respondents in that study mentioning that adaptation strategy

	This study	*Aberdare Range*	*Mt Kenya*	*Central Kenya*	*Kakamega*	*Mau*	*Mt Elgon*	*Taita Hills*
Increased farm size		x						x
Change crop variety	x	x	x	x			x	x
Change new crop			x					
Change coffee variety	x							
Change maize variety	x		x	x				x
Change planting dates	x	x	x	x			x	
Increased use soil conservation techniques	x		x	x (mulching)			x	x
Increased use fertilisers	x			x			x	x
Increased use pesticides	x			x				x (few)
Increased irrigation	x		x	x			x	
Increased agroforestry	x	x		x			x	x
Increased vegetable farming	x	x					x	
Migration to cities to work				x				
Started rearing animals	x			x			x	
Change livestock feeding	x			x				
Reduce stocking of cattle				x (few)				
Off-farm diversification (no details)	x	x		x			x	
Labour	x	x						x
Sell firewood/ charcoal	x	x					x	
Collect wild foods		x						

Jairo and Korir (2019); Karienye and Kamiri (2020); Asayehegn et al. (2017); Saalu et al. (2019); Chepkoech et al. (2018); Kong'ani et al. (2018); Nyberg et al. (2020); Motaroki et al. (2021).

region, Table 22.2), we did not find evidence of human migration due to climate change. Most likely this is linked to place attachment, as we discuss below. Increasing farm size (reported in Taita Hills, Motaroki et al., 2021) was not mentioned in our study, probably due to high population density in both Mt Kenya and the Aberdare Range. In terms of *Arabica* coffee, farmers focused on improved productivity, including increasing shade and inputs, and using improved varieties – all reported by coffee farmers elsewhere in Ethiopia and Tanzania (e.g., Eshetu et al., 2021; Kaganzi et al., 2021). Improving coffee quality and certification to offset unit area yield reductions related to climatic changes was only cited in one Meru village, which had received external support from an NGO.

Contrary to the observations of climatic changes and impacts, important differences were observed in adaptation across ethnic groups, with irrigation being more widespread amongst the Meru, and increasing veterinary care and supplementary feed, sowing seeds earlier or replanting seeds (if they die) being more widespread amongst Kikuyu. Similar findings (greater differences in adaptation than in observations reported by different ethnic groups) were also reported in Mt Oku in Cameroon (Ngute et al., 2021) or Mt Kahuzi in DR Congo (Batumike et al., 2021). In our study, differences in culture and social networks are likely to explain these climate change adaptation differences, as biophysical contexts and access to markets are similar across studied sites. For example, irrigation was more widespread among the Meru than among the Kikuyu. It can be argued that the sense of community and the value of reciprocity is stronger among the Meru, who also have stronger social networks (traditional leadership is strong, they have regular meetings of different groups such as youth and women groups). These characteristics might have help them to create and maintain the water user associations, providing irrigation even to poorer households.

The Meru, who feel they are custodians to the land, also seem to have a stronger attachment to their land than the Kikuyu. This has likely contributed to making water management both sustainable and fair. Notably, thanks to regular meetings (e.g., women association, male association), traditional ecological knowledge and the related traditional management systems might be more shared with younger generations, contributing to the long-term sustainability of cultural values. Another example could be the role 'top farmers' play as change agents in adaptation amongst the Meru. These rich farmers (with better access to information, technology, and inputs) engage with other Meru farmers because of their sense of community, value of reciprocity. It is also possible that as Meru is a smaller ethnic group, they maintain a greater sense of community.

Indeed, the Kikuyu are known to have a weak sense of place-attachment: they often migrate to other areas for work. It could be argued that the Kikuyu place a higher value than the Meru on 'individual' success and cash economy. Another example could be how culture affected the acceptance and uptake of certain adaptation strategies. Amongst the Meru, uptake of cattle rearing for milk production was widespread (a strategy initiated by top farmers as a change agent in a village), while in another village, improved varieties of coffee – and coffee farming in general – were widespread (a strategy related to coffee certification initiated by local NGO). Trust among actors (e.g., standard farmer and 'top farmers') and acceptance of new strategies seemed greater amongst the Meru, possibly due to stronger social networks and values of who can, or should, take action. In areas with strong social networks, one option to foster adaptation could be to strengthen the horizontal sharing and learning of methods where farmers are leading the process through their own groups and associations, as already suggested (e.g., Rosset & Martínez-Torres, 2012). Future work on the differences in social networks between the Meru and the Kikuyu is recommended.

In spite of the above-mentioned differences, some similar effects of Meru and Kikuyu cultures were also observed. Most Meru and Kikuyu farmers still use traditional knowledge to determine sowing dates, which they related to a lack of trust in weather forecasts. Cultural dynamics can act both as a barrier to and an enabler of adaptation (Few et al., 2020). However, it is not all about culture. There are practical limitations to the selection of adaptation options, such as access to credit, or technical solutions available, as highlighted by other studies in Kenya (e.g., Nyberg et al., 2020). Also, Meru farmers explained that there was limited access to improved varieties of banana. Indeed, seed systems for banana receive relatively little attention from development-oriented research and commercial seed sector actors, despite their importance for food security, nutrition, and rural livelihoods in certain areas (Almekinders et al., 2019).

Conclusions

This research has investigated how culture shapes reports of climatic changes, its impacts, and adaptation strategies in two mountain regions in Kenya. Results show little differences between Meru and Kikuyu farmers in terms of climatic changes and impacts reported, but several differences in adaptation strategies. Different values, place attachment, social networks, and social cohesion are likely to explain these differences in climate change adaptation actions. For adaptation actions to be successful, government agencies or NGOs should consider farmers' culture(s), and implement more participatory engagement with them to facilitate their adaptation.

Acknowledgements

We are deeply grateful to our study participants, who graciously shared their time, energy, and stories. Thanks to NACOSTI for granting us a research permit (NACOSTI/P/21/11045), and local authorities from the eight villages we studied.

Author contributions

Conceptualization: L.O. and A.C.-S. Methodology: B.M. and A.C.-S. Data gathering: B.M. Formal analysis: B.M. and A.C.-S. Writing original draft preparation: all co-authors. All authors reviewed the final draft of the manuscript. All authors have read and agreed to the published version of the manuscript.

Funding

The Mountain Research Initiative has provided financial support for this research under its Synthesis Workshops funding programme for MRI Community-led Activities.

We also acknowledge funding from the Mountain Sentinels Fellowship 2021 (allocated to the first author).

Institutional Review Board statement

Ethical review and approval were received for this study by the Ethics Committee of the University of Nairobi.

Informed consent statement

Informed consent was orally obtained from all subjects in the study.

Conflicts of interest

The authors declare no conflict of interest.

References

Abid, M., Ngaruiya, G., Scheffran, J., & Zulfiqar, F. (2017). The role of social networks in agricultural adaptation to climate change: Implications for sustainable agriculture in Pakistan. *Climate*, 5(4), 85.

Adger, W. N., Barnett, J., Brown, K., Marshall, N., & O'brien, K. (2013). Cultural dimensions of climate change impacts and adaptation. *Nature climate change*, 3(2), 112–117.

Amani, R. K., Riera, B., Imani, G., Batumike, R., Zafra-Calvo, N., & Cuni-Sanchez, A. (2022). Climate change perceptions and adaptations among smallholder farmers in the mountains of eastern Democratic Republic of Congo. *Land*, 11(5), 628.

Asayehegn Gebreeyesus, K., Temple, L., Sanchez, B., & Iglesias, A. (2017). Perception of climate change and farm level adaptation choices in central Kenya. *Cahiers Agricultures*, 26(2), 1–11.

Autio, A., Johansson, T., Motaroki, L., Minoia, P., & Pellikka, P. (2021). Constraints for adopting climate-smart agricultural practices among smallholder farmers in Southeast Kenya. *Agricultural Systems*, 194, 103284.

Batumike, R., Imani, G., Bisimwa, B., Urom, C., Mambo, H., Kalume, J., ... & Cuni-Sanchez, A. (2021). From tree species to forest services: Ethnic differences in Lomami, Democratic Republic of the Congo. *Economic Botany*, 1–12.

Bernard, F. E. (1972). East of Mount Kenya: Mem agriculture in transition (No. 75). München.

Chen, Z. (2024). Roads, riding stables, and highland barley: Livelihood diversification as climate change adaptation among Tibetans in Shangri-La, China. In V. Reyes-García, S. Álvarez-Fernández, P. Benyei, L. Calvet-Mir, D. García-del-Amo, A. B. Junqueira, X. Li, V. Porcher, A. Porcuna-Ferrer, A. Schlingmann, & R. Soleymani (Eds.), *Routledge Handbook of Climate Change Impacts on Indigenous Peoples and Local Communities*. Oxfordshire & New York: Routledge. Chp 20, pp 335–345.

Chepkoech, W., Mungai, N. W., Stöber, S., Bett, H. K., & Lotze-Campen, H. (2018). Farmers' perspectives: Impact of climate change on African indigenous vegetable production in Kenya. *International Journal of Climate Change Strategies and Management*. 10(4), 551–579.

Cuni-Sanchez, A., Pfeifer, M., Marchant, R., Pompeu, P. V., & Burgess, N. D. (2018). Harvesting fodder trees in montane forests in Kenya: species, techniques used and impacts. *New forests*, 49, 511–528.

Dapilah, F., Nielsen, J. Ø., & Friis, C. (2020). The role of social networks in building adaptive capacity and resilience to climate change: A case study from northern Ghana. *Climate and Development*, 12(1), 42–56.

Eshetu, G., Johansson, T., Garedew, W., & Yisahak, T. (2021). Determinants of smallholder farmers' adaptation options to climate change in a coffee-based farming system of Southwest Ethiopia. *Climate and Development*, 13(4), 318–325.

Few, R., Spear, D., Singh, C., Tebboth, M. G., Davies, J. E., & Thompson-Hall, M. C. (2021). Culture as a mediator of climate change adaptation: Neither static nor unidirectional. *Wiley Interdisciplinary Reviews: Climate Change*, 12(1), e687.

Francis, M., & Kraybill, D. (2012). Climate change and the future of mountain farming on Mt. Kilimanjaro. *The Future of Mountain Agriculture*, 73–88. Berlin, Heidelberg: Springer Berlin Heidelberg.

Fresque-Baxter, J. A., & Armitage, D. (2012). Place identity and climate change adaptation: a synthesis and framework for understanding. *Wiley Interdisciplinary Reviews: Climate Change*, 3(3), 251–266.

Hamilton, A. (1998). Vegetation, climate and soil: Altitudinal Relationships on the East Usambara Mountains, Tanzania. *Journal of East African Natural History*, 87, 85–89.

Jairo, M., & Korir, E. (2019). Climate knowledge, adaptation and intensity of adaptation strategies among farmers in the slopes of Mount Kenya. *Climatology and Weather Forecasting*, 7(1).

Junqueira, A. B., Fernández-Llamazares, Á., Torrents-Ticó, M., Haira, P. L., Nasak, J. G., Burgas, D., ... & Reyes-García, V. (2021). Interactions between climate change and infrastructure projects in changing water resources: an ethnobiological perspective from the Daasanach, Kenya. *Journal of Ethnobiology*, 41(3), 331–348.

Kaganzi, K. R., Cuni-Sanchez, A., Mcharazo, F., Martin, E. H., Marchant, R. A., & Thorn, J. P. (2021). Local perceptions of climate change and adaptation responses from two mountain regions in tanzania. *Land*, 10(10), 999.

Karienye, D., & Kamiri, H. (2020). Trends of banana production among smallholders' farmers due to rainfall and temperature variations in Mount Kenya Region, Kenya. *Budapest International Research in Exact Sciences (BirEx-Journal)*, 2(2), 213–227.

Kassian, L. M., Tenywa, M., Liwenga, E. T., Dyer, K. W., & Bamutaze, Y. (2017). Implications of climate change and variability on stream flow in Iringa region, Tanzania. *Journal of Water and Climate Change*, 8, 336–347.

Klein, J. A., Tucker, C. M., Nolin, A. W., Hopping, K. A., Reid, R. S., Steger, C., ... & Mountain Sentinels Network. (2019). Catalyzing transformations to sustainability in the world's mountains. *Earth's Future*, 7(5), 547–557.

Kong'Ani, L. N. S., Mutune, J. M., & Thenya, T. (2018). Analysis of climate change knowledge and its implications on livelihood options in Naituyupaki Location, Maasai Mau Forest, Narok County, Kenya. *Asian Journal of Forestry*, 2(2), 62–66.

Leclerc, C., Mwongera, C., Camberlin, P., & Moron, V. (2014). Cropping system dynamics, climate variability, and seed losses among East African smallholder farmers: a retrospective survey. *Weather, Climate, and Society*, 6(3), 354–370.

Meru CIDP. (2018). County integrated development plan 2018-2022. County Government of Meru.

Mgumia, A. H., Mattee, A. Z., & Kundi, B. A. T. (2015). Contribution of innovation intermediaries in agricultural innovation: The case of agricultural R&D in Tanzania. *African Journal of Science, Technology, Innovation and Development*, 7, 151–160.

Middleton, J. (1953). The central tribes of the north-eastern Bantu: (the Kikuyu, including Embu, Meru, Mbere, Chuka, Mwimbi, Tharaka, and the Kamba of Kenya).

Motaroki, L., Ouma, G., & Kalele, D. (2021). 'Conservation agriculture' possible climate change adaptation option in Taita Hills, Kenya (Chapter 65). In: *African Handbook of Climate Change Adaptation*, W. Leal Filho et al. (eds.). Available at: https://doi.org/10.1007/978-3-030-45106-6_184.

Mulinde, C., Majaliwa, J. G. M., Twinomuhangi, R., Mfitumukiza, D., Komutunga, E., Ampaire, E., ... & Jassogne, L. (2019). Perceived climate risks and adaptation drivers in diverse coffee landscapes of Uganda. *NJAS-Wageningen Journal of Life Sciences*, 88, 31–44.

Mwakalila, S. (2014). Climate variability, impacts and adaptation strategies: The case of Mbeya and Makete Districts in Great Ruaha Catchment in Tanzania. *Journal of Water Resource and Protection*, 6, 43–48.

Nyeri CIDP. (2018). County integrated development plan 2018-2022. County Government of Nyeri.

Nkurunziza, A., Intwarinkase Mutaganzwa, D., Ndayitwayeko, W. M., Nkengurutse, J., Kaplin, B. A., Teixidor Toneu, I., Zafra-Calvo, N., & Cuni-Sanchez, A. (2023). Local observations of climate change and adaptation responses: A case study in the mountain region of Burundi-Rwanda. *Land*, 12, 329

Nyberg, Y., Jonsson, M., Ambjörnsson, E.L., Wetterlind, J., & Öborn I. (2020). Smallholders' awareness of adaptation and coping measures to deal with rainfall variability in Western Kenya. *Agroecology and Sustainable Food Systems*, 44(10),1280–1308.

Ouma, J. O., Murithi, F. M., Mwangi, W., Verkuijl, H., Gethi, M., & De Groote, H. (2002). Adoption of maize seed and fertilizer technologies in Embu district, Kenya.

Pauline, N. M., & Grab, S. (2018). Whose knowledge matters in climate change adaptation? Perceived and measured rainfall trends during the last half century in south-western Tanzania. *Singapore Journal of Tropical Geography*, 39, 266–280.

Reyes-García, V., Álvarez-Fernandez, S., Benyei, P., García-del-Amo, D., Junqueira, A. B., Labeyrie, V., Li, X., Porcher, V., Porcuna-Ferrer, A., Schlingmann, A., & Soleymani, R. (2023). Local indicators of climate change impacts described by Indigenous Peoples and local communities: Study protocol. *PLoS One*, 18(1), e0279847.

Rosset, P. M., & Martínez-Torres, M. E. (2012). Rural social movements and agroecology: context, theory, and process. *Ecology and society*, 17(3).

Ruggieri, F., Porcuna-Ferrer, A., Gaudin, A., Faye, N. F., Reyes-García, V., & Labeyrie, V. (2021). Crop diversity management: Sereer smallholders' response to climatic variability in Senegal. *Journal of Ethnobiology*, 41(3), 389–408.

Saalu, F. N., Oriaso, S., & Gyampoh, B. (2019). Effects of a changing climate on livelihoods of forest dependent communities: Evidence from Buyangu community proximal to Kakamega tropical rain forest in Kenya. *International Journal of Climate Change Strategies and Management*. 12(1), 1–21.

Savo, V., Lepofsky, D., Benner, J. P., Kohfeld, K. E., Bailey, J., & Lertzman, K. (2016). Observations of climate change among subsistence-oriented communities around the world. *Nature Climate Change*, 6(5), 462–473.

Schumacher, B. L. (2018). Farmer perceptions of climate change and variability in villages adjacent to the Udzungwa Mountains National Park, Tanzania. Master's Thesis, University of California, Los Angeles, CA, US.

Sébastien, L. (2010). The Chagga people and environmental changes on Mount Kilimanjaro: Lessons to learn. *Climate and Development*, 2, 364–377.

Smale, M., & Jayne, T. S. (2003). Maize in Eastern and Southern Africa: "seeds" of success in retrospect (No. 581-2016-39425).

Thomas, K., Hardy, R. D., Lazrus, H., Mendez, M., Orlove, B., Rivera-Collazo, I., ... & Winthrop, R. (2019). Explaining differential vulnerability to climate change: A social science review. *Wiley Interdisciplinary Reviews: Climate Change*, 10(2), e565.

Tiyo, C. E., Orach-Meza, F. L., & Edroma, E. L. (2015). Understanding small-scale farmers' perception and adaption strategies to climate change impacts: Evidence from two Agro-Ecological Zones bordering national parks of Uganda. *Journal of Agricultural Science*, 7(10), 253.

Velempini, K., Smucker, T. A., & Clem, K. R. (2016). Community-based adaptation to climate variability and change: Mapping and assessment of water resource management challenges in the North Pare highlands, Tanzania. *African Geographical Review*, 6812, 1–22.

Wagner, S., Jassogne, L., Price, E., Jones, M., & Preziosi, R. (2021). Impact of climate change on the production of coffea arabica at Mt. Kilimanjaro, Tanzania. *Agriculture*, 11, 53.

Wambugu, F. M., & Kiome, R. M. (2001). The benefits of biotechnology for small-scale banana producers in Kenya. ISAAA. Book.

Wandiga, S. O., Opondo, M., Olago, D., et al. (2010). Vulnerability to epidemic malaria in the highlands of Lake Victoria basin: The role of climate change/variability, hydrology and socio-economic factors. *Climatic Change*, 99, 473–497.

Zizinga, A., Kangalawe, R. Y., Ainslie, A., Tenywa, M. M., Majaliwa, J., Saronga, N. J., & Amoako, E. E. (2017). Analysis of farmer's choices for climate change adaptation practices in South-Western Uganda, 1980–2009. *Climate*, 5(4), 89.

20

ROADS, RIDING STABLES, AND HIGHLAND BARLEY

Livelihood diversification as climate change adaptation among Tibetans in Shangri-la, China

Zhuo Chen

Introduction

"No rain still," my Tibetan Papa said, looking up at the sky and clouds. He was feeding the two horses some highland barley dough in the front yard of the house. Water that occasionally dropped from the tap was stored in a rubber tire. "Not enough water to drink, not enough grass to eat. Poor horses," he said again, holding the highland barley dough and looking at the horse eating in his palms. I looked at the clear eyes of the horses and sighed. Shangri-la county had been experiencing extreme drought when I arrived in the spring of 2021. Observing the sky while hoping for rain became a new daily routine for local Tibetan villagers. Almost every morning, I followed my Tibetan grandpa to the village pagoda and observed the villagers' morning rituals there. The pagoda stood near the road, just opposite the riding stable in our village. Many villagers drove a tricycle slowly on the road, followed by two horses. They were leading the horses to the riding stable. Some horses were even led by slowly driven cars. Since I'm not a Tibetan Buddhist, I could not join the villagers. So, I typically stood by the road and shared some words with those who had finished the ritual. They would share information about coming activities I might want to attend—the ceremonial ritual of praying for rain, for example. Trucks and cars occasionally passed by me at normal speed, either toward the Shika Snow Mountain in the north or the central town in the east.

Climate change has adversely affected many places on the planet in myriad ways, and climate adaptation has become an increasingly pressing issue for various societies and communities. Scholars argue that developing adaptation starts with decreasing vulnerability (O'Brien et al., 2004), but it is important to recognize that not all communities are equally vulnerable to harm from exposure to climate change because of different natural, political, and sociocultural contexts (Füssel, 2010). Eriksen et al. (2011) further argue that to develop more sustainable climate adaptation in a particular community, the contexts for vulnerability should be thoroughly recognized, differing values and interests that affect adaptation outcomes should be acknowledged, and local knowledge should be integrated into adaptation responses.

The Indigenous communities who mainly reside in less developed rural areas and primarily depend on nature for their livelihoods have been identified as among the most vulnerable to climate variabilities (Salick & Byg, 2007; Xu et al., 2009; Ghimire et al., 2010; Bardsley & Wiseman, 2012). Scholars have found that livelihood diversity is one of the strategies adopted by

DOI: 10.4324/9781003356837-25

Indigenous people to decrease vulnerability, thus helping them cope with local climate change impacts (Osbahr et al., 2008; Aryal et al., 2014). However, previous work analyzing the correlation between diversifying livelihoods and decreasing vulnerability has been mostly quantitative. A more in-depth and qualitative investigation of how diversifying livelihoods affects climate adaptation practices and how diversification is formed and sustained in an Indigenous community remains unexplored. Particularly, the ways Indigenous and local knowledge interact with scientific knowledge and other mainstream interests and values in diversifying livelihoods have rarely been discussed.

Interrogating how local knowledge interacts with national policies to form livelihood diversification is crucial for understanding Tibetan communities' adaptation to climate change impacts in Shangri-la county, China. Research has shown that the Tibetan Plateau, at the east edge of where Shangri-la locates, has been identified as one of the most sensitive areas to global climate change (Liu & Chen, 2000; Yan, 2015). The area's temperature rise is approximately three times the average global warming rate (Qiu, 2008) and the snow cover retreats faster in this area than the global average (Yan, 2015). This reality has brought significant impacts on the local livelihoods and forced Tibetan people to take adapting measures (Yin & Zachary, 2018). Local knowledge makes a great contribution to the adaptation of Tibetan livelihoods (Yin & Zachary, 2018), resonating with a surging scholarly discussion on the importance of integrating Indigenous and local knowledge into climate change adaptation (García-del-Amo et al., 2020; Chakauya et al., 2024, Jungsberg and Wendt-Lucas, 2024, However, the Tibetan livelihoods in Shangri-la county have also been dramatically influenced by national policies that encourage the tourism industry and infrastructure development (Yang et al., 2003; Wu et al., 2015). The way traditional Tibetan knowledge interacts with mainstream values in transforming local livelihoods for climate adaptation has yet to be explored. Thus, in this study, I investigated how local knowledge interacts with national policy, roads, and markets to shape diversifying livelihoods among Tibetans in Shangri-la and how such diversified livelihoods help them adapt to climate change impacts.

Tibetans in Shangri-la county and their livelihoods and land-use transition

The Diqing Tibetan Autonomous Prefecture of Yunnan Province, commonly known as the Shangri-la region, is situated in the south of the Qinghai-Tibet Plateau and east of the Himalayas. It is at the junction of the Yunnan, Tibet, and Sichuan Provinces, with a total area of 23,870 square kilometers and a population of about 387,500 (Yunnan Yearbook, 2022). Nearly 90% of the residents are ethnic minorities, defined in China as non-Han population (State Council of People's Republic of China, 2014). Among them, Tibetans account for 36.18%. The prefecture administers three counties—Shangri-la, Deqin, and Weixi. The four villages where I conducted fieldwork are located in Shangri-la county. More than 99% of villagers are Tibetans and share a common belief in Tibetan Buddhism (Figure 20.1).

The Napahai Wetland Reserve is 7 km northwest of the studied villages. The reserve is a provincial natural reserve that was established in 1984 and listed as a Wetland of International Importance by the secretariat of the Ramsar Convention in 2004. Shika Snow Mountain is located 2 km south of Napahai and 10 km east of the villages. According to folklore, the Napahai Wetland and Shika Snow Mountain have played crucial roles in local Tibetan life for centuries.

Four villages located along the large outer ring of the Napahai Wetland were selected for this study. The villages had 31, 59, 56, and 96 households. Roads surrounding the wetland were built in the 2000s, when the reserve became an ecotourism attraction. Each village built a riding stable on the grasslands surrounding Napahai to attract tourists. A tourism bus line called the "Napahai

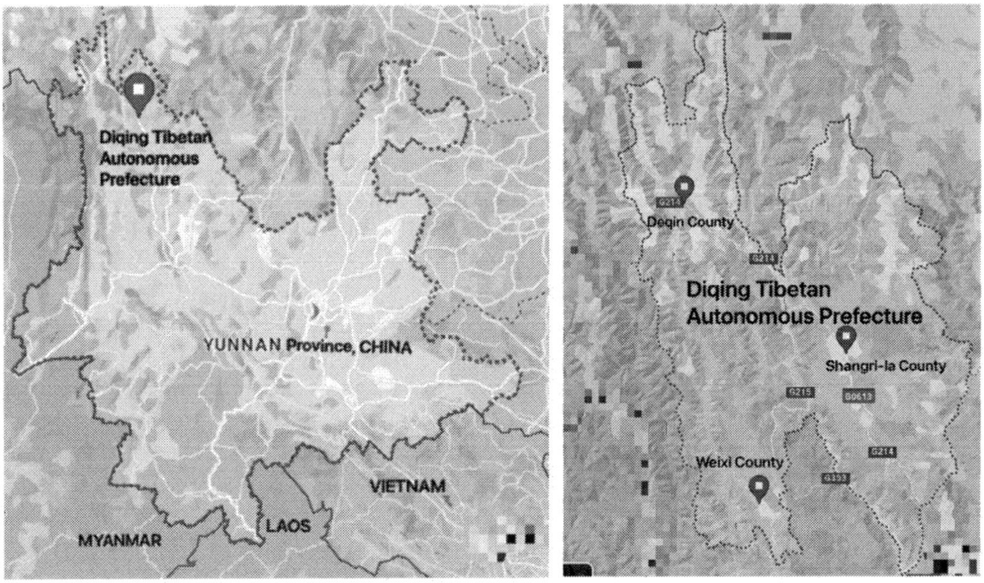

Figure 20.1 Location of Diqing Tibetan Autonomous Prefecture. Reproduced by the author from Google Maps, 2023.

Line" has seven scheduled trips every day from the county center to the wetland. All four villages can be reached by that circuit bus. Most households are located on the outer side of the road surrounding the Napahai Wetland, spreading from right at the edge of the road (Chugu) to farther into the grassland (Yizhai) or up into the mountains (Gedong and Hamugu-Jiaorong). Wooden houses with two or three floors, farmland spots, highland barley shelves, Buddhism pagodas, and riding stables are staggered in between, circling the wetland.

The Tibetans who have lived there for centuries have traditionally relied on agropastoralism and the gathering of non-timber forest products for their livelihood. Yaks and horses are the primary livestock for animal husbandry and grazing, and highland barley is the primary crop for traditional farming. The principal items gathered are various fungi, including matsutake. Local Tibetans have produced rich knowledge of traditional livelihood products. To them, water, highland barley, and yak butter are holy items that have significant material, social-cultural, and spiritual values for local life (Li & Xu, 2007; Li, 2010; Li et al., 2011; Kehoe, 2015; Yang, 2016; Sina, 2019). However, increasingly impacted by national policy on land-use transition, since the 2000s local villagers began to depend more on income from tourism-related businesses.

Chinese national policy started impacting the Shangri-La landscape dramatically in the 1970s. In 1972, the Chinese central government issued a policy encouraging commercial logging to stimulate local economic development. Logging factories spread across the grand Shangri-la region in the following two decades (Yang et al., 2003). As a result, from the 1970s to the 1990s, the forest area decreased by 60%, and the wetland area decreased to only 30% of its original size (Li et al., 2010). After this catastrophic beginning, state and prefecture governments issued a series of policies to restore the ecosystem and landscape in the following decades. In 1998, the Chinese central government implemented the Natural Forest Conservation Program. The program prohibits further logging in the upper and middle reaches of the major Chinese rivers' basin, covering most areas of the Shangri-La region. A variety of governmental programs followed refocusing natural resource policy from a concentration on fast economic development to ecological restoration (Chan, 2002; Cao et al., 2010; Gutiérrez Rodríguez et al., 2016). In Shangri-la, national, provincial, and prefectural policies and programs were implemented to reduce natural landscape degradation (Li et al., 2010). Napahai was listed as one of the priority wetland ecological restoration projects at both the national and provincial levels, and in 2005 grazing was forbidden in the Napahai area.

Policies on tourism development came hand in hand with landscape conservation policies in the late 1990s. Shangri-la became famous as a land of idyllic beauty, for which national and local tourism policies emphasize its ecological richness as a main tourist attraction. In 2002, cross-provincial policies were issued to support ecotourism development in Shangri-la, and Napahai was listed as one of four pilot spots for ecotourism by the Diqing prefecture government (Wen et al., 2003).

In addition to the lasting landscape restoration and ecotourism development policies, a third national policy contributed to land-use changes in Shangri-la. Since the early 2000s, collective assets and property (including land) rights have gone through shareholding reform in rural China (Chen, 2016). Under the compound influence of these policies, local Tibetans have experienced a significant land-use transition. Common lands in the county, such as wetlands, grasslands, and mountain lands, have been developed into ecotourism sites, while maintaining traditional livelihood functions. Since 2003, nine horse-riding stables have been built on the grassland surrounding the Napahai Reserve, each collectively owned by all the villagers. Similarly, Shika Mountain has been transitioned into Shika Snow Mountain Scenic. Tibetan villagers receive compensation via profit sharing from the companies or rural cooperatives that contracted the land for these ecotourism activities (Ma & Liu, 2020). These changes dramatically stimulated infrastructure development in Shangri-La, including the construction of utilities and roads.

Method

This study used an ethnographic method. Fieldwork was conducted from April to June 2021. Participant observations and semi-structured interviews were used to collect the data following the protocol of Local Indicators of Climate Change Impacts project (Reyes-García et al., 2023). Archival research of national and local policies on land use, economic development, and ecological and environmental protection in Shangri-la was conducted as a supplement to fieldwork.

I collaborated with a local research institute that issued a reference letter for me to the Diqen Prefecture Ethnic and Religious Affairs Committee. I introduced the study and its objectives with this letter and oral presentations to the county officers to obtain their permission to conduct research before entering the field. I only conducted research in villages where I obtained local authorities' consent.

To obtain better observations of local life and to gain cooperation from the Tibetan people, I stayed with a Tibetan family, living in their house, eating meals with them, and following them to the riding stables. Each stable had a resting room where villagers gathered, chatting and playing cards, while waiting for their turns to lead the horses. The resting room became a space for me to develop rapport with the villagers and to begin my observations and interviews. After becoming familiar with them, I also participated in other productive activities. Most participant observations were conducted alongside Tibetans operating within their usual livelihoods, such as in the stable while the informants were cleaning the horses, on the grasslands while they were collecting yak dung, on the farmlands while they were working for the vegetable farms, in the temple while they were chanting and circumambulating, or in the backyard while they were making yak butter or washing highland barley.

Semi-structured interviews were conducted at the same lively sites to explore observed climate changes, drivers of those changes, and relevant adaptation and coping measures. Judgmental sampling was used to select the first five informants, each of whom was knowledgeable about farming, pasturing, and gathering, and had devoted more than 30 years to these activities. Then, snowball sampling was used to identify 20 more informants (Bernard, 2017). The informants comprised 13 men and 12 women, spanning age groups from the 20s to the 60s. Their everyday language is Tibetan. However, as a result of the booming ecotourism industry, most of the younger generations can speak Chinese Mandarin, so the interviews with them were conducted in Mandarin. Interviews with older people who were less fluent in Mandarin were conducted with translation help from younger Tibetans.

Content analysis was used for data analysis. The data were coded and analyzed following a codebook containing themes and subthemes on the topics of interest (Saldaña, 2015; Bernard et al., 2017).

Results

Diversifying livelihoods: roads and local knowledge

According to memories of Tibetan villagers in their 80s, interim grazing used to rotate from the Napahai Wetland in the spring, to Shika Snow Mountain in the summer and back to the villages in the winter. Since pasturing in Napahai was forbidden in the early 2000s, villagers then herded only on the summer and winter grazing lands and became more dependent on the grasslands near the villages, a large portion of which has been transformed into riding stables. The management of each stable was done by a specific villager who signed a contract with the county government (the villager then also became the stable manager). In the studied villages, each household had two horses registered in the stable, and the villagers took turns leading the horses for tourists who wanted to try horseback riding. As they were ecotourism sites while still being used as pastures, the riding stables had many special characteristics. For example, none of the stables had fences or walls, and even the "entrances" near the roads were open because the grassland was still a pasture. Led by the villagers, tourists rode horses on the grasslands near the entrance. Meanwhile, yaks and other horses (not for tourist riding) were grazed on the same grassland, but at a further distance from the entrance. Every morning, the villagers pastured the yaks, cattle, and horses to the grasslands from home and got a queuing number from the manager to lead the horse. Then, animals started to wander on the grassland, and people gathered in the resting room or on the land, chatting, drinking yak-butter tea, or making handmade goods. When their turn to lead was about to come, they would run to the far grasslands to get their horses back and wait together at the entrance for the next wave of tourists, most of whom came by tour buses (though a few came by car). The boundary between traditional pasturing and tourism has been blurred in the grassland/riding stable. The Shika Snow Mountain area—the traditional summer grazing land—has experienced similar transitions. In this case, the villagers receive annual profit-sharing dividends and find labor-wage work cleaning the scenic area. Meanwhile, they can still pasture and collect fungi in other areas of the mountains.

With the transition of the common land to commercial uses, the villagers have made a relatively stable income from the mountain and grasslands as tourist sites. Almost all the villagers mentioned "leading the horse in the stable" and "dividends from the Shika Mountain" as the main sources of their annual household income. A similar transition also occurred with farming lands. Compared to other livelihoods, farming requires more time and brings in a lower income. As a result, local

Tibetan villagers started to rent farming lands to people who came from other places in the Yunnan province to grow vegetables for sale.

Although traditional agropastoral livelihoods have been heavily influenced by the market economy, none of the Tibetan households has abandoned crop farming and pasturing, particularly with highland barley planting and yak grazing. In the villagers' narratives, highland barley was a gift from Tibetan deities, who gave seeds as big as an adult's palm thousands of years ago. According to local narratives, it was the highland barley that made the Tibetan people and the yaks strong and healthy despite living in the harsh environment of the plateau. Thus, highland barley is indispensable to local life, both for humans and animals. The villagers' typical breakfast and afternoon tea is yak-butter tea with fried highland barley flour. Almost every household stores piles of highland barley hay on the attic floor of their house, preparing for herding yaks and horses in the winter. Moreover, highland barley plays a crucial spiritual and cultural role in Tibetan life. Symbolizing pristine grains, highland barley is considered holy. Every morning during my stay in the field, I could see the villagers bring highland barley flour to the pagoda and burn it with clean water and branches of conifers as a religious and cultural ritual to begin their day.

Like highland barley, yak butter is also integral to all aspects of Tibetan life. The villagers drink yak-butter tea and worship with yak-butter lamps every day. They also use yak butter as skin cream when the weather is dry. From their point of view, yak-butter tea symbolizes their Tibetan identity. As one of my interlocutors said, "Drinking yak-butter tea makes us Tibetan." (Anonymous. 2021. Interview with villager by author. April 14th. Resting room of Yizhai Riding stable.) Thus, according to local views, the value of highland barley and yak butter is irreplaceable and supersedes their market value, allowing traditional agriculture and pasturing to survive regardless of the profound influence of the market economy in the area.

Road construction has played a crucial role in Shangri-la county's land-use and livelihood transition. Many older villagers remembered that the first roads from the county center to the mountains were built in the 1970s to transport wood. After logging was forbidden, ecotourism continually attracted government investment in road construction. Many young men in the villages have regular jobs as truckers. They leave home for transportation trips about every two weeks, then stay home to lead the horses at the stable or gather fungi on the mountain while waiting for the next long-distance transportation task. Some also choose to work as part-time drivers for tourists between transportation tasks. Meanwhile, the roads provide the villagers with easier connections to the market at the county center. They used to walk half a day there to sell the dairy products and fungi they collected, but now such trips only take ten minutes by bus.

Although the roads were originally built for tourism and the market economy, they also brought benefits to traditional pasturing. Before there were roads in the area, transporting living materials to the mountain mainly depended on yak carts through trails, which were laborious and time-consuming. But now, as one villager in his sixties said, "My son or my friend could drive all the things up there for me within a half day. It's easy and very quick." (Anonymous. 2021. Interview with villager by author. April 20th. Chugu Riding stable.) Moreover, the villagers previously had to make tradeoffs on summer transhumant timing so they would be able to depart *before* the paths were flooded by the wetland but *after* the grass on the mountain had grown. Now, they worry less about the paths and pasture in the upper areas only when there is enough green for the yaks to eat. Pasturing becomes more flexible in schedule with the roads connecting the villages, the grasslands, and the mountain.

In summary, governmental policy and local knowledge have synergistically contributed to the change in local livelihoods. As a result, market-oriented and traditional livelihoods intertwine with each other in space and time, particularly surrounding the roads, and provide the Tibetan villagers diversified economic choices with both stability and flexibility. The following sections show how such diversifying livelihoods has become one of their most crucial strategies in the face of local climate variabilities.

Drought, water shortage, and various explanations

Extreme drought, as the one experienced in 2021, and drier spring seasons were the most significant local climate change impacts identified by Tibetan villagers. According to them, it had not snowed or rained even once in the area since the winter of 2020. When I interviewed or chatted with some people in their sixties or older, they often pointed to Snow Mountain and told me that, when they were young, the snow on the top of the mountain would not melt even on summer days. "You can see the whole mountaintop was white, you know," the 65-year-old former village head recounted the scenery of Shika Snow Mountain in his memory when we were chatting on the grassland. He continued with a soft sigh. "Now it's only April, and you can see almost no snow there. It just doesn't snow." (Anonymous. 2021. Interview with villager by author. April 17th. Chugu Village.) Many villagers provided similar narratives and claimed that less rain and snow threatened their accessibility to water. In the four studied villages, all households had tap water, sourced from meltwater from the snow mountains. Almost all villagers I interviewed claimed that they did not have enough tap water to use, either occasionally or often. Anxiety spread across the studied area from this water shortage. "The tap water is cut off more frequently these days. Horses and yaks don't have enough water to drink. And we (people) have to save it all for eating and drinking, with no water left for washing clothes or showering," one young girl complained. (Anonymous. 2021. Interview with villager by author. April 24th. Resting room of Chugu Riding stable.)

Villagers mentioned that the drought brought difficulties for the horses and yaks in finding water to drink. Worse yet, it was also hard for them to find enough grass to eat. As the drought lasted, the grasslands remained yellow, and the grass did not grow. According to the villagers, various plants and fungi in the grassland grew slower than before. Moreover, many villagers claimed that there had been negative changes in highland barley due to less rain and snow in recent years. People described the highland barley seeds as lower in quality and smaller in size, with leaves growing later and an increase in insect pests. "Some black pets fell down on the highland barley from the air, like a pest rain," three of the villagers mentioned the abnormal number of highland barley insect pests two years ago. "We have never seen such things in our lives," they said (Anonymous. 2021. Interview with villager by author. April 18th. Rain praying ritual site).

Facing these negative impacts, local Tibetans provided explanations of the drivers of such changes derived from both scientific and local perspectives. Many informants claimed that extreme drought happened because of touristic exploitation of the land. Some referred to outsiders renting the land and using too many plastic covers to grow vegetables. "Using plastic is not good, not to mention they burn the waste plastic covers," one 31-year-old woman complained when we walked by her rented farmland. "That definitely makes the climate worse. We have fewer sunny days, you see. Always cloudy, but no rain." (Anonymous. 2021. Interview with villager by author. April 18th.) Meanwhile, their beliefs also played an important role in narratives. For example, the villagers mentioned that Tibetan people practiced water burial for hundreds of years. Only since the end of the 20th century have they turned to cremation, following a national policy. However, they still believe that water burial is the only right way. "You should not burn things or bodies during the winter. The deities will get angry and punish you," one 73-year-old woman said (Interview with villager by author. April 18th 2021. Rain praying ritual site.) Like her, many villagers believed that the extreme drought in 2021 and drier springs in recent years were punishments from the Tibetan deities.

Flexibility as response

As mentioned, local Tibetan people have developed diversified livelihoods combining agriculture, pastoralism, gathering, ecotourism, and wage labor in the transportation industry and tourism-related business. Such livelihoods provide the villagers with diversified sources of income, including the relatively stable income from tourism, transportation business, and renting farming land, and the more flexible income from gathering and selling fungi, selling yak butter to tourists, and selling horses and yaks every few years. This differentiated income portfolio helps them adapt to drought and water shortages by allowing them to schedule their work with flexibility.

In the interviews, the villagers named many adaptations to drought by adjusting the timing of activities. For example, for farming, they postponed sowing highland barley seeds to wait for rain and harvested earlier to avoid damage from the abnormal frost; for herding, they transported up to the mountains later—which became possible with the well-connected roads—if it did not rain or snow enough during the winter and spring; for gathering, they postponed collection in the mountains, waiting for fungi to grow bigger. "Even if we go into the mountains now, we could find nothing; no rain, no fungi. But thank God we have this riding stable. At least we can get money from there," one villager said (Anonymous. 2021. Interview with villager by author. April 28th. Resting room of Chugu Riding stable.) Diversifying sources of income prevents local Tibetans from depending solely on one specific livelihood and provides them with a more flexible schedule for arranging productive activities in the face of irregular climate conditions. When climate variabilities impact one or several living resources, they are able to unhurriedly and comprehensively evaluate the influences on each specific livelihood and make a balanced arrangement. As discussed above, local knowledge and road construction synergistically contribute to forming such scheduling flexibility in the timing of livelihood activities.

Reflecting on the sacredness of water, one of the local Tibetans' most crucial ceremony rituals is to pray for rain. Normally, in early spring, women gather at a particular well at the village entrance. They burn wood and add water and highland barley flour into the fire, then sit in a circle and chant. The ceremony lasts an entire day. If it does not rain in the two weeks or so that follows, another ceremony is held. Many Tibetan women claimed that they had held this ritual praying for rain more frequently due to the drought. Believing in its sacredness, the villagers store water, develop ways of recycling the used water, and are very careful about burning things to avoid intensifying the drought. In addition, some villagers mentioned they transport water from somewhere else when the tap water stopped too long. The roads connecting villages and mountains enabled them to transport water by truck from the Snow Mountain far away to the villages and to use the water for both human and livestock, as well as watering the highland barley seedlings. Moreover, there were situations in which some villagers bought bottled water to drink and cook, which was a luxury choice for most residents, yet still affordable in extreme cases. The Tibetan villagers' response to water shortages further demonstrates that the synergy of roads and local knowledge could also provide flexibility in a broader sense, which is reflected in their various coping strategies—from ritual ceremonies to market behaviors—to a particular climate impact.

Discussion and conclusion

In this study, I explored the ways in which governmental land-use policy and local knowledge work together to form diversified livelihoods among Tibetan people in Shangri-la county and how such diversification is relevant to local climate change adaptation. The findings suggest that diversifying livelihoods decreases local villagers' vulnerability to drought and water shortages through diversifying income, allowing flexibility in scheduling, and enabling complementary and flexible coping strategies.

Some research has revealed that, in nature-dependent societies in developing countries, forms of livelihood diversification are relevant to local communities' vulnerability to climate change impacts. For example, Osbahr et al. (2008) found that in the farming community of Nwadjahane, Mozambique, local people developed a diverse livelihood portfolio of traditional skills in craftwork, herbal medicine, and construction. Such diversification contributed to livelihood resilience and flexibility in response to climate variability. In another case, Aryal et al. (2014) showed that in the Indigenous transhumant community in Himalaya, Nepal, beyond agriculture and livestock income from tourism and tourism-related business contributed to more strategic and diversified livelihoods for local households, thus decreasing their vulnerability to climate change. My study provides one more example to prove the relevance of livelihood diversification to climate change adaptation. Furthermore, through an ethnographic approach, this study provides a more nuanced and thicker analysis of how livelihood diversification works in practice to decrease Tibetan communities' vulnerability to drought and water shortages.

Although Tibetans in Shangri-la county show some similarity in livelihood diversification with Indigenous communities in other areas, their relationship to the new transportation infrastructure is peculiar, as roads have connected all their livelihood spaces, pervaded all aspects of local life, and, to a large extent, driven their livelihood diversification. Li and Hu (2018) studied the Tibetans in another village in the Shangri-la region, Benzilan, and found that roads, as a localized manifestation of national power, stimulated the development of the logging industry and matsutake trade in the 1980s and ecotourism and the transportation industry since the 2000s. The villages I studied experienced almost the same change. Yet, in my case, roads were not merely manifestations of governmental policies or simply stimulated market-oriented livelihoods. Local people also actively used roads to revitalize traditional livelihoods with more flexibility. Roads thus became a key element in their adaptation to drought, both in direct (i.e., enabling flexible transhumant and water transportation) and indirect ways (i.e., through riding stables built surrounding the roads to guarantee tourism). Local knowledge is essential in transforming the roads from a policy expression to a climate change adaptation instrument for local Tibetan people. In his book *Building on Borrowed Time: Rising Seas and Failing Infrastructure in Semarang*, anthropologist Lucas Ley (2021) provides a counterexample showing how infrastructure (the urban sewage system in Semarang) could fail as a "modern" technology to temporarily "fix" the flood problem brought by climate change in Indonesia. On the contrary, in this study, the transportation infrastructure has been integrated into local life in a way that is reconciled with local knowledge. Traditional Tibetan knowledge and governmental policy synergistically repurposed the roads into a sustainable element in local climate change adaptation through villagers' diversifying livelihood practices.

I do not intend to romanticize road construction in remote areas, the negative impact of which has been deliberated on a great deal in both public discourse and academic literature (see Gallice et al., 2019; Pattiselanno & Krockenberger 2021 as the most recent examples). My aim, rather, is to present the complexity of life in communities that have already integrated roads into their livelihoods and the potential of such integration to contribute to local climate change adaptation. In this sense, this study could be a starting point for exploring how roads, or, more broadly speaking, the market economy, could help—rather than impede—local communities in decreasing vulnerability to climate change risks, especially in communities that are under rapid socioeconomic transition and unavoidably affected by the market industry. The Tibetan people's experiences and the essential role of their local knowledge make a valuable case for further exploration.

Acknowledgments

I would like to thank my Tibetan research assistant and my Tibetan family for taking care of me and supporting my fieldwork. I would also like to thank Xiaoyue Li for her generous help throughout my research and for introducing me to the LICCI team. Many thanks to Victoria Reyes-García for her valuable advice and guidance on academic writing.

References

Aryal, S., Cockfield, G., & Maraseni, T. N. (2014). Vulnerability of Himalayan transhumant communities to climate change. *Climatic Change*, 125(2), 193–208. https://doi.org/10.1007/s10584-014-1157-5

Bardsley, D. K., & Wiseman, N. D. (2012). Climate change vulnerability and social development for remote indigenous communities of South Australia. *Global Environmental Change*, 22(3), 713–723. https://doi.org/10.1016/j.gloenvcha.2012.04.003

Bernard, H. R. (2017). *Research methods in anthropology: Qualitative and quantitative approaches*. Sixth Edition. Rowman & Littlefield.

Bernard, H. R., Wutich, A., & Ryan, G. W. (2017). *Analyzing qualitative data: Systematic approaches*. Second Edition. SAGE Publications.

Cao, S., Wang, X., Song, Y., Chen, L., & Feng, Q. (2010). Impacts of the natural forest conservation program on the livelihoods of residents of Northwestern China: Perceptions of residents affected by the program. *Ecological Economics*, 69(7), 1454–1462. https://doi.org/10.1016/j.ecolecon.2009.04.022

Chan, Y. N., & Malm, T. (2002). Livestock and rangeland management in Shangri-la Gorge, Shangri-la county, Northwest Yunnan, China. Master's Thesis. Lund University, *Master's Programme in Environmental Science*.

Chakauya, R., Materechera, S.A., Jiri, O., Chakauya, E., and Machate, M. (2024) Climate change impacts on agriculture, adaptation and resilience: insights from local farmers in South-East Zimbabwe. In V. Reyes-García, S. Álvarez-Fernández, P. Benyei, L. Calvet-Mir, D. García-del-Amo, A. B. Junqueira, X. Li, V. Porcher, A. Porcuna-Ferrer, A. Schlingmann, & R. Soleymani (Eds.), *Routledge Handbook of Climate Change Impacts on Indigenous Peoples and Local Communities*. Oxfordshire & New York: Routledge. Chp 16, pp 273–287.

Chen, A. (2016). The politics of the shareholding collective economy in China's rural villages. *The Journal of Peasant Studies*, 43(4), 828–849. https://doi.org/10.1080/03066150.2015.1078318

Eriksen, S., Aldunce, P., Bahinipati, C. S., Martins, R. D. A., Molefe, J. I., Nhemachena, C.,… & Ulsrud, K. (2011). When not every response to climate change is a good one: Identifying principles for sustainable adaptation. *Climate and Development*, 3(1), 7–20. https://doi.org/10.3763/cdev.2010.0060

Füssel, H. M. (2010). How inequitable is the global distribution of responsibility, capability, and vulnerability to climate change: A comprehensive indicator-based assessment. *Global Environmental Change*, 20(4), 597–611. https://doi:10.1016/j.gloenvcha.2010.07.009

Gallice, G. R., Larrea-Gallegos, G., & Vázquez-Rowe, I. (2019). The threat of road expansion in the Peruvian Amazon. *Oryx*, 53(2), 284–292.

García-del-Amo, D., Mortyn, P. G., & Reyes-García, V. (2020). Including Indigenous and local knowledge in climate research: An assessment of the opinion of Spanish climate change researchers. *Climatic Change*, 160(1), 67–88. https://doi.org/10.1007/s10584-019-02628-x

Ghimire, Y. N., Shivakoti, G. P., & Perret, S. R. (2010). Household-level vulnerability to drought in hill agriculture of Nepal: Implications for adaptation planning. *International Journal of Sustainable Development & World Ecology*, 17(3), 225–230. https://doi.org/10.1080/13504501003737500

Gutiérrez Rodríguez, L., Hogarth, N. J., Zhou, W., Xie, C., Zhang, K., & Putzel, L. (2016). China's conversion of cropland to forest program: A systematic review of the environmental and socioeconomic effects. *Environmental Evidence*, 5(1), 1–22.

Leneisja, J., & Nicola, W. L. (2024). Local indicators, adaptation actions, and resilience efforts for a warming climate in North Greenland. In V. Reyes-García, S. Álvarez-Fernández, P. Benyei, L. Calvet-Mir, D. García-del-Amo, A. B. Junqueira, X. Li, V. Porcher, A. Porcuna-Ferrer, A. Schlingmann, & R. Soleymani (Eds.), *Routledge Handbook of Climate Change Impacts on Indigenous Peoples and Local Communities*. Oxfordshire & New York: Routledge. Chp 22, pp 356–369.

Kehoe, T. (2015). I am Tibetan? An exploration of online identity constructions among Tibetans in China. *Asian Ethnicity*, 16(3), 314–333. https://doi.10.1080/14631369.2015.1015254

Ley, L. (2021). *Building on borrowed time: Rising seas and failing infrastructure in Semarang*. U of Minnesota Press.

Li, Z. & Hu, Q. (2018). Roads, Livelihoods and national identity: Based on Benzilan village field work in Yunnan Tibetan area. *Journal of North Minzu University*, 141(3), 43–52 (In Chinese).

Li, M., & Xu, J. (2007). Simmering practices of Tibetan families in Yunnan – A case study of two Tibetan communities in Diqing Tibetan Autonomous Prefecture. *Ethno-National Studies*, 5, 46–55 (In Chinese).

Li, J., Hu, J., Dong, Y., & Yuan, H. (2010). Study on the landscape change from 1994 to 2006 of Napahai catchment and the Wetlands, Northwest Yunnan mountainous region. *Journal of Mountain Science*, 28(2): 247–256 (In Chinese).

Li, Y., Long, C., Kato, K., Yang, C., & Sato, K. (2011). Indigenous knowledge and traditional conservation of hulless barley (Hordeum vulgare) germplasm resources in the Tibetan communities of Shangri-la, Yunnan, SW China. *Genetic Resources and Crop Evolution*, 58(5), 645–655. https://doi.org/10.1007/s10722-010-9604-2

Liu, X., & Chen, B. (2000). Climatic warming in the Tibetan Plateau during recent decades. *International Journal of Climatology: A Journal of the Royal Meteorological Society*, 20(14), 1729–1742. https://doi.org/10.1002/1097-0088(20001130)20:14<1729::AID-JOC556>3.0.CO;2-Y

Ma, G. & Liu, Q. (2020). Comparative study on the benefits of different land management models in deeply impoverished areas under the background of rural rejuvenation-taking Diqing Tibetan area of Yunnan province as an example. *Journal of Guizhou University of Finance and Economics*, 38(02), 105 (in Chinese).

O'Brien, K., Eriksen, S. E., Schjolden, A., & Nygaard, L. P. (2004). *What's in a word? Conflicting interpretations of vulnerability in climate change research*. CICERO Working Paper.

Osbahr, H., Twyman, C., Adger, W. N., & Thomas, D. S. (2008). Effective livelihood adaptation to climate change disturbance: Scale dimensions of practice in Mozambique. *Geoforum*, 39(6), 1951–1964. https://doi.org/10.1016/j.geoforum.2008.07.010

Pattiselanno, F., & Krockenberger, A. (2021). Road development and indigenous hunting in Tanah Papua: Connecting the facts for future wildlife conservation agendas. *Forest and Society*, 5(1), 181–189.

People's Government of Yunnan Province (2022). Yunnan Yearbook 2022. Diqing Government. http://www.yn.gov.cn/yngk/szmp/dqz/201902/t20190227_99291.html

Qiu, J. (2008). China: The third pole. *Nature*, 454(7203), 393–397. https://doi.10.1038/454393a

Reyes-García, V., Álvarez-Fernandez, S., Benyei, P., García-del-Amo, D., Junqueira, A. B., Labeyrie, V., Li, X., Porcher, V., Porcuna-Ferrer, A., Schlingmann, A., & Soleymani, R. (2023). Local indicators of climate change impacts described by Indigenous Peoples and local communities: Study protocol. *PLoS One*, 18(1), e0279847.

Rumbidzayi, C., Simeon, A. Materechera, Obert, J., Ereck, C., & Machete, M. (2024). Climate change impacts on agriculture, adaptation and resilience: Insights from local farmers in Chiredzi, South-East Zimbabwe. In V. Reyes-García, S. Álvarez-Fernández, P. Benyei, L. Calvet-Mir, D. García-del-Amo, A. B. Junqueira, X. Li, V. Porcher, A. Porcuna-Ferrer, A. Schlingmann, & R. Soleymani (Eds.), *Routledge Handbook of Climate Change Impacts on Indigenous Peoples and Local Communities*. Oxfordshire & New York: Routledge. Chp 16, pp 273–286.

Saldaña, J. (2015). *The Coding manual for qualitative researchers*. Third Edition. London: SAGE.

Salick, J., & Byg, A. (2007). *Indigenous peoples and climate change*. Tyndall centre for climate change research. Oxford: University of Oxford.

Sina, B. (2019). Exploring the symbolic and socio-cultural meanings of Yak Butter Tea in Tibet. Doctoral dissertation, Auckland University of Technology.

State Council of the People's Republic of China (2014, August 26). Ethnic groups in China. State Council of the People's Republic of China. http://english.www.gov.cn/archive/china_abc/2014/08/27/content_281474983873388.htm

Wen, C., Yang, G., & Wang, H. (2003). Environmental effects of eco-tourism in Shangri-la eco-tourism demonstration region, Northwestern Yunnan. *Journal of Agro-Environment Science*, 22(1), 82–85 (In Chinese).

Wu, G., Gao, Y., Wang, Y., Wang, Y., & Xu, D. (2015). Land-use/land cover changes and their driving forces around wetlands in Shangri-la county, Yunnan Province, China. *International Journal of Sustainable Development & World Ecology*, 22(2), 110–116. https://doi.org/10.1080/13504509.2014.915894

Xu, J., Grumbine, R. E., Shrestha, A., Eriksson, M., Yang, X., Wang, Y. U. N., & Wilkes, A. (2009). The melting Himalayas: cascading effects of climate change on water, biodiversity, and livelihoods. *Conservation Biology*, 23(3), 520–530. https://doi.org/10.1111/j.1523-1739.2009.01237.x

Yan, Y., Zhang, Y., Shan, P., Zhao, C., Wang, C., & Deng, H. (2015). Snow cover dynamics in and around the Shangri-la county, southeast margin of the Tibetan Plateau, 1974–2012: The influence of climate change and local tourism activities. *International Journal of Sustainable Development & World Ecology*, 22(2), 156–164. https://doi.org/10.1080/13504509.2014.918909

Yang, Q. (2016). Inheritance and development of Tibetan yak butter culture. *The Border Economy and Culture*, 2016(1), 48–49. https://doi.10.3969/j.issn.16725409.2016.01.021

Yang, X., Li, B., Di, A., & Xu, J. (2003). Land use changes and Tibetan's livelihood strategy in NW Yunnan—A case study on Jisha village, Shangri-la county. *Acta Botanica Yunnanica*, XIV, 72–81 (In Chinese).

Yin, L., & Zachary, M. M. (2018). Traditional knowledge and community-based research in climate change adaptation in East Himalayas. *In Tume, SJP, and Tanyanyiwa, VI (Eds), Climate Change Perception and Changing Agents in Africa & South Asia*. Delaware & Malaga: Vernon Press. Chp 8, pp 133–147.

21

FAITH, RECIPROCITY, AND BALANCE

Inner Mongolian Ovoo offering ritual and its contribution to climate change adaptation

Ouerle Chao, Xiaoyue Li and Victoria Reyes-García

Introduction

There is growing worldwide concern about climate change and how it might impact people, for which adaptation strategies have received considerable attention, from individual to national level (Aguiar et al., 2018; Biesbroek et al., 2010). However, in these efforts, Indigenous Peoples have rarely been considered in public debates until recently (Nyong et al., 2007; Salick & Ross, 2009). Under the current climate crisis, Indigenous Peoples are disproportionately susceptible to environmental stress because of their historical dependence on local biodiversity and cultural landscapes as a source of sustenance and well-being (Adger et al., 2012). However, Indigenous Peoples and local communities are not mere helpless victims of climate change impacts, but have valuable knowledge systems, which are at the roots of their capacity to cope with environmental changes and uncertainties (Hosen et al., 2020; Smith & Sharp, 2012).

The last decade has seen a surge of interest in community-based and participatory climate change adaptation approaches that integrate Indigenous knowledge (Galappaththi, 2024; McNamara & Buggy, 2016). In that sense, the importance of bringing together different forms of knowledge, including Indigenous knowledge, to enhance existing practices is recognized by many climate change researchers (e.g., García-del-Amo et al., 2020). For example, the Sixth Assessment Report of the IPCC (2021) emphasizes the importance of recognizing Indigenous knowledge, as well as integrating insights from different knowledge systems with scientific research to fully understand and address the implications of climate change. Despite this surge, when considering how to bring different knowledge systems together, most of the previous focus has been on management aspects embedded in Indigenous knowledge systems (Reyes-García et al., 2021), with aspects related to Indigenous worldviews and spirituality, as well as their significance for climate change adaptation, having received little attention (Luetz & Nunn, 2020).

Indigenous knowledge (IK) is known by a wide variety of terms, including, local knowledge, traditional knowledge, and traditional ecological knowledge. It has many definitions and there is no consensus on an operational definition applicable across disciplines. Rajasekaran (1993, p.171), for example, fixes the scope of IK to 'the systematic body of knowledge acquired by local people through the accumulation of experiences, informal experiments and intimate

DOI: 10.4324/9781003356837-26

understanding of the environment in a given culture'. In contrast, Fikret Berkes (2017) argues that IK includes not only direct observation and interaction of local people with the environment, but also a broad spectrum of cultural and spiritual knowledge and values that underpin human–environment relationships. This definition of IK is more representative of Indigenous Peoples' cosmologies, which acknowledge that the environment consists of conscious, sentient entities and spirits.

Since Indigenous Peoples are often strongly spiritually engaged, their understanding of environmental changes and responses to them are framed within spiritual beliefs, rationalizations, and elements of their long-standing culture. Under certain circumstances, Indigenous worldviews and spirituality have been described as preserving conservative and fatalistic attitudes that constrain the development of effective climate actions (Wolf & Moser, 2011). However, Indigenous worldviews have also been described as playing positive roles. For example, Quechuan communities in the Potato Park of Cusco region, Peru, have successfully grown and conserved 2,300 potato varieties, focusing on varieties that can withstand climatic extremes. Moreover, they have also rejected the use of modern farming techniques and chemicals in favour of maintaining traditional practices that are better connected with local cycles and can help adapt to climate change impacts. These activities are all propelled by Quechuan communities' traditional values, such as *Ayni*, which refers to the reciprocal process of giving and receiving from elements of nature to maintain the harmony of the world, or *chanincha,* which refers to the solidarity between the human, the natural, and the spiritual worlds (Walshe & Argumedo, 2016).

Indeed, many spiritual rituals practiced by Indigenous Peoples are also directly related to nature, includins some specifically related to weather (e.g., rainmaking rituals). Consequently, it is not much of a leap to expect that the belief system that embeds these rituals might influence individual or community responses to weather variability and/or climate change. To explore the issue, this chapter focuses on the case of Inner Mongolia, a northern autonomous region of China. Through different periods of environmental changes, Inner Mongolians have demonstrated their persistent capability to conserve ecosystem services and co-evolve with the natural world (Tang & Gavin, 2010; Xie & Li, 2008). The role of the Ovoo offering ritual in this adaptation process has not been explored. Thus, the main question we ask in this chapter is 'How do rituals might help communities to cope with the ongoing climate change impacts?' To answer this question, we focus on the Ovoo offering ritual and its potential contribution to transmitting traditional worldviews, bridging networks of reciprocity, and promoting sustainable behaviours.

Methods

This chapter used a qualitative approach based on content analysis of primary and secondary data. The primary data were obtained from semi-structured interviews conducted among herders from East Ujimqin Banner, while the secondary data were obtained from relevant published books and journal articles.

We acknowledge the mixed positionality of the researchers. The first author was born and nurtured in the grasslands of East Ujimqin Banner and she is currently studying in Europe. She self-identifies as an Indigenous Mongolian researcher, for which conducting research in her mother tongue, native land, and with her community has intellectual and personal importance for her. The second author is from China and has completed her Ph.D. and postdoctoral study focusing on Indigenous and local knowledge systems. The senior author is a non-Indigenous scholar who has worked with plural knowledge systems for over two decades.

Description of the study area

Inner Mongolia, located in northern China, has 87 million ha of natural grassland, and is home to over 4.5 million Mongolians (Han et al., 2009). Nomadic pastoralism, dependent on grassland availability, continues to be an important part of the local herder's identity.

Most of the Mongolian grasslands are situated in arid and semi-arid regions. In recent years, the rapidly changing climate, various human activities, and policy implementations, namely land tenure reform, have accelerated grassland degradation in Inner Mongolia (Li et al., 2017), disturbing the well-being of Mongolian herders and threatening their culture (Li & Huntsinger, 2011).

Despite their long history of interactions with the environment, Inner Mongolian herders are often depicted as responsible for land degradation due to overgrazing (e.g., Fang & Wu, 2022). They are also considered as an uneducated group of people who lack scientific knowledge of grassland management, are victims of climate change, and need help (Zhang et al., 2021). Indeed, over the last decades, the Chinese government has designed policies for the development of the region (Li & Huntsinger, 2011), which neglect and undervalue local forms of knowledge and management. For example, in the early 1980s, the Chinese government introduced the 'land tenure reform' to improve grassland management and avoid the 'tragedy of the commons' in Inner Mongolia (Fang & Wu, 2022). This policy, however, resulted in the overall deterioration of the Inner Mongolian pastoral system and grasslands. The distribution of land property rights to individual households, as proposed by the land tenure reform, made it hard for many herders to continue their mobility strategy, resulting in an inability to respond quickly to the unexpected environmental crisis (Conte, 2015). Moreover, the land tenure reform also caused the collapse of the cooperative relationships between herders. Before the land tenure reform, reciprocal relations between herders helped them cope with adverse weather events. However, the land tenure policy transformed the former reciprocal relations into competitive ones (Li & Huntsinger, 2011).

Data collection

Secondary data

This chapter collected secondary data from published books and journal articles. Four different sources of ethnographic research data were used: Bilegsaikhan (2017), Lindskog (2016, 2019), and Huang (2019). Each researcher from which we collected ethnographic data had conducted around two to three years of field research and documented many interview responses from herders related to the Mongolian Ovoo offering ritual. Additionally, the detailed book of 阿尔门德 (2004) about Mongolian religion and the Ovoo offering ritual was used to understand the origin and customary norms of this ritual.

Primary data

The first author conducted semi-structured interviews through *Wechat*, a Chinese social media application. We aimed to interview local people who (1) were representative community members; (2) were or had been herders; (3) were knowledgeable about the Ovoo offering ritual and Mongolian culture in general, and (4) were able and willing to share their knowledge. We started by interviewing a personal contact of the first author and then used the snowball sampling method to identify three other interviewees through the interviewee's recommendations.

In order to capture all the details, such as the interviewee's facial expressions, voices, and emotions, video calls were used. Interviews were conducted in the Mongolian language and explored herders' perspectives about the Ovoo offering ritual and its potential role in (1) transmitting traditional worldviews, (2) bridging networks of reciprocity, and (3) promoting sustainable behaviours.

Data analysis

We used content analysis (Bernard, 2011) to analyse secondary and primary data. After familiarizing ourselves with the data, citations from secondary sources and information from the interview notes were divided into meaningful units. Then, we developed codes (e.g., celebrate collectivity and respect for nature) that are descriptive labels for each meaningful unit and sorted the codes into different categories. Finally, categories were classified into three predetermined themes that guide our analysis: transmission of the traditional worldview, bridging networks of reciprocity, and promoting sustainable behaviours.

Results

The Ovoo offering ritual

In Mongolian, the word *Ovoo* literally means 'pile.' But for Mongolians, the word contains parts of their history and culture. Ovoo is generally an impressive stone feature on the landscape, often in an unusual shape. It can be found throughout the Inner Mongolian region, but it is often located on mountaintops or close to water sources. Mongolians believe in a world populated by spirits and deities that are an organic part of nature. Every creature has a spirit and significance. They feel that everything in their world is connected through their spirits. In that sense, every Ovoo is believed to house a master spirit who often takes a strikingly similar physical form to the landscapes where they are found and are usually distinguished by their unique myths. For example, the spiritual master of the *Arsleng* (lion) Ovoo in East Ujimqin is a lion. In *Arsleng* Ovoo mythology, the lion master protects humans and the environment. In addition to animal deities, there are also human deities, predominantly female figures. The *Eej Nuur* (mother lake) Ovoo in East Ujimqin banner is a spiritual master in the form of a mother who provides milk to her people.

The Ovoo offering rituals aim to protect communities from misfortunes and environmental hazards through communicating and pleasing spiritual masters. Moreover, spiritual masters are not always friendly and can sometimes be wrathful. Anyone who has violated the customary norms concerning the master spirits could be punished by them. Therefore, Ovoo sites carry a set of rules and regulations regarding people's behaviour. For example, following unwritten norms of Mongolian herders, hurting animals, cutting down trees, polluting the sites, and drinking alcohol are all activities banned during offerings (Lindskog, 2019). In addition to these general rules, based on its geographic location and landscape characteristics, each Ovoo has specific rules regarding what kind of food and drink can be offered (阿尔门德, 2004). For example, when the Ovoo is situated near a water source, milk and animal meat cannot be offered to avoid potential pollution of the water.

Almost every offering is followed by an *Ovoon Naadam*, a festival where the 'three games of men' are played: wrestling, horse racing, and archery. *Ovoon Naadam* is held for three successive days after the conclusion of an offering. All local and nearby women, men, and children participate or watch the games. Every family would move their yurt near the festival site and invite each other to eat, drink, sing, and dance together. The elders usually play Mongolian chess, while the younger play *shagai*, a traditional dice game using sheep ankle bones.

The role of Ovoo in transmitting traditional Mongolian worldviews

Mongolian traditional worldview is predominantly spiritual and espouses a minimal separation between the physical and metaphysical world. Mongolians believe that different and legitimate ways of beings, such as humans, animals, stones, and rivers, are conceived as a collection of energies that influence and balance each other mutually. This understanding also can be seen in the Mongolian language. For example, the Mongolian term '*baigal*,' translated as nature, is closely connected to '*baidal*,' which refers to an ideal state of equality and balance between different elements of a unified whole (Humphrey & Sneath, 1999). This perceived notion of balance comprises a fundamental ontological principle for Mongolian people to understand and engage with the spiritual and the natural world. According to a herder's statement, the livelihoods of herders and the welfare of pastures and animals depend on the balance among them. Mongolian nomadism is rooted in the belief of a harmonious balance between all beings (Lindskog, 2016).

The Ovoo offering ritual itself is supported by and embedded in the Mongolian worldview of a harmonious balance between all beings. Mongolians practice this ritual to please the land and ancestral spirits, strengthen the relationship between humans and spirits, and pray for the prosperity of the grassland and animals. One of the interviewees believes that the region's current challenges (i.e., desertification, severe sandstorms, and frequent extreme weather events) are the consequence of a state of imbalance caused by human misbehaviour (Anonymous, 2021. Interview with herder by author. March 9th). For instance, she posits that the primary driver of land degradation in the region is attributed to mining activities. She further stated that the environmental disruptions had inranged the spirits, resulting in a series of unfortunate events upon the grassland.

Additionally, Ovoo-related myths depict significant events of nature, human (mostly female) and animal deities, and how they protect the grassland and local people. They are the foundational building blocks and tools for communicating the Mongolian worldview. As one of the interviewees, a researcher of the Ovoo offering ritual stated, these stories work as an old teaching method to tell children the importance of nature (Anonymous, 2021. Interview with herder by author. March 3rd.2021).

The role of Ovoo in bridging networks of reciprocity

The Ovoo offering ritual further helps build networks and relationships of trust and reciprocity that define social capital. In Mongolian pastoral society, herders rely on relations and social networks to share resources such as pastures, gain information, build new institutions, and create collective norms to increase their adaptive capacity to climate variability or other environmental changes (Fernandez-Gimenez et al., 2011; Humphrey & Sneath, 1999; Zhang et al., 2013). The Ovoo offering ritual, particularly the festival afterward, provides a joyful platform for Mongolians to come together to socialize, celebrate their tradition, and strengthen their collective cultural identity, altogether leading to strong and extensive social networks. As a herder said, there is little chance for relatives and friends to meet each other because of their daily work, and Ovoo offering rituals provide such opportunities (Anonymous 2021. Interview with herder by author. March 22nd.). In a similar note, another herder commented that the families she met through this event had provided her family their land for practicing *'otor'* movement, a traditional mobility strategy that helps cope with severe weather events (Anonymous, 2021. Interview with herder by author. March 9th.). In the case of adverse climatic conditions in one area, arrangements are made for herder families to use neighbouring pasture areas. Some families are renting seasonal pastures from other herders. Some herders combined their pastures by establishing cooperatives. This strategy provides an important means of overcoming localized drought or winter disasters (Fernández-Giménez et al., 2015). This kind of collaborative strategy is increasingly common in Inner Mongolia after the land tenure reform. For example, in Xilingol, it has been found that herders whose contracted pastures are close to each other, or who have kinship or friendship relations, have joined together to reactivate the *'otor'* movement on their allocated winter, summer, spring and autumn pastures based on an oral or written agreement among them. Moreover, encounters facilitated by the practice of the Ovoo ritual also create the best opportunity for herders to share and gain information concerning the weather, the way of managing rangeland and herds, and the experiences in handling extreme weather events, among many others.

Maintaining a social network of family, friends, and neighbours is critical to the success of herding, and the Ovoo ritual contributes to creating and strengthening such social capital. These networks, based primarily on generalized reciprocity, can allow herders to gain access to needed labour and receive essential resources (Janes & Chuluundorj, 2015). In that sense, the Ovoo offering ritual can help reduce herders' vulnerability to climate change through bridging networks of reciprocity.

The role of Ovoo in promoting sustainable behaviours

Ovoo offering rituals is critical in encouraging people to lead more sustainable lifestyles by directing individual behaviours. The grassland ecosystem provides crucial resources such as fodder and water to pastoral societies and supports herders in coping with highly variable and uncertain environments. Thus, sustainable management is embedded in Mongolian culture. For centuries, Mongolians have viewed the Ovoo, and nature in general, as equipped with agency and power to treat or punish the human world. Many customary rules for the protection of Ovoo sites have been established as a result of the respect and fear that they feel for it. As documented in the book Ujimqin Ovoo worship, human misbehaviours, including digging holes, leaving trash behind, cutting young trees, and killing young and female animals during the Ovoo offering will anger the spirits, which will bring misfortunes to the land and the people (阿尔门德, 2004).

New rules and taboos have also been created due to the region's changes, such as modernization and biodiversity loss. The killing of the steppe marmot, for example, is forbidden in certain areas (Anonymous, 2021. Interview with herder by author. March 22nd.). Several interviewees also mentioned the new restrictions on driving cars and a call for horseback riding during the Ovoo offering (Anonymous, 2021. Interview with herder by author. March 3rd and March 27th.). These new emerging rules prove that Mongolians are not merely following customary rules but are also creating rules for the challenges they face, but according to traditional beliefs. Instead of viewing these rules just as Ovoo rules, many herders view them as a guide to a more sustainable future. One interviewee stated that he uses a mouthful of water to wash his face in order to save water (Anonymous, 2021. Interview with herder by author. March 27th.) Another herder chooses to collect and burn dried dung for fuel instead of coal (Anonymous, 2021. Interview with herder by author. March 9th.) The household use of local sustainable energy sources, such as dried dung, can be one of the ways to mitigate the throughput of energy use.

Discussion

This chapter explored the Ovoo offering ritual and its potential to transmit traditional worldviews, bridge networks of reciprocity, and promote sustainable behaviour. The findings suggest that the Inner Mongolian Ovoo offering ritual can contribute to local climate change adaptation in at least the following three ways.

Firstly, the Ovoo offering ritual can convey and reinforce the Mongolian holistic worldview, which in turn may prioritize adaptation strategies that recognize local knowledge and preserve local culture. According to Hulme (2014), the 'thick' accounts of moral reasoning behind most spiritual beliefs and values can give people a motivational force not mirrored by economics or science and other secular messaging, which helps develop culturally situated responses to climate change. In Inner Mongolia, herders are reviving the traditional mobility strategy, '*otor*,' to cope with grassland hazards. The '*otor*' movement is a traditional mobility strategy to manage the grassland and to cope with grassland disasters. However, since the implementation of the land tenure reform policy, the practice of this strategy has become harder. The return to the '*otor*' movement rests on the fact that, according to the Mongolian worldviews, sticking to natural rules and getting along with nature is the best way to address the current environmental crisis. In contrast to this strategy, the central government in China has implemented different top-down adaptation strategies, including a grazing ban where the land will be fenced and forbidden for grazing from three months to one year for grassland restoration in Inner Mongolia. Although this grassland management strategy is often claimed as a success in alleviating land degradation (Qiu et al., 2020), some scholars have noticed its long-term negative impacts on grassland, including biodiversity losses and numerous negative social and economic impacts on local communities (e.g., Li & Huntsinger, 2011). This exemplifies how different beliefs and values may lead to different adaptation strategies.

Secondly, Ovoo offering rituals provide a community platform to bridge social capital and to increase the community's cohesion and solidarity, which increases the resilience capacity of the whole community. The social networks that have been strengthened or built during Ovoo offering rituals help herders collaboratively overcome environmental hazards. The use of social capital has largely been documented in the literature to deal with many shocks. For example, the Mazungunye community in Zimbabwe relies on social capital to lend their livestock to their relatives and friends who have better grazing lands through the system called *kuronzera*, the destocking of livestock strategy (Nyahunda & Tirivangasi, 2021). Such a strategy allows them to respond to climate change impacts collectively. Reciprocal use of pastures and underlying social networks can also allow herders to use pastures efficiently and survive in regions with frequent climate hazards. In the same line of argument, Fernández-Giménez and her colleagues (2015) argue that social capital can play a crucial role in coping with harsh winters collectively among Mongolians. However, it should be noted that little attention has been paid to how spiritual rituals amongst Indigenous communities may foster social capital and its role in local climate change adaptation. For example, the collective ritual of Tungus in Siberia is believed to connect not only the local nomads but also very different actors, such as the regional governors. Through this gathering, different actors can discuss the reasons behind climate change and to find a solution to the changing environment (Lavrillier, 2013).

Thirdly, the Ovoo offering ritual can promote sustainable practices of local people, which can help them mitigate and adapt to climate change. The ritual encourages the conservation of the grassland ecosystem, and this may result in a build-up of nature towards climate change adaptation and resilience. In the face of climate change, the role of conservation becomes more critical to enhancing the adaptive capacity of local people (Reyes-García et al., 2021). For example, sacred sites of Indigenous Peoples can provide ecosystem services such as drinking water, carbon storage, and soil stabilization (Simms et al., 2006). Similarly, the findings of this case study show that customary norms of Inner Mongolia regulate and guide individual behaviours and thus protect grassland ecosystems, which is essential for providing material resources such as fodders and water to local animals and herders. In addition, the availability of well-preserved pasture lands is related to practices like the '*otor*' movement. More importantly, these norms have been internalized by Mongolians and seem to regulate the way in which they interact with nature on a daily basis. For example,

Mongolians, like many other pastoral societies around the world, have used dung as a fuel source until now, which provides an option for sustainable energy production (Ramos-Suárez et al., 2019). In some societies, rituals can direct and encourage collective acts to balance the ecosystem. For example, the Kaiko ritual of the Tsembaga Maring of New Guinea plays a vital role in maintaining the animal–human–environmental balance to respond to environmental instabilities. During this ritual, local people will plant trees and hunt overpopulated pigs to maintain ecosystem balance.

Conclusion

This chapter demonstrates the importance of Indigenous knowledge to climate change adaptation discourses and the need for further research on the topic. Findings from this work are valuable for future researchers because they dived into one largely understudied part of Indigenous knowledge systems, spiritual ritual, trying to connect it to climate change adaptation research. Rather than presenting the spiritual dimensions of Indigenous knowledge as barriers to climate change discourses, the case study shows that these dimensions are one of many crossing pieces of both individual and community identities that come together in various ways to influence community adaptive capacity, adaptation decisions, and outcomes.

Based on the results of this chapter, we suggest that policymakers recognize the potential for cultural responses to climate change that are locally meaningful and morally compelling. Different environments require different adaptations, and not every culture works towards a similar 'norm' of adaptation. As such, each culture has a unique pattern of ecological adaptation. To ensure effective local adaptation, the interventions of climate change policy and planning should reflect and respond to the local culture, including its values and beliefs. For example, instead of implementing a grazing ban policy to help the grassland regenerate, making traditional mobile grazing practice possible or integrating this idea into the policy planning might be a more culturally appropriate option for the ecosystem and the livelihood of the herders.

In Inner Mongolia, there is often a lack of communication between Indigenous Peoples and the government, which makes the integration of local knowledge and scientific knowledge very difficult. The Ovoo offering ritual can be a platform for the local government to participate and create a dialogue with Indigenous Peoples to discuss and plan for future adaptation actions collectively. Climate communication should not be exclusive and pluralistic approaches to climate change should be the future.

Acknowledgments

We would like to thank the contribution of four interviewees who collaborated in all possible ways to ensure that this article would be delivered on time and with proper research done regarding their opinion. We thank funding from the European Research Council (ERC) under grant agreement No 771056-LICCI-ERC-2017-COG. This work contributes to the "María de Maeztu" Programme for Units of Excellence (CEX2019-000940-M).

References

Adger, W. N., Barnett, J., Brown, K., Marshall, N., & O'Brien, K. (2012). Cultural dimensions of climate change impacts and adaptation. *Nature Climate Change*, 3(2), 112–117. https://doi.org/10.1038/nclimate1666

Aguiar, F. C., Bentz, J., Silva, J. M. N., Fonseca, A. L., Swart, R., Santos, F. D., & Penha-Lopes, G. (2018). Adaptation to climate change at local level in Europe: An overview. *Environmental Science & Policy*, 86, 38–63. https://doi.org/10.1016/j.envsci.2018.04.010

Berkes, F. (2017). *Sacred ecology*. Fourth edition. New York: Routledge. https://doi.org/10.4324/9781315114644
Bernard, H. R. (2011). *Research methods in anthropology: Qualitative and quantitative approaches*. Lanham, MD: Rowman Altamira.
Biesbroek, G. R., Swart, R. J., Carter, T. R., Cowan, C., Henrichs, T., Mela, H., Morecroft, M. D., & Rey, D. (2010). Europe adapts to climate change: Comparing national adaptation strategies. *Global Environmental Change*, 20(3), 440–450. https://doi.org/10.1016/j.gloenvcha.2010.03.005
Bilegsaikhan, T. (2017). Some remarks on Ovoo worship among the Dariganga Mongol. *Rocznik Orientalistyczny/Yearbook of Oriental Studies*, T. LXX(2), 261–273.
Conte, T. (2015). The effects of China's grassland contract policy on Mongolian herders' attitudes towards grassland management in northeastern Inner Mongolia. *Journal of Political Ecology*, 22(1). https://doi.org/10.2458/v22i1.21079
Fang, X., & Wu, J. (2022). Causes of overgrazing in inner Mongolian grasslands: Searching for deep leverage points of intervention. *Ecology and Society*, 27(1), 8. https://doi.org/10.5751/ES-12878-270108
Fernandez-Gimenez, M. E., Batbuyan, B., Batkhishig, B. (2011). Understanding resilience in Mongolian pastoral social-ecological systems: Adapting to disaster before, during and after 2010 dzud-year 1 report. World Bank, Washington, DC. https://openknowledge.worldbank.org/handle/10986/2199
Fernández-Giménez, M. E., Batkhishig, B., Batbuyan, B., & Ulambayar, T. (2015). Lessons from the Dzud: Community-based rangeland management increases the adaptive capacity of Mongolian Herders to winter disasters. *World Development*, 68, 48–65. https://doi.org/10.1016/j.worlddev.2014.11.015
Galappaththi, E. (2024). How can co-management support adaptation to climate change? The case of co-existing fisheries from Pangnirtung, Nunavut, Canada. In V. Reyes-García, et al. (Ed.), *Routledge Handbook of Climate Change Impacts on Indigenous Peoples and Local Communities*. Routledge: Oxfordshire & New York. Chapter 23, 370–388.
García-del-Amo, D., Mortyn, P. G., & Reyes-García, V. (2020). Including Indigenous and local knowledge in climate research: An assessment of the opinion of Spanish climate change researchers. *Climatic Change*, 160(1), 67–88. https://doi.org/10.1007/s10584-019-02628-x
Han, X., Owens, K., Wu, X. B., Wu, J., & Huang, J. (2009). The grasslands of Inner Mongolia: A special feature. *Rangeland Ecology & Management*, 62(4), 303–304. https://doi.org/10.2111/09-002.1
Hosen, N., Nakamura, H., & Hamzah, A. (2020). Adaptation to climate change: Does traditional ecological knowledge hold the key? *Sustainability*, 12(2), 676. https://doi.org/10.3390/su12020676
Huang, H. (2019). Afterthoughts: Nature, culture, and shamanism in Inner Mongolia, PRC. *EnviroLab Asia*, 3(3), 1–13. https://doi.org/10.5642/envirolabasia.20190303.02
Humphrey, C., & Sneath, D. A. (1999). *University lecturer in social anthropology and fellow Caroline. The end of nomadism? Society, state, and the environment in Inner Asia*. Durham, NC: Duke University Press.
IPCC. (2021). Climate change 2021*: The physical science basis. Contribution of working group I to the sixth assessment report of the intergovernmental panel on climate change*. Cambridge University Press, Cambridge and New York, in press. https://doi.org/10.1017/9781009157896
Janes, C., & Chuluundorj, O. (2015). Making disasters: Climate change, neoliberal governance and livelihood insecurity on the Mongolian Steppe. *Nomadic Peoples*, 21(1), 161–164. https://doi.org/10.3197/np.2017.210111
Lavrillier, A. (2013). Climate change among nomadic and settled Tungus of Siberia: Continuity and changes in economic and ritual relationships with the natural environment. *Polar Record*, 49(03), 260–271. https://doi.org/10.1017/s0032247413000284
Li, W., & Huntsinger, L. (2011). China's grassland contract policy and its impacts on herder ability to benefit in Inner Mongolia: Tragic feedbacks. *Ecology and Society*, 16(2). https://doi.org/10.5751/es-03969-160201
Li, Z., Wu, W., Liu, X., Fath, B. D., Sun, H., Liu, X., Xiao, X., & Cao, J. (2017). Land use/cover change and regional climate change in an arid grassland ecosystem of Inner Mongolia, China. *Ecological Modelling*, 353, 86–94. https://doi.org/10.1016/j.ecolmodel.2016.07.019
Lindskog, B. V. (2016). Ritual offerings to ovoos among nomadic Halh herders of west-central Mongolia. *Études mongoles et sibériennes, centrasiatiques et tibétaines* (47). https://doi.org/10.4000/emscat.2740
Lindskog, B. V. (2019). Managing uncertainty, beckoning security: Ritual offerings to a local Ovoo in Mongolia. *Ethnos*, 1–19. https://doi.org/10.1080/00141844.2019.1699143
Luetz, J. M., & Nunn, P. D. (2020). Climate change adaptation in the Pacific islands: A review of faith-engaged approaches and opportunities. *Climate Change Management*, 293–311. https://doi.org/10.1007/978-3-030-40552-6_15

McNamara, K. E., & Buggy, L. (2016). Community-based climate change adaptation: A review of academic literature. *Local Environment*, 22(4), 443–460. https://doi.org/10.1080/13549839.2016.1216954

Nyahunda, L., & Tirivangasi, H. M. (2021). Harnessing of social capital as a determinant for climate change adaptation in Mazungunye communal lands in Bikita, Zimbabwe. *Scientifica*, 1–9. https://doi.org/10.1155/2021/8416410

Nyong, A., Adesina, F., & Osman Elasha, B. (2007). The value of Indigenous knowledge in climate change mitigation and adaptation strategies in the African Sahel. *Mitigation and Adaptation Strategies for Global Change*, 12(5), 787–797. https://doi.org/10.1007/s11027-007-9099-0

Qiu, H., Su, L., Feng, X., & Tang, J. (2020). Role of monitoring in environmental regulation: An empirical analysis of grazing restrictions in pastoral China. *Environmental Science & Policy*, 114, 295–304. https://doi.org/10.1016/j.envsci.2020.08.023

Rajasekaran, B. (1993). A framework for incorporating indigenous knowledge systems into agricultural research and extension organizations for Sustainable Agricultural Development in India (dissertation). Iowa State University Digital Repository, 171.

Ramos-Suárez, J., Ritter, A., Mata González, J., & Camacho Pérez, A. (2019). Biogas from animal manure: A sustainable energy opportunity in the Canary Islands. *Renewable and Sustainable Energy Reviews*, 104, 137–150. https://doi.org/10.1016/j.rser.2019.01.025

Reyes-García, V., Fernández-Llamazares, Á., Aumeeruddy-Thomas, Y., Benyei, P., Bussmann, R. W., Diamond, S. K., García-del-Amo, D., Guadilla-Sáez, S., Hanazaki, N., Kosoy, N., Lavides, M., Luz, A. C., McElwee, P., Meretsky, V. J., Newberry, T., Molnár, Z., Ruiz-Mallén, I., Salpeteur, M., Wyndham, F. S., … Brondizio, E. S. (2021). Recognizing indigenous peoples' and local communities' rights and agency in the post-2020 biodiversity agenda. *Ambio*, 51(1), 84–92. https://doi.org/10.1007/s13280-021-01561-7

Salick, J., & Ross, N. (2009). Traditional peoples and climate change. *Global Environmental Change*, 19(2), 137–139. https://doi.org/10.1016/j.gloenvcha.2009.01.004

Simms, A., Maldonado, M. J., & Reid, H. (2006). *Up in smoke? Latin America and the Caribbean: The threat from climate change to the environment and human development (International Development)*. Oxfam Publishing, London: New Economics Foundation (NEF).

Smith, H. A., & Sharp, K. (2012). Indigenous climate knowledges. *Wiley Interdisciplinary Reviews: Climate Change*, 3(5), 467–476. https://doi.org/10.1002/wcc.185

Tang, R., & Gavin, M. C. (2010). Traditional ecological knowledge informing resource management: Saxoul conservation in Inner Mongolia, China. *Society & Natural Resources*, 23(3), 193–206. https://doi.org/10.1080/08941920903375503

Walshe, R., & Argumedo, A. (2016). Ayni, ayllu, Yanantin, and Chanincha: The cultural values enabling adaptation to climate change in communities of the Potato Park, in the Peruvian Andes. *GAIA – Ecological Perspectives for Science and Society*, 25(3), 166–173. https://doi.org/10.14512/gaia.25.3.7

Wolf, J., & Moser, S. C. (2011). Individual understandings, perceptions, and engagement with climate change: Insights from in-depth studies worldwide. *WIREs Climate Change*, 2(4), 547–569.

Xie, Y., & Li, W. (2008). Why do herders insist on otor? Maintaining mobility in Inner Mongolia. *Nomadic Peoples*, 12(2), 35–52. https://doi.org/10.3167/np.2008.120203

Zhang, C., Li, W., & Fan, M. (2013). Adaptation of herders to droughts and privatization of rangeland-use rights in the arid alxa left banner of Inner Mongolia. *Journal of Environmental Management*, 126, 182–190. https://doi.org/10.1016/j.jenvman.2013.04.053

Zhang, R., Yeh, E. T., & Tan, S. (2021). Marketization induced overgrazing: The political ecology of neoliberal pastoral policies in Inner Mongolia. *Journal of Rural Studies*, 86, 309–317. https://doi.org/10.1016/j.jrurstud.2021.06.008

阿尔门德拉. (2004). 乌珠穆沁敖包祭祀. 内蒙古人民出版社.

22

LOCAL INDICATORS, ADAPTATION ACTIONS, AND RESILIENCE EFFORTS FOR A WARMING CLIMATE IN NORTH GREENLAND

Leneisja Jungsberg and Nicola Wendt-Lucas

Introduction

During the last four decades, the Arctic has been warming nearly four times faster than the rest of the globe (Rantanen et al., 2022). Indigenous Peoples living in the Arctic (e.g., Sami, Inuit, Nenet) are among the first groups to face the direct consequences of climate change, due to their dependence upon and close relationship with nature (United Nations, n.d.). Indeed, many Indigenous communities living in the Arctic region depend on hunting for seals, whales, polar bears, musk ox, caribou, fishing, and gathering not only for food to support the household economy, but also as a cultural practice that represents their social identity and heritage (Pearce et al., 2010; Poppel, 2015; Sharma, 2010). Some of the concerns facing Indigenous Peoples in the Arctic include changes in species presence and abundance, availability of traditional food sources, reduction in weather predictability, and the safety of traveling in changing ice and weather conditions. Together these changes pose serious challenges to human health and food security in the region (Pearce et al., 2010; Poppel, 2015; Sharma, 2010). Moreover, climate change impacts exacerbate the difficulties already faced by Indigenous communities, including political and economic marginalization and loss of land and resources (United Nations, n.d.).

In this context, there is a large interest in developing adaptation strategies with and for Indigenous communities. A big share of the international funding for adaptation actions is allocated to address technical actions, ranging from information and communication technology to early warning systems and new improved infrastructure. However, these actions do not always produce satisfying results. This is partly due to adaptation being understood and approached as a technical response to a technical problem (Biagini et al., 2014), which contrasts with research in the field illustrating how adaptation barriers also involve a lack of human and financial resources (Piggott-McKellar et al., 2019). Consequently, critical Arctic voices are increasingly pointing to the need for more deep-rooted and socially embedded approaches to build the necessary resilience to meet the climate challenges affecting Arctic livelihoods (Fedele et al., 2019).

On a fundamental level, adaptation seems simple: adjusting behaviours, actions, and decisions within biological, social, and built systems in response to climatic changes (Owen, 2020). However, adaptation is always contingent upon the events or conditions to which a person or a community is responding to (Owen, 2020), which implies that adaptation further hinges on the multiple

DOI: 10.4324/9781003356837-27

and intersecting ways in which people know, experience, and deal with climate change and its impacts. People have diverse conceptualizations and multiple ways of understanding reality, including climatic changes, and consequently multiple ways to adapt behaviours accordingly. These ontological and epistemological differences suggest that adaptation scholars must grapple with coexistent realities of, experiences with, and ways of gaining knowledge about climate change (Owen, 2020).

Much literature points to the importance of empirically grounded studies to grasp the ongoing, climate-induced changes (Baker et al., 2012; Biagini et al., 2014; Biesbroek et al., 2015). Previous findings have highlighted the value of local knowledge to explore, understand and foster adaptation practices that enhance resilience (e.g., Engle, 2014). One goal of the development of adaptation strategies is to increase people's adaptive capacity, or their ability – in terms of resources and practices – to respond effectively to changing stresses and shocks to manage or reduce risk, and to prevent the loss of future development options (Engle, 2011). Adaptation strategies also aim to increase resilience, or the ability of a system to continue functioning when confronted with shock and stress. The concepts of resilience and adaptation can hence be seen as complementary to one another (Nelson, 2011). Livelihood diversification, community development, governance, and public health programmes are all aspects that are relevant to understand interventions that build resilience and support adaptation actions (Engle et al., 2014).

Indicators and proxies can provide transparent syntheses of knowledge about representative aspects of resilience, and to some extent, they can aid in making comparisons and examining the sources and impacts of adaptation and resilience (Engle et al., 2014). So far, few efforts exist to establish a system of local indicators of climate change in a resilience context for North Greenland, building on the collection and analysis of locally observed changes linked to global warming. Existing studies addressing climate change impacts in Qaanaaq and surrounding areas have focused on specific aspects of climate change, for instance changes in the sea-ice, glaciers or economic activities (e.g., Andersen & Flora, 2019; Nuttall, 2020; Sugiyama, 2020). By suggesting a resilience framework consisting of climatic and physical, biological, and socio-economic systems, this chapter establishes a structure to capture and analyse local indicators of climate change impacts. The contribution of this study is hence threefold: (1) it provides novel empirical insights into ongoing environmental changes affecting the inhabitants of Qaanaaq, (2) it offers reference points for future research into the progression of climate change impacts in Qaanaaq through the categorization of the observed changes into local indicators of change (LIC), and (3) it allows the identification of bottom-up adaptation actions that could inform future adaptation design to help local communities tackle the implications of a warming climate.

Resilience framework

The concept of resilience has emerged from the ecological sciences and in that context, it "captured the ability of the system to fluctuate within the domain of attraction without being pushed over the boundary" (Holling, 1973, p. 3). Later, the concept has expanded to various disciplinary backgrounds, under the argument of the need to study human and environmental systems and their interactions, since the human element is one of the most important drivers of ecosystem change (Engle et al., 2014).

The Arctic Council has also worked with the concept of resilience suggesting the following definition: "The capacity of people to learn, share and make use of their knowledge of social and ecological interactions and feedbacks, to deliberately and effectively engage in shaping adaptive or transformative social-ecological change" (Arctic Council, 2016, p. 8). The emphasis on

Table 22.1 Resilience framework presenting three systems important for local resilience in Northern Greenland

System changes	*Local and Indigenous knowledge*	*Adaptation actions*
Climatic and physical	Local & Indigenous knowledge referring to atmospheric and physical aspects such as weather conditions and associated factors such as (sea) ice and ground stability.	The ability of community members to resist the stress from changes in the atmospheric system and to adapt practices that rely on elements of these systems (e.g., the accessibility and stability of sea-ice).
Biological	Local & Indigenous knowledge referring to biological factors such as migratory patterns of animal populations, animal behaviour, composition of animal populations and animal welfare.	The ability of community members to withstand biological changes to sustain themselves by adapting hunting and fishing practices.
Socio-economic	Local and Indigenous knowledge referring to socio-economic aspects such as employment and housing.	The ability of community members to manage and cope with changes in the socio-economic system which might affect their everyday lives through unstable income or unsafe housing conditions.

knowledge as a facilitator to engage in change is a relevant aspect that implicitly leans towards the concept of adaptation and the enabling of adaptation actions.

Drawing on the definition proposed by the Arctic Council, we suggest a more explicit linkage between local and Indigenous knowledge and adaptation capabilities and processes for resilience. We do so by presenting local and Indigenous knowledge as an enabling factor in the context of climate change impacts, as it allows the local population to draw on the knowledge of past generations and current observations to dynamically adapt to changing environmental conditions. We present three systems that are relevant for adaptation actions: (1) the climatic and physical, (2) the biological, and (3) the socio-economic systems (see Table 22.1).

By foregrounding the value of local and Indigenous knowledge systems, this study also spotlights the collectivist underpinning of the concept and advancement of resilience in Indigenous communities. According to the work of Berkes and Berkes (2009), in its holistic approach towards the lived environment, Indigenous knowledge is better situated to encompass the various factors forming the complex system of environmental change. The authors compare the collection and exchange of mainly qualitative, observational data on environmental change to a fuzzy logic system. Placing more emphasis on these nexuses of bottom-up knowledge can help to gain a more complete understanding of climate change-related impacts and of the necessary adaptive actions on a collective and individual level that can help to preserve livelihoods but also cultural identity and heritage.

Methods

Study area: Qaanaaq, North Greenland

Research was conducted in Qaanaaq, North Greenland (Figure 22.1 Territorial division of Greenland). Climate in the area is classified as cold tundra and the winter is dark and cold, with three winter months of polar nights and three summer months with 24-hour daylight. Peak temperature

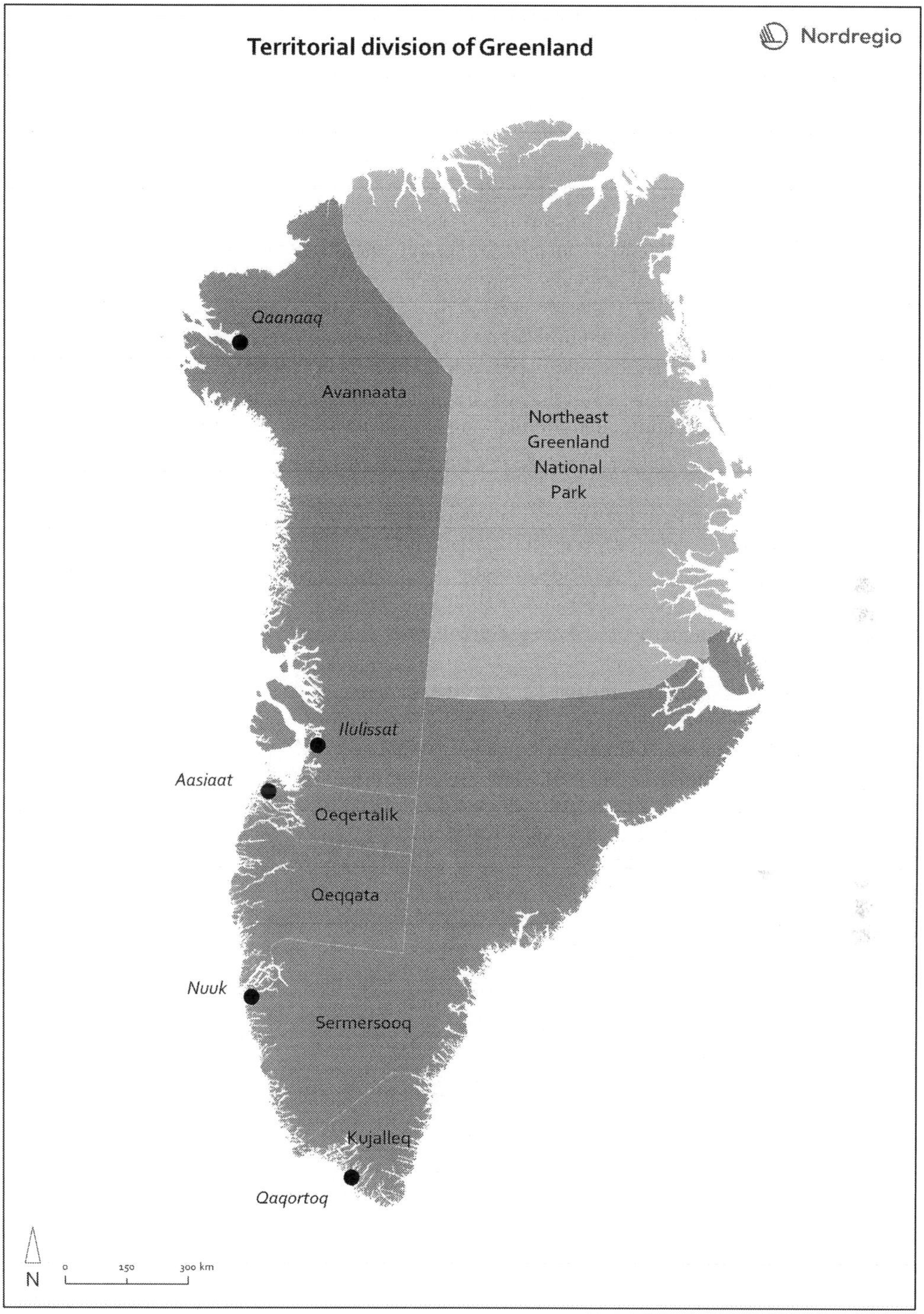

Figure 22.1 Territorial division of Greenland

in July seldom exceeds 14°C, but occasionally it can get as warm as 20°C. During winter, the average temperature ranges from –20 to –30°C, with a measured record of –40°C.

The people of Qaanaaq migrated from Canada around 1100 CE and belong to the Inughuit group (Hastrup, 2018). They have a unique dialect and their own specific cultural traits, for example, the usage of harpoons for hunting. Qaanaaq was established in 1953 as part of a relocation from Thule (Old Thule/Pittufik).[1] The United States Air Force constructed an air defence site near the village and the Danish government forcibly relocated "Old Thule", with about 130 inhabitants, to a new village approximately 97 km north. In a Danish Supreme Court judgement of 28 November 2003, the move was considered an expropriative intervention and an unlawful act, for which the Thule tribe was granted an economic compensation.

Qaanaaq is part of Avannaata municipality, and the administrative centre is Ilulissat, where most residents travel approximately once a year as a transit to other parts of Greenland or for health care purposes. During the long winters, the sea-ice covers the Northern Greenlandic coast, which means that only two Royal Arctic Line ships (one in July and one in September) supply the town with materials for the built environment, food, and other imported equipment. There are one to two weekly passenger flights depending on the season. Water in town is supplied directly from the river during the four summer months, from tanks filled during summer during the four fall and onset of the winter months, and from a water plant that melts sea-ice during winter. About half of the households have a water tank, while approximately a quarter gets water through pipes, and the other quarter needs to get it from a public pump. Common livelihood activities are wage labour in the public and private sectors.[2] Most households in Qaanaaq go leisure hunting and fishing to supplement their income with local fresh food resources.

Empirical data and methods

This study is based on four weeks of fieldwork in Qaanaaq during February–March 2020. The first author collected data in collaboration with a local field-assistant who acted as translator and gatekeeper. Most communication between the first author and the field assistant was in Danish, and besides a few interviews, the field assistant translated the questions to Greenlandic and facilitated translations between Danish and Greenlandic. The field assistant had an extensive social network in town and was also chairing the local hunter and gather association (KNAPK), which facilitated approaching potential informants.

We conducted 20 qualitative interviews. The study participants were to a large extent part of the local network of the field assistant. However, we also used the snowball sampling technique and got referred to other community members by study participants. Table 22.2 describes the sample based on age, gender, and occupation. While only five participants were full-time fishers/hunters, most participants engaged in hunting and fishing as a leisure activity.

Fewer women than men participated in the qualitative interviews. Throughout the fieldwork, some individuals stated that they were not interested in participating; most of them were women.

1 Today, Thule is still a military base operating with soldiers from the US, Canada and Denmark. The base also hosts a weather station and a number of satellite networks (21st space Wings global network of warnings of space surveillance and space control) collaborating with American Aerospace Defence command (NORAD) and Air Force Space Command (AFSPC).

2 Common employers include the municipal office, local school with first to tenth grade, small hospital/health reception, old people's home, fish factory, grocery shop, electricity and water supply centre Nukissiorfiit, polar oil, town's dump and sanitarian care, Tele Greenland contact point, youth club, craftsman as well as the local hotel.

Table 22.2 Age, gender, and occupations profile of study participants in the qualitative interviews

Age	*Gender*	*Fisher/ hunter*	*Public sector*	*Private sector*	*Students in education*	*Total*
25–80	Men	5	2	4		11
	Women		6	1		7
18–25	Men				2	2
	Women					
Total		5	8	5	2	20

For instance, none of the female students at Majoriaq[3] wanted to participate in the study. Lower women's participation might be due to women feeling shy and thinking they did not have much to contribute with to the topic of nature observations, environment, and climate trends, because male members of the community are more likely to engage in close-to-nature activities such as hunting and fishing. An alternative explanation is that the field assistant was a man, for which he could more easily appeal to other men in terms of convincing them to invest their time answering interview questions.

We collected observations of climate change with qualitative interviews. The semi-structured interviews focused on observed changes and often started with an opening question "Compared to when you were young, have you noticed any changes in nature, in the climate, the plants, the animals, or the local community?" The identified changes occurring three or more times were validated in two focus group discussions of five to seven participants. In the focus groups, the discussion centred around 20 observations that had emerged from the interviews, although participants only agreed on 15 observations upon discussion. The methodology used draws on the protocol developed by Reyes-García et al. (2023).

The question posed to understand the adaptation mechanism behind local changes came as a follow-up question after the study participant had described observed changes. Specifically, we asked "Have these changes led to new routines, habits, adjustments, adaptations etc. in your life"? The transcripts were coded based on actions with the purpose of coping with local environmental changes. Both the local indicators of change and the adaptation actions were collected through semi-structured interview questions to understand how the study participants were experiencing and coping with local climatic and physical, biological, and socio-economic changes. The methodology to connect local indicators with adaptation actions is partly inspired by the methodological guidelines from Braun and Clarke (2006).

To understand whether the individually observed changes were more broadly observed by a broader representation of the community, an additional questionnaire was conducted with a larger sample. The local field assistant carried out the survey in Qaanaaq. The population of Qaanaaq was 646 (as of 2020 when the fieldwork was conducted) of whom 450 are above the age of 18. The number of respondents ($n=45$), therefore, represents 10% of all adult inhabitants in Qaanaaq. Both men and women were represented in the sample, with 21 female and 24 male survey respondents. Survey questions focused on individual observations of a set of changes attributed to climate change, adaptation actions, household economy, subsistence, and other livelihood activities. The survey was conducted during February and March 2020.

3 Majoriaq in Greenland deals with three main tasks: Job placement – Guidance for education and work – Upskilling for education and work for young adults.

Results

Assessing local observations of climate change

Local and Indigenous knowledge is a valuable source of information to understand community's observations of climate change impacts (Table 22.3).

The climatic changes reported by the participants refer to wind, sea-ice, snow, temperature, and storms. Sea-ice and snow are reported to decrease, whereas temperature, wind, and storms are reported to increase.

Sea-ice is essential for the livelihood and culture in North Greenland as it serves as the primary hunting and (ice) fishing ground. Qaanaaq is located by the North Water (Pikialasorsuaq), which is a polynya – an area of open sea, which never freezes completely, but is circumscribed by sea-ice. Every summer, the sea-ice melts, and the polynya disappears, to reappear later, in fall or winter. However, informants reported that the predictability of these sea-ice changes has diminished, and it has become hard to say exactly when the sea-ice will freeze. According to the study participants, in 2019, there was open water until the end of December, which is much later compared to 20–30 years ago, when the water froze around late September. Participants also reported that summers are getting longer and that there are more warm days compared to 20 or 30 years ago. One participant shared that when she was a child, the sea-ice stayed until early July, while today, the ice often disappears already in late May or early June.

Besides a decrease in sea-ice, participants also reported that there is less snow and noted changes in the amount and frequency of snowfall. The observations and experiences regarding snow were strongly debated in the focus groups and some participants argued that there has always been less snow in Qaanaaq than in other areas of Greenland. In some winter periods, the participants had to wait for the snow to arrive, calling these periods *black winter.* Snow is essential for dog sledding and makes a difference during the three months of polar nights, as the snow brightens up the atmosphere and gives a sense of light.

Participants further reported a negative synergy between the retard of the snowfall and the increasing frequency of strong winds. Often, the arrival of a thin layer of snow would now be followed by wind gusts that blow the snow away. This is a big concern, as strong winds and storms reduce time for hunting, fishing, and other outdoor activities. The winds are particularly concerning during fall, winter, and spring.

The biological changes reported by the study participants relate to changes in the migration of animals on land and to changes in the composition of marine animal and plant species. In the summer, there is 24/7-hour daylight, and it is common to go out fishing on a boat and stay overnight in wooden cabins. From their fishing trips, 67% of the participants reported having seen new fish species. These include cod and capelin. These two species have been observed during the last ten years in the waters surrounding Qaanaaq, but have historically been associated with the southern Greenland waters. Another change in the biological system is that polar wolves have been observed closer to town. Approximately 22% of the participants explained that it may be due to changes in caribou patterns or to the general lack of food. So far, it is mainly the men who have been out hunting who have observed polar wolves in the vicinity of Qaanaaq and the wolves have not posed any danger to the community.

Changes in the temporal distribution of snow also affect the dogs' paws when sledding. The study participants explained that during the periods with less snow, it is rough on the dog paws to

Table 22.3 Reported observations and corresponding Local Indicators of Change (LIC)

System	*Observation*	*Local indicator of change (LIC)*	*% survey participants confirming the observation*
Climatic and physical systems	The wind is stronger in certain periods.	Changes in wind strength or speed	69
	Sea-ice cover that is safe to travel on with dog sled or snowmobile appears later than in previous years (e.g., it did not appear before January in 2020).	Changes in the extent of sea-ice surface	93
	The snow cover decreases in some periods – as wind blows it away – while there is an increase in other periods.	Changes in the amount of snowfall	84
	There are more periods with thin sea-ice.	Changes in the thickness of sea-ice	84
	The temperature in winter is fluctuating.	Changes in temperature fluctuations	84
	Summers are getting longer.	Changes in the mean temperature in a given season	89
	More water is flowing out from the mountains behind Qaanaaq town.	Changes in the thawing or melting of permafrost	29
	More storms with strong gusts have been observed.	Changes in the intensity of windstorms	49
	There has been less snow.	Changes in the frequency of snowfall	84
	The number of warm days has increased.	Changes in the frequency of hot/ warm days	84
	The soil is sinking in certain areas.	Changes in the extent of permafrost surface	58
Biological system	New fish species (cod and caplin) have been found in the sea water.	Changes in the species composition of marine animal and plant species	67
	Polar wolves are coming closer to the town.	Changes in terrestrial fauna migration areas and routes	22
	The paws of the sled dogs are getting hurt more easily due to a bad snow cover.	Changes in temporal distribution of snowfall	73
Socio-economic system	Houses in the town are becoming damaged and skewed.	Changes in the depth of the permafrost layer	73

run on the ground leading to a decrease in performance and an increase in injuries. Without the snow, the ground is rocky and stony and hence not optimal for the sledding dogs' paws. When the snow-covered period during winter gets shorter, the benefit of having dogs diminishes as they need to be fed all year round, regardless of whether they are useful in subsistence activities or not. Therefore, more people in Greenland choose to have snowmobiles rather than dogs and a sled, as the snow-mobile only needs gasoline when being used and does not require any care in summer or the snow-free seasons.

The socio-economic changes reported mostly concerned impacts on housing. In Qaanaaq, most houses are built on sedimentary deposits and are therefore prone to deformations due to permafrost thawing. Most built structures are at risk of becoming skewed. However, it was debated in the interviews whether the deformations and tilted structures are caused by the deepening of the active layer of the permafrost or by poor-quality construction work in the 1960s and 1970s. During this period, many houses were constructed on the principle "as cheap and fast as possible" and the houses' long-term endurance to cope with the annual active layer changes might not have been accounted for. Other minor housing issues, such as doors that do not close, small cracks in the walls, and tilting floors, were also reported. For these minor damages, participants usually decide to live with them, while in the case of bigger deformations and degrading ground, there is a need to engage in reconstruction. One participant explained how two different house owners were fixing the foundation themselves, using a jack to lift the foundation and adding soil and rocks to make it more stable.

Another topic that indirectly occurred but was not validated in the focus group discussions was an indirect socio-economic impact related to the prolonged ice-free season, which provides touristic cruise ships easier access to Qaanaaq. Participants explained that last summer seven large ships arrived at the town. The tourists went off the ship and walked around the town for a few hours. This sudden and short influx of tourists was critically perceived by the participants, as cruise ship tourists only spend a limited amount of time in town. Participants mentioned their preference for tourists who come and stay for a couple of nights in Qaanaaq and hence spend money on accommodation, food etc. Cruise tourism is one example of how climatic changes also bring about socio-economic changes.

Assessing adaptation actions

Most of the observed climate change impacts bring about a set of adaptation actions that enable the community to better cope with them. In this section, we map reported adaptation actions onto the previously discussed observed changes, which are here described as drivers of such responses. We also analyse the potential implications of these actions.

We start by noting that not all observed changes require or have resulted in changes of local practices. For example, so far, the changes in the behaviour of polar wolves, which are now observed being closer to town, have not resulted in any behavioural change. However, most of the climatic changes have led to new behaviours that can also be characterized as adaptation practices. Table 22.4 illustrates the climate-driven observed changes and the corresponding adaptation actions, or the effects on the community and the local culture. The expected effects of the adaptations are based on the authors' interpretation and analysis of the empirical data.

Generally, the changing weather circumstances have far-reaching repercussions for the hunters. As a result, hunters meet at least once a month to discuss their observations and how they interpret the weather conditions' implications for hunting and fishing. *Changes in the sea-ice* have resulted in a shorter season for hunters. An adaptation strategy shared by hunters

Table 22.4 Climate-driven changes, adaptation actions, and adaptation impact

Driver	*Adaptation action*	*Expected effect of the adaptation*
Changes in the extent of sea-ice surface.	Do an English course to diversify income possibilities.	Better knowledge of English, stronger engagement with the tourists.
Changes in wind-induced snow/ice erosion and loss.	Shorter dog sledge trips.	Less engagement in dog sledding and dog breeding more generally.
Changes in wind-induced snow/ice erosion and loss| Changes in the incidence of animal injuries.	Change hunting routes (on inland ice) because of bad snow cover	Less predictability of circumstances and certain paths becoming inaccessible in some seasons.
Changes in the thickness of sea-ice| Changes in wind-induced snow/ ice erosion and loss| Changes in temperature fluctuations| Changes in the mean temperature in a given season	Observe weather, discuss options among other hunters/people.	Enhanced information and knowledge exchange amongst Qaanaaq's hunters/fishermen.
Changes in wind strength or speed	Stay home.	Less engagement in traditional and subsistence practices, increasingly sedentary lifestyle.
Changes in the mean temperature in a given season| Changes in wind-induced snow/ice erosion and loss| Changes in temperature fluctuations	Work with tourism.	Growing tourism sector and income diversification of local population.
Changes in wind-induced snow/ice erosion and loss| Changes in the incidence of animal injuries| Changes in the extent of sea-ice surface| Changes in the thickness of sea-ice	Fish halibut.	Increased halibut fishing, shorter hunting trips, additional halibut supply to local factory.
Changes in the thickness of sea-ice	Ask for help.	Broad information, material, and knowledge exchange amongst community members in Qaanaaq's.
Changes in temperature fluctuations| Changes in wind strength or speed	Make savings (for period November–January).	Different household management and economic planning.

participating in the study has been to prepare to diversify. They have done so by signing up for an English course to be able to communicate with tourists. According to study participants, being able to speak English can help diversify income, as they can offer their services to tourists who want to go dog sledding. It is common that hotel-owners would contact local hunters when they have visiting guests who want to go dog-sledding. The English courses were provided by the Greenlandic public authorities that are aware of the economic impact from the shorter ice-season. However, dog sledding trips are also affected by climatic shifts, as *changes in wind-induced snow loss* can lead to changes in the incidence of animal injuries. The adaptation strategy is then that in those periods, the hunters go on shorter trips to avoid hurting dogs' paws.

Hunters reported *changes in the wind strength or speed*. The main adaptation strategy to cope with strong winds and storms is to stay at home to avoid life-threatening weather conditions. This strategy further links to the strategy of making savings for the period November-January, as less catches are expected during those months due to the challenging weather conditions.

Having a volatile income can be particularly difficult for those hunters who do not live in a household in which a member has a monthly income. Women are often occupied in wage labour and can secure a steady income for those months of the year where hunting and fishing opportunities are limited. This gives safety for the household. However, some men in the community on their own. These men are more dependent on private savings and receiving help from their closest relatives and friends in town.

In case of bad weather, when hunters cannot go out on long trips to hunt the big catches, it is common that they go on short trips to fish halibut. For several years, the prices of this fish have been favourable and many local hunters have provided for themselves and their families with halibut fishing.

When changes in the wind, snow, or sea-ice conditions are extraordinarily rough, another strategy locally pursued is to ask for help from the extended family or closest social network. The help can be to share food, to loan money, or to borrow reserve parts or extra material to make necessary repairs to equipment. In the local hunter association, KNAPK, local hunters also discuss and support each other and have joined efforts in their protest about limiting certain hunting quotas to show their voice to the Greenlandic Government.

Discussion

What enables resilience in the climatic and physical system?

According to our informants, changes in the sea-ice are the most important climate change impact altering existing livelihood practices, followed by the increase of periods of strong winds. Both factors were reported as hindering participants' engagement in their usual subsistence practices. Changes in sea-ice and their impact on Inuit communities have been widely documented in the literature. For instance, Nuttall (2020) discusses the changing sea-ice in northwest Greenland in relation to Indigenous rights and communities' efforts to engage in community-based monitoring of on-going change, rather than presenting local communities as victims passively accepting the limitation following the climatic and physical system changes.

Some of the popular adaptation actions include the involvement in other livelihood activities such as tourism and halibut fishing. Both the tourism business and halibut fishing activity have already made their way into the seasonal calendars of many of the vocational hunters (see, e.g., Tay, 2020). These adaptations build on collective knowledge of halibut stocks, cooperation with local hotels, and continuing training efforts. While these adaptations offer an alternative income source to previous activities, they are also associated with new risks and dependencies. For instance, both halibut sales and income from the tourism sector are dependent on broader market forces and make the local population of Qaanaaq more reliant on external factors. However, together, these adaptation actions enable wider community resilience.

Besides income diversification, other actions in the climatic and physical system build on risk avoidance to ensure resilience. Interviewees described how they would shorten or change their hunting or sledding routes to avoid the risk of injury to both humans and dogs. In case of strong gusts, some individuals would even stay in Qaanaaq. Therefore, the observed climatic changes seem to drive income and activity diversification on the one hand, while restricting the exercise of other activities, including long-standing Indigenous traditions, such as frequent week-long hunting trips. In the case of adaptation actions that avoid a certain activity, the value and need of conserving Indigenous knowledge become more prominent. This observation relates to scholarly debates on the decline of Indigenous knowledge in the context of environmental changes (e.g., Reyes-García et al., 2013) by highlighting that this loss can stem from safety-critical adaptation actions.

What enables resilience in the biological system?

Observed biological changes primarily relate to the local fauna. While new species, including cod and caplin, have made their way to Greenlandic waters, other native species have altered their migratory routes or disappeared. Also, behavioural changes have been reported, including the presence of polar wolves closer to human settlements.

In the light of these developments, participants described a shift towards the capture of other species both for hunting and fishing. Further alterations to the local fauna would, however, not only imply the potentially irreversible loss of ecological diversity, but also the need for further adaptation by the local population. Indigenous communities' breadth and depth of knowledge on the living environment offer a foundation for the shift towards other species to ensure food security. Yet, the role of new species and the extent to which existing knowledge is transferable to new activities, as well as the effect on communities' resilience needs to be addressed in future research.

What enables resilience in the socio-economic system?

Regarding the socio-economic implications of climate change, an increasingly need of adapting local housing infrastructures has become apparent. Adaptive capacity in this system can stem from individual affordances and measures (e.g., adding additional gravel to support the foundation of the house). However, our results also show that the ability to adapt to climate change in Qaanaaq is not solely reliant on an individual's inherent traits, capabilities, or affordances, but also on a collective process based on local and Indigenous knowledge. Moreover, many of the adaptation actions are enabled through local networks. For instance, in the example of the house-owner struggling with permafrost thaw, there were two friends helping to fill in gravel to adjust the foundation. Through collaboration, local knowledge is shared among community members and the techniques for dealing with deformation due to a thawing ground becomes collectively refined. Against the backdrop of global warming impacts, Indigenous knowledge should hence not only be seen as a well of knowledge that requires increased conservation efforts, but also as a tool to dynamically collect, exchange and renegotiate insights to better inform adaptive practices across the climatic, physical, biological, and socio-economic systems in order to build up community resilience.

Conclusion

Climate change has had a continuous and encompassing impact on everyday life practices of Qaanaaq's inhabitants, affecting and altering hunting and fishing activities, local ecosystems, and infrastructures. These changes might not only undermine local food security but also pose a potentially existential threat to a unique set of Indigenous knowledge and traditions which are closely intertwined with lived experiences in the Arctic environment. Results from this study stress the close connection between Indigenous knowledge and resilience to climate change impacts on climatic, physical, biological, and socio-economic resilience systems. We show that many of the adaptive *actions* are based on local and Indigenous knowledge and, together, these actions create what we identify as adaptive *capacity.* A high adaptive capacity enables adaptation actions to adjust to environmental changes. By discussing the findings in a resilience framework, this chapter also made clear that Indigenous knowledge can serve as a dynamic tool to further develop adaptive capacities and advance community resilience in Northern Greenland. This can further the development of a more holistic resilience approach that builds on Indigenous knowledge developed and practised by communities most affected by climate change.

Acknowledgement

The research leading to this study received support from the ERC Consolidator Grant (agreement No. 771056-LICCI-ERC-2017-COG) and contributed to the 'María de Maeztu Unit of Excellence' (CEX2019-000940-M), the Nunataryuk project, funded by the European Union's Horizon 2020 Research and Innovation Program under grant agreement No. 773421, Belmont Forum project SeMPER-Arctic (Swedish Research Council Grant Agreement 2019-02326). We would like to thank our reviewers for their insightful and valuable comments.

References

Andersen, A. O., & Flora, J. (2019). Puzzling pieces and situated urgencies of climate change and globalisation in the High Arctic: Three stories from Qaanaaq. In *Climate, Capitalism and Communities: An Anthropology of Environmental Overheating* (pp. 115–132). Pluto Press. Astrid Bredholt Stensrud, Thomas Hylland Eriksen.

Arctic Council (2016). In Carson, M. and Peterson, G. (Eds). *Arctic Resilience Report*. Stockholm Environment Institute and Stockholm Resilience Centre, Stockholm. 1–240.

Baker, I., Peterson, A., Brown, G., & McAlpine, C. (2012). Local government response to the impacts of climate change: An evaluation of local climate adaptation plans. *Landscape and Urban Planning*, *107*(2), 127–136.

Berkes, F., & Berkes, M. K. (2009). Ecological complexity, fuzzy logic, and holism in Indigenous knowledge. *Futures*, *41*(1), 6–12.

Biagini, B., Bierbaum, R., Stults, M., Dobardzic, S., & McNeeley, S. M. (2014). A typology of adaptation actions: A global look at climate adaptation actions financed through the Global Environment Facility. *Global Environmental Change*, *25*, 97–108.

Biesbroek, R., Dupuis, J., Jordan, A., Wellstead, A., Howlett, M., Cairney, P., ... & Davidson, D. (2015). Opening up the black box of adaptation decision-making. *Nature Climate Change*, *5*(6), 493–494.

Braun, V., & Clarke, V. (2006). Using thematic analysis in psychology. *Qualitative Research in Psychology*, 3, 77–101.

Engle, N. L. (2011). Adaptive capacity and its assessment. *Global Environmental Change*, *21*(2), 647–656.

Engle, N. L., de Bremond, A., Marlone, E. L., & Moss, R. H. (2014). Towards a resilience indicator framework for making climate-change adaptation decisions. *Mitigation and Adaptation Strategies for Global Change, 19*(8): 1295–1312.

Fedele, G., Donatti, C. I., Harvey, C. A., Hannah, L., & Hole, D. G. (2019). Transformative adaptation to climate change for sustainable social-ecological systems. *Environmental Science & Policy*, *101*, 116–125.

Hastrup, K. (2018). A history of climate change: Inughuit responses to changing ice conditions in North-West Greenland. *Climatic Change*, *151*(1), 67–78.

Holling, C. S. (1973). Resilience and stability of ecological systems. *Annual Review of Ecology and Systematics*, 4:1–23.

Nelson, D.R. (2011), Adaptation and resilience: responding to a changing climate. WIREs Clim Change, 2: 113–120. https://doi.org/10.1002/wcc.91.

Nuttall, M. (2020). Water, ice, and climate change in northwest Greenland. *Wiley Interdisciplinary Reviews: Water*, *7*(3), e1433.

Owen, G. (2020). What makes climate change adaptation effective? A systematic review of the literature. *Global Environmental Change*, *62*, 102071.

Pearce, T., Smit, B., Duerden, F., Ford, J. D., Goose, A., & Kataoyak, F. (2010). Inuit vulnerability and adaptive capacity to climate change in Ulukhaktok, Northwest Territories, Canada. *Polar Record*, *46*(2), 157–177.

Piggott-McKellar, A. E., McNamara, K. E., Nunn, P. D., & Watson, J. E. (2019). What are the barriers to successful community-based climate change adaptation? A review of grey literature. *Local Environment*, *24*(4), 374–390.

Poppel, B. (Ed.) (2015). *SLiCA: Arctic Living Conditions: Living Conditions and Quality of Life among Inuit, Saami and Indigenous Peoples of Chukotka and the Kola Peninsula*. Nordic Council of Ministers.

Rantanen, M., Karpechko, A. Y., Lipponen, A., Nordling, K., Hyvärinen, O., Ruosteenoja, K., ... & Laaksonen, A. (2022). The Arctic has warmed nearly four times faster than the globe since 1979. *Communications Earth & Environment*, *3*(1), 1–10.

Reyes-García, V., Álvarez-Fernandez, S., Benyei, P., García-del-Amo, D., Junqueira, A. B., Labeyrie, V., Li, X., Porcher, V., Porcuna-Ferrer, A., Schlingmann, A., Soleymani, R. (2023). Local indicators of climate change impacts described by Indigenous peoples and local communities: Study protocol. *PLoS One*. Submitted September 2022. 18(1): e0279847. https://doi.org/10.1371/journal.pone.0279847.

Reyes-García, V., Guèze, M., Luz, A. C., Paneque-Gálvez, J., Macía, M. J., Orta-Martínez, M., ... & Rubio-Campillo, X. (2013). Evidence of traditional knowledge loss among a contemporary Indigenous society. *Evolution and Human Behavior*, *34*(4), 249–257.

Sharma, S. (2010). Assessing diet and lifestyle in the Canadian Arctic Inuit and Inuvialuit to inform a nutrition and physical activity intervention programme. *Journal of Human Nutrition and Dietetics*, *23*, 5–17.

Statistics Greenland. (2022). Greenland in Figures 2022. ISBN 978-87-998113-7-3.

Sugiyama, S. (2020). Through the Japanese field research in Greenland: A changing natural environment and its impact on human society. *Polar Record, 56*, E8. doi:10.1017/S003224742000011X.

Tay, V. (2020). The possibilities and limitations of tourism development in Greenland impacting self-efficiency and socio-economic wellbeing of coastal communities. In *Arctic Marine Sustainability* (pp. 345–363). Springer, Cham. Eva Pongrácz, Victor Pavlov, Niko Hänninen.

United Nations Department of Economic and Social Affairs, Indigenous Peoples (n.d.). Climate change. Retrieved October 12 2022, from https://www.un.org/development/desa/indigenouspeoples/climate-change.html

23

HOW CAN CO-MANAGEMENT SUPPORT ADAPTATION TO CLIMATE CHANGE? THE CASE OF CO-EXISTING FISHERIES FROM PANGNIRTUNG, NUNAVUT, CANADA

Eranga K. Galappaththi

Introduction

Small-scale fisheries are crucial for Arctic Indigenous and local communities, where they sustain economies, food security, health, and cultures (Pearce *et al.*, 2015). However, in the Arctic, climate change is occurring four times faster than in the global south, largely impacting the livelihoods and lifeways of coastal communities (Divine *et al.*, 2021). Climate change impacts in this region are well documented. They include the loss of sea ice, melting glaciers and permafrost, and species shifts (Arctic Council, 2016). Responding to these climate change impacts requires dynamic relationships among diverse actors, based on mutually agreed-upon environmental management objectives. Among the range of management responses, co-management is identified as an applied governance approach for small-scale fisheries to better handle the complexities and uncertainties resulting from climate stressors (Butler *et al.*, 2016).

Co-management (or collaborative management) refers mainly to the "sharing of power and responsibility between the government and the local resource users" (Carlsson & Berkes, 2005: 66). Some of the important features of co-management are the sharing of authority, partnerships between local resource users and the government, decentralised decision-making, and linkages for governance (vertical or horizontal) (Galappaththi & Berkes, 2015; Plummer *et al.*, 2012). Adaptive co-management builds on co-management, adding to it the active, continued, and dynamic learning of complex systems' dynamics (Kristofferson & Berkes, 2005; Plummer & Baird, 2013). Olsson *et al.* (2004) identified adaptive co-management as a flexible community-based resource management process tailored to specific situations and places and supported by, and working with, various organisations at multiple scales. For instance, Galappaththi & Berkes (2015) examined how local aquaculture farmers partnered with government institutions to develop a collaborative crop calendar system to respond to shrimp disease challenges in northwestern Sri Lanka.

Scholars recognise (adaptive) co-management's ability to build resilience in a social-ecological system setting. Co-management is a collaborative process of social learning that seeks to

DOI: 10.4324/9781003356837-28

increase the coherence between social processes, management, and ecosystem dynamics, which is a prerequisite for climate adaptation by fostering sustainable resource use and building resilience within Indigenous fisheries systems (Armitage *et al.*, 2011; Galappaththi *et al.*, 2022). For example, Armitage *et al.* (2011) recorded the important role of co-management institutions in creating conditions for social learning and climate adaptation in rapidly changing Canadian Arctic environments. Further, Galappaththi *et al.* (2022) proposed resilience-based steps to practice adaptive co-management of Arctic char (*Salvelinus alpinus*) fisheries with Inuit of the Canadian Arctic. These steps are (1) dialogue through a discussion forum, (2) place-based participatory research, (3) resilience-building management actions, (4) collaborative monitoring, and (5) joint process evaluation.

This chapter examines how the Pangnirtung Inuit community of the Canadian Arctic co-manages co-existing fisheries in a climate change adaptation context. The three objectives of this chapter are (1) to examine the ways in which Inuit fishers experience climate change, (2) to investigate the co-management and co-management-associated responses that Inuit fishers use to adapt to such change, and (3) to assess the characteristics of Pangnirtung fisheries co-management that foster adaptation to climate change. As part of the third objective, this chapter will investigate the structural and functional characteristics of Pangnirtung co-management, which include some of the characteristics already featured in co-management literature [as documented by Armitage *et al.* (2011), Carlsson and Berkes (2005), Kristofferson and Berkes (2005), Olsson *et al.* (2004), Plummer and Baird (2013)]. The study reveals various means by which Inuit fisheries systems build resilience and minimise vulnerability (i.e., adapt) to climate change impacts (Galappaththi *et al.* 2019a, 2021).

Study area

Archaeological research indicates that Inuit have continuously occupied the Nunavut region for over 4,000 years. The history of Canadian Inuit goes beyond the Nunavut Land Claim Agreement in 1993 and the confederation of Nunavut as a separate territory in 1999 (Government of Nunavut, 2016; Priest & Usher, 2004). Inuktitut, the language of Inuit, is still alive and growing with various dialects. It is widely used throughout the region and Inuit from across the Arctic can understand one another. Hunting, trapping, and fishing comprise Inuit lifestyle. For instance, Inuit had been whaling bowhead whales in the Cumberland Sound region of Nunavut long before Europeans arrived in the 1800s. Each year they caught between eight and 12 whales to feed the population of about 1,000 Inuit in the community (Inuit Elder, personal communication, July 4, 2017). However, with the European invasion and their extractive quest (e.g., whales, fur), Inuit faced changes in their way of life. Additionally, Christianity has dramatically impacted Inuit spiritual beliefs and the Roman Catholic and Anglican churches took every opportunity to destroy Inuit beliefs (Laugrand and Oosten, 2010).

Pangnirtung (population 1,481 in 2016, 94% Inuit; Statistics Canada, 2017) is an isolated Inuit fishing community located in the Canadian territory of Nunavut (Figure 23.1). Pangnirtung Inuit have a long history. Pangnirtung is the present location of the community, although traditionally Inuit lived in various places in the Cumberland Sound region, named outpost camps. In 1962, many Inuit living in Cumberland Sound started moving to Pangnirtung partly due to a distemper epidemic that killed most of their sledge dogs. Later, the government forced most Inuit who lived in the outpost camps across the current Cumberland Sound region to move to the current Pangnirtung area. The last Inuit from the outpost camps moved to Pangnirtung in 2008. Even after Inuit moved to Pangnirtung, they continued hunting and fishing in the Cumberland Sound region as the maintained their historical connection to the land and waters. Pangnirtung Inuit no longer live in Igloos (since 1950s–1960s) (Galappaththi, 2020).

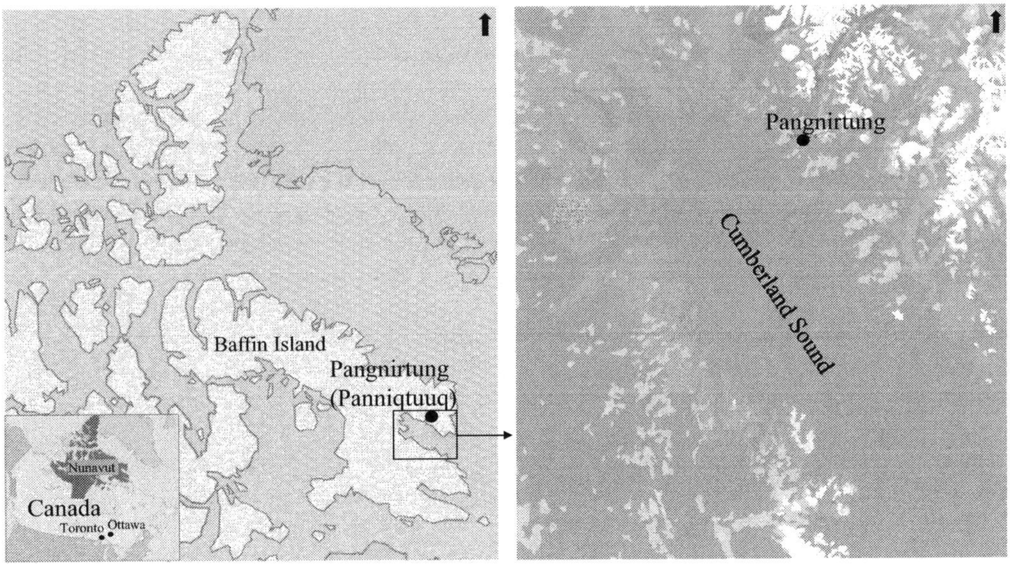

Figure 23.1 Location of Pangnirtung (the community) and the Cumberland Sound (water body) in Baffin Island, Canada. The red-coloured region in the map indicates the Nunavut territories in Canada. Pangnirtung Inuit use the surrounding lakes for winter Arctic char fishing for both exploratory and commercial purposes (CC BY-SA 4.0).

The community is accessible only by aircraft. Flying via the nearest commercial (Toronto) and administrative (Ottawa) cities is expensive partly because of the limited availability of flight services and infrastructure. Residents get supplies via boat only in the summer and delivery is highly reliant on the prevailing sea ice conditions. Residents must cope with unique challenges, including high rates of suicide, food insecurity, housing shortages, tuberculosis outbreaks, and low high school graduation rates, comparable to other small Nunavut settlements. Lack of economic opportunities is a common characteristic across the Nunavut region. Pangnirtung Inuit still have an intimate connection to the surrounding Cumberland Sound area for fishing and hunting and detailed knowledge of the species.

Fishing and hunting for local "country food" is a critical part of the Inuit culture and way of life (Galappaththi *et al.*, 2019b). The community has two grocery stores (Northern and store co-op) that provide alternative food sources, yet Inuit still consider country food to be their main food source as opposed to the expensive, less nutritious, and processed foods from the store. Caribou (*Rangifer tarandus*), seal (*Pusa hispida*), and Arctic char (*Salvelinus alpinus*) are the most important food sources for the community because of their year-round availability (Table 23.1). Several other seasonal resources, such as turbot (*Reinhardtius hippoglossoides*), ptarmigan (*Lagopus muta*), eiders (*Somateria mollissima*), polar bear (*Ursus maritimus*), kelp (*Laminaria solidungula* and *Saccharina latissimi*), Arctic hare (*Lepus arcticus*), clam (*Spisula solidissima*), and beluga (*Delphinapterus leucas*), are also important to health, culture, and well-being.

Finding country food is a laborious process affected by many factors. Short-term fishing and hunting are family activities. Long-term hunting trips are led mainly by male adults. Fishing and hunting are mostly individual or small-group activities. However, large operations like beluga hunting take place in large groups (e.g., 10–20 Inuit). Inuit fishers/hunters go out "on the land" and spend days outside the community, around the Cumberland Sound area. Yet, the availability

Table 23.1 Inuit seasonal food calendar

Month	*Season*	*Activities people do for a living*	*Locations where livelihood activities take place*
December January February	Winter	Fishing and hunting (e.g., turbot, arctic hare, ptarmigan, fox, wolf, kelp, seal, arctic char)	*Turbot*: Fishing takes place on frozen sea ice in the Cumberland Sound area.
March April May	Early spring	Fishing and hunting (e.g., turbot, arctic hare, ptarmigan, seal pup, seal, arctic char)	*Arctic char*: Fishing takes place on surrounding frozen lakes during the winter and in the Cumberland Sound
June	Spring	Fishing and hunting (e.g., goose, beluga, eggs, eider duck, walrus, clam, seal, arctic char)	and Pangnirtung fjord waters during the summer. *Marine animals*: Hunting takes
July August September	Summer	Tourism (selling arts and crafts to summer cruise ship visitors and people who visit nearby Auyuittuq National Park) Fishing and hunting (e.g., kelp, narwhal, berries, eider duck, clam, seal, arctic char)	place during the summer in the Cumberland Sound and Pangnirtung fjord waters. *Land animals*: Pangnirtung and the surrounding Cumberland Sound areas. *Tourism*: During the summer, cruise
October November	Fall	Fishing and hunting (e.g., kelp, narwhal, seal, arctic char)	ships will visit the community only if the ice clears up. Also, the community is the entrance for visitors of Auyuittuq National Park.

of country food is determined by many factors, including changing access to ice, land, and water, which nowadays is highly reliant on climate and environmental change (Ford *et al.*, 2019). For instance, the Pangnirtung area was a good caribou hunting ground before the caribou out-migrated to western Arctic lands, resulting in an increased reliance on the ocean for food security (Galappaththi *et al.*, 2019b). Thus, changing climate can have a significant impact on the Inuit diet.

Pangnirtung is one of the few communities in Nunavut territory that has significant commercial and subsistence fishing activities. The community started commercial Arctic char fisheries in the late 1980s and established a community fish processing plant in 1992, specifically for turbot fisheries. This commercial fishery expanded into winter and summer turbot fisheries. About 90% of turbot products are exported to eastern Asia, including South Korea, Japan, Taiwan, Vietnam, and China. These commercial fisheries co-exist with subsistence fisheries, as both Arctic char and turbot fisheries are key to Pangnirtung Inuit livelihood and food security (Galappaththi *et al.*, 2019b).

Arctic char is the most important fish species for Pangnirtung Inuit (Galappaththi *et al.*, 2022). It is an anadromous species, feeding in the sea and overwintering in lakes. This species migrates to the ocean in the summer to feed on prey from lipid-rich Arctic marine food webs. Its growth and survival are thus linked to marine ecosystem dynamics. Evidence has accumulated about climate-driven transformations in Arctic marine ecosystems and the possibility of rapid, unexpected cascading effects on fisheries (Arctic Council, 2016; Huntington *et al.*, 2020; Ulrich & Tallman, 2021; Yurkowski *et al.*, 2018). Arctic char is obtained from subsistence and commercial fisheries and is also considered a staple food supporting healthy diets, cultures, and revenue. During the summer, when the Pangnirtung fjord is clear of ice, fishers start boating into Cumberland Sound waters for char, using gill nets. During the winter and spring, after the formation of strong sea ice, fishers travel via snowmobile to surrounding lakes to fish Arctic char using a short stick and a line with bait (referred to as "*jigging*"). Inuit use Arctic char as bait for turbot longlines. Pangnirtung has an Arctic char commercial fishery, available mainly during the summer, that involves issuing char licences using a lottery system (i.e., randomly allocated among licence applicants on an annual basis) coordinated by the local Hunters and Trappers Association.

Commercial turbot (also known as Greenland halibut) fishing is popular among Pangnirtung Inuit because of its cash-earning potential. Each year, it brings in over CAD$1 million (or US$780,000) worth of benefits in terms of fish processing jobs and fish sales to the plant. During the spring and winter turbot fishing seasons, fishers drive snowmobiles for about 100 km on the frozen ocean and spend several days on the sea ice in the Cumberland Sound area. This requires travelling to the Cumberland Sound sea ice, for which strong and thick sea ice conditions are essential for this fishery. Inuit bring their shacks to fishing spots and travel with the Qamutik, which allows them to transport the turbot harvest and other fishing gear/equipment such as longlines, flashlights, and hauliers. Turbot fishing takes place on the frozen ocean among a few (one to four) turbot ice holes (fishing spots). This is a high-risk fishing operation due to continuous darkness, extreme cold (<−40°C with wind chill), and the fact that the Greenland shark is a potential by-catch for turbot longlines (Galappaththi *et al.*, 2019b).

Pangnirtung has a system to co-manage Arctic char and turbot co-existing fisheries (Galappaththi *et al.*, 2019b, 2021). As outlined in the Nunavut Agreement Article 5, fisheries co-managers are both the Hunters and Trappers Association (HTA) and the Department of Fisheries and Oceans (DFO), along with the Nunavut Wildlife Management Board (NWMB) and other designated Inuit organisations (Figure 23.2). The aim of the Hunters and Trappers Association is to co-manage fisheries with the DFO and the NWMB, which includes selecting licence holders for the char commercial fishery. The HTA belongs to Inuit of Pangnirtung and decision-making is carried out by a board of directors. The DFO is the federal government representative and contributes to co-management through community consultations. Some of the roles include issuing fisheries licences and closure notices, monitoring quotas, and monitoring compliance concerns. The NWMB is the representative of the Nunavut territorial government and closely monitors resource use; it also plays an extension role. The NWMB is the main institution for wildlife management, but the DFO has the power to reject or amend the NWMB's decisions if a clear conservation concern exists. The HTAs can also implement management measures in response to local concerns and issues, such as momentary fishery closures regarding both time and space (DFO, 2021).

Fish quotas are based on the fiscal year and the HTA decides when the water bodies are fished (summer versus winter) (Galappaththi *et al.*, 2021). The Pangnirtung HTA has been the licence holder for the exploratory char fisheries and designates a quota for its membership. Also, the Pangnirtung HTA has been the licence holder for the Cumberland Sound turbot fishery and advises the DFO on which members shall be added to the licence. The HTA should advise the fish plant and the DFO on the fishers who will be fishing under the issued licences. As part of the fisheries co-management process, feedback about fisheries activities is transmitted to the DFO, HTA, and NWMB. The fish plant is the community's largest employer in terms of fishery activities. It facilitates the co-existence of Arctic char and turbot commercial fisheries and it purchases fish from Inuit fishers, then processes and ships the fish to local and international markets. The fish plant is a subsidiary of Nunavut Development Corporation. This private institution is managed by an Inuit board of directors.

The co-management process is strengthened by the Regional Wildlife Organization (RWO) and Nunavut Tunngavik Incorporated (NTI). For instance, the RWO oversees the HTA's harvesting practices and represents "Inuit rights," while the NTI advocates and makes decisions as an Inuit stakeholder to protect "Inuit rights." Both the RWO and NTI are linked to this co-management process, building on the provisions of Article 5 of the Nunavut Land Claim Agreement. Additionally,

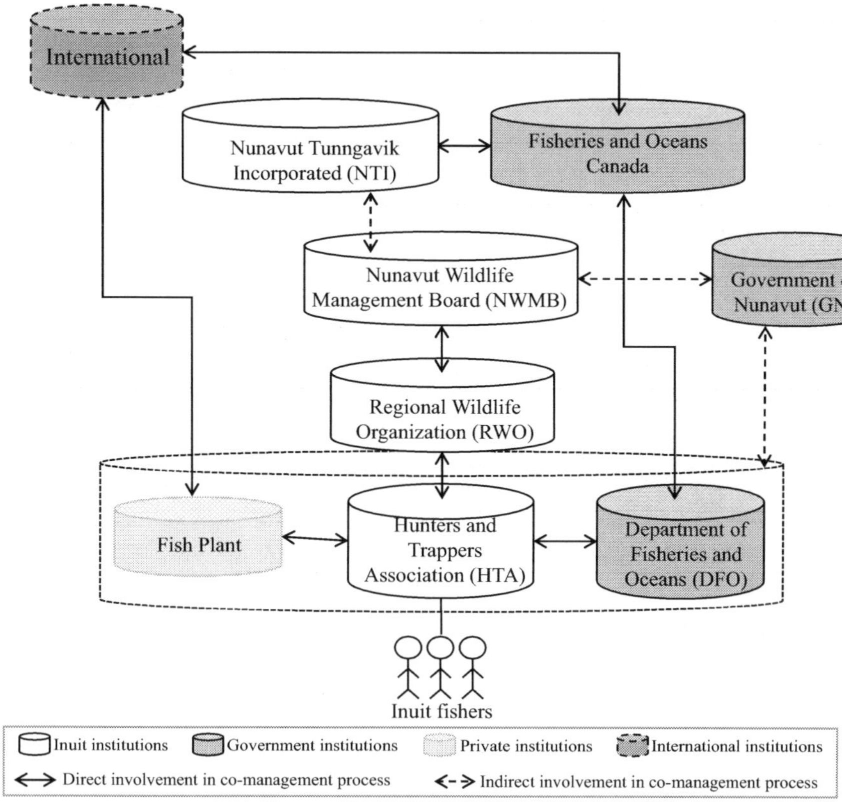

Figure 23.2 Co-management structure for Pangnirtung Arctic char and turbot fisheries (Adapted from Galappaththi *et al.*, 2021).

the on-site fish plant plays a facilitation role that includes buying fish and providing seasonal job opportunities in processing and shipping, as well as contributing to community events and supporting the community soup kitchen (Galappaththi *et al.*, 2019b).

Data collection methods

To examine the ways in which Inuit fishers in Pangnirtung experience and respond to climate change, a mixed-method approach was used. The approach included participant observation, semi-structured interviews, focus group discussions, key informant interviews, and surveys. From May 2016 to November 2022, five field visits were made to the community, and over 16 weeks of fieldwork were conducted through participation and observation of Inuit fisheries' way of life to obtain contextual knowledge (Collings, 2009). Participant observation included an time spent with Inuit fishers in the form of attending community events, drinking tea and chatting with elders, attending meetings, visiting local institutions, and making six fishing trips to the Cumberland Sound to experience winter turbot fishing and summer Arctic char fishing. The researcher maintained a field diary to record participant observation.

Semi-structured interviews ($n = 62$) were conducted with Inuit fishers to document the changes being observed in the region and to identify and characterise ways of responding to climate change (Longhurst, 2016). An interview guide featuring key themes (e.g., change, place, human agency, collective action, institutions' knowledge systems, learning) was used to maintain the focus. A snowball sampling technique was used to choose participants. Interviews were conducted, audio-recorded, and transcribed in the community of Pangnirtung from May 2017 through April 2018. The semi-structured interview questioning focused on "environmental change" in general (rather than climate change) to avoid biassing answers and to keep interviews open-ended, thus covering the issues and changes that Inuit viewed as most important. Interview questions relating to "change" and "co-management" referred to "about 30 years back" in time, aiming for the social-ecological memory of fishers in the geographical area of Pangnirtung and the surrounding Cumberland Sound area (Rodríguez et al., 2019). Regardless of age, Inuit possess a cumulative body of Indigenous and local knowledge, practice, and belief, evolving through the adaptive process and handed down through generations by cultural transmission. See Table 23.2 for sample profiles.

Two focus group discussions were carried out to validate data and build thematic areas related to changes that fishers experience and the key ways in which co-management responds to such changes (Carey & Asbury, 2016). Participants were invited using a community radio announcement. Inuit fisher groups of four to eight individuals participated in the focus group discussion (FGD) organised during the latter stage of the data collection process. The first FGD was conducted in spring 2018 with eight participants. Six of them were men (three elders and three middle age). Two of them were women from the same extended family (one elder and one middle age). The second FGD was conducted remotely in spring 2021. Two Inuit research assistants led this discussion. Five participants were selected based on purposive sampling. The five participants consisted of three Inuit women and two men. They were active in both Arctic char and turbot fisheries and represented the age ranges of 30–40 ($n = 1$) and 40–50 ($n = 4$).

Thirty-eight key informant interviews were conducted (Taylor & Blake, 2015). Key informants included individuals who possess specific skills, knowledge, experience, and specialised background on the research project and project participants. Their assistance enabled me to obtain

Table 23.2 Sample profiles of semi-structured interview respondents ($n = 62$)

Variable	*Number of respondents (%)*
Gender	
Female	18 (29)
Male	44 (71)
Age*	
<20	1 (2)
20–39	24 (39)
40–59	22 (35)
60–79	14 (23)
80–99	1 (2)
Occupation	
Fishing	62 (100)
Hunting/trapping	36 (58)
Art-related work (e.g., carving, painting, crafts)	27 (44)
Tourism-related work (e.g., translating, outfitting services)	7 (11)
Income support (government income assistance program)	18 (29)
Work for other entities (e.g., construction, fish plant, northern store, daycare, hamlet office)	13 (21)

* Regardless of age, Inuit possess a cumulative body of Indigenous and local knowledge, practice, and belief, evolving through the adaptive process and handed down through generations by cultural transmission.

specific information that is difficult to access through other methods. The study interviewed representatives of co-managers of the Pangnirtung co-existing fisheries, including officers from the DFO, NWMB, HTA, GN, fish plant, and RWO. The nature of these interviews was very similar to that of the semi-structured interviews. This method was used mainly to gather information related to the co-management institution, and particularly its process and characteristics.

Forty-five surveys were conducted (Dalati *et al.*, 2018) among individuals related to Inuit fisheries to further examine how Inuit experience change, as well as their perception of fisheries co-management as an adaptive response. Surveys were conducted at the latter stage (2020–2022) following LICCI project guidelines to gather individual (n = 45) and household (n = 40) level climate change impacts and adaptation data (Reyes-García *et al.*, 2023). These surveys helped validate and create an understanding of the connection among data gathered using other methods.

Data gathered from the semi-structured interviews, key informant interviews, and field diary were analysed using textual content analysis (Krippendorff, 2018). With the help of Inuit translators, recorded data were translated into English and then transcribed for analysis. Key techniques were manifest and latent content analysis to develop common themes and patterns related to the ways in which fishers experience and respond to climate change. The theme-building process was monitored and members were checked with community participants to ensure that the results were relevant and meaningful for Inuit. Survey data were entered into MS Excel data entry formats, following the LICCI protocols (Reyes-García *et al.*, 2023). Additionally, direct quotes from the transcriptions and cropped versions of the field photos were used as evidence to support themes and descriptive results related to community responses to climate change impacts.

Results

How do Inuit experience climate change?

The study identified five areas in which climate change is experienced, according to the perspectives documented from Inuit of Pangnirtung (Table 23.3). The five areas are changes related to (1) sea ice conditions, (2) landscape, (3) seascape, (4) fish, and (5) weather conditions. In terms of the direction of change based on the Inuit perspective (derived from FGD), these five changes show an increasing trend. Based on Inuit observations, the study also identified various local indicators for climate

Table 23.3 How do Inuit fishers experience climate change?

Change	*Local indicators of climate change impacts*	*Quotes from Pangnirtung Inuit*
Sea ice	Sea ice melting speed Changes in ice melting Ice breaking time Thickness of sea ice	• *Fishing season gets shorter each year. Ice breaks up faster now. Last year ice was weak … once we boat in December … so strange … ice doesnt break at the right time.* • *Back then [when I was a kid], snow melts slower than now. Snow melts faster now.* • *Ice crack early. Once, I lost my shack and fishing gear because ice broke earlier than expected.* • *Ice doesn't break at the right time … now ice forms slowly and breaks up early … We used to have more room to fish.* • *In the 1990s, sea ice was 5-6' thinner but now [2021] it's 3-4' or thicker.*

(*Continued*)

Table 23.3 (Continued)

Change	*Local indicators of climate change impacts*	*Quotes from Pangnirtung Inuit*
Landscape	Thawing or melting of permafrost Productivity of wild plant species Changes in river water levels Terrestrial animal species migration areas and routes Speed of melting glaciers	• *...We live nearby the river and mountains up there ... our view is moving, and I think our land is moving ... our house is not level anymore.* • *Our old bridge had to be replaced due to permafrost collapse ... It was there since I was a baby.* • *River has drastically changed as well, as land has drastically changed. Our river has moved a lot in the last couple of years ... the river had erosion.* • *Lately, not many blue and blackberries now. Not even a couple of cups. I used to harvest buckets of berries when I was a teenager.* • *During the spring, we see more water than before from melting glaciers. After they melt, we see more water running all over the place.* • *This community was known for its name 'Pangniqtuuq,' which is the name for bull caribou; it was known to have plenty of caribou. Then apparently, the herds migrated to the western part of the Canadian Arctic.*
Seascape	Changes in pollution levels in marine water bodies Direction of ocean currents	• *I found some plastic bags in the sea while fishing ... it can damage my motor... wind can bring plastic anywhere.* • *Ice moving to unusual areas of the Cumberland Sound ... we are not used to that.*
Fish	Quality of marine food sources Abundance or occurrence of marine species stated as invasive Changes in marine food webs	• *Arctic char meat is white now ... its not red anymore, not sure why ... most of them are smaller than back then* • *The[re] were no Capelin back then, it showed up lately, now there are many ... grandmother said that the reason for Arctic char flesh turning white.* • *Seal meat taste, texture, and color are different now, not like back then.*
Weather	Amount of snowfall	• *Less snow now ... I used to go to Iqaluit every year in April through the mountains ... Now when we Skidoo, we hit rocks because of less snow in April.* • *In January, people from other communities came here, and they wore parkas ... they are saying it is warm here in Pang now.* • *Spring is getting warmer and warmer.* • *We get more Greenland sharks [Somniosus microcephalus] for turbot [Reinhardtius hippoglossoides] long lines ... When the water gets warmer, you get more Greenland sharks as a bycatch.* • *Water seems hotter these days.*

Adapted from Galappaththi *et al.* (2019b).

Note: Exact quotes are not included to protect confidential information, as requested by Inuit participants.

change impacts. Many of these changes refer to visible changes in the biophysical environment (e.g., sea ice, fish texture, and colour). These changes are interrelated but partly distinct. Inuit experience multiple stressors related to climate change simultaneously, resulting in integrative vulnerability to climate change. For example, mobility for hunting and fishing activities is highly reliant on sea ice conditions. During the summer, floating icebergs restrict boating to the Cumberland Sound for beluga and seal hunting. During the winter and spring seasons, the nature of sea ice (e.g., thickness, strength, texture) determines the safety of Inuit who travel using snowmobiles. Warm winds and water currents from the south can change sea ice conditions within a very short time, which leads to major safety and food security issues for the community. Climate impacts combined with other non-climatic changes (e.g., lack of economic opportunities, COVID pandemic, markets and fish price changes in the Canadian Arctic and beyond, and high suicide rates) affect Inuit in mixed ways.

Co-management-related responses to climate change impacts

The study found co-management is a primary response to the systemic climate change impacts identified in Pangnirtung fisheries. The aim of the Pangnirtung co-management is the sustainable use of fisheries to ensure Inuit food security and well-being in the rapidly changing Arctic conditions, while developing government and community partnerships. The co-management response is associated with other identified responses such as country food sharing within and beyond extended family members, use of technology, use of different knowledge systems for decision-making, in-house fish plant establishment and co-existing fisheries, and food system transformations (land- to sea-based) (Table 23.4 and Figure 23.3).

Table 23.4 Co-management-related responses to climate change

Responses to climate change	*Description*
Country food sharing	Inuit organised food-sharing events (the HTA has some government subsidies; when people bring seal and char, they purchase it from hunters, and then the HTA makes it available to the community). Also, unorganised food sharing (going on the radio and saying, "*Look, I got a seal; come on over and help yourself*") minimises the uneven distribution of food (including fish) among the community. Such food sharing minimises the vulnerabilities related to food insecurity and improves social cohesion. (Plate C, Figure 3)
Use of technology	Fishers have adopted technology such as global positioning systems (GPSs), VHF radios, and advanced rifles for fishing/hunting activities. Yet some prefer not to rely on technology, as they have limited access to service/repairs due to the community's isolated nature. (Plate B, Figure 3)
Use of different kinds of knowledge	Pangnirtung Inuit possess various kinds of knowledge that have accumulated and evolved over the generations, and that is shared among friends and peer groups. This includes place-specific knowledge of Arctic char, turbot fishing techniques, local environmental knowledge, and on-site fish processing and marketing knowledge. This body of knowledge is used for joint decision-making in co-management. "*We [HTA, DFO, and NWMB] use the best available scientific and local knowledge … Community consultations are one way of incorporating Indigenous and local knowledge into co-management decision-making process…*" – DFO officer (Plates A & E, Figure 3)

(*Continued*)

Table 23.4 (Continued)

Responses to climate change	*Description*
In-house fish plant establishment and co-existing fisheries	The fish plant is the primary reason for the co-existing commercial fisheries. It brings significant financial benefits to the community (over 1 million Canadian dollars per year). The fish plant is also involved in operational aspects of co-management, including record keeping of yield data, resource facilitation, and coordination. (Plate D, Figure 3)
Food system transformations (land- to sea-based)	Pangnirtung Inuit food systems are shifting from land-based (caribou) to sea-based food systems for food security. With the outmigration of caribou, Inuit are relying more on the ocean for country food (Arctic char, seal, turbot, beluga). This puts more pressure on fisheries co-management aimed at the community's food security. "*We [Inuit] are heavily relying on our fish for food … now more than ever after the caribou out-migration.*" – Inuit elder

Figure 23.3 Photo evidence of co-management-related responses to climate change. Photo plate-A illustrates Inuit women involved in on-site turbot fish processing. Turbot fishing is practised mostly as a family (or group) activity and Inuit use their knowledge about on-site turbot fish processing. Also, Inuit use knowledge about survival skills in this operation, for instance, building a snow wall to create a wind barrier. Photo plate-B illustrates the use of advanced technology in fishing and hunting practices to improve productivity. For example, on this hunting trip, Inuit used advanced rifles for seal hunting and very-high-frequency radio for more efficient communication. The Government of Nunavut has specific programs to support fishers and hunters in purchasing these kinds of equipment to improve harvest productivity. Photo plate-C illustrates a summer country food-sharing activity. Most organised country food-sharing activities are supported by co-management institutions such as the Government of Nunavut, fish plant, or Hunters and Trappers Association. Photo plate-D illustrates the interior of the fish plant. The fish plant brings significant amounts of financial resources and seasonal job opportunities to the community. Photo plate-E illustrates a community consultation event related to fisheries co-management, which fosters efforts to integrate Inuit and western knowledge systems for decision-making.

Co-management intersects and brings together various responses, such as the fish plant, the use of different kinds of knowledge, country food sharing, and the use of technology for fishing and hunting. Such responses collectively respond to systems-level challenges created and complicated by climate change. For example, according to semi-structured interview data, in 2008, Inuit first spotted capelin on the Pangnirtung fjord shore. Capelin is a warm water invasive species that is moving towards global north waters due to global warming. The same interview data support changes related to the main country food, for example, the changing flesh colour, taste, and texture of Arctic char and seal. Some elders believe that the emergence of capelin has caused such changes (see the section on 'Abundance or occurrence of marine species stated as invasive' in Table 23.3). This has food security concerns because most elders do not like changes in country food. For instance, elders do not like to eat whiter and softer Arctic char. This condition is further aggravated by the limited access to caribou due to out-migration. Co-management supports and ensures economic opportunities for Inuit through commercial fisheries. This allows Inuit to supplement their food sources, for instance, ordering country food from neighbouring communities or buying food from the two stores in the community.

Characteristics of co-management foster climate adaptation

The study found five structural characteristics of co-management in the Pangnirtung fisheries system that foster climate adaptation: (1) partnerships, (2) mixed regimes, (3) vertical and horizontal linkages, (4) sharing of responsibility, authority, and power, and (5) on-site fish processing plant (Figure 23.2; Table 23.5).

First, the DFO, HTA, and NWMB directly co-manage Arctic char and turbot fisheries, while NTI, GN (Government of Nunavut), and RWO are partners in the decision-making process. Informally, an Inuit-owned private-entity fish plant has a significant influence on the co-management process.

Second, mixed regimes represent government (e.g., DFO), private (e.g., fish plant), and communal (e.g., HTA) actors. These partnerships and mixed regimes will foster adaptation by increasing the range/richness of information available for effective decision-making in co-management. Diverse stakeholder interests will improve the flexibility of the fisheries co-management process to adjust to changing conditions.

Third, both vertical and horizontal linkages are active within the mixed regime. For example, the federal government (i.e., DFO) and community organisations (i.e., HTA), with the support of private-sector industry organisations (i.e., fish plant), horizontally connect for fisheries management while provincial government entities (i.e., GN, NWMB, RWO) vertically connect to support decision-making. This vertical and horizontal connectivity will improve the feedback-spreading process that improves the productivity of the fisheries management system. For example, co-management meetings allow all stakeholders to discuss lessons learned from the previous fishing seasons to incorporate into future fisheries management plans (e.g., weather, fish population dynamics, and safety issues).

Fourth, the sharing of responsibility, authority, and power with community organisations is another characteristic that might foster climate adaptation. For example, the HTA is the co-management licence holder for Arctic char and turbot fishing. It uses a lottery system to issue licences for commercial char fishing. This kind of power-sharing improves the sense of belongingness (or place attachments) within the fisheries management process. Also, the active inclusion of Inuit (as knowledge holders) will improve the effectiveness of decision-making and, thereby, the ability to adapt to community-level changes.

Table 23.5 Quotes to support co-management characteristics that foster adaptation to climate change

Characteristics	*Quotes*	*Source*
Structural		
Partnerships Mixed regimes	*Both DFO and the HTA along with the Nunavut Wildlife Management Board and other Designated Inuit Organizations are co-managers of the fisheries in Nunavut as outline in the Nunavut Agreement Article 5*	DFO officer
Vertical and horizontal linkages	*...federal, provincial and Inuit organizations are connected and contribute at different levels to make the final decision in co-management... [] all bring different expertise to the table...*	GN officer
Sharing of responsibility, authority, and power	*...the HTA is a co-managers of the fisheries in Nunavut as outlined in the Nunavut Agreement Article 5. With respect to Pangnirtung in particular the HTA has been the license holder for the Exploratory Char Fisheries and designates a quota to its membership. Also, the Pangnirtung HTA has been the license holder for the Cumberland Sound Turbot Fishery and advises DFO on what members shall be added onto the license.*	DFO officer
On-site fish processing plant	*The fish plant is the main reason for the Arctic char and turbot commercial fisheries, and purchases fish from locals [Inuit fishers], then processes and ships the fish to local [Canada] and international markets.*	Fish plant officer
Functional		
Community consultations and knowledge integration for joint decision-making	*...inclusion of traditional knowledge and community perspectives are key in co-management... community consultations help us [co-managers] to combine best available scientific advice and Traditional ecological knowledge for decision-making...*	DFO officer
Continuous monitoring of fishery resources	*...real-time monitoring of land and wildlife is a traditional practice of our [Inuit] way of life DFO and NWMB also do their own way of monitoring [e.g., stock assessments] ... we [co-managers] are strong together...*	HTA officer
Evaluations of the co-management process	*...annual meetings are held before and after each fishing season to assess how the fishery went regarding sustainability goals and to ensure that local priorities are met. This [meetings] reassess each fishing season and incorporate lessons for the following season...*	GN officer
Learning and knowledge production	*...this [co-management] is a long journey ... sometimes it works ... sometimes it doesn't ... sometimes we [co-managers] can easily agree ... sometimes disagree ... but we all can agree to disagree... It [co-management] is a co-learning process...*	HTA officer

Note: Exact quotes are not included to protect confidential information, as requested by Inuit participants.

Fifth, the fish processing plant's contribution to the co-management of Pangnirtung's co-existing fisheries is a unique structural characteristic that might foster climate adaptation. Arctic char and turbot commercial fisheries are highly reliant on the fish plant for research, industry development, fish purchasing, processing, and marketing and selling activities. For instance, the fish plant is diversifying its fishery portfolio, thus reducing vulnerability to market fluctuations, as the fish plant is involved in both Arctic char and turbot commercial fisheries. The fish plant has recently been able to pay more attention to turbot, as its Arctic char markets have been shrinking, due in part to competitive aquaculture char production from the global south. Furthermore, the fish plant is piloting new fisheries projects (e.g., summer turbot fishing). The fish plant maintains direct administrative connections with the community HTA and DFO for commercial fish quota allocations (e.g., coordination support of licences issued by the HTA) and data management (e.g., harvest data collection and record-keeping).

The study also found four functional characteristics of co-management that foster climate adaptation in the Pangnirtung fisheries system (Table 23.5). They are (1) community consultations and knowledge integration for joint decision-making; (2) continuous monitoring of fishery resources; (3) evaluations of the co-management process; and (4) learning and tacit knowledge.

Community consultations and knowledge integration for joint decision-making are important features of Pangnirtung fisheries co-management. Community consultations are an important part of co-management processes; quotas are based on a combination of the best available science advice and traditional knowledge and must be approved by the NWMB and DFO. For example, Indigenous knowledge is used with scientific knowledge for more inclusive fisheries decision-making and in the development of scientific research and fishery management plans. Although Indigenous knowledge holders are regularly consulted, the DFO's stock assessments are based mostly on scientific methods, such as morphometric measurements and modelling. The most recent published stock assessment for Pangnirtung was in 2015. However, knowledge plurality in management is rapidly evolving and the Government of Nunavut is working to ensure that Inuit Qaujimajatuqangit directly feeds into research and management. In Inuit Nunangat, Inuit knowledge is called Inuit Qaujimajatuqangit and is a *way of knowing* transcending knowledge which also encompasses customs and values (Pedersen *et al.*, 2020). The commercial fisheries are operated by Inuit fishers in partnership with the local fish processing plant and both fishers and managers are involved in consultation processes.

Continuous monitoring of fishery resources is conducted collaboratively through tight partnerships among managers, scientists, fishers, and fish plants. The DFO conducts annual programs to monitor fishery data and plans to ultimately measure catch-per-unit-effort (CPUE), although CPUE data is currently limited. In Pangnirtung, both government scientists and Inuit are involved in ecosystem monitoring activities. The DFO carries out formal stock assessments, while the HTA, NWMB, fish plant, and DFO collectively manage the processes of harvesting data and quota monitoring related to Arctic char commercial fisheries. Monitoring is moving towards more community-based, transdisciplinary approaches based on Inuit Qaujimajatuqangit principles, which could provide, among other things, temporal information about changes in fish and their social-ecological implications with a scale of detail that science-based DFO stock assessments do not capture. Indeed, real-time ecosystem monitoring is a traditional practice of Inuit through regular observations of the land and wildlife (e.g., char habitats, fishing spots, weather and ice changes, Arctic char taste/colour/size, meat texture).

Co-management process evaluation is implemented to learn how to improve fishery sustainability. The evolution of the co-management approach over the past decades towards better responses to fishery issues illustrates this shared willingness of co-managers and stakeholders to collaboratively learn and improve. Before and after each fishing season, multi-stakeholder meetings are held to assess how the fishery aligns with short- and long-term sustainability objectives and to ensure that local priorities are considered. At the decision-making level, stakeholder meetings reassess each fishing season and incorporate lessons for the following season. Regular decision-making meetings occur among government entities (DFO, NWMB, Government of Nunavut) and Indigenous knowledge holders. Frequent process evaluation is one of the key places where the "adaptive" part of management occurs: when objectives are iteratively updated or strategies to achieve objectives are tuned according to the latest observations.

Learning and tacit knowledge are characteristics of the Pangnirtung fisheries system. Inuit learn from various activities in daily life and update their knowledge. They rely highly on such knowledge systems during fishing, including survival skills on the ice, ecological knowledge of important fish species, and tides and water currents. At the co-management level, this is practised in many ways. For instance, considering the size of fish populations and migratory patterns, the fish quota will be reviewed annually based on the best available science and Indigenous and local knowledge. Community fishers are part of the fish population monitoring program. Continuous application of learning at individual and collective levels will allow for learning-by-doing, reassessing present knowledge, and constantly co-producing new knowledge for use in coping with and adapting to new conditions.

Discussion and conclusion

This chapter assesses how Inuit experience climate change and respond to its impacts using co-management in the Arctic fisheries context. The Arctic in general, and the Baffin Island region in particular, is experiencing rapid climate change and associated impacts, including changing access to hunting and fishing lands due to unexpected variations in sea ice conditions (Ford *et al.*, 2019; Jungsberg & Wendt-Lucas, 2024). Understanding climate change impacts and their implications is important in planning adaptation interventions using existing community responses to climate change (Termeer *et al.*, 2016).

Many of the climate change impacts are characterised by clearly noticeable changes in Arctic biophysical systems, but generate complexity and uncertainty in Indigenous fisheries. Climate change-related impacts such as sea-ice conditions, landscape, seascape, fish, and weather conditions are documented for the Arctic region (AMAP, 2018; Arctic Council, 2016; Giles *et al.*, 2013; Idrobo & Berkes, 2012). However, how people experience these impacts at the social-ecological systems level is understudied (see Galappaththi *et al.* 2019b for exception). Indigenous fisheries systems are undergoing multiple stressors simultaneously (i.e., integrative vulnerability) (Debortoli *et al.*, 2019). This is also documented for tropical fisheries systems of the global south that are undergoing climate change impacts (Galappaththi *et al.*, 2021). To further understand the complexities of integrative vulnerabilities, locally relevant indicators of climate change impacts are important (e.g., sea ice melting speed, changes in river water levels, and the amount of snowfall).

In terms of responses to climate change, co-management is an important response for Pangnirtung Inuit because of their high reliance on fisheries for food security, nutrition, and well-being. However, co-management is not the only response to climate change impacts, because various other community responses are simultaneously happening, such as the culture of country food sharing, the use of technology, and the use of different kinds of knowledge (Galappaththi *et al.*, 2019b). The chapter offers evidence for co-managing co-existing fisheries to generate broader insights about sustainable ways of managing Indigenous fisheries towards climate change adaptation. Both the federal government's Department of Fisheries and Oceans and the community Hunters and Trappers Association, along with the Nunavut Wildlife Management Board and other designated Inuit organisations, such as Nunavut Tunngavik Incorporated, are co-managers of the Pangnirtung fishery. Building on the case of Pangnirtung, the study argues that co-management is a viable climate adaptation strategy because of its ability to support adaptation to climate change in Indigenous fisheries.

In the case of Pangnirtung, co-management supports climate adaptation in three ways. First, co-management supports community's food security and livelihood. Fishing and hunting for "country food" is an essential part of the Inuit way of life (Pearce *et al.*, 2015). Seal, Arctic char, and caribou are key country food sources throughout the year, though Pangnirtung caribou migration to western lands and changes in fish quality and abundance have resulted in a higher reliance on the ocean for Inuit food security. In this setting, co-managing Arctic char and turbot fisheries supports Inuit livelihoods and food security in many ways. For example, a wide range of food, income, and market options can improve the adaptive capacity of the fisheries system mainly through (a) year-round distributed income-generating activities that allow Inuit to afford alternative food sources (purchased from the store), (b) access to a wide range of country food that will minimise vulnerability in terms of health issues and food insecurity, and (c) multiple markets that will improve the resilience of the local fishing industry in terms of adapting to changes in global trade.

Second, under favourable conditions, co-management can foster social learning. Co-management facilitates co-learning at various levels as a part of the collaborative management process. Learning related to climate change at the Inuit fishers' level could be turned into social learning under conditions such as the presence of community-based institutions, collective action, and partnerships (April, 2008). Social learning is an important aspect of adaptation to climate change and reducing vulnerability (Collins & Ison, 2009). When fisheries management is characterised by the use of collective action and collaboration through local institutions (Karlsson & Hovelsrud, 2015), it could also unlock social learning into social action that leads to problem-solving and builds adaptive capacity.

Third, co-management can foster knowledge co-production under favourable conditions. Given the vibrant Inuit Qaujimajatuqangit available regarding fishery social-ecological systems (e.g., GN, 2016; Idrobo & Berkes, 2012; Keith *et al.*, 2021), additional efforts to interweave knowledge through co-management processes could help promote a better understanding of Arctic change, allowing fisheries co-managers to assess stocks and, more broadly, examine how stocks are and will be affected by climate-related changes in social-ecological systems. The use of diverse kinds of knowledge systems for fisheries decision-making is an important characteristic that builds the resilience of co-management (Armitage *et al.*, 2011). Co-managers use the best available scientific and Indigenous knowledge for collective decision-making, for example, determining fish quotas and fishing lakes (Armitage *et al.*, 2011). Galappaththi *et al.*, (2019b) identified five types of knowledge possessed by Pangnirtung Inuit (1) place-specific knowledge of Arctic char, (2) turbot fishing techniques, (3) turbot fish processing and marketing knowledge, (4) local environmental knowledge, and (5) co-produced knowledge.

The structural and functional characteristics of co-management are important in creating favourable conditions for co-management to support adaptation to climate change. However, these favourable conditions will change based on specific contexts, including the fishery, the people, and the place. Many of the structural characteristics of Pangnirtung fisheries co-management are featured in other co-management systems (partnerships, mixed regimes, vertical and horizontal linkages, and sharing of responsibility and power) (Armitage *et al.*, 2011; Berkes & Armitage, 2010; Fidelman *et al.*, 2017; Galappaththi & Berkes, 2015; Kocho-Schellenberg & Berkes, 2015) and are sometimes identified as adaptive co-management (Fidelman *et al.*, 2017; Galappaththi *et al.*, 2022; Plummer *et al.*, 2012). However, the presence of an on-site fish processing plant to run co-existing fisheries and support co-management put Pangnirtung fisheries in a stronger position to adapt to the integrated vulnerabilities produced by climate change.

Some of the functional characteristics are also important to support climate adaptation. For instance, community consultations as a part of the co-management process are characterised by the use of collective action through local institutions. It could also activate knowledge co-creation and (individual) learning, and, going beyond, unlock individual learning into social learning. Further, continuous real-time monitoring of land and wildlife is a usual practice for Inuit. However, scientific stock assessments led by the DFO are also important to bridge knowledge systems for joint decision-making. Based on the assessment of Nunavut fisheries, Galappaththi *et al.* (2022) proposed steps that could build resilience in the adaptive co-management process of Arctic char, which relates to the functional characteristics identified in this chapter.

Overall, co-management can support climate change adaptation that builds resilience underpinning the adaptive capacity of fisheries' social-ecological systems. This chapter provides empirical evidence from Inuit fisheries co-management in the Canadian Arctic to examine the feasibility of co-management to foster climate change adaptation. Co-management is not a single response to climate change but, rather, is associated with many other responses (e.g., the use of technology) supporting adaptation. Co-management is widely applied in the Canadian Arctic region for natural resources management. Its success (as defined by co-management partners) and directions can depend on many conditions (e.g., structural/functional characteristics) (Berkes & Armitage, 2010; Galappaththi *et al.*, 2022; Kocho-Schellenberg & Berkes, 2015). Yet, this does not mean that the co-management approach could replace all conventionally-managed Indigenous fisheries systems.

Acknowledgements

I sincerely appreciate the support of Inuit fishers and community members of the Pangnirtung fishing community in Baffin Island, Nunavut, Canada. Fieldwork is funded primarily by multiple sources (the Canadian Institute of Health Research-CIHR, the Social Science and Humanities Research Council-SSHRC, and ArcticNet). I acknowledge the financial support of the SSHRC doctoral fellowship and field support from the Northern Scientific Training Program (NSTP) and a McGill North Engagement Grant through McGill University. I also acknowledge funding support from the LICCI project (ERC Consolidator Grant #771056). Further, I acknowledge Anna Schlingmann, Yorgos Stratoudakis, and Victoria Reyes-García for providing feedback on an earlier version of the chapter.

References

AMAP, 2018. *Adaptation Actions for a Changing Arctic: Perspectives from the Baffin Bay/Davis Strait Region.* Arctic Monitoring and Assessment Programme (AMAP), Oslo, xvi–354pp.

April, W.I., 2008. Experiential learning from the Khoi-Khoi (Namibia) and Maori (New Zealand) communities. *The International Journal of Entrepreneurship and Innovation* 9, 103–110.

Arctic Council, A., 2016. *Arctic Resilience Report*. Arctic Council, Stockholm.

Armitage, D., Berkes, F., Dale, A., Kocho-Schellenberg, E., Patton, E., 2011. Co-management and the co-production of knowledge: Learning to adapt in Canada's Arctic. *Global Environmental Change* 21, 995–1004.

Berkes, F., Armitage, D., 2010. Co-management institutions, knowledge, and learning: Adapting to change in the Arctic. *Etudes/Inuit/Studies* 34, 109–131.

Butler, J.R.A., Suadnya, W., Yanuartati, Y., Meharg, S., Wise, R.M., Sutaryono, Y., Duggan, K., 2016. Duggan K: Priming adaptation pathways through adaptive co-management: Design and evaluation for developing countries. *Climate Risk Management* 12, 1–16.

Carey, M.A., Asbury, J.-E., 2016. *Focus Group Research*. Routledge, London, p. 117.

Carlsson, L., Berkes, F., 2005. Co-management: Concepts and methodological implications. *Journal of Environmental Management* 75, 65–76.

Collings, P., 2009. Participant observation and phased assertion as research strategies in the Canadian Arctic. *Field Methods* 21, 133–153.

Collins, K., Ison, R., 2009. Jumping off Arnstein's ladder: Social learning as a new policy paradigm for climate change adaptation. *Environmental Policy and Governance* 19, 358–373.

Dalati, S., Marx Gómez, J., Marx Gómez, J., Musaiman, S., 2018. Surveys and questionnaires. In: *Modernizing the Academic Teaching and Research Environment: Methodologies and Cases in Business Research.* Cham. Eds: Jorge Marx Gómez and Sulaiman Mouselli Publisher: Springer. pp. 175–186.

Debortoli, N.S., Clark, D.G., Ford, J.D., Sayles, J.S., Diaconescu, E.P., 2019. An integrative climate change vulnerability index for Arctic aviation and marine transportation. *Nature Communications* 10, 1–15.

Divine, L.M., Pearce, T., Ford, J., Solovyev, B., Galappaththi, E.K., Bluhm, B., Kaiser, B.A., 2021. Protecting the future Arctic. *One Earth* 4, 1649–1651.

DFO 2021. Proceedings of the regional peer review on the estimated abundance and sustainable harvest levels for the Jayko and Halokvik (30 Mile) rivers in the Cambridge Bay commercial fishery, 2010–2015 DFO Can. Sci. Advis. Sec. Proceed. Ser. 2020/029 (24–26 January 2017).

Fidelman, P., Van Tuyen, T., Nong, K., Nursey-Bray, M., 2017. The institutions-adaptive capacity nexus: Insights from coastal resources co-management in Cambodia and Vietnam. *Environmental Science & Policy* 76, 103–112.

Ford, J.D., Clark, D., Pearce, T., Berrang-Ford, L., Copland, L., Dawson, J., New, M., Harper, S.L., 2019. Changing access to ice, land and water in Arctic communities. *Nature Climate Change* 9, 335–339.

Galappaththi, E.K., 2020. *Opportunities for Adaptation to Climate Change: A Comparative Analysis of Indigenous Fisheries Systems in the Canadian Arctic and Eastern Sri Lanka*. Department of Geography, Montreal, Canada.

Galappaththi, E.K., Berkes, F., 2015. Can co-management emerge spontaneously? Collaborative management in Sri Lankan shrimp aquaculture. *Marine Policy* 60, 1–8.

Galappaththi, E.K., Falardeau, M., Harris, L.N., Rocha, J.C., Moore, J.-S., Berkes, F., 2022. Resilience-based steps for adaptive co-management of Arctic small-scale fisheries. *Environmental Research Letters* 17, 083004.

Galappaththi, E.K., Ford, D.J., Bennett, E.M., 2019a. A framework for assessing community adaptation to climate change in a fisheries context. *Environmental Science and Policy* 92, 17–26.

Galappaththi, E.K., Ford, D.J., Bennett, E.M., Berkes, F., 2019b. Climate change and community fisheries in the Arctic: A case study from Pangnirtung, Canada. *Journal of Environmental Management* 250, 109534.

Galappaththi, E.K., Ford, D.J., Bennett, E.M., Berkes, F., 2021. Adapting to climate change in small-scale fisheries: Insights from Indigenous communities in the global north and south. *Environmental Science and Policy* 116, 160–170.

Giles, A.R., Strachan, S.M., Doucette, M., Stadig, G.S., Municipality of Pangnirtung, 2013. Adaptation to aquatic risks due to climate change in Pangnirtung, Nunavut. *Arctic* 66, 207–217.

GN, 2016. *Nunavut Fisheries Strategy, Department of Environment, Fisheries and Sealing Division, 2016–2020*, The Territory of Nunavut, Government of Nunavut (GN), Department of Environment.

Government of Nunavut, 2016. *Nunavut Fisheries Strategy, Government of Nunavut* (GN), Department of Environment, Fisheries and Sealing Division. 2016. 47 pp.

Huntington, H.P., Danielson, S.L., Wiese, F.K., Baker, M., Boveng, P., Citta, J.J., De Robertis, A., Dickson, D.M., Farley, E., George, J.C., 2020. Evidence suggests potential transformation of the Pacific Arctic ecosystem is underway. *Nature Climate Change* 10, 342–348.

Idrobo, C.J., Berkes, F., 2012. Pangnirtung Inuit and the Greenland shark: Co-producing knowledge of a little discussed species. *Human Ecology* 40, 405–414.

Jungsberg, L., Wendt-Lucas, N., 2024. Chapter 22: Local indicators, adaptation actions, and resilience efforts for a warming climate in North Greenland. In V. Reyes-García, S. Álvarez-Fernández, P. Benyei, L. Calvet-Mir, D. García-del-Amo, A. B. Junqueira, X. Li, V. Porcher, A. Porcuna-Ferrer, A. Schlingmann, & R. Soleymani (Eds.), *Routledge Handbook of Climate Change Impacts on Indigenous Peoples and Local Communities*. Oxfordshire & New York: Routledge. Chp 22, pp 356–369.

Karlsson, M., Hovelsrud, G.K., 2015. Local collective action: Adaptation to coastal erosion in the Monkey River Village, Belize. *Global Environmental Change* 32, 96–107.

Kitikmeot Heritage Society, 2021. A guidebook for research with Nunavut communities, Iqaluktuuttiaq (Cambridge Bay).

Kocho-Schellenberg, J.-E., Berkes, F., 2015. Tracking the development of co-management: Using network analysis in a case from the Canadian Arctic. *Polar Record* 51, 422–431.

Krippendorff, K., 2018. *Content Analysis: An Introduction to Its Methodology*. Sage, London.

Kristofferson, A.H., Berkes, F., 2005. Adaptive co-management of Arctic Char in nunavuit territory. In: Berkes, F., Rob, H., Fast, H., Manseau, M., Diduck, A. (Eds.), *Breaking Ice: Renewable Resource and Ocean Management in the Canadian North*. Calgary, Alberta, University of Calgary Press. p. 249.

Laugrand, F.B., Oosten, J.G., 2010. *Inuit Shamanism and Christianity: Transitions and Transformations in the Twentieth Century*. McGill-Queen's Indigenous and Northern Studies, Montreal, Vol. 58, p. 467.

Longhurst, R., 2016. Semi-structured interviews and focus group. In: Clifford, N., Cope, M., Gillespie, T., French, S. (Eds.), *Key Methods in Geography*. London: Sage, pp 103–115.

Olsson, P., Folke, C., Berkes, F., 2004. Adaptive comanagement for building resilience in social–ecological systems. *Environmental Management* 34, 75–90.

Pearce, T., Ford, J., Willox, A.C., Smit, B., 2015. Inuit traditional ecological knowledge (TEK), subsistence hunting and adaptation to climate change in the Canadian Arctic. *Arctic* 68, 233.

Pedersen, C., Otokiak, M., Koonoo, I., Milton, J., Maktar, E., Anaviapik, A., Milton, M., Porter, G., Scott, A., Newman, C., 2020. ScIQ: An invitation and recommendations to combine science and Inuit Qaujimajatuqangit for meaningful engagement of Inuit communities in research. *Arctic Science* 6, 326–339.

Plummer, R., Baird, J., 2013. Adaptive co-management for climate change adaptation: Considerations for the Barents Region. *Sustainability* 5, 629–642.

Plummer, R., Crona, B., Armitage, D.R., Olsson, P., Tengö, M., Yudina, O., 2012. Adaptive comanagement: A systematic review and analysis. *Ecology and Society* 17, 1–17.

Priest, H., Usher, P., 2004. *The Nunavut Wildlife Harvest Study: Final Report*. Nunavut Wildlife Management Board, Iqaluit.

Reyes-García, V., S. Álvarez-Fernandez, P. Benyei, D. García-del-Amo, A.B. Junqueira, V. Labeyrie, X. Li, V. Porcher, A. Porcuna-Ferrer, A. Schlingmann, R. Soleymani. 2023. Local indicators of climate change impacts described by Indigenous Peoples and local communities: Study protocol. *PLoS One* 18(1): e0279847. doi:10.1371/journal.pone.0279847.

Rodríguez, M.V., Davidson-Hunt, I., Berkes, F., 2019. Social-ecological memory and responses to biodiversity change in a Bribri Community of Costa Rica. *Ambio-Biodiversity Change and Human Adaptation* 48, 1470–1481.

Taylor, G.A., Blake, B.J., 2015. Key informant interviews and focus groups. In: Chesnay, M.D. (Ed.), *Nursing Research Using Data Analysis: Qualitative Designs and Methods in Nursing*. New York, Springer. pp. 153–165.

Termeer, C.J., Dewulf, A., Biesbroek, G.R., 2016. Transformational change: governance interventions for climate change adaptation from a continuous change perspective. *Journal of Environmental Planning and Management* 60, 558–576.

Ulrich, K.L., Tallman, R.F., 2021. Multi-indicator evidence for habitat use and trophic strategy segregation of two sympatric forms of Arctic char from the Cumberland Sound region of Nunavut, Canada. *Arctic Science, 7* (2) 512–544.

Yurkowski, D.J., Hussey, N.E., Ferguson, S.H., Fisk, A.T., 2018. A temporal shift in trophic diversity among a predator assemblage in a warming Arctic. *Royal Society Open Science* 5, 180259.

24

CLIMATE CHANGE ADAPTATION IN FIJI

Local adaptation strategies to enhance national policy

Priyatma Singh, Daniela Schulman, Dhrishna Charan, Victoria Reyes-García, Anna Schlingmann, Ashneel A. Singh and Kelera Railoa

Introduction

Human activity has already warmed the planet by 1.0°C above pre-industrial levels, with severe consequences for climate-vulnerable communities around the world (IPCC, 2018). Among climate-vulnerable communities, those living in Pacific Island countries face large-scale climate change risks ranging from sea level rise to extreme weather events, warming temperatures, and changing rainfall patterns (Oppenheimer et al., 2019). These persistent and worsening risks present a need for climate change adaptation planning, finance, and implementation across all levels of government and society.

Following the Intergovernmental Panel on Climate Change (IPCC), in this study adaptation is understood as "the process of adjustment to actual or expected climate change and its effects" (IPCC, 2014). To reduce risks from any specific environmental changes, individuals and communities often adapt by implementing reactive actions that are mostly driven by local experience and not guided by governmental or other institutional programs or planning (Adger et al., 2003; Mersha and Van Laerhoven, 2018). This is referred to as 'autonomous adaptation', implying that local communities undertake actions to adapt to climate-related risks based on local resources and traditional knowledge (Bawakyillenuo et al. 2016; Eakin et al., 2014; Fenton et al., 2017). The local perception of actual or potential climate change impacts is essential in the undertaking of autonomous adaptation measures. In contrast, 'planned adaptation' refers to strategic actions taken by government, development agencies, or other actors to reduce the vulnerability of a target population (IPCC, 2014). Whether planned or autonomous, adaptive actions and strategies may range from short-term coping mechanisms to long-term incremental adaptation and even transformation (Fedele et al., 2019; Marijn et al., 2023; Pelling et al., 2015; Termeer et al., 2017), and may be effective and successful (Adger et al., 2005; Owen, 2020; Singh et al., 2021) or maladaptive (Juhola et al., 2016; Maruna & Bajec, 2015; Schipper, 2020; Webber & O'Neill, 2010).

The United Nations Environment Programme (UNEP) Adaptation Gap Report (AGR) states that it is relatively easy to track large-scale adaptation projects implemented by international

DOI: 10.4324/9781003356837-29

donors and development partners (UNEP Adaptation Gap Report, 2021). Although sub-national and local projects play a significant role in mitigating local climate risks, they are often autonomous efforts that lack documentation (Kinoshita et al., 2018; Rahman et al., 2021). Indeed, the UNEP AGR (2021) highlights a gap in local data availability that hampers the development of sub-national and sectoral adaptation planning instruments. Moreover, even less is known of how adaptation at different levels can complement one another or negatively interfere (d'Armengol et al., 2018; Galappaththi, 2024; Gero et. al., 2011; Mersha & Van Laerhoven, 2018).

With the intention to enhance policy approaches that provide better responses to local needs, in this chapter, we document current autonomous adaptation measures in local villages and compare them with planned national adaptation measures, identifying tensions and synergies. Our study was conducted in Fiji, an archipelagic country in the South Pacific consisting of over 300 islands. Fiji represents an excellent case study for comparison of local and national adaptation plans for at least two reasons. First, the Fijian government has been playing an active role in the global climate change arena (Betzold & Nunn, 2019) as Fiji was the first country in the world to ratify the Paris Agreement in 2015, the first small island state to assume presidency of the United Nations Framework Convention on Climate Change (UNFCCC) in 2017, and one of the only 20 countries in the world that has submitted a National Adaptation Plan (NAP). The NAP process was established under the Cancun Adaptation Framework and allows parties to adapt it to national context as means for achieving the global adaptation goals under the Paris Agreement. The Fijian government published its first NAP in 2018, which is intended to serve as a holistic guide to comprehensively address climate change in Fiji. The NAP was formulated with the support of the NAP Global Network and resulted from national consultation workshops and key informant interviews with civil society, regional bodies, and leaders of pioneering adaptation programmes.

The NAP is aligned with pre-existing national policy plans, including the National Development Plan (NDP), the National Climate Change Policy (NCCP), the Green Growth Framework, the Disaster Risk Reduction Policy, and the Nationally Determined Contributions (NDCs) to the Paris Agreement. Moreover, the NAP was thought of as a "continuous, progressive, and iterative" approach to adaptation in government decision-making. The second reason why the island nation of Fiji is an excellent case to study interactions between local and national adaptation measures is because residents of Fiji, including the three villages selected for this study, face extreme risks from the impacts of climate change (Janif et al., 2016; Nurse et al., 2014). In view of this, Fiji's administrative structure allows for vertical integration of climate change adaptation initiatives, thus enabling practical opportunities to implement actions at various scales. Primary data collected through our study is a timely intervention to support the next phase of Fiji's adaptation plans, which focuses on data collection by any entities that have implemented NAP actions at the local level.

Study site

There are 14 provinces and 1 dependency in Fiji (Sharma et al., 2021), divided into 197 districts. Ba Province, in the northwest of Viti Levu, is the largest province in Fiji, comprising 21 districts and 107 villages (Census of Population and Housing – Fiji Bureau of Statistics, 2017). This study was conducted in three *iTaukei* villages in the district of Ba, in the Province of Ba. The Indigenous villages selected for the study site are Votua, Nawaqarua, and Natutu (Figure 24.1), which have a combined population of approximately 1,000 individuals. These villages are positioned on the

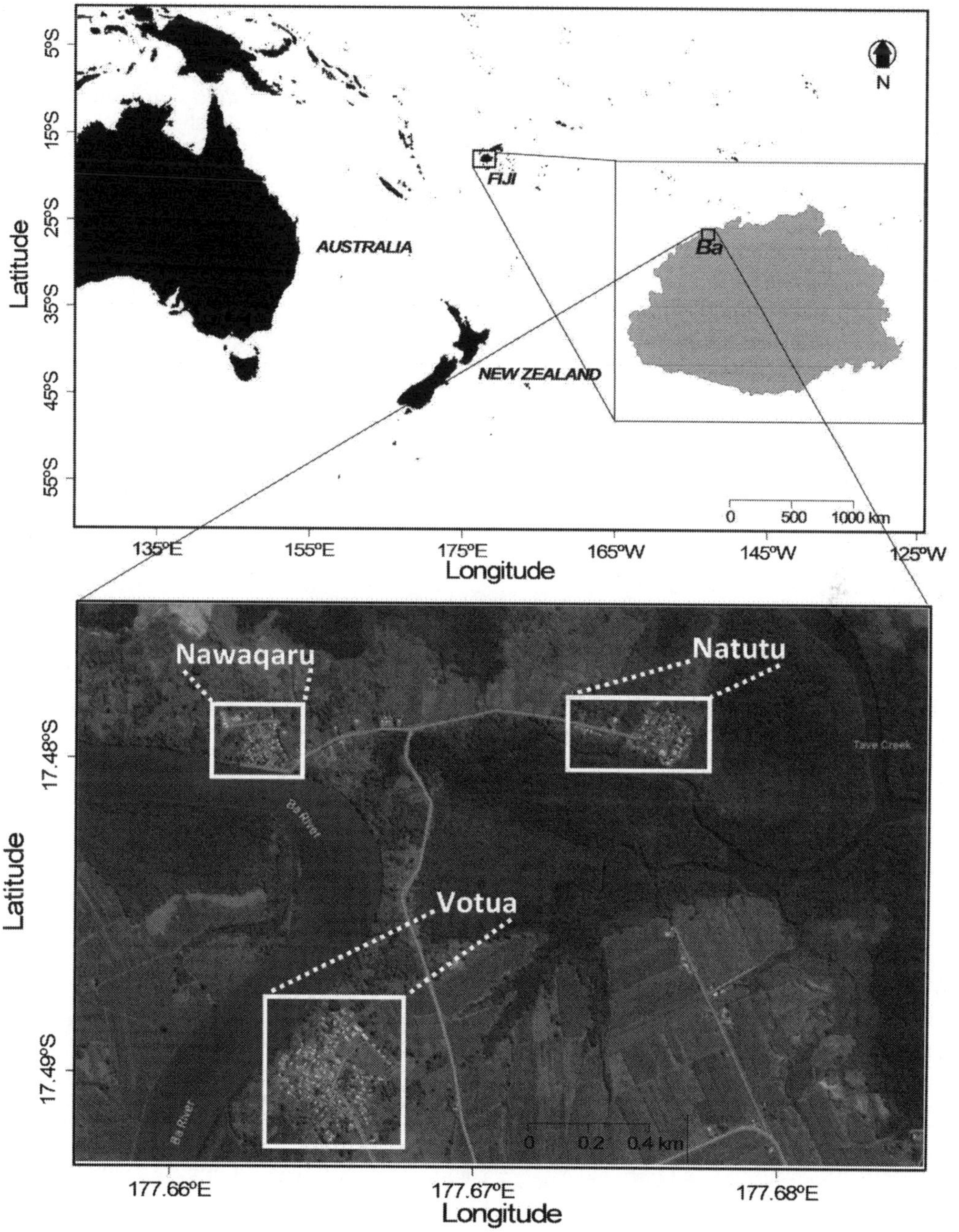

Figure 24.1 Location of Votua, Nawaqarua, and Natutu village in the Ba Province on the main island of Viti Levu, Fiji.

banks of the Ba River that connects to the seas of the Yasawa Islands, an archipelago of about 20 volcanic islands off the coast of Western Viti Levu. The main sources of income in the studied villages are fishing and employment in the nearby towns. A few households are engaged in sugarcane farming, while others practice small-scale subsistence farming of yam, *dalo* (taro), and cassava. Some villagers also engage in pig and cattle farming. Votua is the largest village in size and has an extensive *qoliqoli* (traditional fishing ground) boundary, which is shared amongst those in the Ba Delta villages, including Nawaqarua and Natutu. *Qoliqoli* in Fiji presents a remarkably well-established system of traditional fishing grounds, officially referred to as customary fishing rights area, which extend to the outer reef slope. These areas benefit from some legal recognition and community custodianship, enabling local villagers to oversee fisheries management policy implementation and monitoring.

Aquatic ecosystems such as the Ba River and the coastal and marine areas provide a great diversity of resources that sustain local livelihoods. These resources provide more than just food supply. They are integral parts of the *iTaukei* culture, particularly related to recreation and unique traditions valuable to the *iTaukei* way of life. Today, the rich cultural and environmental diversity in Fijian coastal villages is at risk of succumbing to changing climate and an increasing number of natural hazards. According to a survey report by Brown et al. (2017), flooding is the greatest local threat in the Ba River catchment and the most common natural disaster after tropical cyclones. The Ba River basin is located on the leeward side of Viti Levu, which also makes it vulnerable to prolonged periods of drought. Climate change-induced temperature rise and ocean acidification further harm marine ecosystems already suffering from varied sources of local pollution and overharvesting (Baxter et al., 2016; Oppenheimer et al., 2019). The villages have a low average income, which, when coupled with property damages, decline of agricultural crops, and loss of fishing days due to natural disasters, affect the capacity of community members to adapt to the changing climatic conditions (Singh et al., 2022).

Methods

We used an ethnographic research approach to better understand the natural setting of the *iTaukei* coastal fishing community. Culture shapes the way people experience and respond to the world around them. Thus, it is important to understand the impacts of global and climate change through cultural lenses (Kaijser and Kronsell, 2014; Marks et al., 2022). Prior to the commencement of field activity, a research permit was acquired from the Ministry of *iTaukei* Affairs, the government department responsible for overseeing development activities related to the welfare of the *iTaukei* people. In addition, the authors followed the ethical guidelines of the LICCI protocol as outlined in Reyes-García et al. (2023), including free, prior, and informed consent with village leaders and study participants who kept the right to withdraw their consent at any moment.

We used both primary and secondary data sources. Primary data on climate change impacts and autonomous adaptation measures were collected via in-depth semi-structured interviews and focus group discussions (FGD) following the protocol developed by Reyes-García et al. (2023). Through semi-structured interviews, we tried to understand local climate change impacts and the applied response measures. The FGD served to understand group consents on the direction and drivers of experienced impacts (beyond climate change). Semi-structured interviews and FGD were audio recorded with the permission of the interviewees. The secondary data used for this study consist of the Republic of Fiji's NAP (2018).

Semi-structured interviews

Twenty-five in-depth semi-structured interviews were carried out. Quota sampling was applied to select participants across three categories: gender, age, and livelihood. The first five participants included village leaders and chiefly family members who had good knowledge of people's livelihoods and of the recent local events. The rest of the interviewee list included ten women and ten men participants ranging from 25 to 75 years of age. Some selected women were engaged in subsistence farming and fishing, while some sold handicrafts and food products in the local market. The selected men were involved in offshore fisheries, subsistence farming, and paid employment. Most of the participants had attended secondary school, whereas very few young women and men had a trade or tertiary qualification. Elderly participants from age 55 to 75 had completed primary education.

In semi-structured interviews, we asked informants to report changes and impacts observed in their local environment (hereafter environmental changes and impacts). The participants were encouraged to think of atmospheric, physical, and biological changes they had observed since they were young and to provide examples of changes in their livelihood system, particularly with regard to fishing, livestock rearing, crops, and diseases. Each observation of change described by a participant was followed by questions on the potential drivers of the reported change. For changes and impacts that were attributed to elements of the atmospheric system, for example, temperature and precipitation (hereafter climate change impacts), the participants were then asked to describe how the community, or the affected households, were responding to the changes (hereafter local adaptation). Responses were noted verbatim.

Focus group discussion

Two FGD were conducted to validate the changes reported during semi-structured interviews. Due to the patriarchal nature of the villages within the study site, there was a strong preference for gender-based group discussions. One group consisted of nine women and the second group consisted of ten men. All participants ranged from 25 to 75 years of age. We used convenience sampling to recruit participants for FGD. The participants were selected from all three villages where we had conducted semi-structured interviews. The groups included participants engaged in fishing, farming, paid employment, and other professional activities, such as church or women's groups.

We used a guide to keep the discussion focused. The focus group discussion turned into a *talanoa*, an informal open conversation that shares stories, thoughts, and feelings. This seemed effective for the *iTaukei* because of the group dynamics that exist within the closely knitted communities. The participants felt at ease and shared comprehensive answers in a storytelling mode. The environmental changes extracted from the interviews were validated through the group discussion. We also asked the participants to explain how they respond to those changes. For instance, we wanted to know if people have changed or are in the process of changing any activities and what they are doing in response to the changes reported. This helped us to better clarify interlink between adaptation measures and local observations of climate change impacts.

National Adaptation Plan

The Republic of Fiji's NAP is a high-level action plan for climate change adaptation. This document is to be reinforced in three additional stages to ensure effectiveness: development of a monitoring and evaluation system, formation of communication strategy, and development of financing plans. Based on climate change projections and impacts, as well as on resolvable adaptation barriers, NAP stakeholders (i.e., government, private sector, civil society, faith-based organizations, professional and academic institutions, and actors representing low-income and otherwise vulnerable groups) reached consensus on around 160 priority adaptation action points (thereafter NAP planned adaptation). These adaptation action points are sorted into ten categories throughout the NAP, called "systems and sectoral components" (Table 24.1). Systems component actions are designed to enable and increase investment flows for adaptation as well as climate-resilient development. Sectoral component actions are relevant to vulnerable components of society and the economy. Detailed definitions of each component are provided in the NAP, which is available on a public government website (https://bit.ly/3m2wGXN).

Data analysis

Information on environmental changes reported during semi-structured interviews and FGD were tabulated and classified according to the hierarchical classification proposed by Reyes-García et al. (2023) that distinguishes between changes observed in the atmospheric, the physical, and the life systems. Each coping or adaptation action described by the villagers was grouped into Local Adaptation to Climate Change Impacts (LACCI) identified using the classification proposed by Schlingmann et al. (2021) that distinguish between changes in timing, location, livelihood product, practices and techniques, capacity building, resource input, and practiced (main) livelihood system. For the categorization of the NAP, we used the existing organizational structure of the document, considering all 160 adaptation measures across all 10 systems and sectoral components.

Using a hybrid deductive and inductive technique for thematic analysis (Fereday & Muir-Cochrane, 2006), three researchers independently compared the reported local adaptations (i.e., LACCIs) with corresponding categories from the NAP. After discussion between the coders, each distinct LACCI identified in the primary data was linked in a spreadsheet to all possible relevant adaptation measures from the NAP. The results were compared to identify the linkages between LACCIs and NAP and common themes where synergies and gaps have manifested. The comparison constitutes the basis for our discussion.

Results

Local observations of environmental change

The villagers at our study site observe numerous changes in the atmospheric, physical, and life systems and attribute many of those changes to climate change. We categorized local observations in 31 indicators of environmental change: 9 referred to changes in elements of the atmospheric system, 8 to elements of the physical system, and 14 to elements of the life system (Table 24.2).

Observations of changes in elements of the atmospheric system referred to changes in temperature, precipitation, and air masses. Nearly all 25 participants reported higher temperatures in the day and at night (100%), fewer cold days (92%), higher humidity (100%), longer lasting periods of droughts (92%), irregular rainfall patterns (96%), frequent flooding (100%), and stronger cyclones (100%). Regarding the physical system, participants observed changes in marine, freshwater, and terrestrial subsystems. All of the participants described widening of the Ba River (100%) and frequent river flooding (72%). Approximately 60% of the participants reported an increase in sea water temperature and a rise in sea level. Most participants noted that the village is increasingly being inundated with saltwater, gradually contaminating the soil (92%). About 80% of the participants expressed their concern with regards to soil salinity (76%) and almost all the participants reported changes in soil fertility and productivity (92%). A few participants also stated an increase in frequency of king tides (32%) (Table 24.2).

Table 24.1 Breakdown of prioritized adaptation actions in Fiji's National Adaptation Plan (2018) – system and sectoral components (adopted from Fiji's National Adaptation Plan, 2018)

Systems component	*Definition*	*Number of adaptation actions*
Climate information services and management	Actions for climate information services and management will improve capacity to generate, manage, disseminate, and use climate change information.	10
Horizontal integration	Actions for horizontal integration will mainstream climate change issues into national-level development planning processes.	11
Vertical integration	Actions for vertical integration will integrate climate change issues into subnational development planning processes.	10
Climate change awareness and knowledge	Actions for climate change awareness and knowledge will enhance understanding of climate change by increasing the flow of relevant information to relevant adaptation stakeholders.	11
Resource mobilization	Actions for resource mobilization will improve the amount of resources available and the way available resources are utilized.	13

Sectoral component	*Definition*	*Number of adaptation actions*
Food and nutrition security	Actions for food and nutrition security will improve capacity to anticipate and reduce environmental and climate risks and support sustainable food production efforts.	23
Health	Actions for health will improve systems and infrastructure to manage the negative impacts caused by future climate variability and change.	10
Human settlements	Actions for human settlements will reduce vulnerability to major assets, infrastructure, and population centres, providing the ingredients for resilient growth.	12
Infrastructure	Actions for infrastructure will help to ensure the full life span of investments can be reached by addressing environmental and climate risks.	44
Biodiversity and the natural environment	Actions for biodiversity and the natural environment will support the maintenance of vital ecosystems and the services they provide.	16

Table 24.2 Local indicators of environmental change reported by iTaukei villagers ($n = 25$)

Observation [N° of interviewees]	Local Indicator of Climate Change Impact [direction of change]	Impacted Element	Subsystem	System
'Higher temperatures as compared to 10 years ago' [25] 'Increase in temperature at night as compared to 10-15 years ago' [25] 'Fewer cold days as compared to 10 years ago' [23]	Changes in mean temperature (not further specified) [increase] Changes in the temperature during the night [increase] Changes in the frequency of cold days [decrease]	Mean temperature	Temperature	Atmospheric system
'Increase in humidity as compared to 10 years ago' [25]	Changes in air moisture/humidity [increase]	Moisture/humidity, dew and frost	Precipitation	Atmospheric system
'Increase in drought events as compared to 10-15 years ago' [25] 'Increase in length of drought season' [23]	Changes in the frequency of drought events [increase] Changes in the length/duration of drought [increase]	Drought	Precipitation	Atmospheric system
'Increase in the number of flash floods and higher flood water levels' [25]	Changes in the intensity of flash floods [increase]	Precipitation Extremes	Precipitation	Atmospheric system
'Decrease in rainfall' [24]	Changes in mean rainfall [decrease]	Mean precipitation	Precipitation	Atmospheric system
Increase in strength of tropical cyclones' [25]	Changes in the intensity of cyclones [increase]	Cyclons	Air masses	Atmospheric system
'Increase in temperature of sea water' [15]	Changes in the sea surface temperature [increase]	Sea temperature	Marine physical systems (ocean & sea)	Physical system
'Increase in sea level' [15] 'Increase in frequency of king tides' [8]	Changes in the sea level [increase] Changes in the level of tides [increase]	Sea-level rise	Marine physical systems (ocean & sea)	Physical system
'Widening of the river' [25] 'A lot of soil erosion along the river bank' [15]	Changes in the frequency of river or pond bank erosion [increase] Changes in the intensity of river or pond bank erosion [increase]	River bank/pond erosion and sedimentation	Freshwater physical systems (continental waters)	Physical system
'Frequent river flooding' [18]	Changes in the frequency of river/lake floods [increase]	River and lake floods	Freshwater physical systems (continental waters)	Physical system
'Soil quality is deteriorating, increase in difficulty of crop cultivation, e.g. cassava' [23]	Changes in soil fertility [decrease] Changes in soil productivity [decrease]	Soil fertility, structure and biology	Terrestrial physical systems (soil & land)	Physical system
'Increase in soil salinity' [19]	Changes leading to soil degradation [increase]	Soil degradation	Terrestrial physical systems (soil & land)	Physical system
Observation [N° of interviewees]	**Local Indicator of Climate Change Impact [direction of change]**	**Impacted Element**	**Subsystem**	**System**
'Marine fish breeding season is changing' [25]	Changes in the timing of mating or reproduction of marine animal species [more irregular]	Marine spp reproduction	Marine biological system	Biological system
"Decreasing numbers of some fish species' [16]	Changes in the abundance of marine fish [decrease]	Marine spp abundance	Marine biological system	Biological system
'Decline in productive crop growing seasons' [25] 'Decrease in crop yield, e.g. breadfruit' [23] 'Decrease in size of food crops' [18] 'Some crops are less tasty as compared to 20 years ago' [9]	Changes in the frequency of successful cropping seasons [decrease] Changes in crop productivity / yield [decrease] Changes in cultivated species' fruit size [decrease] Changes in the taste of crop species/products [modified]	Cultivated spp productivity and quality	Cultivated plant spp (crops, orchards)	Human system
'Flowering times of important crops are changing' [18] 'Flowering time of some fruits and crops has changed, e.g. mangoes' [18]	Changes in crop flowering time [more irregular]	Cultivated spp phenology and reproduction	Cultivated plant spp (crops, orchards)	Human system
'Decrease in farmland' [15]	Changes in crop suitable cultivation areas [decrease]	Cultivated spp suitable areas	Cultivated plant spp (crops, orchards)	Human system
'Increase in cases of crop weeds' [14]	Changes in the frequency or occurrence of weed species states as invasive[increase]	Crop Weeds	Cultivated plant spp (crops, orchards)	Human system
'Food shortages after natural disasters and flooding' [25]	Changes in the frequency of famine/food shortage episodes [increase]	Hunger	Human health	Human system
'Increase in infectious diseases' [10]	Changes in the incidence of human vector borne and waterborne diseases [increase]	Diseases	Human health	Human system
'Increase livestock parasites' [10]	Changes in the frequency of parasites in livestock [increase]	Livestock disease/pest/ mortality	Livestock	Human system
'Decline in reproduction rate of livestock' [8]	Changes in the number of pups or offspring in livestock/raised animals [decrease]	Livestock/raised animals phenology, reproduction and behaviour	Livestock	Human system

Reports of changes in elements of the life system included changes observed in marine biological subsystems, as well as changes in health, agricultural, and socioeconomic systems. Thus, participants reported a decrease in the numbers of some fish species (64%) and changes in fish breeding season (100%). Informants also reported an increase in infectious disease (40%) in both humans and pests, a decline in crop productivity (100%), with a consequent loss of income. The informants also reported change in the flowering time of some fruits and crops (72%), as well as a decrease in livestock reproduction rate (32%). An increase in the frequency and severity of cyclones and flooding events reportedly led to increasing cases of infectious diseases, destruction of crops, and disruptions to fishing schedules that adversely affect livelihood security.

The synergistic effects of climate and other drivers of change were also highlighted by the participants. Most of the observations were directly linked to climate change, such as a decline in productive crop growing seasons and changes in fish breeding patterns. However, others were put in the context of multiple drivers, including climate change, socioeconomic, and technical drivers. For example, flooding was associated both to increases in heavy rainfall events and to poor drainage systems. Similarly, villagers noted a decline in freshwater and marine shellfish, which they linked to pollution from a nearby sugar mill industry, and they attributed the decline in marine fish catch to both climate change and overharvesting activities over the past four decades.

Local adaptation to climate change

The villagers anticipate, to a great extent from their own observations, that climate change will continue to cause multiple and potentially compounding changes. In response to the reported climate change impacts, *iTaukei* villagers are implementing changes in their agriculture, livestock farming, fishing, and ways of living. These changes are often based on their Indigenous knowledge as well as on their local social networks. The reported measures include efforts that range from short-term coping strategies, including sharing of products that help cover basic necessities, such as food and water, during and in the aftermath of disaster, but also incremental long-term adaptation measures, such as planting trees around riverbanks and shifting plantations away from coastal areas, and even transformative adaptations, such as livelihood diversification (Table 24.3).

Agriculture and food security appear to be key areas in which respondents have implemented most autonomous adaptation measures. Local autonomous adaptations mostly rely on local resources. For example, incremental adaptation measures to prolonged periods of dry weather include planting drought-resistant crop varieties and adopting agronomic measures such as multicropping. Villagers also reported the use of natural pesticides such as the addition of ash and worms to counter pest infestation, measures related to reciprocity and sharing and those related to livelihood diversification were implemented autonomously. In contrast, most of the local adaptation measures related to climate information and services (e.g., access to weather forecast from Fiji Meteorological Service) were planned adaptation by the government. Adaptation measures in the study site that were implemented by the government were termed as local planned adaptation.

Some of the local adaptation measures could not be purely classified as either autonomous or planned adaptation, but they resulted from a collaborative effort in which the adaptive measures initiated by the communities were, in some form, assisted by government's planned activities. So, while not a direct result of government interventions, these local adaptation measures by the communities were informed to some degree by government programmes, especially those related to prior awareness campaigns. We classified these as multiscalar adaptations. Most common forms of multiscalar adaptation measures include technological changes such as diversification of crop grown, including drought-resistant varieties, changes to the farming techniques, and

Table 24.3 Local adaptations to climate change impacts by iTaukei villagers

Sector domain type		*Adaptation measure reported*	*LACCI*
Agriculture			
Livelihood product Composition	1	Planting of drought-resistant crop varieties in consultation with Dept of Agriculture	Changes in crop variety/landrace composition
	2	Planting two or more crops at the same time within same farm area in consultation with Dept of Agriculture	Changes in multiple cropping practices, incl. Agroforestry
Resource input (pesticide) Demand (natural pesticide)	3	Addition of ashes to prevent pest infestation	Changes in the quantity of pesticide (natural pesticide)
Practices and techniques Methods & technology	4	Addition of worms to prevent pest infestation	Changes in soil (structure) conservation practices [not further specified]
Location Relocation (close-by)	5	Soil management: Use of fallowing practices	Changes in the rotation of cultivated plots (incl. fallow practices)
Livestock			
Practiced livelihood system Livelihood systems	6	Mixed farming approach – backyard gardening, livestock farming, and yaqona plantation	Changes in practicing livestock husbandry as a livelihood
Fishing			
Time management Timing	7	Change in harvesting/gleaning times to allow for mussels and prawns to rejuvenate	Changes in the timing of seasonal fishing activities Changes in the sustainable use of natural resources (e.g., access regulation) in fishing activities
Living			
Practiced livelihood system Income/financial capital	8	Income diversification – women engaged in selling of traditional handicrafts, herbal medicines, tailoring, and bakery business.	Changes in other (low) natural resource-dependent income
	9	Some households are engaged in paid employment as laborers, e.g., Carpenters/security officers to ensure consistent income	Changes in other non-natural resource-dependent income, including diversifying [not further specified]
	10	Villagers have been receiving remittances from relatives working abroad for quite some time now	Changes in the financial support, remittance, loans, credits, borrows
Practices and techniques Built construction	11	New houses are built on raised platforms	Changes in the housing/shelters
Natural ecosystems/ ecosystem-based	12	Riverbank protection – planting of woody trees and mangrove rehabilitation to fortify coastal areas	Changes in re-/afforestation/tree & grass plantings, other than Agroforestry
Location relocation (close-by)	13	Relocation of plantations to higher grounds	Changes in the relocation (close-by) of other activities, persons, or materials

(*Continued*)

Table 24.3 (Continued)

Sector domain type	*Adaptation measure reported*	*LACCI*
Resource input (water) Storage	14 Investing in tanks for storing safe drinking water	Changes in the storage of water in other activities, incl. rainwater harvesting
Resource input (food) Storage	15 Preservation of food	Changes in the storage of processed/purchased food
Practices & techniques Observation, monitoring, & planning	16 Use of traditional knowledge to forecast weather and relying on past experiences	Changes in the use of weather observation, monitoring, & forecast, incl. early warning
	17 Villagers have access to weather forecast from Fiji Meteorological via their mobile phones/TV/radio	Changes in the use of weather observation, monitoring & forecast, incl. early warning
	18 Disaster management – evacuation strategy prepared in consultation with local government	Changes in the planning & preparation of future actions/ decision
	19 Formation of community fisheries committee to monitor fisheries activities in the area led by Dept of Fisheries	Changes in other observation, monitoring & mapping
Methods & technology	20 Behavioral changes in livelihood practices – Destroying mosquito breeding sites and boiling water, especially during times of disaster in consultation with Ministry of Health	Changes in protective measures against diseases (e.g., use of mosquito nets, sun cream. filtering water)
Capacity building Human capital (Knowledge & gender role)	21 Capacity building exercises – climate change awareness workshops and conservation projects in consultation with NGOs and relevant government agencies	Changes in education, information & knowledge sharing, awareness training
Social capital (social networks & relations)	22 Sharing of items such as food during times of disaster/aftermath of disaster	Changes in social relationships & household/individual-based reciprocity

ecosystem-based adaptation. There are fewer numbers of local planned adaptation measures in these communities compared to local autonomous and multiscalar adaptation.

Linkages between local and NAP adaptation measures

The comparison of local adaptation measures by coastal *iTaukei* communities (numbered from 1 to 22 in Table 24.3) and the measures proposed in Fiji's NAP (Table 24.1) revealed key synergies, but also gaps (Table 24.4). The local adaptation measures are classified under the system and sectoral components of NAP in Table 24.4. The number of local adaptation measures under the systems component of the NAP is noticeably lower than under the sectoral component. For example, NAP sectoral component "Food and nutritional security" is well covered by local adaptations, while there is a lack of evidence of vertical or horizontal integration as well as resource mobilization in the local adaptation measures. Adaptation measures identified under the system's components are important for climate-resilient development to occur. We find that several important sectors such

Table 24.4 Comparison of synergies, discrepancies and gaps between local adaptations (i.e., autonomous, planned and multiscalar adaptation) in the study site and Fiji's NAP, structured around the different system and sectoral components of the NAP

	Local autonomous adaptation[a]	*Local planned adaptation*[b]	*Local multiscalar adaptation*[c]
National Adaptation Plan systems components			
Climate information services and management	16	17, 18, 21	
Horizontal integration			
Vertical integration			
Climate change awareness and knowledge		18	
Resource mobilization			
National Adaptation Plan: sectoral components			
Food and nutrition security	3, 4, 7, 13, 15	19	1, 2, 5, 6, 12, 20
Health		21	
Human settlements	11		12
Infrastructure		14	12
Biodiversity and the natural environment		21	12
Gaps in the local and NAP planned adaptation			
Reciprocity and sharing	22		
Reducing dependency from environment	8, 9, 10		

The numbers stated here refer to numbers listed beside adaptation measures in Table 24.3.

[a] Local autonomous adaptation: adaptation measures implemented by the *iTaukei* community members uniquely.
[b] Local planned adaptation: adaptation measures implemented by the government in the study site.
[c] Multiscalar adaptation: Local adaptation measures implemented in cooperation by both actors, *iTaukei* villagers and the government.

as health, human settlements, infrastructure, biodiversity, and natural environment are also inadequately addressed by local adaptation.

The results from interviews also indicate that income diversification (LACCI 8, LACCI 9, LACCI 10) is an important strategy for some villagers to cope with climate-related shocks. In addition, *iTaukei* communities have a communal way of living, according to which resources are shared between families, which often supersede individual interests. This is reflected in the way communities are choosing to adapt to food shortages, with many communities sharing food during and in the aftermath of a disaster (LACCI 22). Sharing, reciprocity, and income diversification, especially pertaining to non-nature-based jobs were identified as important local adaptation measures but were found to be lacking in the NAP adaptation actions. Food and nutrition security appear to be a key focus of adaptive action for the local communities in this study and could easily be regarded as one of the main criteria for evaluating adaptation efficiency at local levels.

Discussion

While villages at the study site are prone to sea-level rise, coastal inundation, and riverbank erosion, these climate-induced threats have not diminished the spirit of the villagers who are making efforts to reinforce their coastal areas, protect food security, take precautionary health measures, and safeguard their place of birth. The autonomous adaptation activities documented cover sectors that are vulnerable to the impacts of climate change in an effort to reduce risk and enable villagers to cope with the changing conditions. The autonomous adaptation strategies documented for *iTaukei* villagers largely focus on improving food and nutrition security as well as practiced livelihoods through initiatives ranging from changes in crop-livestock systems to changes in fishing behavior and storage of food. For instance, an increase in the occurrence of coastal flooding has prompted some villagers to relocate their plantations to higher farmlands. However, such autonomous adaptations are not without risks. In the case of farm relocation, for instance, if the new plantation site falls within the village's customary boundaries, then there are no complications. However, if the plantation is moved to a land that belongs to a different clan, then tensions may arise if proper negotiations and legal procedures are not followed (Gharbaoui & Blocher, 2016, 2018). This example shows that autonomous adaptations might face a myriad of challenges derived from different prioritization of place-based attachment and new livelihood opportunities, signaling the need for a robust participatory decision-making process (Singh et al., 2020).

Many autonomous adaptation measures are also oriented to income diversification, as communities move away from reliance on natural elements that are being threatened by climate change impacts and towards non-nature based jobs and remittances. The community members have diversified their income through low resource-dependent activities with women selling handicrafts and establishing small businesses such as bakeries and tailoring businesses. Some members are even increasing their household income by getting paid jobs. But there appears to be a gap in planned adaptation with regards to how community income can be diversified. For instance, Fiji's NAP livelihood diversification measures focus on enhancing communities' capacity within practiced livelihoods and through the use of local resources available. Examples of these strategies include promoting the diversification of agricultural products for subsistence consumption and market sales or encouraging aquaculture and non-extractive cultured fisheries. In this line, NAP prioritizes increasing support to enterprises, including microenterprises, to enable people to better address and manage environmental and climate risks. The climate change awareness and knowledge and resource mobilization system components of NAP include strategies that can further increase the adaptive capacity by increasing flow of climate information and resources. If implemented, these planned adaptation strategies, which include increasing access of local communities to greater levels of public consultation, improving financial literacy, and access to financial services products can greatly support locals in their attempts to diversify income.

This study also demonstrated the deep connection villagers have with their environment and the importance of this connection in dealing with climate change impacts. For instance, villagers use their local knowledge to predict weather and to plant crops or to fish. Women also use traditional food preservation skills to ensure food is available during times of disaster. Similar findings have been reported by McNamara and Prasad (2014). While the essence of this knowledge system is appreciated by all villagers, there are concerns regarding its potential loss, in particular, traditional ways of growing food, managing pests, preserving food, and conserving fish and other marine resources (Janif et al., 2016). The threat of this potential loss may be exacerbated by the location of our study sites on Viti Levu (the largest island in Fiji and economic hub of the Pacific region), where the influence of globalization may interrupt intergenerational transmission of Indigenous knowledge.

The community also displayed strong kinship relations, which became prominent during the COVID-19 crisis. One of the important frameworks of Fiji's planned adaptation strategies through NAP is the promotion of Indigenous knowledge and practices and, while this national strategy has identified several options that focus on the use of these (e.g., traditional ways of protecting freshwater aquifers from saltwater intrusion, traditional use of food preservation, traditional seed varieties), social networks and kinship components are largely unaddressed. This neglect contrasts with autonomous adaptation measures documented, among which social networks were prominent. We have identified this as a gap in the NAP and currently implemented planned adaptation. There are opportunities for NAP to learn and incorporate the autonomous adaptation options which are being carried out by the communities to improve the likelihood of sustaining livelihoods during difficult times. Indigenous knowledge and lived experience of adaptation solutions that are, or are not, working at the community level must flow into a feedback loop that connects with and informs policy, programs, and projects at the sub-national and national levels. These feedback loops must be continuous and iterative.

While most of the autonomous adaptive measures were initiated and completed by the community members, some were also influenced to a certain extent by previous awareness and training sessions conducted by government agencies or through assistance from them, which is also reflected in the NAP's strong focus on awareness raising. For instance, the villagers have been practicing ecosystem-based adaptation approaches (also a priority in the NAP), such as mangrove rehabilitation and planting of trees around the riverbank to increase their adaptive capacity and to continue living at their village site. In this aspect, the NAP is aligned with local priorities and recognizes managed retreat as a last resort. The NAP focuses on strengthening coastal boundaries through hybrid or nature-based solutions to reduce risk and stave off the need to relocate communities and infrastructure. This observation also aligns with the timescale of the Fijian government's increased participation in national, regional, and international climate change dialogues (Government of Fiji, 2018). Ecosystem-based approaches have the potential to build resilient coastal systems, but local adaptation measures must closely align with NAP to evade maladaptation (Mfitumukiza et al., 2020). Ecosystem-based adaptation measures must fully account for social complexity, such as community leadership, kinship and hierarchy, implementation, monitoring, and/or evaluation, in their vulnerability analysis to be able to offer adequate protection.

Villagers in the study site have been planting more drought-resistant crop varieties. This measure was adopted after the Department of Agriculture created awareness and the village committee consulted the Department about potential options. The NAP promotes the integration of climate-smart agriculture practices into farming, training, and extension services. In view of this, the extension and advisory services embedded within the agriculture and even the fisheries sector can play a major role in boosting the autonomous actions to improve food security and connect the villagers with established local and export markets. The extension officers are best positioned to act as intermediaries and create connections between local communities, local markets, and various agencies, including government departments. However, to achieve this goal, the role of extension officers needs revamping to enhance monitoring and evaluation of autonomous and planned adaptation strategies and actions at different levels. This is particularly important as the Fiji government works on implementing an effective communications strategy to support the awareness, adoption and implementation of the NAP principles by all relevant stakeholders (Hidalgo et al., 2022). The communities are one of the major stakeholders of the NAP.

Recommendations and way forward

The adaptation measures considered at the national level can have their largest impact at the community level, thus making stakeholder participation and consultation in the identification and prioritization of adaptation options necessary. To be successful, planned adaptation measures must be acceptable to those who are to implement them, including local government and communities.

The exclusion of community members from the decision-making process means that community buy-in will be difficult to achieve. While engaging every individual member of an entire community is logistically complex, community consultation processes may be strategically designed to carefully take into consideration which types of interactions carry the most benefit for specific communities. For instance, traditional leaders have significant influence over community development plans and local governance policies and programmes. So, do religious leaders, such as church pastors, who bring a powerful voice to issues of importance (Singh et al., 2022). Engaging these authoritative community members in all local adaptation planning processes up to the national level could offer a strong foundation for vertical integration in the evolution of the NAP. We recommend in-depth local consultation and the preservation of Indigenous knowledge to avoid the erosion of climate resilience from cultural loss. This would also ensure that communities' voices are heard and their experiences incorporated in the way national adaptation plans are designed and implemented (Nunn et al., 2014). A participatory approach is also important for understanding the traditional and local knowledge and kinship that forms the basis of any adaptation strategy at the community level (Remling & Veitayaki, 2016).

International donors, development partners, and national governments play important roles in channelling adaptation resources and technical support from the top down, but effective adaptation will eventually take place via the dynamics of local government, non-government organizations (NGO), and communities. Apart from gathering quantitative data on climate change and its impacts, qualitative social research can play a crucial role in autonomous adaptation to understand local observations of risk, explore the adaptive capacity of individuals and groups, harness local knowledge, and enable tailored engagement. Investing in qualitative social research may enable the government to garner support for and participation in its adaptation activities. This resonates with findings from McConney et al. (2024) who emphasize the role of local knowledge in creating and operating complete, multi-level sargassum policy cycles in the affected countries. We recommend further research in important dimensions that influence adaptation in Pacific Island Countries, namely food security, income diversification, resource use (or non-use) and implementation, local knowledge, participatory policymaking, and monitoring and evaluation.

Fiji's NAP is a broad, high-level climate change adaptation policy that lays out an overarching action plan for the country. Several actions outlined in the NAP are beyond the scope of what villagers can do, for example, building climate-resilient infrastructure, establishing flood management activities, and managing waterway resources. However, such actions can be enhanced to incorporate co-management and be integrated at different levels. Successful co-managed adaptation by actors at different levels is found for example in a fishery context in the Canadian Arctic (Galappaththi, 2024). D'Armengol et al. (2018) identified a supportive legislative framework with clear operational rules and power symmetries as key conditions for successful co-management in small-scale fisheries.

For the Fijian government to mobilize resources to avoid maladaptation, it must establish mechanisms to successfully implement the actions outlined in the NAP. The revised version of the NAP should have some key performance indicators. Climate change adaptation is a complex process that requires equal effort by communities, civil society, private sector, and the government. Thus far, the lack of coordination among these different actors has resulted in an ad hoc implementation of NAP at the local level. Our findings suggest a clear delineation of roles and responsibilities for the monitoring and evaluation plan that is stipulated as the next stage in the NAP implementation. This includes responsibilities for sharing monitoring information with relevant stakeholders, such as NGOs and academic institutions that could take responsibility for conducting community-based surveys and communicating the data to sub-national authorities.

We also recommend formulation of a local or sub-national adaptation plan that may offer a pragmatic way to evaluate what people practice in response to the opportunities and constraints they face in the context of climate change, and to integrate adaptive measures into local planning. Autonomous adaptation alone might not be sufficient to address future climate change and meet sustainable development goals. Thus, a good interplay between autonomous and planned adaptation is crucial. Local or sub-national adaptation plans may bridge the gaps observed between local and national adaptation measures, affording a pathway to reconcile autonomous and planned adaptation and perhaps even to encourage communities to continue implementing multiscalar adaptation measures. Considering that many Pacific Island Countries are still developing NAPs or have only recently developed one, it may take some time before a subnational adaptation plan materializes. Nonetheless, it is a way forward and necessitates further research to establish what may work (or not) and why (not). Several coastal communities in PICs are highly vulnerable to climate risks. These communities do not have time to wait. Prioritizing the development of a local or a sub-national adaptation plan may bridge central planning and local priorities, fast-tracking implementation processes by channelling resources into the hands of those who need them most.

Acknowledgements

Research was funded by an ERC Consolidator Grant (FP7-771056-LICCI). We are grateful to all the participants, village leaders, and Ba Provincial Council who made important contributions to this study.

Declaration of interest None.

References

Adger, W. N., Arnell, N. W., & Tompkins, E. L. (2005). Successful adaptation to climate change across scales. *Global Environmental Change-Human and Policy Dimensions, 15*(2), 77–86. https://doi.org/10.1016/j.gloenvcha.2004.12.005

Adger, W. N., Huq, S., Conway, D., & Hulme, M. (2003). Adaptation to climate change in the developing world. *Progress in Development Studies, 3*(3), 179–195. https://doi.org/10.1191/1464993403ps060oa

Bawakyillenuo, S., Yaro, J. A., & Teye, J. K. (2016). Exploring the autonomous adaptation strategies to climate change and climate variability in selected villages in the rural northern savannah zone of Ghana. *Local Environment, 21*(3), 361–382. https://doi.org/10.1080/13549839.2014.965671

Baxter, J., Laffoley, D. & Simard, F. (2016). Marine protected areas and climate change, IUCN: International Union for Conservation of Nature. Retrieved 25 July, 2022, from https://policycommons.net/artifacts/1373208/marine-protected-areas-and-climate-change/1987430/

Betzold, C., & Nunn, P. D. (2019). Adaptation to climate change in small Island developing states: A systematic literature review of academic research. *The Journal of Environment & Development, 28*(2), 196–218. https://doi.org/10.1177/1070496519835895

Brown, P., Daigneault, A., & Gawith, D. (2017). Climate change and the economic impacts of flooding on Fiji. *Climate and Development, 9*(6), 493–504. https://doi.org/10.1080/17565529.2016.1174656

Census of Population and Housing – Fiji Bureau of Statistics. (2017). Fiji Bureau of Statistics. Retrieved July 22, 2022, from http://www.statsfiji.gov.fj/statistics/2007-census-of-population-and-housing

D'Armengol, L., Castillo, M. P., Ruiz-Mallén, I., & Corbera, E. (2018). A systematic review of co-managed small-scale fisheries: Social diversity and adaptive management improve outcomes. *Global Environmental Change-Human and Policy Dimensions, 52*, 212–225. https://doi.org/10.1016/j.glo envcha.2018.07.009

Eakin, H., Tucker, C. M., Castellanos, E., Diaz-Porras, R., Barrera, J. F., & Morales, H. (2014). Adaptation in a multi-stressor environment: Perceptions and responses to atmospheric and economic risks by coffee growers in Mesoamerica. *Environment, Development and Sustainability 16*(1), 123–139. https://doi.org/10.1007/s10668-013-9466-9

Fedele, G., Donatti, C. I., Harvey, C. A., Hannah, L., & Hole, D. J. (2019). Transformative adaptation to climate change for sustainable social-ecological systems. *Environmental Science & Policy, 101*, 116–125. https://doi.org/10.1016/j.envsci.2019.07.001

Fenton, A., Paavola, J., & Tallontire, A. (2017). Autonomous adaptation to riverine flooding in Satkhira District, Bangladesh: Implications for adaptation planning. *Regional Environmental Change, 17*(8), 2387–2396. https://doi.org/10.1007/s10113-017-1159-8

Fereday, J., & Muir-Cochrane, E. (2006). Demonstrating rigor using thematic analysis: A hybrid approach of inductive and deductive coding and theme development. *International Journal of Qualitative Methods, 5*(1), 80–92. https://doi.org/10.1177/160940690600500107

Galappaththi, E.K. (2024) How can co-management support adaptation to climate change? The case of co-existing fisheries from Pangnirtung, Nunavut, Canada. In V. Reyes-García et al. (Ed.), *Routledge Handbook of Climate Change Impacts on Indigenous Peoples and Local Communities*. Routledge: Oxfordshire & New York. Chp 23, pp 370–388.

Gero, A., Méheux, K., & Dominey-Howes, D. (2011). Integrating community-based disaster risk reduction and climate change adaptation: examples from the Pacific. *Natural Hazards and Earth System Sciences, 11*(1), 101–113. https://doi.org/10.5194/nhess-11-101-2011

Gharbaoui, D., Blocher, J. (2016). The Reason Land Matters: Relocation as Adaptation to Climate Change in Fiji Islands. In: Milan, A., Schraven, B., Warner, K., Cascone, N. (eds) Migration, Risk Management and Climate Change: Evidence and Policy Responses. *Global Migration Issues*, vol 6. Springer, Cham. https://doi.org/10.1007/978-3-319-42922-9_8

Gharbaoui, D., Blocher, J. (2018). Limits to Adapting to Climate Change Through Relocations in Papua-New Guinea and Fiji. In: Leal Filho, W., Nalau, J. (eds) Limits to Climate Change Adaptation. *Climate Change Management*. Springer, Cham. https://doi.org/10.1007/978-3-319-64599-5_20

Government of Fiji. 2018. Republic of Fiji National Climate Change Policy. Available online: http://fijiclimatechangeportal.gov.fj/sites/default/files/documents/National%20Climate%20Change%20Policy%202018%20-%202030.pdf

Hidalgo, D. M., Nunn, P. D., Beazley, H., Burkhart, S., & Rantes, J. (2022). Adaptation, sustainable food systems and healthy diets: An analysis of climate policy integration in Fiji and Vanuatu. *Climate Policy, 22*(9–10), 1130–1145. https://doi.org/10.1080/14693062.2022.2095969

IPCC (2014). *Climate Change 2014: Synthesis Report*. Contribution of Working Groups I, II and III to the Fifth Assessment Report of the Intergovernmental Panel on Climate Change [Core Writing Team, R.K. Pachauri and L.A. Meyer (eds.)]. IPCC, Geneva, 151 pp.

IPCC (2018). Summary for Policymakers. Global warming of 1.5°C. An IPCC Special Report on the impacts of global warming of 1.5°C above pre-industrial levels and related global greenhouse gas emission pathways, in the context of strengthening the global response to the threat of climate change, sustainable development, and efforts to eradicate poverty. In V. Masson-Delmotte, P. Zhai, H. O. Pörtner, D. Roberts, J. Skea, P.R. Shukla, A. Pirani, W. Moufouma-Okia, C. Péan, R. Pidcock, S. Connors, J. B. R. Matthews, Y. Chen, X. Zhou, M. I. Gomis, E. Lonnoy, T. Maycock, M. Tignor, & T. Waterfield (eds.). World Meteorological Organization, Geneva, Switzerland, 32 pp.

Janif, S. Z., Nunn, P. D., Geraghty, P., Aalbersberg, W. G., Thomas, F. M., & Camailakeba, M. (2016). Value of traditional oral narratives in building climate-change resilience: Insights from rural communities in Fiji. *Ecology and Society, 21*(2). https://doi.org/10.5751/es-08100-210207

Juhola, S., Glaas, E., Linnér, B., & Neset, T. S. (2016). Redefining maladaptation. *Environmental Science & Policy, 55*, 135–140. https://doi.org/10.1016/j.envsci.2015.09.014

Kaijser, A., & Kronsell, A. (2014). Climate change through the lens of intersectionality. *Environmental Politics, 23*(3), 417–433. https://doi.org/10.1080/09644016.2013.835203

Kinoshita, Y., Tanoue, M., Watanabe, S., & Hirabayashi, Y. (2018). Quantifying the effect of autonomous adaptation to global river flood projections: Application to future flood risk assessments. *Environmental Research Letters, 13*(1), 014006. https://doi.org/10.1088/1748-9326/aa9401

Marijn, Z., Schlingmann, A., Reyes-García, V., & García-del-Amo, D. (2023). Incremental and transformational adaptation to climate change among Indigenous Peoples and local communities: A global review. [Manuscript submitted for publication].

Marks, D., Bayrak, M. M., Jahangir, S., Henig, D., & Bailey, A. (2022). Towards a cultural lens for adaptation pathways to climate change. *Regional Environmental Change, 22*(1). https://doi.org/10.1007/s10113-022-01884-5

Maruna, M., & Bajec, N. L. (2015). Planned and autonomous actions: Belgrade waterfront adaptation to climate change. *Environmental Engineering and Management Journal, 14*(1), 1–10. https://doi.org/10.30638/eemj.2015.001

McConney, P., Cumberbatch, J., Hinds, C., Oxenford, H. A. & Pena, P. (2024) Sargassum seaweed challenges from local to national level in the Caribbean: a policy cycle perspective. In V. Reyes-García, S. Álvarez-Fernández, P. Benyei, L. Calvet-Mir, D. García-del-Amo, A. B. Junqueira, X. Li, V. Porcher, A. Porcuna-Ferrer, A. Schlingmann, & R. Soleymani (Eds.), *Routledge Handbook of Climate Change Impacts on Indigenous Peoples and Local Communities*. Oxfordshire & New York: Routledge. Chp 3, pp 61–73.

McNamara, K. E., & Prasad, S. (2014). Coping with extreme weather: Communities in Fiji and Vanuatu share their experiences and knowledge. *Climatic Change, 123*(2), 121–132. https://doi.org/10.1007/s10584-013-1047-2

Mersha, A. A., & Van Laerhoven, F. (2018). The interplay between planned and autonomous adaptation in response to climate change: Insights from rural Ethiopia. *World Development, 107*, 87–97. https://doi.org/10.1016/j.worlddev.2018.03.001

Mfitumukiza, D., Roy, A.S., Simane, B., Hammill, A., Rahman, M. F., & Huq, S. (2020). Scaling local and community-based adaptation. Global Commission on Adaptation Background Paper. Rotterdam and Washington, DC. Available online at www.gca. org/global-commission-on-adaptation/report/papers

Nunn, P. D., Aalbersberg, W. G., Lata, S., & Gwilliam, M. (2014). Beyond the core: community governance for climate-change adaptation in peripheral parts of Pacific Island Countries. *Regional Environmental Change, 14*(1), 221–235.

Nurse, L.A., McLean, R.F., Agard, J., Briguglio, L.P., Duvat-Magnan, V., Pelesikoti, N., Tompkins, E., & Webb, A. (2014). Small Islands. In *Climate Change 2014: Impacts, Adaptation, and Vulnerability. Part B: Regional Aspects. Contribution of Working Group II to the Fifth Assessment Report of the Intergovernmental Panel on Climate Change*. Cambridge University Press, Cambridge and New York, pp. 1613–1654. https://www.ipcc.ch/site/assets/uploads/2018/02/WGIIAR5-Chap29_FINAL.pdf

Oppenheimer, M., Glavovic, B. C., Hinkel, J., van de Wal, R., Magnan, A. K., Abd-Elgawad, A., Cai, R., Cifuentes-Jara, M., DeConto, R. M., Ghosh, T., Hay, J., Isla, F., Marzeion, B., Meyssignac, B., & Sebesvari, Z. (2019). Sea level rise and implications for low-lying Islands, coasts and communities. In H.-O. Pörtner, D.C. Roberts, V. Masson-Delmotte, P. Zhai, M. Tignor, E. Poloczanska, K. Mintenbeck, A. Alegría, M. Nicolai, A. Okem, J. Petzold, B. Rama, & N.M. Weyer (Eds.), *IPCC Special Report on the Ocean and Cryosphere in a Changing Climate.* Cambridge University Press, Cambridge and New York, pp. 321–445. https://doi.org/10.1017/9781009157964.006.

Owen, G. (2020). What makes climate change adaptation effective? A systematic review of the literature. *Global Environmental Change-Human and Policy Dimensions, 62*, 102071. https://doi.org/10.1016/j.gloenvcha.2020.102071

Pelling, M., O'Brien, K. D., & Matyas, D. (2015). Adaptation and transformation. *Climatic Change, 133*(1), 113–127. https://doi.org/10.1007/s10584-014-1303-0

Rahman, H. M. T., Albizua, A., Soubry, B., & Tourangeau, W. (2021). A framework for using autonomous adaptation as a leverage point in sustainable climate adaptation. *Climate Risk Management, 34*, 100376. https://doi.org/10.1016/j.crm.2021.100376

Remling, E., & Veitayaki, J. (2016). Community-based action in Fiji's Gau Island: a model for the Pacific? *International Journal of Climate Change Strategies and Management, 8*(3), 375–398. https://doi.org/10.1108/ijccsm-07-2015-0101

Republic of Fiji National Adaptation Plan. (2018). Fiji climate change & national designated authority portal. Fiji_National-Adaptation-Plan.pdf (fijiclimatechangeportal.gov.fj).

Reyes-García, V., Álvarez-Fernández, S., Benyei, P., García-Del-Amo, D., Junqueira, A. B., Labeyrie, V., Li, X., Porcher, V., Porcuna-Ferrer, A., Schlingmann, A., & Soleymani, R. (2023). Local indicators of climate change impacts described by indigenous peoples and local communities: Study protocol. *PLoS One, 18*(1), e0279847. https://doi.org/10.1371/journal.pone.0279847

Schipper, E. L. F. (2020). Maladaptation: When adaptation to climate change goes very wrong. *One Earth, 3*(4), 409–414. https://doi.org/10.1016/j.oneear.2020.09.014

Schlingmann, A., Graham, S., Benyei, P., Corbera, E., Sanesteban, I. M., Marelle, A., Soleymani-Fard, R., & Reyes-García, V. (2021). Global patterns of adaptation to climate change by Indigenous Peoples and local communities. A systematic review. *Current Opinion in Environmental Sustainability, 51*, 55–64. https://doi.org/10.1016/j.cosust.2021.03.002

Sharma, K., Sachan, H. K., & Krishna, D. (2021). Ginger production constraints and future perspectives in Fiji. *Reviews in Agricultural Science, 9*, 260–270. https://doi.org/10.7831/ras.9.0_260

Singh, C., Iyer, S., New, M., Few, R., Kuchimanchi, B., Segnon, A. C., & Morchain, D. (2021). Interrogating 'effectiveness' in climate change adaptation: 11 guiding principles for adaptation research and practice. *Climate and Development, 14*(7), 650–664. https://doi.org/10.1080/17565529.2021.1964937

Singh, P., Charan, D., Kaur, M., Railoa, K., & Chand, R. (2020). Place attachment and cultural barriers to climate change induced relocation: Lessons from Vunisavisavi Village, Vanua Levu, Fiji. *Climate Change Management.* https://doi.org/10.1007/978-3-030-40552-6_2

Singh, P., Tabe, T., & Martin, T. (2022). The role of women in community resilience to climate change: A case study of an Indigenous Fijian community. *Women's Studies International Forum, 90,* 102550. https://doi.org/10.1016/j.wsif.2021.102550

Termeer, C. J., Dewulf, A., & Biesbroek, R. (2017). Transformational change: Governance interventions for climate change adaptation from a continuous change perspective. *Journal of Environmental Planning and Management, 60*(4), 558–576. https://doi.org/10.1080/09640568.2016.1168288

UNEP. (2021). Adaptation Gap Report 2020. In *United Nations Environment Programme (UNEP).* https://www.unep.org/resources/adaptation-gap-report-2020

Webber, M., & O'Neill, S. L. (2010). Maladaptation. *Global Environmental Change-Human and Policy Dimensions, 20*(2), 211–213. https://doi.org/10.1016/j.gloenvcha.2009.11.004

POSTFACE

Carla Roncoli

Drawing on the Local Indicators of Climate Change Impacts (LICCI) project, this edited volume illustrates the synergistic value of integrative work across disciplines, institutions, geographies, and positionalities. While each chapter makes a significant contribution in its own unique way, as a whole, this collection and the rich diversity and granularity of the case studies it comprises offer readers a wide-ranging, multidimensional picture of how Indigenous People and Local Communities (IPLC) around the world perceive, experience, and adapt to the climate crisis. In doing so, this book marks notable advances in empirical knowledge, methodological approaches, and policy propositions around the critical role that Indigenous and local ways of knowing can play in building climate resilience, particularly among marginalized and/or underserved communities. Through the individual chapters and the synthesizing introduction, we learn a great deal about knowledge as a dimension of self, place, interests, and, especially, power.

Before delving into such learnings, I recognize that, as a white European woman living and working in an American university, I cannot speak for – or even claim to fully grasp – the perspectives and experiences of Indigenous people. Rather, I write from the standpoint of an environmental anthropologist who, for the last three decades, has been involved in research partnerships with rural communities and national collaborators in Africa, developing meaningful and enduring friendships in the process. Drawing on such engagements, my work has aimed to support policy advisory and research programs at national and regional scales in Africa as well as global assessments, such as the Intergovernmental Panel on Climate Change. This postscript inevitably reflects this positionality, as the diverse identities and social locations of this collection's authors also inform the representation of their research findings. Some of them are Indigenous people, while others are non-Indigenous allies who have been and/or are partnering with Indigenous and local communities in pursuit of an engaged and liberatory vision of scholarly research. They do so, not only by deploying innovative participatory methodologies and/or ethnographic "thick descriptions" of Indigenous and local knowledge systems, but also by elucidating the distinctive nature and adaptation potential of those ways of knowing.

As a whole, the case studies highlight holism as a key feature of Indigenous and local knowledge, often in juxtaposition to the fragmentation of knowledge in techno-scientific disciplines. The authors of various chapters detect overlaps and disconnects between these different ways of

DOI: 10.4324/9781003356837-30

knowing, but they would also, more or less explicitly, recognize the incommensurability of those systems; in that, they cannot be defined by the same categories and parameters. Indigenous and local knowledge is holistic because it accounts for ecosystem interaction, understood as relationships among humans, animals, plants, water, soil, and terrain. Rather than reducing complexity, Indigenous knowledge embraces it, acknowledging and analyzing the multiple causes, modes, and scales of environmental change. Most of all, Indigenous and local knowledge is holistic because it is acquired through sensorial, embodied, emotional, experiential, and spiritual ways of learning. It reflects what knowledge holders see, feel, value, desire, practice, and believe, as a community. Thus, it cannot and should not be extracted from histories, contexts, meanings, and relationships that define it to be codified, catalogued, quantified, or validated in terms of external instruments and standards. Environmental features and phenomena that local people rely upon to understand and anticipate climate variability and change are not generic, objective, and impersonal "indicators", but are known and valued as living beings and agents in everyday life. Taken together, the chapters in this volume offer a compelling account of this embedded and epistemic holism in all its situatedness.

An intrinsic connection to place characterizes the ways of knowing explored by this collection. Disruptions and vulnerabilities related to climate change are experienced in everyday life at the local level, in geographies that are intimately familiar to the Indigenous and local people that inhabit them. At the same time, these localities and communities have been profoundly altered – and often destabilized – by pressures stemming from the state and the market and by influences emanating from modernity and globalization. Climate change interacts with these broader forces in ways that translate into localized shocks and losses. But Indigenous lands and local spaces are also sites of resilience, recovery, resistance, and renewal. They are stages in which environmental resources and collective action are mobilized to adapt to climate shocks as well as to respond to new opportunities opened by a changing climate. This collection exemplifies the varied ways Indigenous and local people draw upon their cultural knowledge and collective experience to exercise their agency in addressing a myriad of intersecting stressors, uncertainties, potentialities, and innovations.

Indigenous knowledge – as expressed in interpretations and expectations of environmental change – infuses such agency, guiding analyses and decisions relative to tradeoffs across production systems, livelihood needs, natural resource uses, and the broader economy. Of course, recognizing the adaptive role of Indigenous knowledge and the resilience of local lifeways does not necessarily imply sustainable and equitable outcomes at all scales and for all groups. A simplistic overstatement of such capacities obfuscates the unjust reality – so well articulated in the chapters and introduction to this volume – that the climate vulnerability of Indigenous and local people stems from a historical process of dispossession, disenfranchisement, and displacement. It also rationalizes unfairly attributing to Indigenous and local communities most of the responsibility of coping with a climate crisis that they have had no role in causing. Taken together, the case studies achieve a sensible balance of exposing how the negative impacts of climate change fall disproportionately and inequitably upon Indigenous people, without confining them to a hapless victim narrative. Rather, even as they document the colonial and capitalist roots of those disparities, the authors never fail to account for the creative and skillful ways local communities respond to their changing contexts and climates. In doing so, they convincingly advocate for scientists and policymakers to recognize the importance of Indigenous knowledge, not just in academic publications and official pronouncements, but by realigning research agendas and protocols, national and global adaptation planning, and the allocation of authority and resources with the interests and priorities of Indigenous and local communities.

Yet I am positive that editors and authors would also agree that well-intentioned calls for "consideration", "recognition", "consultation", "integration", and "preservation" fall short of achieving accountability unless Indigenous and local people are themselves meaningfully and legitimately represented amongst science leaders and policymakers. The LICCI project makes strides in this direction by supporting training and research opportunities for Indigenous scientists and those allied with them and by fostering the creation of multi-actor networks and data storage and sharing platforms, to promote solidarity, global dialogue, and social learning. By providing young scientists of both Indigenous and non-Indigenous backgrounds with opportunities to conduct and publish research (including this important volume), the project also facilitates their work to be included in the IPCC and other global bodies. The project also developed a rigorous and comprehensive protocol to guide and engage in mutually beneficial, respectful, and sustainable research partnerships with Indigenous and local communities, while also investing in research ethics staffing. The project leaders and authors of this volume's introduction likewise advocate for the development of research standards for reducing or eliminating project-related carbon emissions, as exemplified in their travel policies and communication practices. I believe that the authors of this collection are well-aware that much remains to be done to revert instrumentalizing, extractive, and colonial modes of scientific research, to rebuild trust and relationships between scientists and Indigenous people, and, especially, to ensure that the latter have a transformational presence and influential role in scientific and policymaking circles. Deep change must especially occur at higher levels of engagement, such as grant requirements by donor agencies, promotion criteria in academia, publication conventions of scholarly journals, and parameters for the inclusion of content and authors in assessments by the IPCC and other authoritative scientific bodies. In the meantime, we can be encouraged and inspired by conscious efforts (albeit inevitably imperfect and incomplete) to develop and sustain relationships grounded in intentionality and mutuality such as those by the authors of this collection.

GLOSSARY[1]

Adaptation The process of adjustment to an actual or expected climate and its effects. In human systems, adaptation seeks to moderate or avoid harm or exploit beneficial opportunities. In some natural systems, human intervention may facilitate adjustment to the expected climate and its effects. Adaptations can be incremental (i.e., modifying livelihood strategies) or transformative (i.e., adopted at large scales, that are new to a particular region or resource system, or that transform places or shift locations). Adaptations are also differentiated between planned (i.e., deliberate policy decisions on the part of public agencies), and autonomous (i.e. initiatives that occur naturally by private actors without the intervention of public agencies).

Adaptive capacity The general ability of institutions, systems, and individuals to adjust to potential damage, to take advantage of opportunities, or to cope with the consequences.

Adaptive management A formal procedure for learning by doing that is particularly amenable to sequential decision problems.

Agrobiodiversity The biological diversity that sustains key functions, structures, and processes of agricultural ecosystems. It includes the variety and variability of animals, plants, and microorganisms, at the genetic, species, and ecosystem levels.

Barriers to adaptation The absence of key enabling conditions, or conditions that facilitate approaches to addressing social and ecological challenges.

Biodiversity The variability among living organisms from all sources including terrestrial, marine, and other aquatic ecosystems and the ecological complexes of which they are a part. This includes variation in genetic, phenotypic, phylogenetic, and functional attributes, as well as changes in abundance and distribution over time and space within and among species, biological communities, and ecosystems.

Cascading effect (from a climate event) The dynamics in which an initial impact in the atmospheric system exceeds certain thresholds generating a sequence of cascading impacts in elements of the physical and the life systems. Cascading effects are complex and

1 This glossary includes terms commonly used in different chapters of the book. Most terms are taken from IPBES (https://www.ipbes.net/glossary) and IPCC (https://www.ipcc.ch/report/sr15/glossary/) online glossaries, and adapted for consistency with the work presented here when necessary.

multi-dimensional, and the combined effect of the resulting impacts is larger than the initial one.

Cascading impact (from a climate event) The sequence of secondary impacts in physical and life systems generated by an initial impact in the atmospheric system.

Climate change As defined in Article 1 of the UNFCCC, a change of climate that is attributed directly or indirectly to human activity that alters the composition of the global atmosphere and which is in addition to natural climate variability observed over comparable time periods.

Climate justice The fair treatment and meaningful involvement of all people regardless of race, color, national origin, or income with respect to the development, implementation and enforcement of climate laws, regulations, and policies.

Climate governance The diplomacy, mechanisms, and specific response measures towards preventing, mitigating or adapting to potential climate-related risks.

Climate policy Policy that regulates institutional infrastructure, pledges, plans, and programs to mitigate greenhouse gasses and respond to climate change impacts.

Co-management Process of management in which the government shares power with resource users, with each given specific rights and responsibilities relating to information and decision-making. **Adaptive co-management** refers to the means of sustaining socio-ecological systems by building their resilience and adaptive capacity and establishing sustainable pathways as well as providing a novel institutional arrangement to generate adaptive responses.

Coping The use of existing resources to achieve various desired goals during and immediately after unusual, abnormal, and adverse conditions of a hazardous event or process.

Co-production A relationship between service provider and service user that draws on the knowledge, ability, and resources of both to develop solutions to issues that are claimed to be successful, sustainable, and cost-effective for both parties involved, changing the balance of power from the service provider towards the service user. The approach is used in work with both individuals and communities.

Drivers of change (also referred to as stressors) All the factors that, directly or indirectly, cause changes in nature, anthropogenic assets, nature's contributions to people, and a good quality of life. Direct drivers of change can be both natural and anthropogenic. Direct drivers have direct physical (mechanical, chemical, noise, light, etc.) and psychological (disturbance, etc.) impacts on nature and its functioning, and on people and their interaction. Direct drivers unequivocally influence biodiversity and ecosystem processes. They are also referred to as 'pressures' or 'stressors'. Direct drivers include, inter alia, climate change, pollution, land use change, resource extraction, invasive alien species, and zoonoses, including their effects across regions. Indirect drivers are drivers that operate diffusely by altering and influencing direct drivers as well as other indirect drivers (also referred to as 'underlying causes'). Interactions between indirect and direct drivers create different chains of relationship, attribution, and impacts, which may vary according to type, intensity, duration, and distance. These relationships can also lead to different types of spill-over effects. Global indirect drivers include economic, demographic, governance, technological, and cultural ones, among others.

Ecosystem A dynamic complex of plant, animal, and micro-organism communities and their non-living environment interacting as a functional unit.

Exposure The nature and degree to which a system experiences environmental or socio-political stress. The characteristics of these stresses include their magnitude, frequency, duration, rapidity of onset, and areal extent of the hazard.

Indigenous and local knowledge systems (also referred to as traditional knowledge or traditional ecological knowledge or local ecological knowledge). Indigenous knowledge systems and local knowledge systems are dynamic bodies of integrated, holistic, social and ecological understandings, know-hows, practices and beliefs pertaining to the relationship of living beings, including people, with one another and with their environment. Indigenous and local knowledge is grounded in territory, is highly diverse and is continuously evolving through the interaction of experiences, skills, innovations, and different types of wisdom expressed in multiple ways (written, oral, visual, tacit, practical, and scientific). Such knowledge can provide information, methods, theory, and practice for sustainable ecosystem management. Indigenous and local knowledge systems have been, and continue to be, empirically tested, applied, contested, and validated through different means in different contexts. However, in recognition of the growing political sensitivities and terminology distinctions between Indigenous knowledge (IK) and local knowledge (LK), authors used the most adequate terminology in each chapter.

Indigenous Peoples and local communities We use the definition proposed by The Intergovernmental Science-Policy Platform on Biodiversity and Ecosystem Services (IPBES), which considers that Indigenous peoples and local communities are, typically, ethnic groups who are descended from and identify with the original inhabitants of a given region, in contrast to groups that have settled, occupied or colonized the area more recently. However, in recognition of the growing political sensitivities and terminology distinctions between Indigenous Peoples (IP) and local communities (LC), authors used the most adequate terminology in each chapter.

Intersectionality The interweaving of oppressions on people based on structural inequalities in the global system around categories such as race, class, and gender.

Livelihood diversification The process by which rural families construct a diverse portfolio of activities and social support capabilities in their struggle for survival and in order to improve their standards of living.

Livelihood security Adequate and sustainable access to income and resources to meet basic needs (including adequate access to food, potable water, health facilities, educational opportunities, housing, time for community participation, and social integration).

Local Indicators of Climate Change Impacts Indigenous Peoples' and local communities' first-hand observations of changes in local weather and climatic variability as well as changes in local social-ecological systems which are attributed to climate change and grounded on people's interaction with their direct environment. These qualitative place-based observations of climate change impacts can then be categorized as indicators, or more general descriptions of observations—what we call 'local indicators of climate change impacts' (LICCI). In turn, LICCI can be grouped based on the natural element or process reportedly being impacted, which can be then further grouped into sub-systems ultimately corresponding to the four main systems: climatic, physical, biological, and socioeconomic.

Maladaptation A trait that is, or has become, more harmful than helpful, in contrast with an adaptation, which is more helpful than harmful.

Perceptions The first stage of the human cognitive process. Perceptions are not neutral as they pass through rational and emotional filters which assess and interpret the relevancy of what people see. These filters are conditioned by individual experience, education, and collective worldviews.

Resilience The level of disturbance that an ecosystem or society can undergo without crossing a threshold to a situation with different structures or outputs. Resilience depends on factors such as ecological dynamics as well as the organizational and institutional capacity to understand, manage, and respond to these dynamics.

Social-ecological system A concept used in a variety of analytical approaches intended to examine the relationship between people and nature as interlinked, recognizing that humans should be seen as a part of, not apart from, nature, and nature as interlinked to social systems.

Stressors See Drivers of change.

Sustainability A characteristic or state whereby the needs of the present and local population can be met, with the successful application of justice and equity, without compromising the ability of future generations or populations in other locations to meet their needs.

Sustainable livelihood approach A theoretical approach that organizes the factors that constrain or enhance livelihood opportunities and shows how they relate. Different households have different access to livelihood assets, among which households often make trade-offs and choices. Assets comprise a combination of financial, human, natural, physical, and social capitals.

Scientific knowledge Western academic knowledge systems (also called modern science, Western scientific knowledge or international science) are used as a broad term to refer to knowledge typically generated in universities, research institutions and private firms following paradigms and methods typically associated with the 'scientific method' consolidated in Post-Renaissance Europe based on wider and more ancient roots. It is typically transmitted through scientific journals and scholarly books. Some of its central tenets are observer independence, replicable findings, systematic skepticism, and transparent research methodologies with standard units and categories.

Synergies Situation where the enhancement of one desirable outcome leads to enhancement of another.

Trade-off A situation where an improvement in the status of one aspect of the environment or of human well-being is necessarily associated with a decline in or loss of a different aspect. Trade-offs characterize most complex systems and are important to consider when making decisions that aim to improve environmental and/or socio-economic outcomes.

Triggering impact Initial impact in the atmospheric system generated by weather/climate events which produces the amplification of the hazard through impacts in other elements of the social-ecological system by cascading effects.

Vulnerability The propensity or predisposition to be adversely affected. Vulnerability encompasses a variety of concepts and elements including sensitivity or susceptibility to harm and lack of capacity to cope and adapt.

INDEX

Note: **Bold** page numbers refer to tables and *italic* page numbers refer to figures.

FSC
www.fsc.org
MIX
Papier | Fördert gute Waldnutzung
FSC® C083411

Kolibri 360 GmbH
Ferdinand-Jühlke-Straße 7
99095 Erfurt, Deutschland
produktsicherheit@kolibri360.de